PIMLICO

268

THE SECOND WORLD WAR

John Keegan was for many years Senior Lecturer in
Military History at the Royal Military Academy,
Sandhurst, and is now Defence Editor of the *Daily
Telegraph*. He is the author of many books, including *The
Face of Battle*, *The Mask of Command*, *Six Armies in
Normandy*, *The Battle for History*, *Battle at Sea* and *A
History of Warfare*, which was awarded the Duff Cooper
Prize. *The Second World War* is his sixth book to be
published by Pimlico.

John Keegan is a Fellow of the Royal Society of
Literature and received the OBE in the Gulf War Honours
List.

THE SECOND
WORLD WAR

JOHN KEEGAN

PIMLICO

PIMLICO

An imprint of Random House
20 Vauxhall Bridge Road, London SW1V 2SA

Random House Australia (Pty) Ltd
20 Alfred Street, Milsons Point, Sydney
New South Wales 2061, Australia

Random House New Zealand Ltd
18 Poland Road, Glenfield
Auckland 10, New Zealand

Random House South Africa (Pty) Ltd
Endulini, 5A Jubilee Road, Parktown 2193, South Africa

Random House UK Ltd Reg. No. 954009

First published by Century Hutchinson 1989
Pimlico edition 1997

1 3 5 7 9 10 8 6 4 2

Papers used by Random House UK Limited are natural,
recyclable products made from wood grown in sustainable forests.
The manufacturing processes conform to the environmental
regulations of the country of origin

Maps designed by John Sutherland and Glenn Swann
Printed and bound in Great Britain by
Mackays of Chatham, PLC

ISBN 0-7126-7348-2

CONTENTS

FOREWORD

The Second World War is the largest single event in human history, fought across six of the world's seven continents and all its oceans. It killed fifty million human beings, left hundreds of millions of others wounded in mind or body and materially devastated much of the heartland of civilisation.

No attempt to relate its causes, course and consequences in the space of a single volume can fully succeed. Rather than narrate it as a continuous sequence of events, therefore, I decided from the outset to divide the story of the war into four topics – narrative, strategic analysis, battle piece and 'theme of war' – and to use these four topics to carry forward the history of the six main sections into which the war falls: the War in the West, 1939-43; the War in the East, 1941-3; the War in the Pacific, 1941-3; the War in the West, 1943-5; the War in the East, 1943-5; and the War in the Pacific, 1943-5. Each section is introduced by a piece of strategic analysis, centring on the figure to whom the initiative most closely belonged at that time – in order, Hitler, Tojo, Churchill, Stalin and Roosevelt – and then contains, besides the appropriate passages of narrative, both a relevant 'theme of war' and a battle piece. Each of the battle pieces has been chosen to illustrate the nature of a particular form of warfare characteristic of the conflict. They are air warfare (the Battle of Britain), airborne warfare (the Battle of Crete), carrier warfare (Midway), armoured warfare (Falaise), city warfare (Berlin) and amphibious warfare (Okinawa). The 'themes of war' include war supply, war production, occupation and repression, strategic bombing, resistance and espionage, and secret weapons.

It is my hope that this scheme of treatment imposes a little order for the reader on the chaos and tragedy of the events I relate.

Acknowledgements

My thanks are due above all to the colleagues and pupils among whom I spent twenty-six years at the Royal Military Academy Sandhurst. When I joined the academic staff of the Academy in 1960, many of the military instructors were veterans of the Second World War and it was from conversation with them that I first began to develop an understanding of the war as a human event. I also learnt a great deal from my pupils; because of the Sandhurst method of instruction, which requires cadets to prepare 'presentations' of battles and campaigns, I was often almost as much a listener as a teacher in the Sandhurst Halls of Study and found a great deal of illumination in hearing those episodes described by embryo officers too young to have taken part in them. A number of my pupils have subsequently become professional military historians themselves, including Charles Messenger, Michael Dewar, Anthony Beevor and Alex Danchev. Of all Sandhurst influences, however, none was stronger than that of the Reader in Military History, Brigadier Peter Young, DSO, MC, FSA, a distinguished Commando soldier of the war, the founder of the War Studies Department and an inspiration to generations of officer cadets.

The Sandhurst Library contains one of the most important collections of Second World War literature in the world, and I was fortunate enough to be able to use it almost daily for many years. I would particularly like to thank the present Librarian, Mr Andrew Orgill, and his staff; I would also like to thank Mr Michael Sims and his staff at the Staff College Library, Mr John Andrews and Miss Mavis Simpson at the Ministry of Defence Library and the staff of the London Library.

Friends, and colleagues past and present, at Sandhurst and *The Daily Telegraph* whom I would particularly like to thank include Colonel Alan Shepperd, Librarian Emeritus of Sandhurst, Mr Conrad Black, Mr James Allan, Dr Anthony Clayton, Lord Deedes, Mr Jeremy Deedes, Mr Robert Fox, Mr Trevor Grove, Miss Adela Gooch, Mr Nigel Horne, Mr Andrew Hutchinson, Mr Andrew Knight, Mr Michael Orr, Mr Nigel Wade, Dr Christopher Duffy and Professor Ned Willmott. I owe warmest thanks of all to Mr Max Hastings, the Editor of *The Daily Telegraph* and a distinguished historian of the Second World War. Among others I would like to thank are Mr Andrew Heritage and Mr Paul Murphy.

The manuscript was typed by Miss Monica Alexander and copy-edited by Miss Linden Stafford and I thank them warmly for their professional help. I would also like to thank my editor, Mr Richard Cohen of Hutchinson, and the team he assembled to see the manuscript through production, particularly Mr Robin Cross, Mr Jerry Goldie and Miss Anne-Marie Ehrlich. I owe much gratitude, as always, to my literary agent, Mr Anthony Sheil, and Miss Lois Wallace, my former American literary agent. I am especially indebted to the scholars who read the manuscript: Dr Duncan Anderson, Mr John Bullen, Mr Terry Charman, Mr Terence Hughes, Mr Norman Longmate, Mr James Lucas, Mr Bryan Perrett, Mr Antony Preston, Mr Christopher Shores and Professor Norman Stone. For the errors which remain I alone am responsible.

My thanks finally to friends at Kilmington, particularly Mrs Honor Medlam, Mr Michael Gray and Mr Peter Stancombe, to my children, Lucy Newmark and her husband Brooks, Thomas, Rose and Matthew, and my darling wife, Susanne.

John Keegan
Kilmington Manor
June 6, 1989

PROLOGUE

ONE

Every Man a Soldier

The First [World] War explains the second and, in fact, caused it, in so far as one event causes another,' wrote A. J. P. Taylor in his *Origins of the Second World War*. 'The link between the two wars went deeper. Germany fought specifically in the Second War to reverse the verdict of the first and to destroy the settlement that followed it.'

Not even those who most vehemently oppose Mr Taylor's version of inter-war history will take great issue with those judgements. The Second World War, in its origin, nature and course, is inexplicable except by reference to the First; and Germany – which, whether or not it is to be blamed for the outbreak, certainly struck the first blow – undoubtedly went to war in 1939 to recover the place in the world it had lost by its defeat in 1918.

However, to connect the Second World War with the First is not, if the former is accepted as the cause of the latter, to explain either of them. Their common roots must be sought in the years preceding 1914, and that search has harnessed the energies of scholars for much of this century. Whether they looked for causes in immediate or less proximate events, their conclusions have had little in common. Historians of the winning side have on the whole chosen to blame Germany, in particular Germany's ambition for world power, for the outbreak of 1914 and hence to blame Germany again – whatever failing attaches to the appeasing powers – for that of 1939. Until the appearance of Fritz Fischer's heretical revision of the national version in 1967, German historians generally sought to rebut the imputation of 'war guilt' by distributing it elsewhere. Marxist historians, of whatever nationality, have overflown the debate, depicting the First World War as a 'crisis of capitalism' in its imperialist form, by which the European working classes were sacrificed on the altar of competition between decaying capitalist systems; they are consistent in ascribing the outbreak of the Second World War to the Western democracies' preference for gambling on Hitler's reluctance to cross the brink rather than accept Soviet help to ensure that he did not.

These views are irreconcilable. At best they exemplify the judgement that 'history is the projection of ideology into the past'. There can indeed be no common explanation of why the world twice bound itself to the wheel of mass war-making as long as historians disagree about the logic and morality of politics and whether the first is the same as the second.

A more fruitful, though less well-trodden, approach to the issue of causes lies along another route: that which addresses the question of *how* the two World Wars were made *possible* rather than why they came about. For the instances of outbreak are themselves overridingly important in neither case. It was the enormity of the events which flowed from the upheavals of August 1914 and September 1939 that has driven historians to search so long for reasons to explain them. No similar impetus motivates the search for the causes of the Austro-Prussian War of 1866 or the Franco-Prussian War of 1870, critical as those conflicts were in altering the balance of power in nineteenth-century Europe. Moreover, it is safe to say that had Germany won the critical opening battle of the First World War, that of the Marne in September 1914, as she might well have done – thereby sparing Europe not only the agony of the trenches but all the ensuing social, economic and diplomatic embitterment – the libraries devoted to the international relations of Germany, France, Britain, Austria-Hungary and Russia before 1914 would never have been written.

However, because it was not Germany but France, with British help, who won the Marne, the First – and so the Second – World War became different from all wars previously fought, different in scale, intensity, extensiveness and material and human cost. They also came, by the same measure, closely to resemble each other. It is those differences and those similarities which invest the subject of their causation with such apparent importance. But that is to confuse accident with substance. The causes of the World Wars lay no deeper and were no more or less complex than the causes of any other pair of conjoined and closely sequential conflicts. Their nature, on the other hand, was without precedent. The World Wars killed more people, consumed more wealth and inflicted more suffering over a wider area of the globe than any previous war. Mankind had grown no more wicked between 1815, the terminal date of the last great bout of hostilities between nations, and 1914; and certainly no sane and adult European alive in the latter year would have wished, could he have foreseen it, the destruction and misery that the crisis of that August was to set in train. Had it been foretold that the consequent war was to last four years, entail the death of 10 million young men, and carry fire and sword to battlefields as far apart as Belgium, northern Italy, Macedonia, the Ukraine, Transcaucasia, Palestine, Mesopotamia, Africa and China; and that a subsequent war, fought twenty years later by the same combatants over exactly the same battlefields and others besides, was to bring the death of 50 million people, every individual and collective impulse to aggression, it might be thought, would have been stilled in that instant.

That thought speaks well for human nature. It also speaks against the way the world had gone between 1815 and 1914. A sane and adult European alive in the latter year might have deplored with every fibre of his civilised being the prospect foretold to him of the holocausts that were to come. To do so, however, he would have had to deny the policy, ethos and ultimately the human and material nature of the state – whichever state that was – to which he belonged. He would even have had to deny the condition of the world which surrounded him. For the truth of twentieth-century European civilisation was that the world it dominated was pregnant with war. The enormous wealth, energy and population increase released by Europe's industrial revolution in the nineteenth century had transformed the world. It had created productive and exploitative industries – foundries, engineering works, textile factories, shipyards, mines – larger by far than any at which the intellectual fathers of the industrial revolution, the economic rationalists of the eighteenth century, had guessed. It had linked the productive regions of the world with a network of communications – roads, railways, shipping lanes, telegraph and telephone cables – denser than even the most prescient enthusiast of science and technology could have foreseen. It had generated the riches to increase tenfold the population of historic cities and to plant farmers and graziers on millions of acres which had never felt the bite of the plough or the herdsman's tread. It had built the infrastructure – schools, universities, libraries, laboratories, churches, missions – of a vibrant, creative and optimistic world civilisation. Above all, and in dramatic and menacing counterpoint to the century's works of hope and promise, it had created *armies*, the largest and potentially most destructive instruments of war the world had ever seen.

The militarisation of Europe

The extent of Europe's militarisation in the nineteenth century is difficult to convey by any means that catch its psychological and technological dimensions as well as its scale. Scale itself is elusive enough. Something of its magnitude may be transmitted by contrasting the sight Friedrich Engels had of the military organisation of the independent North German city-states in which he served his commercial apprenticeship in the 1830s with the force which the same German military districts supplied to the Kaiser of the unified German Reich on the eve of the First World War. Engels's testimony is significant. A father of Marxist theory, he never diverged from the view that the revolution would triumph only if the proletariat succeeded in defeating the armed forces of the state. As a young revolutionary he pinned his hopes of that victory on the proletariat winning the battle of the barricades; as an old and increasingly dispirited ideologue, he sought to persuade himself that the proletariat, by then the captive of Europe's conscription laws, would liberate itself by subverting the states' armies from within. His passage from the hopes of youth to the doubts of old age can best be charted by following the transformation of the Hanseatic towns' troops during his lifetime. In August 1840 he rode for three hours from his office in Bremen to watch the combined manoeuvres of the armies of Bremen, Hamburg, Lübeck free city and the Grand Duchy of Oldenburg. Together they formed a force a regiment – say, to err on the side of generosity, 3000 – men strong. In the year of his death in 1895 the same cities provided most of the 17th and part of the 19th Divisions of the German Army, together with a cavalry and artillery regiment – at least a fourfold increase. That accounts for only first-line troops, conscripts enrolled and under arms. Behind the active 17th and 19th Divisions stood the 17th and 19th Reserve Divisions to which the Hanseatic cities would contribute an equal number of reservists – trained former conscripts – on mobilisation. And behind the reserve divisions stood the *Landwehr* of older ex-conscripts who in 1914 would provide half of another division again. Taken together, these units represent a tenfold increase in strength between 1840 and 1895, far outstripping contemporary population growth.

This enormous multiplication of force was nevertheless in the first instance a function of demographic change. The population of most states destined to fight the First World War doubled and in some cases tripled during the nineteenth century. Thus the population of Germany, within the boundaries of 1871, increased from 24 million in 1800 to 57 million in 1900. The British population increased from 16 million in 1800 to 42 million in 1900; but for the Irish famine and emigration to the United States and the colonies, producing a net outflow of about 8 million, it would have tripled. The population of Austria-Hungary, allowing for frontier changes, increased from 24 million to 46 million; of Italy, within the 1870 frontiers, from 19 million to 29 million, despite a net outflow of perhaps 6 million emigrants to North and South America. Belgium's population grew from 2.5 to 7 million; that of European Russia between the Urals and the western frontier of 1941 nearly tripled, from 36 to 100 million. Only two of the combatant states, France and the Ottoman empire, failed to show similar increases. The French population, once the largest in Europe, rose only from 30 to 40 million and chiefly through extended longevity; the birth-rate remained almost static – the result, in Professor William McNeill's view, of Napoleon's returning warriors bringing home techniques of birth control learned on campaign. The population of Turkey within its present frontiers scarcely increased at all; it was 24 million in 1800 and 25 million in 1900.

The French and Turkish cases, though falling outside the demographic pattern, are nevertheless significant in explaining it. The increased longevity of the French was due to improved standards of living and public health, the outcome of the application of science to agriculture, medicine and hygiene. The failure of the Turkish population to increase had an exactly contrary explanation: the poor yields of traditional farming and incidence of disease in a society without doctors ensured that population, despite high birth-rates, remained at a static level. Whenever increased agricultural output (or input) combined with high birth-rates and improved hygiene, as they did almost everywhere in Europe in the nineteenth century, the effect on population size was dramatic. In England, the centre of the nineteenth-century economic miracle, it was spectacular. Despite a massive emigration of the population from the countryside to the towns, overcrowded and often jerry-built, the number of the English increased by 100 per cent in the first half and by 75 per cent in the second half of the century. Sewer-building, which ensured the elimination of cholera from 1866 and of most other water-borne diseases soon after, and vaccination, which when it was made compulsory in 1853 eliminated smallpox, sharply reduced infant mortality and lengthened the life expectancy of the adult population; death from infectious disease declined by nearly 60 per cent between 1872 and 1900. Improved agricultural yields from fertilised and fallowed fields, and, in particular, the import of North American grain and refrigerated Australasian meat, produced larger, stronger and healthier people. Their intake of calories was increased by the cheapening of luxuries such as tea, coffee and especially sugar, which made grain staples more palatable and diet more varied.

The combined effect of these medical and dietary advances on growing populations was not only to increase the size of the contingents of young men liable each year for conscription (*classes*, as the French labelled them) – by an average of 50 per cent, for example, in France between 1801 and 1900 – but to make them better suited, decade on decade, for military service. There is an apparently irreducible military need for a marching soldier to bear on his body about 50 lb of extraneous weight – pack, rifle and ammunition. The larger and stronger the soldier, the more readily can he carry such a load the desirable marching norm of twenty miles a day. In the eighteenth century the French army had typically found its source of such fit men among the town-dwelling artisan class rather than the peasantry. The peasant, physically undernourished and socially doltish, rarely made a suitable soldier; he was undisciplined, prone to disease and liable to pine to death when plucked from his native heath. It was these shortcomings which prompted Marx a hundred years later to dismiss the peasantry as 'irredeemable' for revolutionary purposes. By the mid-nineteenth century, however, the peasant populations of Germany, France, Austria-Hungary and Russia had so much improved in physique that they were regularly supplying to their national armies a proportion of new conscripts or *classes* large enough to give Marx the lie. His analysis may have been skewed by his standpoint in England, where large-scale emigration to the towns left only the least enterprising under the thumb of squire and parson. In the continental lands, which were industrialising more slowly than England – the German rural population in 1900 was still 49 per cent of the total – it was the countryside which yielded the *classes* of large, strong young men out of which the great nineteenth-century armies were built.

If the new population surplus yielded by better diet, drugs and drains increased the European armies' recruiting pool, it was the nineteenth-century states' enhanced powers of head-counting and tax-gathering which ensured that recruits could be found, fed, paid, housed, equipped and transported to war. The institution of regular census-taking – in France in 1801, Belgium in 1829, Germany in 1853, Austria-Hungary in 1857, Italy in 1861 – accorded recruiting authorities the data they needed to identify and docket potential recruits; with it died the traditional expedients of haphazard impressment, cajolery, bribery and press-ganging which had raised the *ancien régime* armies from those not fleet enough of thought or foot to escape the recruiting sergeant. Tax lists, electoral registers and school rolls documented the conscript's whereabouts – the grant of the vote and the introduction of free education for all entailed a limitation as well as an enlargement of the individual's liberties. By 1900 every German reservist, for example, was obliged to possess a discharge paper specifying the centre at which he was to report when mobilisation was decreed.

The enormous enlargement of European economies was meanwhile creating the tax base by which the new armies of conscripted recruits were supported; the German economy, for example, expanded by a quarter between 1851 and 1855, by a half between 1855 and 1875 and by 70 per cent between 1875 and 1914. From this new wealth the state drew, via indirect and direct revenue, including the resented institution of income tax, an ever-increasing share of the gross domestic product. In Britain, for example, the government's share of consumption rose from 4.8 per cent in 1860-79 to 7.4 per cent in 1900-14 and in Germany from 4 per cent to 7.1 per cent; rises were proportionate in France and Austria-Hungary.

Most of this increased revenue went to buy military equipment – in the broadest sense. Guns and warships represented the costliest outlay; barracks the more significant. The *ancien régime* soldier had been lodged wherever the state could find room for him, in taverns, barns or private houses. The nineteenth-century conscript was housed in purpose-built accommodation. Walled barracks were an important instrument of social control; Engels denounced them as 'bastions against the populace'. The sixteenth-century Florentines similarly regarded the building of the Fortezza de Basso inside the gates of their city as a symbol of the curtailment of their liberties. Barracks were certainly a principal means of guaranteeing that ready availability of force by which the Berlin revolt of 1848 and the Paris Commune of 1871 were put down.* However, barracks were not only the precinct-stations of the contemporary riot police. They were also the fraternity houses of a new military culture in which conscripts learnt habits of obedience and forged bonds of comradeship which would harden them against a battlefield ordeal more harrowing than any which soldiers had known before.

The new-found wealth of the nineteenth-century state enabled the conscript not only to be housed and equipped but also to be transported to the battlefield and fed amply when he arrived. The soldier of the *ancien régime* had been

---

*It was not only continentals who opposed barrack building. Field Marshal Wade, the eighteenth-century British general, put the British attitude thus: 'the people of this Kingdom have been taught to associate the idea of Barracks and Slavery so closely that, like darkness and the Devil, though there be no manner of connection between them, yet they cannot separate them'.

scarcely better supplied than the Roman legionary; flour ground in the regimental hand-mills, supplemented by a little beef driven on the hoof, was his staple. The nineteenth-century conscript was fed in the field on preserved food; margarine and canning were both the products of a competition founded by Napoleon III to invent rations that would not rot in the soldier's pack. However, the necessity for him to carry his own supply of rations was in any case sharply diminished by the subordination of the burgeoning railway system to military uses. Troops were transported by rail as early as 1839 in Germany. By 1859, when France fought Austria in northern Italy, deployment by rail seemed commonplace. In 1866 and 1870 it underlay Prussia's victories against Austria and France. In the latter year the German rail network, only 469 kilometres in 1840, had increased to 17,215; by 1914 it would total 61,749 kilometres, the greater part of it (56,000 kilometres) under state management. The German government, heavily prompted by the Great General Staff, had early grasped the importance for defensive – and offensive – purposes of controlling the railway system; much of it, particularly in such sectors of low commercial use as Bavaria and East Prussia, had been financed by state-raised loans and laid out at the direction of the General Staff's railway section.*

Railways supplied and transported the soldier of the steam age (at least as far as the railhead; beyond, the old marching and portering imperatives persisted). The technology that built the railways also furnished the weapons with which the soldiers of the new mass armies would inflict mass casualties on each other. The development of such weapons was not deliberate, at least not at the outset; later it may have been. Hiram Maxim, the inventor of the first successful machine-gun, is alleged to have given up experiments in electrical engineering in 1883 on the advice of a fellow American, who said: 'Hang your electricity! If you want to make your fortune, invent something which will allow those fool Europeans to kill each other more quickly.' Initially, however, the reason for the appearance of the faster-firing, longer-range and more accurate weapons that equipped the conscript armies between 1850 and 1900 was the particular conjunction of human ingenuity and industrial capability which made their production feasible.

Four factors were significant. The first was the spread of steam power, which supplied the energy to manufacture weapons by industrial process. The second was the development of the appropriate process itself, originally called 'American' by reason of its origin in the 1820s in the factories of the Connecticut Valley, which were chronically short of skilled labour. This industrial process resulted in 'interchangeable parts', machined by a refinement of the ancient pantographic principle, and achieved an enormous surge of

---

*It is evidence of the military importance the German state and army attached to the free use of the railways that the personnel of the *Reichsbahn* were not allowed to unionise. Understandably so; the word 'sabotage' derives from the Belgian railwaymen's practice of unseating rails from their shoes (*sabots*) during the great strike of 1905.

output. The Prussian manufacturer, Dreyse, inventor of the revolutionary 'needle-gun' (in which a bolt-operated firing-pin struck a metal-jacketed cartridge), managed to turn out only 10,000 units a year by traditional methods in 1847, despite holding a firm contract from the Prussian government to re-equip its whole army. By 1863, in contrast, the British Enfield armoury, rejigged with automatic milling machines, turned out 100,370 rifles, and in 1866 the French government re-equipped the armoury at Puteaux with 'inter-changeable parts' machinery capable of producing 300,000 of the new Chassepot rifles each year.

Advances in metal engineering would have been pointless without improvements in the quality of the metal to be worked; that was assured by the development of processes for smelting steel in quantity – notably by the British engineer Bessemer after 1857 (he also was encouraged by a prize offered by Napoleon III). Bessemer's 'converter' marked the third significant advance. With similar furnaces, the German cannon-founder, Alfred Krupp, began in the 1860s to cast steel billets from which perfect cannon-barrels could be machined. His breech-loading field-guns, equivalents on a larger scale of the rifles with which all contemporary infantrymen in advanced armies were now issued, proved the decisive weapons of the Franco-Prussian War of 1870-1. The fourth ingredient of the firepower revolution was supplied shortly afterwards by European chemists, notably the Swede Alfred Nobel, who developed propellants and bursting-charges which drove projectiles to a greater distance and detonated them with more explosive effect than ever before. The effective range of infantry weapons, for example – a function equally of engineering and propellant developments – increased from a hundred to a thousand yards between 1850 and 1900. When the recuperation of chemical-energy discharges was applied to the mechanism of small arms and artillery in the period 1880-1900, it produced the machine-gun and the quick-firing artillery piece, the ultimate instruments of mass death-dealing at distance.

Surplus and war-making capacity

Long-range, rapid-fire weapons constituted the threat by which all the increments of offensive force assembled by the industrial and demographic revolutions of the nineteenth century were to be negated. There lay an irony. The material triumph of the nineteenth century had been to break out of the cycle of recurrent lean and plenty which had immemorially determined the condition of life even in the richest states, and to create permanent surplus – of food, energy and raw materials (though not of capital, credit or cash). Market fluctuations perpetuated boom and recession in the peaceful life of states. Surplus transformed their war-making capacity. War at any level above the primitive ritual of raid and ambush had always required surplus for its waging. However, accumulated surpluses had rarely been large enough historically to fund wars that culminated in the decisive victory of one side over another; self-funding wars, in which the spoils of conquest sustained the impetus of a victorious campaign, had been rarer still. Extraneous factors – gross disparity in the opposed technologies of war-making or in the dynamism of opposed ideologies, or, as Professor William McNeill has suggested, susceptibility to unfamiliar germ strains transported by an aggressor – had usually explained one society's triumph over another; and they certainly underlay such military sensations as the Spanish destruction of the Aztec and Inca empires, the Islamic conquests of the seventh century and the American extinction of Red Indian warriordom.

In the warfare of Europe between the Reformation and the French Revolution, waged between states occupying a level plateau of war-making skills, will to war and resistance to common disease, such extraneous factors had played no decisive part; while the surpluses available for offence had been heavily offset by the diversion of funds into means of defence, particularly siege engineering. A great deal of such siege engineering had been dedicated to the destruction of the feudal strongholds from which local magnates had defied central authority once the fashion for castle-building seized the European land-holding class in the eleventh century. It was extremely costly; and to the costs had been added those of replacing local with national fortifications in the frontier zones throughout the sixteenth, seventeenth and eighteenth centuries. Investment in siegecraft, destructive and constructive, had the collateral effect of securing under-investment in civil infrastructures – roads, bridges and canals – which might otherwise have made the passage of armies on offensive campaigns swift and decisive. As late as 1826, for example, while the British road network – much of it in Scotland deliberately built for military purposes after the Jacobite revolt of 1745 – extended to over 21,000 miles, that of France (three times the size) was no greater, while Prussia, which occupied much of the most strategically significant terrain in northern Europe, had a road network of only 3,340 miles, most of it in her Rhineland provinces. Her eastern lands were virtually roadless, as Poland and Russia were to remain – to Napoleon's and then Hitler's cost – well into the twentieth century.

The surplus created by the economic miracle in nineteenth-century Europe cancelled out the effects of under-investment in road-building and over-investment in frontier fortification. Mass armies, transported and supplied along the new infrastructure of railways, swamped strategically significant territory as if by tidal force in an era of changed sea levels. In 1866 and 1870 the armies of Prussia overflowed the frontier regions of Austrian Bohemia and French Alsace-Lorraine without hindrance by the costly fortifications that guarded them. Strategic movement in Europe achieved a fluidity equivalent almost to that which had characterised the western campaigns of the American Civil War, fought by mass armies in a landscape free from artificial obstructions of any sort. Regions disputed by Habsburg and Bourbon generals in two hundred years of toothpick campaigning for advantage in each cavity and

crevice of each other's borderlands went under the hammer of steam power in
a few weeks of brutal resculpturing. It seemed that a second 'military
revolution', equivalent to that brought about by gunpowder and mobile cannon
at the dawn of the Renaissance and Reformation, stood at hand. Blood, iron and
gold – available in quantities more copious than any of which the richest king
had ever disposed – promised victories swifter and more total even than those
which had been achieved by Alexander the Great or Genghis Khan.

Such victories were promised but could not necessarily be delivered; for the
greatest material riches do not avail if the human qualities necessary to animate
them are lacking. But here too the nineteenth century had wrought a sea-
change. The eighteenth-century soldier had been a poor creature, the liveried
servant of his king, sometimes – in Russia and Prussia – an actual serf delivered
into the state's service by his feudal master. Uniform was, indeed, a livery,
which reigning monarchs conspicuously did not wear. Those who did bore it as
a mark of surrendered rights. It meant that they had succumbed to 'want or
hardship', the most common impulse to enlistment; that they had changed sides
(turncoat prisoners of war formed large contingents in most armies); that they
had accepted mercenary service under foreign colours (as tens of thousands of
Swiss, Scots, Irish, Slavs and other highlanders and backwoodsmen did
throughout the *ancien régime*); that they had 'plea-bargained' out of imprisonment for petty crime or attachment for civic debt; or simply that they had failed
to run fast enough from the press-gang. The volunteer was almost the rarest if
the best of soldiers. Because so many of his comrades-in-arms were unwilling
warriors, the penalties for desertion were draconian and the code of discipline
ferocious. The eighteenth-century soldier was flogged for infractions of duty
and hanged for indiscipline, both sorts of offence being loosely interpreted.

The nineteenth-century soldier, by contrast, was a man who wanted to be
what he was. A willing, often an enthusiastic, soldier, he was usually a conscript
but one who accepted his term of (admittedly short) service as a just subtraction
from his years of liberty, to be performed with cheerfulness as well as
obedience. This was the case at least from mid-century onwards and in the
armies of the most advanced states – Prussia first and foremost, but also France
and Austria, with the smaller and more backward hurrying to follow suit. Such
a change of attitude is difficult to document but real enough nevertheless.
Perhaps its most tangible manifestation was the appearance of the regimental
souvenir which began to be manufactured in tens of thousands towards the end
of the nineteenth century. The souvenir, typically in Germany a china drinking
mug, decorated with pictures of regimental life, usually bore the names of the
conscript's fellow platoon members, some couplets of doggerel verse, a
salutation to the regiment – 'Here's to the 12th Grenadiers' – and the universal
superscription 'In memory of my service time'. The young soldier who had
been sent off garlanded with flowers by his neighbours – a strikingly different
farewell from that given to the Russian serf conscript of the eighteenth century,

for whom the village priest said a requiem mass – bore back his souvenir when his service time was over to stand in a place of honour in the family home.

This remarkable change of attitude was literally revolutionary. The roots of the change were manifold, but the three most important led directly to the French Revolution and the principal slogans of its ideology: Liberty, Equality, Fraternity.

Military service became popular in the nineteenth century first because it was an experience of *equality*. 'Cook's son – Duke's son – son of a belted Earl,' Rudyard Kipling wrote of the army Britain sent to fight the Boers in 1900, with some accuracy. Popular enthusiasm for the war did sweep all classes into the ranks as common soldiers; but they were, of course, volunteers. Universal conscription in the European armies took all classes willy-nilly – in Prussia from 1814, in Austria from 1867, in France from 1889 – and bound them to service for two or three years. There were variations in the proportion of annual *classes* enlisted and fluctuations in the length of service. There were alleviations of obligation for the better educated; typically, for example, high-school graduates served only one year and were then transferred to the reserve as potential officers. Yet the principle of universal obligation that generally held good was also accepted as persisting. Reservists during their early years of discharge returned annually to the colours for retraining; as they grew older they moved to a wartime reserve (*Landwehr* in Germany, Territorial Army in France); and their final years of able manhood were spent on the list of the Home Guard. Reserve training was borne with good humour, even regarded as a sort of all-male holiday. Freud, a reserve medical officer in the Austrian army, writing to a friend from manoeuvres in 1886, observed that 'it would be ungrateful not to admit that military life with its inescapable "must" is good for neurasthenia. It all disappeared in the first week.'

Conscription was also relatively egalitarian in its outreach. Jews, like Freud, were as liable as Gentiles and in the Habsburg army automatically became officers if educationally qualified; in the German army, Jews could become reserve officers but were barred by regimental anti-Semitism from holding regular commissions, though Bismarck's financier, Bleichroder, managed to get his son a regular commission in the household cavalry. The officer who recommended Hitler for his Iron Cross 1st Class was a Jewish reserve officer. This was 'emancipation' in its military aspect, and it applied not only to Jews. The universality of conscription swept up every nationality in the Habsburg lands, Poles and Alsace-Lorrainers in Germany, Basques, Bretons and Savoyards in France. All, by being soldiers, were also to be Austrians, Germans or Frenchmen.

Conscription was an instrument not only of equality but also of *fraternity*. Because it applied to all at the same moment of their lives and in principle treated all in the same way, it forged bonds of brotherhood young Europeans had never before felt. Universal compulsory education, a simultaneous innovation, was currently taking children outside their families and plunging them into a common experience of learning. Conscription took young adults from their locality and plunged them into the experience of growing up – confronting them with the challenge of separation from home, making new friends, dealing with enemies, adjusting to authority, wearing strange clothes, eating unfamiliar food,* shifting for themselves. It was a genuine *rite de passage*, intellectual, emotional and, not least of all, physical. Nineteenth-century armies, told that they were 'schools of the nation', took on many of the characteristics of contemporary schools, not only testing and heightening literacy and numeracy but also teaching swimming, athletics and cross-country sports as well as shooting and the martial arts. Turnvater Jahn, the pioneer of physical education in Germany, was a potent influence on Prussian military training; his ideas were propagated in France through the specialist athletics instructors of the Bataillon de Joinville, while in Italy Captain Caprilli founded a school of military horsemanship which was to transform the art of riding throughout the Western world. The healthy outdoorsmanship of military life, lived round the campfire and under canvas, would eventually develop into the ideals of the German youth movement and the code of the Boy Scouts and so make its way back into social and military life by a convergent route.

The *rite de passage* of universal conscription was not a liberating experience for all. As Professor William McNeill has pointed out, individuals drafted into the army from a society which was rapidly urbanising and industrialising, marching them away from the plough and the village pump,

> found themselves in a simpler society than the one they knew in civil life. The private soldier lost almost all personal responsibility. Ritual and routine took care of nearly every working hour. Simple obedience to the orders that punctuated that routine from time to time, and set activity off in some new direction, offered release from the anxieties inherent in personal decision-making – anxieties that multiplied incontinently in urban society, where rival leaders, rival loyalties and practical alternatives as to how to spend at least part of one's time competed insistently for attention. Paradoxical as it may sound, escape from freedom was often a real liberation, especially for young men living under very rapidly changing conditions, who had not yet been able to assume fully adult roles.

Even when allowance is made for the force of this percipient observation, however, the ultimate importance of universal conscription in changing attitudes to military service was that it ultimately connected with *liberty*, in its

---

*Often very much better in the army than at home. In the 1860s the French national intake was 1.2 kilograms, the army intake 1.4. The contemporary Flemish conscripts' refrain, reflecting the hardship of peasant life, ran: 'Every day in the army meat and soup without working.'

political if not its personal sense. The old armies had been instruments of oppression of the people by kings; the new armies were to be instruments of the people's liberation from kings, even if that liberation was to be narrowly institutional in the states which retained monarchy. The two ideas were not mutually contradictory. The French National Convention had decreed in 1791 that 'the battalion organised in each district shall be united under a banner bearing the inscription: "The French people united against tyranny".' That decree encapsulated the idea inherent in the United States Constitution that 'the right to bear arms', once made common, was a guarantee of direct freedoms. Two years earlier the revolutionary leader, Dubois-Crance, had articulated the congruent proposition: 'Each citizen should be a soldier, and each soldier a citizen, or we shall never have a constitution.'

The tension between the principles of winning freedoms by revolutionary assault and extracting them in legal form by performance of military duty was to transfix European political life for much of the nineteenth century. The excess of freedom won by force of arms in France provoked the reaction of Thermidor and diverted the fervour of the extremist *sans-culottes* into conquest abroad. The victories of the 'revolutionary' armies (after 1795 firmly under the control of their officers, many of them, ironically, returned monarchists) then had the effect of provoking their enemies, particularly the Prussian and Austrian kings, into decreeing a variation of the *levée-en-masse* or general conscription, the original manifestation of the French Revolution in its military form. Such conscription produced popular forces – *Landwehr, Landsturm, Freischützen* - to oppose the French on their home territories.

Landwehr and *Freischützen* became an embarrassment as soon as their work was done. With Napoleon safely on St Helena, Prussia and Austria consigned these popular forces, with their liberal-minded bourgeois officers, to the status of reserve contingents, and intended never to call on their services again. Nevertheless they survived until 1848, 'year of revolutions', when their members actively participated in the street battles for constitutional rights in Vienna and Berlin – where the uprising was put down by the Prussian Guard, the ultimate bastion of traditional authority. They had meanwhile been replicated in France, whose National Guard would keep alive the 'liberal' principle in military life under the Second Empire and, after the withdrawal of the Prussians from Paris in 1871, rise against the regular army of the conservative Third Republic in a bloody Commune which would cost the lives of 20,000 of its members.

'No conscription without representation'
The struggle of these citizen forces with the armies of reaction, though ending in physical defeat, nevertheless indirectly exerted the pressure which extracted constitutional and electoral rights from the conservative European regimes. The demand for such rights was in the air; and the *impôt du sang* - 'blood tax',

as conscription laws were called in France – could not be levied if constitutional rights continued to be refused, particularly when neighbour states were enlarging their armies and reserves through the process of conscription. Prussia, the military pace-setter, granted a constitution in 1849, as a direct result of the fright it was caused by armed revolutionaries the previous year. By 1880 both France and the German Empire had introduced universal male suffrage, and France would institute a common three-year term of service as a *quid pro quo* in 1882. Austria extended the vote to all males in 1907; even Russia, most autocratic of states and most exigent in its conscription laws, which imposed a term of four years, had created a representative assembly in 1905, following the defeat of its army by the Japanese in Manchuria and the subsequent revolution of that year.

'No conscription without representation' had, in short, become an unspoken slogan of European politics in the half-century before the First World War; since conscription is indeed a tax, on the individual's time if not money, it exactly echoed the American colonists' challenge to George III in 1776. Paradoxically, in the states where votes were granted to all, or most, free men but where military service was still restricted to those fettered by 'want or hardship' – the United States and Britain – a strange passion for volunteer soldiering seized their citizenry during the great era of military expansion through conscription in nineteenth-century Europe. The opening stages of the American Civil War could not have been fought without the prior existence of a network of entirely amateur regiments, with names like the Liberty Rifles of New Jersey, the Mechanic Phalanx of Massachusetts, the Republican Blues of Savannah, Georgia, and the Palmetto Guard of Charleston, South Carolina. In 1859 a nationwide war scare caused by French naval expansion had brought into being a similar though much larger network in Britain. Tennyson's stirring verses, *Form, Riflemen, Form*, had helped to call 200,000 civilians into amateur military service. This was a serious embarrassment to the government, which could not stop them designing and buying their own uniforms but was reluctant to see or help them arm.

They did so none the less; and the government, which like all others in Europe since the establishment of public order at the beginning of the eighteenth century had energetically carried out the disarmament of its population, was eventually obliged to issue them with rifles from the state arsenals. The issue of the modern rifle, rather than the obsolete musket, was crucially significant. The musket, like the uniform livery of the dynastic armies that used it, was a mark of servitude. So short was its range that its effect could be harnessed to battle-winning purposes only by massing the musketeers in dense rank, and keeping them 'closed up' at pike point. The rifle, by contrast, was a weapon of individual skill. It could kill a common soldier, without much discrimination by its user, at 500 yards; in the hands of a marksman it could kill a general at 1000 yards. Hence the Paris Communards were convinced, as

Thomas Carlyle put it, that 'the rifle made all men tall'. A rifleman was as good as any man. The British Rifle Volunteers, in token of the status their weapons gave them, chose to dress not in the tight scarlet of the soldiers of the line, enlisted from 'want and hardship', but in the loose tweed shooting-suits of country gentlemen; to that garb some added 'Garibaldi' shirts or the 'wideawake' hats of the 1848 revolutionaries. In different varieties of cut and colour – field-grey or khaki – this grousemoor or deerstalker garb would come to clothe all the armies of Europe (with the exception of the French) by 1914, just as the long-range, high-velocity rifle would arm them. No badge of military proficiency would be worn with more ostentation than the marksman's; and those units which had carried the rifle earliest – designated as *Schützen* in Germany, *Jäger* in Austria, *chasseurs* in France, greenjackets in Britain – would arrogate to themselves a particular *esprit de corps* as soldiers of modernity.

In truth, however, all the soldiers who marched to war in 1914 formed a badge of the modernity of the states to which they belonged. They were fit, strong, faultlessly clothed and equipped, armed with weapons of unparalleled lethality, and inspired by the belief that they were free men who, in free activity on the battlefield, would win prompt and decisive victories. Above all they were numerous. No society on earth had ever proportionately put forth soldiers in such numbers as Europe did in August 1914. The intelligence section of the German Great General Staff had evolved a rule of thumb that every million of a nation's population could support two divisions of soldiers, or some 30,000 men. The rule of thumb was narrowly borne out on mobilisation: France, with 40 million population, mobilised 75 infantry divisions (and 10 of cavalry); Germany, with 57 million, 87 divisions (and 11 of cavalry); Austria-Hungary, with 46 million, 49 divisions (and 11 of cavalry); and Russia, with 100 million, 114 divisions (and 36 of cavalry). Since each was formed from a particular locality – the German 9th and 10th Divisions, for example, from Lower Silesia, the French 19th and 20th from the Pas de Calais, the Austrian 3rd and 5th from the vicinity of Linz (Hitler's home town), the Russian 1st, 2nd and 3rd from the Baltic states – their departure denuded their home districts of their young manhood overnight. In the first fortnight of August 1914 some 20 million Europeans, nearly 10 per cent of the populations of the combatant states, donned military drab and shouldered rifles to take the train to war. All had been told and most believed that they would be back 'before the leaves fell'.

It would be four years and five autumns before the survivors returned, leaving on the battlefields some 10 million dead. The vast crop of fit and strong young men which formed the fruit of nineteenth-century Europe's economic miracle had been consumed by the forces which gave them life and health. The original divisions which had mobilised in 1914 had 'turned over' their personnel at least twice and in some cases three times. War-raised divisions had suffered comparable losses, for the conscription machine drove on throughout the war's course, not only consuming new *classes* as each came annually of military age but also spreading its jaws to swallow the older, younger and less fit whom it would have rejected in peacetime. Ten million Frenchmen passed through the military machine between 1914 and 1918; out of each nine enlisted, four became casualties. German fatal casualties exceeded 3 million, Austrian a million, British a million, those of Italy, which entered the war only in May 1915 and fought on the narrowest of fronts, over 600,000; the dead of the Russian army, whose collapse in 1917 permitted the Bolsheviks to seize power, have never accurately been counted. The graves of the Russian dead, and those of the Germans and Austrians who opposed them, were scattered from the Carpathians to the Baltic; those of the French, British, Belgians and Germans who fell on the Western Front were concentrated in a narrow belt of frontier territory forming cemeteries which have become major and permanent land-marks in that countryside. Those constructed by the British – for which Edwin Lutyens, the great neo-classicist, designed the architecture and Rudyard Kipling, himself a bereaved Great War parent, wrote the funerary inscriptions, 'Their name liveth for evermore' and, on the tombs of the unidentified dead, 'A soldier of the Great War, known unto God' – are places of heartrending beauty.

'Cities of the dead' they have been called, though 'gardens of the dead' is more apt; they are supreme achievements of that romantic landscape art which is one of England's donations to world culture. But they were filled from zones which in their time were cities of the living, foci of activity, emotional and intellectual as well as physical, more intense than any Europe had known since the French Revolution. 'The front cannot but attract us,' the French Jesuit philosopher, Teilhard de Chardin, had written, 'because it is in one way the *extreme* boundary between what you are obviously aware of and what is still in the process of formation. Not only do you see these things that you experience nowhere else but you also see emerge from within yourself an underlying stream of clarity, energy and freedom that is to be found hardly anywhere else in ordinary life.' Teilhard de Chardin's rhetoric harks back directly to that of the barricades, those of 1871, 1848, ultimately of 1789; and with good reason. The trenches of the Western Front were indeed barricades. Alan Seeger, a poet and victim of the trenches, called them 'disputed barricades' – across which the emancipated youth of Europe levelled their rifles, symbols of their status as free citizens, in defence of the values of liberty, equality, fraternity. The nineteenth century had given these values to all, but nationalism had persuaded each citizen that they inhered meaningfully only in the state to which he belonged. Revolution, its fathers had quite genuinely believed, would be a gift freely given to all, a gift whose effect would be to foster a fraternity of nations as well as of people. It had, none the less, never been successfully internationalised. Even at its dawn it had manifested itself as the dynamic of a single nationality alone; when its values came to be more widely diffused, their transmission, by a bizarre perversion, succeeded only in reinforcing the *amour-propre* of each nation among which they rooted. The French Revolution persuaded the

French – as it still does – that they were unique in their devotion to equality; its influence reinforced the Germans' commitment to fraternity; its proclamation of liberty convinced the British that they already possessed it more fully than latecomer claimants to their freeborn rights ever could.

The fruits of victory

The states to which the First World War brought both victory and its fruits – France and Britain foremost – were able to adjust the sense of suffering they had undergone to their belief in the higher values that had animated their war-making without grave damage to their national psyches. For each of them, in a real but unexpressed material dimension, the First World War had been worth the sacrifice. Despite the human and, in the case of France, material cost, the war had re-energised and expanded their home economies, even if much overseas investment had been liquidated to purchase raw materials and finished goods in the process; more important, it had greatly expanded their overseas possessions. Britain and France, in that order, remained in 1914 the most important of the world's imperial powers (a major factor in motivating Germany to attack them); by 1920, after the distribution of the defeated powers' possessions under League of Nations mandate, their empires had become larger still. France, already dominant in North and West Africa, added Syria and Lebanon to its Mediterranean holdings. Britain, head of the largest imperial association the world had ever seen, extended it by the addition to its East African colonies of German Tanganyika, thus making the dream of an Africa British 'from Cairo to the Cape' a reality; at the same time it acquired the mandates for Palestine and Iraq, ex-Turkish territories, and so established its power over a 'fertile crescent' running from Egypt to the head of the Persian Gulf.

Crumbs from the table of the German and Turkish empires fell elsewhere; South-West Africa and Papua to South Africa and Australia, Rhodes to Italy, Germany's Pacific islands to Japan – a sop which only time would reveal as ill considered. Italy and Japan believed they deserved more, particularly since the greater allies picked up crumbs too. Their sense of being skimped would feed dangerous rancours in the years to come. But the rancour of these unfavoured victors was as nothing compared to that of the vanquished. Both Austria and Turkey, ancient contestants for mastery in Europe's middle lands, would develop the resignation to adapt to reduced circumstances. Germany would not. Its sense of humiliation bit deep. Not only had it lost the trappings of an embryo colonial power as well as the marches of its historic advance into central Europe in West Prussia and Silesia. It had also lost command of a strategic zone so extensive and central that as late as July 1918 its possession had promised victory, and thereby control of a new empire in the European heartland.

On 13 July 1918, the eve of the Second Battle of the Marne, German armies occupied the whole of western Russia up to a line which touched the Baltic outside Petrograd and the Black Sea at Rostov-on-Don, enclosed Kiev, capital of the Ukraine and historic centre of Russian civilisation, and cut off from the rest of the country one-third of Russia's population, one-third of its agricultural land and more than one-half of its industry. The line, moreover, was one not of conquest but of annexation, secured by an international treaty signed at Brest-Litovsk in March. German expeditionary forces operated as far east as Georgia in Transcaucasia and as far south as the Bulgarian frontier with Greece and the plain of the Po in Italy. Through her Austrian and Bulgarian satellites Germany controlled the whole of the Balkans and, by her alliance with Turkey, extended her power as far away as northern Arabia and northern Persia. In Scandinavia, Sweden remained a friendly neutral, while Germany was helping Finland to gain its independence from the Bolsheviks – as Latvia, Lithuania and Estonia were also shortly to do. In distant south-east Africa a German colonial army kept in play an Allied army ten times its size. And in the west, on the war's critical front, the German armies stood within fifty miles of Paris. In five great offensives, begun the previous March, the German high command had regained all the territory contested with France since the First Battle of the Marne fought four years earlier. A sixth offensive promised to carry its spearheads to the French capital and win the war.

Five months later the war had indeed been won, but by the French, British and Americans, not by Germany. Her soldiers, beaten back to the Belgian frontier by the Allied counter-offensives of July, August and September, had learned in November of the armistice their leaders had accepted, had marched back across the Rhine to home territory and had there demobilised themselves. Within days of their return, the largest army in the world, still numbering over 200 divisions, had returned its rifles and steel helmets to store and dispersed homeward. Bavarians, Saxons, Hessians, Hanoverians, Prussians, even the immortals of the Imperial Guard, decided overnight, in defiance of every imperative by which the German Empire and the European military system had been built over the preceding fifty years, to stop their ears to superior orders and resume civilian life. Cities, towns and villages which since 1914 had been empty of young men suddenly repossessed them in cohorts; but the Berlin government, which had counted unreflectively on the availability of boundless military force for a hundred years, disposed of none whatsoever.

The Freikorps phenomenon

States cannot survive in a military vacuum; without armed forces a state does not exist. This truth was soon discovered by the socialists who came to power after the fall of the Kaiser, committed though they were to popular instead of autocratic government. Confronted by armed communist insurrection and Russian Bolshevik intervention – in Bavaria, in the Baltic and North Sea ports, in Berlin itself – the German Social Democratic government took military help wherever they could find it. It was not a time to be choosy and the choice was not delicate. Friedrich Ebert, Chancellor of the new republic and a lifelong socialist, announced, 'I hate the Social Revolution like sin!'; but he can scarcely have liked the soldiers whom the crisis threw his way. 'War had taken hold of them', Ernst von Salomon wrote of the young republic's first protectors, 'and would never let them go. They would never really belong to their homes again.' The men of whom he spoke – and he was one of them – were a type thrown up by any great military convulsion. They had congregated at Cape Taenarum in the Peloponnese in the fifth century BC after the wars of the Greek city-states – landless men looking for mercenary hire. Germany had been full of them during the Thirty Years War, as had the whole of Europe after the fall of Napoleon, when many had made a living by going to fight for the Greeks in the war of independence against the Turks. In November and December 1918 they called themselves *Frontkämpfer* – 'front fighters', men who had learned in the trenches a way of life from which the onset of peace could not wean them. General Ludwig von Maercker, organiser of the first of the republic's *Freikorps*, spoke of forming 'a vast militia of bourgeoisie and peasants, grouped around the flag for the re-establishment of order'. His vision harked back to a pre-industrial military system in which artisans and farmers united to repress anarchy and sedition. In truth no such system had ever existed. The *Freikorps* were a manifestation of a much more modern principle – the post-1789 belief that a political being was a citizen armed with a rifle which he was trained to use in defence of the nationality to which he belonged and the ideology that nationality embodied.

It was significant that Maercker's original *Freikorps*, the Volunteer Territorial Rifle Corps (*das Freiwillige Landesjägerkorps*), included a tier of 'trusted men' (*Vertrauensleute*) intermediate between officers and rank-and-file, and that its disciplinary code stipulated that 'the leader of a volunteer corps must never inflict a punishment capable of touching a man's honour'. The *Landesjägerkorps*, in short, embodied the idea that statehood was ultimately military in origin, that citizenship was validated by military service, that service should be freely given and that the serviceman's duty of obedience should always be mitigated by the honour owed to him as a warrior. Here was the ultimate realisation of the political philosophy proclaimed by the fathers of revolution in France 130 years earlier.

Maercker's original *Freikorps* was rapidly replicated all over the new German republic; in addition, *Freikorps* sprang up in the regions over which *Deutschtum* ('Germanness') had historic claims to dominate, in the borderlands disputed with the new state of Poland, in the Baltic lands winning their independence from Russia and in the German-speaking remnants of the Habsburg Empire. The titles adopted by such *Freikorps* – the word was itself a direct reference to the popular units raised in Prussia against Napoleon in 1813-14 – were indicative of their ethos: the German Rifle Division, the Territorial Rifle Corps, the Border Rifle Brigade, the Guard-Cavalry Rifle Division, the Yorck von

Wartenburg Volunteer Rifle Corps. There were many others, and some would go to form brigades, regiments or battalions of the 'hundred thousand men' army that Versailles would eventually allow the German republic. Others would naturally disband but take on clandestine existence as the political militias of the parties of the extreme right in Weimar Germany; their defeated left-wing equivalents would survive as the camouflaged street-fighting units of the Red Front.

The *Freikorps* phenomenon was not confined to the German lands alone. Wherever peoples were divided by ideology, as they were in Finland and in Hungary, to say nothing of Russia in the era of Civil War, it appeared, and often hydra-headed. The post-war world was awash with rifles, with rootless and rancorous men and with freebooting officers who knew how to lead them; but it was in Italy that it took its most purposive form. Italy seethed with rancours, diplomatic and domestic. It had benefited little by its blood sacrifice; the acquisition of Trieste, the South Tyrol and the Dodecanese islands was little recompense for 600,000 dead. The survivors benefited from victory not at all. The costs of the war drove post-war Italy into an economic crisis with which the traditional parties, liberal and religious alike, were unable to deal. The only leader to promise salvation was the *Freikorps*-type Benito Mussolini, who advocated military-style solutions to the country's problems. His *Fascio di Combattimento* drew its activists from ex-servicemen, among whom former *arditi* (stormtroopers) were foremost. Their programme, proclaimed on the eve of the 'March on Rome' which delivered the government to the Fascists in October 1922, was 'to hand over to the King and Army a renewed Italy'.

The idea of the army as a social model – centrist, hierarchical and supremely nationalist – was to energise politics over a wide area of Europe throughout the post-war years. It took no root in the great victor nations, France and Britain, nor in the settled bourgeois democracies of northern Europe and Scandinavia. But it proved deeply attractive in the defeated nations, in the successor states of the dismembered empires and in the underdeveloped countries on the European fringe, particularly Portugal and Spain. There the strains of adaptation to democracy or self-government and to the unfamiliar market forces of a suddenly unstable international economy seemed best solved by calling a halt to competition between classes, regions and minorities and consigning authority to a militaristic and often uniformed political high command. The polarisation of politics between the military and political principles would even manifest itself in Bolshevik Russia, where much of the victorious revolutionaries' bureaucratic energies after the defeat of the Whites in 1920 would be devoted to emasculating the Red Army as an alternative political force.

Uniforms and titles of rank were pushed to the margin of political life in Lenin's and then Stalin's Russia. In Italy they dominated the centre; in Austria and Germany they hovered in the wings, poised to occupy the stage at the moment the drama of events gave them their cue. Elsewhere – in Hungary, in Poland, in Portugal, in Spain – career colonels and generals took over and exercised power without the hesitations that their equivalents in states of liberal tradition felt they owed to the conventions of representative rule. A strange transvaluation of the ideal of 1789 took possession of the public life of these countries. Military service was seen no longer as the token by which the individual validated his citizenship but as the form in which the citizen tendered his duty to the state and took part in its functions. 'Every citizen a soldier and every soldier a citizen' had borne a creative and even beneficent meaning in a society like that of France before the Revolution, where the two states of being were historically and sharply separate. In societies where they had become undifferentiated, soldierly obedience all too easily supplanted civic rights in the relationship between masses and government. So it came to be in Italy after 1922; so it would be, comprehensively and fatally, in Germany after 1933.

No European of his time had more potently imbibed the soldierly ethic than Adolf Hitler. As a subject of the Habsburg Empire, he had evaded conscription into its army because that entailed service with the non-Germans – Slavs and Jews – whom he despised. August 1914 offered him the chance to enlist as a volunteer in a unit of the German army and he eagerly seized it. He quickly proved himself a good soldier and served bravely throughout the war, an event that produced in him 'a stupendous impression – the greatest of all experiences. For that individual interest – the interest of one's own ego – could be subordinated to the common interest – that the great heroic struggle of our people demonstrated in overwhelming fashion.' The defeat of November 1918 outraged him as intensely as any of those who joined a *Freikorps* - as he might himself have done. Instead he found a position which better suited his talents and exactly encapsulated that interpenetration of political by military principles of which he would eventually make himself the supreme practitioner. In the spring of 1919 he was appointed a *Bildungsoffizier* in the Weimar Republic's VII District Command, with the task of instructing soldiers of the new army in their duty of obedience to the state. It was a propagandist's job, created by the army for the purpose of inoculating the men against contagion by socialist, pacifist or democratic ideas. *Bildung* is a word of manifold meanings ranging from 'formation' through 'education' to 'culture' and 'civilisation'. The self-taught and dreamily romantic Hitler would have been aware of all of them and conscious of his responsibility not merely to warn against dangerous influences but also to form minds and attitudes. It can have surprised him not at all that the army command in Munich simultaneously encouraged him to join an embryo nationalist movement, the German Workers' Party, nor that his superior, Captain Ernst Röhm, not only fed it with members drawn from the *Freikorps* but also joined it himself; so too did other veterans of Hitler's wartime regiment, Lieutenant Rudolf Hess and Sergeant-Major Max Amann. Röhm quickly organised the toughest ex-soldiers and *Freikorps* men into a party street-fighting force, the *Sturmabteilung* (SA). By 1920 the essential elements of the Nazi Party were in place.

Like its communist antithesis, the Red Front, and its Italian equivalent, the *Fascio di Combattimento*, the Nazi Party was military in ethos, organisation and appearance from the outset. It chose brown as its uniform colour, from that of the victorious British army, whose Sam Browne belt it also adopted; from the elite mountain rifle regiments it borrowed the peaked ski-cap; and its members wore knee-length boots, an age-old symbol of the rough-riding warrior. On parade it formed ranks behind legionary banners; on the march it stepped out to the beat of the drum. Only the absence of rifles differentiated it from an army proper; but, in Hitler's vision, political victory would bring it weapons also. The triumph of the National Socialist revolution would abolish the distinction between party and army, citizen and soldier, and subordinate every German and everything in Germany – parliament, bureaucracy, courts, schools, business, industry, trades union, even churches – to the *Führerprinzip*, the principle of military leadership.

TWO

Fomenting World War

Military leadership implies military action. The first public act of Hitler's political life was to lead a *Putsch* – an attempted military coup – against the constitutional government of the German republic. He had been contemplating it for five years. 'I can confess quite calmly', he disclosed at Munich in 1936, 'that from 1919 to 1923 I thought of nothing else than a *coup d'état.*' During those years Hitler had led a double life. As the leader of a party seeking members and support he had spoken constantly, tirelessly – and electrifyingly – to any audience that he could command throughout the area of his political base in Bavaria. He spoke of the 'criminals of Versailles', of Germany's sufferings in the World War, of her losses of territory, of the iniquity of the disarmament terms, of the presumptions of the new states – Poland most of all – which had been raised on historic German soil, of the extortion of reparations, of the national shame, of the part played by the enemies within – Jews, Bolshevists, Jewish Bolshevists and their liberal republican puppets – in bringing Germany to defeat in 1918. On 25 January 1923, at the first Nazi 'Party Day' in Munich, in a speech which might stand for all his others, he proclaimed: 'First of all, the arch-enemies of German freedom, namely, the betrayers of the German Fatherland, must be done away with. . . . Down with the perpetrators of the November crime [the signing of the armistice]. And here the great message of our movement begins. . . . We must not forget that between us and those betrayers of the people [the republican government in Berlin] . . . there are two million dead.' This was the central theme of his message: that German manhood had fought honourably, suffered and died in a war that had ended by denying the succeeding generation the right to bear arms. As a result, 'Germany disarmed was prey to the lawless demands of her predatory neighbours.' Those neighbours included the Poles, against whom the *Freikorps* had fought a frontier campaign to defend the Reich territory in 1920, and behind them the Bolshevik Russians and the new Slav states, Czechoslovakia and Yugoslavia, as well as the unstable remnants of the Habsburg Empire, Hungary and Austria, which had been threatened by communist takeover and might be again. They also included the French, the most rapacious of the victors, who had not only taken back the Reich provinces of Alsace-Lorraine, but maintained an army in the Rhineland and openly threatened to use military force to back up their demands for full payment of the costs of the war, which had been determined by the Allies at

Versailles in the form of reparations. The menace of these threats and demands, Hitler endlessly reiterated, could only be set aside when Germany had an army once again, not the paltry 100,000-man force allowed it under the Treaty of Versailles, stripped of tanks and aeroplanes and almost of artillery, but a true national army commensurate in size with that of the largest and most populous state on the continent.

This was a message that magnetised Hitler's audiences, which grew steadily in size throughout 1919-23. He had become a brilliant speaker and, as his power with words increased, so did the numbers who heeded them. 'I cast my eyes back', he was to say in 1932, 'to the time when with six other unknown men I founded [the Nazi Party], when I spoke before eleven, twelve, thirteen, fourteen, twenty, thirty, fifty persons. When I recall how after a year I had won sixty-four members of the movement, I must confess that that which has today been created, when a stream of millions is flowing into our movement, represents something unique in German history.' The stream of millions had not yet begun to flow in 1923; his followers were still only numbered in thousands. They responded ecstatically, however, to his call for revenge. 'It cannot be', he said at Munich in September 1922, 'that two million Germans should have fallen in vain and that afterwards one should sit down at the same table as friends with traitors. No, we do not pardon, we demand – vengeance!' Some of them also responded to his call for violent action; for the other side of Hitler's double life was as an organiser of a 'parallel' army within the Weimar Republic as a conspirator against it. By 1923 the *Sturmabteilung* (SA) numbered 15,000 uniformed men, with access to an ample store of hidden arms, including machine-guns; moreover, Hitler believed it had the promise of support by the legitimate army of the state, the Bavarian division of the *Reichswehr*. Hitler had been encouraged in that belief by many of the division's officers, most importantly by Captain Ernst Röhm, the future head of the SA, who until 1923 was also a serving soldier. Through him, but also because of the attitude of the army commander in Bavaria, General Otto von Lossow, Hitler had formed the impression that if the SA and its associated militias, together forming the extreme right-wing *Kampfbund* (Battle League), were to stage a *Putsch* the army would not oppose it. What such a *Putsch* needed was leadership and a pretext for action. Hitler would supply the leadership – though he conceded the role of figurehead to General Erich Ludendorff, the retired First World War chief of staff (technically First Quartermaster General), who had put the *Kampfbund* under his patronage. The pretext was provided by the French. In January 1923, in order to force the German government to sustain its reparations payments, which it insisted it was incapable of meeting, the French government sent troops to occupy the Ruhr, Germany's industrial heartland, to extract payment at source.

This intervention intensified a currency crisis within Germany, in part engineered by its own treasury to substantiate the payment difficulties, and it had the effect of fuelling an inflation that destroyed both the working man's purchasing power and the middle classes' savings. The value of the mark, which stood at 160,000 to the dollar in July (in 1914 it had exchanged at four), declined to a million to the dollar in August and 130,000 million in November. Gustav Stresemann, the German Chancellor, at first declared a campaign of passive resistance in the Ruhr, but this did nothing to deter the French, while the example of illegality it gave encouraged communists in Saxony and Hamburg, separatists in the Rhineland and former *Freikorps* men in Pomerania and Prussia to threaten civil disobedience. When, after quelling these disorders, Stresemann announced the end of the passive resistance campaign, Hitler decided his moment had come. On 8 November, at a prearranged public meeting in the Bürgerbräu Keller in Munich, which General von Lossow and the Bavarian Commissioner of State had unwisely agreed to attend, Hitler arrived armed, with armed men outside, put Lossow and the other notables under arrest and announced the formation of a new German regime: 'The government of the November criminals and the Reich President are declared to be removed. A new National Government will be nominated this very day, here in Munich. A German National Army will be formed immediately. . . . The direction of policy will be taken over by me. Ludendorff will take over the leadership of the German National Army.'

Next day, 9 November 1923, the nucleus of the National Army, the *Kampfbund*, set out to march on the former Bavarian War Ministry building, with Hitler and Ludendorff at its head. Röhm and the SA had taken possession of the War Ministry and were awaiting their arrival; interposed between were armed policemen, barring Hitler's way across the Odeonsplatz. Hitler bargained his way through the first cordon. The second held its ground, opened fire, killed the man at Hitler's side (who pulled Hitler to the ground in his dying grasp), put a bullet into Goering, the future commander of the Luftwaffe, but left Ludendorff untouched. He marched ahead, indifferent to the bloodshed about him, but reached the War Ministry to find only one other at his side. The German National Army had disintegrated.

The immediate consequences of the 'Beer Hall *Putsch*' were banal: nine of the conspirators were tried; Ludendorff was acquitted and Hitler was sentenced to five years' imprisonment, of which he served only nine months, just long enough to dictate to Rudolf Hess (an old comrade of Hitler's regiment) the text of his political manifesto, *Mein Kampf.* The long-term consequences of the trial had a deeper significance. In his closing speech to the court, a speech reported throughout Germany and which made him, for the first time in his career as a demagogue, a national figure, Hitler expressed his relief that it was the police and not the *Reichswehr*, the army, which had fired on him and the *Kampfbund*. 'The *Reichswehr*', he said, 'stands as untarnished as before. One day the hour will come when the *Reichswehr* will stand at our side, officers and men. . . . The army we have formed is growing from day to day. . . . I nourish the proud hope that one day the hour will come when these rough companies will grow to battalions, the battalions to regiments, the regiments to divisions, that the old cockade will be taken from the mud, that the old flags will wave again, that there will be a reconciliation at the last great divine judgement which we are prepared to face.'

This was both Hitler's public and his private verdict on his *Putsch* tactics. 'We never thought to carry through a revolt against the army,' he disclosed at Munich in 1933. 'It was *with it* we believed we should succeed.' After the Munich *Putsch* he changed tactics decisively. He never again undertook illegal action against the state but sought instead to achieve power constitutionally through the ballot box. The point of seeking power, however, though he did not disclose this aim publicly, was to acquire constitutional command of the army and the War Ministry and budgetary authority to vote military credits for rearmament. In the ten years that followed the failure of the *Putsch* Hitler did nothing to discourage the growth of the SA, which on the eve of his seizure of power in 1933 had reached a strength of 400,000, four times the size of the *Reichswehr*. Nor did he discourage the stormtroopers from believing that, when the day came, they would put off their brown, put on field-grey and emerge as soldiers of the 'National Army' he had promised to bring into being in Munich in 1923. He did, however, take care to see that the SA was kept under strict discipline, that its boasts of being ready to seize power by force were silenced, that its pretensions to be a replacement rather than a reinforcement for the *Reichswehr* were deflated, and that its leaders were dissuaded from representing themselves as military rather than political figures. After Munich Hitler remained in no doubt that the generals, with their creed of *Überparteilichkeit* ('being above party'), were a power in the land he could not afford to alienate.

Hitler and the Nazi revolution

Economic crisis had provided Hitler with a false opportunity in 1923. Economic crisis again provided him with opportunity in 1930, and between then and his assumption of the German chancellorship in January 1933 he used it with discreet and consummate skill. In the six years after the catastrophic inflation of 1923, Germany had made a good recovery. The currency had been stabilised, credit restored, industry revitalised and unemployment successfully contained. The sudden world crisis of 1929, which destroyed credit across central Europe, brought much of that achievement to naught. Unemployment in Germany, a nation of 60 million people, rose from 1,320,000 in September 1929 to 3 million a year later, 4.5 million the year after that and over 6 million in the first two months of 1932. Hardship once again spread through the land and the moderate parties of the Weimar Republic, committed to orthodox, pre-Keynesian policies of budget balancing, could find no means to redress it. The parties of the extreme right and left benefited accordingly at the parliamentary elections called as one government after another collapsed under the pressure of events.

In the election of September 1930 the Nazi Party polled 18.3 per cent of the vote, but in July 1932 it increased its share to 37.3 per cent, winning 230 seats and becoming the largest party in the Reichstag. In the words of Alan Bullock, 'with a voting strength of 13,700,000 electors, a party membership of over a million and a private army of 400,000 SA and SS . . . [Hitler] was the most powerful political leader in Germany, knocking on the doors of the Chancellery at the head of the most powerful political party Germany had ever seen.' The parallel success of the Communist Party positively reinforced Hitler's appeal to those voters who were terrified by the spectre of Bolshevism, which they believed had been laid by the violent defeat of the Spartacists in 1919; the Communist Party enormously increased its support in 1930 and again in 1932, when it won 6 million votes and a hundred seats.

The communists too had their private army, the Red Front, which fought street battles with the SA that frequently ended in death. Nazi street violence tainted the Nazi cause; communist street violence – which in July 1932 alone caused the deaths of thirty-eight Nazis and thirty communists – raised the prospect of communist revolution. Though that could not win Hitler a parliamentary majority – which he failed to achieve by 6.2 per cent even after the seizure of power in 1933 – it could and did frighten the moderate politicians into accepting Hitler as a counterweight who might be used to offset revolutionary with merely radical extremism, as they believed Nazism to be. In January 1933, after a number of makeshift ministries had fallen, President Paul von Hindenburg, the war hero, was advised by his ministers to offer Hitler the chancellorship. On 30 January he was installed.

What followed was one of the most remarkable and complete economic, political and military revolutions ever carried through by one man in a comparable space of time. Between 30 January 1933 and 7 March 1936 he effectively restored German prosperity, destroyed not only opposition but also the possibility of opposition to his rule, re-created, in a spectacularly expanded German army, the principal symbol of the nation's pride in itself, and used this force to abrogate the oppressive treaties which defeat had imposed on the nation while he was still a humble soldier. He had luck, notably in the timely death of Hindenburg in August 1934, and in the incendiary attack on the Reichstag building in February 1933. The Reichstag fire allowed him to conjure the fiction of a communist threat to parliamentary institutions, and so panic the moderates into voting with the Nazis for a suspension of parliamentary powers: the Enabling Bill they enacted conferred on Hitler the right to pass binding laws by appending his signature to the necessary document. Hindenburg's death opened the way for him to combine the office of the presidency with his own as Chancellor under the title of Führer, a position in which he exercised the authority of both head of government and head of state. But Hitler did not succeed between 1933 and 1936 purely by luck. His economic policy was not based on theory, certainly not Keynesian theory; but it amounted to a programme of deficit budgeting, state investment in public works and state-guaranteed industrial re-equipment of which Keynes would have approved. This was accompanied by a calculated destruction of the trade-union movement, which removed at a blow all restrictions on free movement of labour between jobs and workplaces, and the effect on unemployment was startling: between January 1933 and December 1934 the number of unemployed declined by more than half, many of the 3 million new workers finding jobs in the construction of the magnificent network of motorways (*Autobahnen*) which were the first outward symbol of the Nazi economic miracle.

Moreover, he succeeded in his plan to rearm Germany not by rushing bullheaded at the disabling clauses of the Versailles Treaty but rather by waiting until the victor nations gave him pretext. Thus he did not announce the reintroduction of conscription until March 1935, when the French, beset as before the First World War by a falling birth-rate, themselves announced that they were doubling the length of their conscripts' military service. Hitler was able to represent this move as a threat to German security which justified the enlargement of the 100,000-man army; on 17 March he also announced the creation of an air force – another breach of the Versailles Treaty. Even so he blurred his intentions by offering France a pact which would limit the size of his army to 300,000 men and that of his new air force to 50 per cent of hers. France's refusal permitted him to fix larger totals.

Hitler and the generals

The reintroduction of conscription gave him by 1936 an army with a skeleton strength of thirty-six divisions, a fivefold increase from the seven of the *Reichswehr*. Few were as yet fully equipped or manned, and, as his generals warned him, he certainly lacked the strength to resist any armed reaction to his anti-Versailles policies. In seeking to realise his deeply held ambition to remilitarise the Rhineland, therefore, he waited once again until he could find the semblance of a legal cause, which he claimed to see in the French parliament's ratification of a mutual-assistance pact with the Soviet Union in March 1936. Since the pact bound France to take action against Germany in the event of German aggression against the USSR, Hitler was able to represent it as a unilateral violation of the provision that France would never make war on Germany except by resolution of the League of Nations – a creation of Versailles from which he had withdrawn in 1933 – and to allege that such a violation justified his taking measures to improve Germany's defence of its frontier with France. On 7 March 1936 he accordingly ordered the reoccupation of the Rhineland, where no German soldier had been stationed since November 1918, correctly confident that the French would not move to expel the force he sent, even though it numbered not even one division but a mere three battalions.

Although Hitler's generals had been apprehensive about the Rhineland adventure, they were not fundamentally disposed to argue with his diplomatic or strategic judgements, since the armed forces, among all the other institutions of state, had up to that moment been the principal beneficiaries of the National Socialist revolution. They had been spared *Gleichschaltung*, the process by which every organ of German life was brought directly under Nazi control; moreover, the leaders of the body which had threatened them with *Gleichschaltung*, the SA, had been summarily and brutally killed in June 1934. Hitler's half-formulated promise that the stormtroopers would one day become soldiers of the new Germany had been made good only in the sense that after March 1935 the younger of them received their call-up papers and found themselves embodied in the Wehrmacht as conscripts among hundreds of thousands of others who had never worn the brown uniform. The armed forces had also benefited more generously than any other body from the programme of state investment. Tanks and aeroplanes – enough to equip a Panzer force of six divisions (soon to be raised to ten) and a Luftwaffe of 2000 combat aircraft – were now coming out of the new armaments factories in a steady stream. The design work which underlay their development had been done in Russia during the brief period of Russo–German friendship in the 1920s. In an ill-calculated act of appeasement in 1935 the British Admiralty had agreed that the German navy should also be partially liberated from the provisions of the Versailles Treaty, and it had begun to acquire capital ships and even U-boats, in numbers equivalent to 33 and 60 per cent respectively of the Royal Navy's fleets. This material largesse enormously enhanced the institutional *amour propre* of the Wehrmacht which, after fifteen years in which it had starved for both men and equipment, suddenly found itself advanced to the front rank of the armed forces of Europe, almost as strong as the largest and better armed than any. Professionally, moreover, Hitler's rearmament programme transformed the career prospects of individual officers: in 1933 the average age of a colonel was fifty-six; by 1937 it had been reduced to thirty-nine, while many in the *Reichswehr* who had reconciled themselves to retirement found themselves by 1937 commanding regiments, brigades, even divisions.

Hitler's seduction of his professional officers was as calculated as any other part of his programme, though he rightly attached more importance to it than the rest. His attitude to the SA had always been duplicitous; though he had needed and been glad to use the political fighting force it had given him in the 'time of struggle' before 1933, he was himself too much the true veteran, the seasoned 'front fighter', to reckon its street bullies proper military material. Hitler was, in many respects, a military snob – and with reason: he had fought in the First World War from beginning to end, suffered wounds and won a high decoration for bravery. The army he wished to re-create would be a model of the one in which he had served, not a disorderly political militia reclothed in field-grey. The Blood Purge of June 1934, when Hitler organised the murder of Röhm and the rest of the paramilitary radicals who had thought to leap to general's rank by political hopscotch, had ensured that he had his way. One consequence of the purge of the SA was the rise of the rival military arm of the Nazi Party – the blackshirted *Schutzstaffel* (SS), a highly disciplined elite corps led by Heinrich Himmler.

Although the generals had been careful to know nothing of the 1934 murders, the results had none the less put Hitler high in their favour; but the reverse was not the case. There was a strict limit to Hitler's military snobbery. He was a combat snob, not a worshipper of rank or title. As he well knew, many of the Wehrmacht's elite, the Great General Staff officers who were now senior commanders, had not fought at the front in the First World War, their brains being thought too valuable to be risked beyond headquarters. Their military as well as their social *hauteur* therefore grated with him. One of the innumerable rancours that he nursed dated back to the Munich trial, when General von Lossow, his fainthearted ally, had testified that he regarded him as no more than 'a political drummer boy'; the wound had been salted by the state prosecutor's statement that the drummer boy had 'allowed himself to be carried beyond the position assigned to him'. It was Hitler who now assigned positions everywhere – except within the army, which retained control of its own promotion structure. However, since the generals continued to choose officers as timorous as they themselves had been over the remilitarisation of the Rhineland, Hitler decided to end the system. He wanted a war army, led by commanders determined to take revenge on the victors of 1918 and their creature states erected on the back of Germany's defeat.

Werner von Fritsch, the army commander-in-chief, was a particular bugbear among the fainthearts; in November 1937 he sought a private interview with Hitler to warn against policies that might provoke war. Two months later, the indiscreet remarriage of the Minister of War, General Werner von Blomberg, provided Hitler with an opportunity to get rid of both men: Blomberg's young bride was discovered to have been a prostitute; while the unmarried Fritsch, his obvious successor, fell speechless when confronted by trumped-up charges of homosexual behaviour. Their enforced retirement did not immediately bring him generals of the bellicose temper he wanted; but it provided him with the pretext to establish a new supra-service command in place of the War Ministry, the *Oberkommando der Wehrmacht* (OKW), of which Hitler made himself the head, and the OKW was given responsibility for the highest level of strategic planning. This was a crucial move, for 1938 was to be the year in which Hitler moved from rearmament to the diplomatic offensive. He had already outlined his intentions to his service commanders on 5 November 1937, when he had argued that Britain and France were unlikely to oppose with military force German moves to strengthen its military position in the east. His first priority was to take advantage of the enthusiasm among German nationalists in Austria for union (*Anschluss*) with the Reich; his second was to attempt the annexation of the German-speaking parts of Czechoslovakia, the Sudetenland. Further, he hoped that Italy, Austria's protector, would shortly be brought to Germany's side by a formal alliance with Mussolini, his fellow dictator. Poland, on which he had longer-term designs, he believed would be immobilised by the speed of Germany's action.

In November 1937 Mussolini did indeed accept a German alliance, the Anti-Comintern Pact against the Soviet Union (originally signed by Germany and Japan a year earlier), thus reinforcing the 'Rome-Berlin Axis' agreement of October 1936. By March 1938 Hitler felt free to act against Austria. He first demanded that Austrian Nazis should be installed in key government posts. When Kurt von Schuschnigg, the Austrian Chancellor, refused, Arthur Seyss-Inquart, the Austrian Nazi leader, was instructed to declare himself the head of a provisional government and request German intervention. On 12 March German troops marched in, *Anschluss* was declared the following day, and on 14 March Hitler made a triumphal entry into Vienna, where he had spent his unhappy and aimless youth. Britain and France protested but did no more. Their inactivity was the confirmation Hitler needed that he could safely proceed to his diplomatic offensive against Czechoslovakia. In April he ordered OKW to prepare plans for a military operation, meanwhile instructing the Nazi groups among the Sudetenland Germans to sustain demands for secession. In August he fixed October as the date for military action and on 12 September, when he delivered a fiery anti-Czech speech at Nuremberg, German troops moved to the frontier.

This 'Czech crisis' seemed to threaten war, even though it was not clear who would fight it. The Czechs were not powerful enough to resist the rearmed Wehrmacht without help, but the Red Army, the only nearby source of assistance, could come to their aid only by crossing Polish territory (or Romanian, but the Romanians were pro-German), a manoeuvre which the Poles, with their deep hostility to and well-founded suspicion of the Russians, were not disposed to permit. The British and the French were also disinclined to see Russia intervening in central Europe and, though France had a treaty with Czechoslovakia, and both Britain and France recognised that honour and prudence demanded that they should not allow Czechoslovakia to be dismembered, they could see no way of protecting her except by military action of their own in the west, from which government and people in both countries shrank. Neither had yet modernised their forces, though they had begun reluctantly to rearm; more to the point, neither had yet developed the will to back protest with force, as was lamentably demonstrated by their succession of failures to implement collective action against aggressors through the League of Nations machinery – against Japan for its aggression in Manchuria in 1931 and in China in 1937, against Italy for its aggression against Ethiopia in 1936. Edouard Daladier and Neville Chamberlain, the French and British Prime Ministers, therefore counselled President Eduard Beneš of Czechoslovakia to acquiesce in Hitler's demands, even though the cession of the Sudetenland meant the cession also of the country's frontier fortifications; once surrendered, Czechoslovakia would have no protection whatsoever against further German demands. Nevertheless Beneß felt obliged to agree, since the Western democracies would not stand by him. The crisis seemed to be settled, but on 22 September Hitler decided to harden his terms. Instead of waiting for an international commission to delimit the revised frontier, he demanded the Sudetenland at once. It was this turning of the screw which provoked the crisis called 'Munich', since it was there that Chamberlain and Daladier went to treat with Hitler again on 29-30 September, in a series of craven meetings that conceded him even more than he had initially demanded.

Munich, it is generally said, marked 'the end of appeasement'; certainly it sent Daladier and Chamberlain home, superficially relieved, but convinced – Chamberlain more strongly than Daladier – that rearmament must henceforth proceed apace. More accurately, however, Munich marked the moment when Hitler abandoned caution in his campaign of aggressive diplomacy and began to take the risks which would stiffen the will of the Western democracies to meet challenge with firm response and eventually force with force. The turning-point was Hitler's treatment of browbeaten Czechoslovakia. Having seized the Sudetenland only six months before, on 11 March 1939 he arranged for the pro-German separatist party in the Slovakian half of what remained of the country to announce their secession and request that he become their protector. When the new Czech President, Emil Hacha, arrived in Berlin to protest, he was physically bullied into requesting a German protectorate over the whole of Czechoslovakia. The following day, 15 March, German troops marched into Prague just in time to form a guard of honour and a protective screen for Hitler when he entered the city on their heels.

The rape of Czechoslovakia drove the democracies to act. The French cabinet agreed that when Hitler next moved he must be stopped. On 17 March Chamberlain publicly announced that if there were further attacks on small states Britain would resist 'to the utmost of its power', a clear warning that Hitler now risked war. Hitler did not believe or did not fear the threat. Since January he had been menacing Poland, to which belonged the largest slice of territory that had been German before 1918, in particular the 'corridor' which divided East Prussia and the German-speaking Free City of Danzig from the Reich heartland. The Poles doggedly resisted his threats and continued to do so even when on 23 March, as an earnest of intentions, he occupied the port of Memel, a former League of Nations territory on Poland's border which had been German until 1918. They were chiefly sustained by the knowledge that Britain and France were now preparing to extend them a guarantee of protection; and on 31 March, eight days after publicly announcing that they would defend Belgium, Holland or Switzerland against attack, Britain and France issued a joint declaration guaranteeing the independence of Poland. Two weeks later, on 13 April, to demonstrate the general hardening of their attitude, they issued similar guarantees to Romania and Greece after Mussolini, in imitation of Hitler, annexed Albania.

Poland, however, was the focus of the growing crisis, which France and Britain now hoped best to solve by drawing the Soviet Union into a protective agreement, even though they knew the Poles were reluctant to accept any help from their traditional enemy. The French and British were themselves mistrustful of the Soviets, besides harbouring a deep dislike of their political system, feelings which were exactly reciprocated. Without Polish resistance, however, an agreement might have been reached; but the Poles adamantly refused to contemplate the Red Army operating on their soil, since they rightly suspected that the Russians desired to annex large parts of Polish territory and might hold these under occupation as their reward for intervention. The British and French could offer Stalin no compensatory inducement to act with them in a hypothetical crisis; during the summer of 1939 the negotiations between the Western democracies and Stalin hung fire.

Hitler, on the other hand, could offer powerfully tempting inducements. He too had been negotiating desultorily with Stalin during the spring and summer, encouraged by hints that Russia had no taste for risking war, even over the future of a country as important to the security of its western border as Poland. The discussions seemed to make no progress, since neither side would reveal its hand. Then in late July Hitler decided to gamble with a thinly veiled offer to let Stalin take a slice of eastern Poland if he agreed not to impede a German invasion of the country from the west. The Russians responded with keen interest and on 22 August the two Foreign Ministers, Molotov and Ribbentrop, signed a non-aggression pact in Moscow. Its secret clauses effectively permitted the Soviet Union, in the event of a German-Polish war, to annex eastern Poland up to the line of the Vistula and the Baltic states of Latvia, Lithuania and Estonia.

Poland was now doomed. On 15 June the German army staff, the *Oberkommando des Heeres* (OKH), had settled on a plan which provided for two army groups, North and South, to attack simultaneously with their objective as Warsaw. Because northern Poland was dominated by the German province of East Prussia, while southern Poland bordered on Czechoslovakia, now an extension of German territory (as the Protectorate of Bohemia-Moravia and the puppet state of Slovakia), Poland was deeply outflanked across the whole length of its two most vulnerable frontiers. Its fortified zone lay in the west, covering the industrial region of Lower Silesia, and since Germany's annexation of Czechoslovakia time had not allowed for new fortifications to be built. The Polish government was naturally concerned to protect the richest and most populous region of the country; it remained ignorant of the Molotov-Ribbentrop Pact, and so of the Russian threat to its army's rear; and it counted on the French, with British assistance, to attack Germany's western border in order to draw off German divisions from the east as soon as the Wehrmacht marched.

Hitler's calculations were different. He believed, correctly as it turned out, that the French would not move against him in the west, which he left defended by only forty-four divisions – to oppose the nominal hundred of the French army – and that the British could do little to hurt Germany during the brief span of time he intended the Polish campaign to fill. He had the advantage of being mobilised, whereas the British and French were not. He had the even more important advantage of deploying superior numbers and immeasurably superior equipment against the Poles. The German Army Group North and South together numbered some sixty-two divisions, of which six were armoured and ten mechanised, supported by 1300 modern combat aircraft. Although the Poles had begun to mobilise in July as war became imminent, they had not fully deployed all their men by 1 September. Together they formed forty divisions, of which none was armoured; the few Polish tanks were old, light models, sufficient to equip only a single brigade; and half the 935 aircraft of the air force were obsolete.

The campaign in Poland

Hitler nevertheless still needed a pretext to attack. He was briefly deterred on 25 August by the news that Britain had entered into a formal alliance with Poland which guaranteed protection against aggression by a third party, and a few days of inconclusive diplomatic sparring followed. On 28 August, however, he formally abrogated the 1934 non-aggression pact with Poland, signed at a time when her army far outnumbered the Wehrmacht, and on the evening of 31 August received news of Polish aggression near the Silesian border town of Gleiwitz; the incident had in fact been carefully staged by his own SS. Next morning, at 4.45 am, his tanks began to cross the frontier. Since it was Hitler's pretence that Germany had been attacked by Poland, he issued no declaration of war.

By the end of 1 September the Polish air force had largely ceased to exist, many of its aircraft having been caught on the ground and destroyed by the Luftwaffe, which also bombed Polish headquarters, communications and cities. All the Wehrmacht ground forces made rapid progress. On 3 September the French and British governments delivered separate ultimatums demanding the withdrawal of German troops from Poland; both ultimatums expired that day and a state of war therefore existed between them and Germany. By that date, however, the Fourth Army advancing from Pomerania had made contact with the Third advancing from East Prussia and had cut off the 'Polish Corridor' to Danzig and Gdynia, Poland's outlet to the sea. By 7 September, after a Polish attempt to stand on the line of the river Warta, west of Warsaw, had failed, the Tenth Army had advanced from the south to within thirty-six miles of the capital, while the Third Army, driving down from the north, was on the river Narew, twenty-five miles away. There was now a German change of plan. It had been expected that most of the Polish army would be entrapped west of the

Vistula, on which Warsaw stands. By rapid disengagement, however, large numbers of troops got across the river and marched to concentrate on the capital to fight a defensive battle there. The German commanders therefore ordered a second and deeper envelopment, aimed at the line of the river Bug, a hundred miles east of Warsaw. While it was in progress, the one and only crisis for the Germans occurred. The Polish Poznan Army, one of those entrapped west of the Vistula, turned and attacked the German Eighth and Tenth Armies from the rear, inflicting heavy casualties on the surprised 30th Division in the first impact. A bitter encirclement battle ensued, ending with the capture of 100,000 Polish troops on 19 September.

Warsaw had been encircled by 17 September; in an effort to reduce its garrison's resistance by terror, it was heavily bombed until 27 September, when the defenders finally capitulated. All hopes of escaping eastward into the remote and difficult country bordering the Pripet Marshes were ended when the Red Army, after appeals for assistance from the Germans on 3 and 10 September, finally moved its White Russian and Ukrainian Fronts across the frontier on 17 September. Some 217,000 of the 910,000 Poles taken prisoner in the campaign fell into Russian hands. By 6 October all Polish resistance had ended. Some 100,000 Poles escaped into Lithuania, Hungary and Romania, whence many would make their way to France and later Britain, to form the Polish armed forces in exile and continue the struggle – as infantrymen in the Battle of France, as pilots in the Battle of Britain, and later on other fronts – until the last day of the war.

At the conclusion of the campaign the Wehrmacht, which had suffered 13,981 fatal casualties in Poland, immediately began to turn its victorious divisions westward to man the Siegfried Line or West Wall and prepare for a campaign against the British and French, who had made no attempt at all to divert German forces, except for a small flurry of activity between 8 September and 1 October known as the 'Saar Offensive'. The only immediate military outcome of the Polish campaign lay not in the west but in the east. There Russia at once capitalised on the terms of the Molotov-Ribbentrop Pact to demand basing rights for its troops in Lithuania, Latvia and Estonia, a manoeuvre which eventually led to the annexation of all three countries to the Soviet Union in June 1940.

The Winter War

Stalin also moved against Finland, though with altogether less convenient results. Finland had been Russian territory between 1809 and 1917; when it won its independence after fighting against Russian and local Bolsheviks during the Russian Civil War, it had obtained a frontier demarcation which Stalin decided ran too close for strategic comfort to Leningrad and the Soviet Baltic ports. On 12 October 1939, a week after Latvia had signed its dictated treaty, the Soviet Union confronted the Finnish government with demands for naval basing rights and the cession of a large strip of Finnish territory in the Karelian isthmus leading to Leningrad. The Finns stonewalled until 26 November, when the Soviet Union staged a border incident. On 30 November the Russians attacked with four armies, deploying thirty divisions; for this blatant act of aggression they were expelled from the League of Nations on 14 December. The Soviet Union was eventually to commit a million men to the campaign. The Finns, though their total mobilised strength never exceeded 175,000, fought back with skill and success. Perhaps the most warlike of all European peoples, and certainly the hardiest, the Finns made circles around their Russian attackers in the snowbound wastes of their native forests, employing so-called *motti* or 'logging' tactics to cut off and encircle their enemies, who were regularly disorientated and demoralised by a style of warfare for which their training had not prepared them. While the main strength of the Finnish army defended the Karelian isthmus on the Mannerheim Line, named after the country's commander-in-chief, who had won the war of independence in 1918, independent units attacked, encircled and destroyed Soviet divisions on the long eastern flank between Lake Ladoga and the White Sea.

In December the Finns actually counter-attacked from the Karelian isthmus, after a series of operations by the Soviets described by Mannerheim as 'similar to a performance by a badly directed orchestra'. By January, however, the Russians had taken the measure of their opponents, recognised their own underestimation of the Finns' military prowess, and brought up sufficient forces to overwhelm them. During February they broke their way through the Mannerheim Line by main force, inflicting casualties which the Finnish government recognised its tiny population could not bear. On 6 March it treated for peace and on 12 March signed a treaty which conceded the demands Russia had made in October; they had lost 25,000 dead since the war had begun. The Red Army, however, had lost 200,000, of whom perhaps the majority had died of exposure while surrounded or out of touch with base. The experience of the 'Winter War', which would be renewed as the 'Continuation War' after June 1941, conditioned the Soviet Union's carefully modulated policy towards Finland when the issue of peace came round again.

Finland had briefly been an inspiration to all enemies of the Axis powers, with which, following the Molotov-Ribbentrop Pact, the Soviet Union was identified during 1940. Britain and France had even considered affording her military assistance, and winter-warfare units from both countries were earmarked to join the Finnish army; fortunately for the future of Soviet-Western relations, the Finns had sued for peace before they were sent.

The Scandinavian campaign
The end of the Winter War did not, however, terminate Anglo-French military involvement in northern Europe. According to the German navy, which kept a

close watch on Scandinavian affairs, Western military assistance for Finland would most probably have passed through Norway, and in doing so would not only have violated Norwegian neutrality but menaced German access to the Kiruna-Gällivare iron ore fields in Sweden which supplied Germany's war economy with a vital commodity. Hitler's Grand Admiral, Erich Raeder, was in any case anxious to acquire north Norwegian bases from which to operate against the Royal Navy, and therefore urged Hitler throughout the autumn and winter of 1939 to pre-empt the Allies by authorising an intervention in Norway. Preoccupied by his plans for the forthcoming attack in the west, Hitler would not allow his interest to be aroused, though in December, after Raeder had arranged for the Norwegian Nazi leader, Vidkun Quisling, to be brought to Berlin, he did authorise OKW to investigate whether Norway would be worth occupying. In mid-February his indifference was dissipated by a blow to his pride.

At the outbreak of the war the *Graf Spee*, one of Germany's 'pocket battleships', had undertaken a commerce-raiding campaign against British merchant shipping in the South Atlantic but had eventually been cornered off the coast of Uruguay by three British cruisers. Its commander had been forced to scuttle it at Montevideo after the Battle of the River Plate on 13 December 1939. The British people were heartened and Hitler consonantly infuriated by this humiliation of the German surface fleet. On 16 February Hitler was even more outraged when the *Altmark*, a supply ship which had tended the *Graf Spee* during its cruise, was intercepted by HM Destroyer *Cossack* in Norwegian territorial waters and 300 British merchant seaman taken by the *Graf Spee* were liberated. He at once decided that Norwegian territorial waters must be denied to the British for good, preferably by invasion and occupation, and instructed General Nikolaus von Falkenhorst, a mountain-warfare expert, to prepare a plan. Falkenhorst quickly concluded that it would be desirable also to occupy Denmark as a 'land bridge' to Norway, and by 7 March Hitler had assigned eight divisions to the operation. Intelligence then indicated that Allied plans to intervene in Norway, providing the legal pretext for aggression on which Hitler normally insisted, had been called off. Raeder nevertheless succeeded in persuading him that the operation was strategically necessary and on 7 April the transports sailed.

Denmark, quite unprepared for war, almost unarmed and with no suspicion that Germany harboured hostile intentions against her, surrendered under the threat of an air bombardment of Copenhagen on the morning of the troops' landing on 9 April. The Norwegians were also taken by surprise. They were, however, ready to fight and at Oslo the ancient guns of the harbour fort held the invaders at bay – sinking the German cruiser *Blücher* – long enough for the government and royal family to escape and make their way to Britain. The survivors of the small Norwegian army then gathered as best they could to oppose the German advance up the coast towards the central cities of

Andalsnes, Trondheim and Namsos, and to counter the German landing in the far north at Narvik. They did not, however, have to fight alone. Because of the preparations made to intervene in Finland, both the British and the French had contingents ready to move and debark. Between 18 and 23 April 12,000 British and French troops were put ashore north and south of Trondheim and advanced to meet the Germans who were making their way north from Oslo up the great valleys of the Gudbrandsdal and the Osterdal. The Germans defeated the leading British brigade in the Gudbrandsdal on 23 April and compelled it to withdraw by sea from Andalsnes, then made contact with their own landing party at Trondheim and forced the evacuation of the rest of the Allied troops through Namsos on 3 May.

In the north the fortunes of war swung the other way. The German navy suffered a serious defeat in the two battles of Narvik, fought on 10 and 13 April between a superior British force and the destroyers transporting General Eduard Dietl's mountain troops. Ten of the destroyers, with a high proportion of Dietl's force, were sunk in the Narvik fiords. Dietl escaped ashore with only 2000 mountain infantry and 2600 sailors with whom to oppose 24,500 Allied troops, including the resolute Norwegian 6th Division. He found himself besieged in Narvik from 14 April onwards and was eventually forced to break out and retreat to the Swedish border, which he reached at the end of May. The collapse of the Allied front in France, however, then brought the campaign to an end, since both the French and the British ordered their troops home through Narvik to replace the losses suffered in the *Blitzkrieg* battles with the Wehrmacht which began on 10 May.

Dietl, though in many respects the least successful of the German generals of 1939-40, was to become Hitler's favourite; his death in an aeroplane crash in June 1944 was regarded by the Führer as a wounding personal tragedy. By then he had come to regard Dietl as irreplaceable and he attempted to conceal the news of his death from the Finns, among whom Dietl had established a towering reputation during the Finnish 'Continuation War' of 1941-4, lest it discourage them further at a time when defeat by the Russians again stared them in the face. Hitler liked Dietl because he argued with him in an explosive, soldierly way that perhaps reminded the Führer of his own army service. He liked him even more because at Narvik he had rescued him from humiliation. So alarmed had Hitler been by the miscarriage of the landing that he had been on the point of ordering Dietl to escape into Sweden and intern his soldiers rather than risk having to surrender them to the British. He had eventually been dissuaded from sending the signal, and in any case Dietl's dogged conduct of the siege and retreat made it unnecessary. Dietl was the model of what Hitler wished every German soldier to be, the type he had looked forward to recruiting and training in thousands from the moment he embarked on the creation of the Wehrmacht. The proof of his quality was his snatching of victory from the jaws of defeat in the mountains of north Norway in June 1940, and so sustaining unblemished the record of German military success since the beginning of the war. To the campaign simultaneously unfolding in the west, however, not even a Dietl could have added a jot to the dimensions of German victory. There *Blitzkrieg* seemed a magic which had taken possession of the army itself.

PART I
THE WAR IN
THE WEST
1940–1943

BLITZKRIEG IN THE WEST, 1940

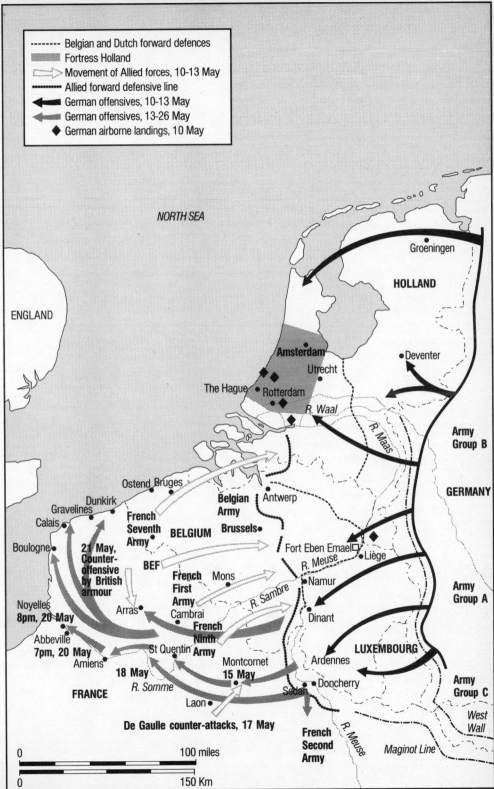

Legend:
- ---- Belgian and Dutch forward defences
- Fortress Holland
- ⇨ Movement of Allied forces, 10–13 May
- •••••• Allied forward defensive line
- ← German offensives, 10–13 May
- ← German offensives, 13–26 May
- ◆ German airborne landings, 10 May

NORTH SEA

ENGLAND

HOLLAND

Groeningen

Deventer

Amsterdam
Utrecht
The Hague • Rotterdam
R. Waal

Army Group B

GERMANY

Ostend Bruges

Belgian Army

Antwerp

Dunkirk
Gravelines
Calais

French Seventh Army

BELGIUM

Brussels

Boulogne

BEF

21 May, Counter-offensive by British armour

Fort Eben Emael
R. Meuse   Liège
R. Meuse
Namur

Army Group A

French First Army

Mons

Noyelles
8pm, 20 May

Arras

Cambrai

R. Sambre

Dinant

Abbeville
7pm, 20 May
Amiens

French Ninth Army

St Quentin

18 May
R. Somme

Montcornet
15 May

Ardennes

LUXEMBOURG

FRANCE

Laon

Sedan • Doncherry

Army Group C

De Gaulle counter-attacks, 17 May

French Second Army

R. Meuse

West Wall

Maginot Line

0   100 miles

0   150 Km

THREE

The Triumph of *Blitzkrieg*

Blitzkrieg – 'lightning war' – is a German word but not known to the German army before 1939. A coining of Western newspapermen, it had been used to convey to their readers something of the speed and destructiveness of German ground-air operations in the three-week campaign against the ill-equipped and outnumbered Polish army. However, as the German generals themselves readily conceded, the Polish campaign had not been a fair test of the army's capabilities. Despite allegations by some of them that the Wehrmacht had not shown itself the equal of the old imperial army – allegations which drove Hitler to a frenzy of rage against General Walther von Brauchitsch, the commander-in-chief, at a meeting in the Reich Chancellery on 5 November – the plodding Polish infantry divisions had offered no match to the mechanised spearheads of Guderian and Kleist. *Blitzkrieg* aptly described what had befallen Poland.

Would *Blitzkrieg* avail against the West? Hitler persisted into October in hoping that its spectacle would persuade France and Britain to accept his Polish victory; but their rejection, on 10 and 12 October respectively, of his peace tentatives, offered in a speech to the Reichstag on 6 October, persuaded him that Germany must make war again. His ambitions required at least the defeat of France, which might persuade Britain to sue for separate terms and inaugurate that accommodation of her maritime with his continental empire for which his upbringing as a subject of the old landlocked Danubian empire led him unrealistically to hope. On 12 September he had told his Wehrmacht adjutant, Rudolf Schmundt, that he believed France could be conquered quickly and Britain then brought to negotiate; on 27 September he warned the commanders-in-chief of the three services that he intended to attack in the west shortly; and on 9 October, even before France and Britain had rejected his peace offer, he issued Führer Directive No. 6 for a western offensive.

In an accompanying memorandum, which accused France and Britain of having kept Germany weak and divided since the Peace of Westphalia in 1648, he announced that nothing less was at stake than 'the destruction of the predominance of the Western powers in order to leave room for the expansion of the German people'. Führer Directive No. 6 described how that destruction was to be achieved:

An offensive will be planned . . . through Luxembourg, Belgium and Holland

[and] must be launched at the earliest possible moment [since] any further delay will . . . entail the end of Belgian and perhaps of Dutch neutrality, to the advantage of the Allies. The purpose of this offensive will be to defeat as much as possible of the French army and of the forces of the Allies fighting on their side, and at the same time to win as much territory as possible in Holland, Belgium and northern France to serve as a base for the successful prosecution of the air and sea war against England and as a wide protective area for the economically vital Ruhr.

The plan of attack, codenamed *Fall Gelb* ('Case Yellow'), was to be worked out in detail by the high command of the army, the *Oberkommando des Heeres* (OKH). Although Hitler as Supreme Commander laid down broad strategic aims, he did not as yet involve himself in technical military affairs. Hitler nevertheless had strong if not clear ideas of what he wanted 'Case Yellow' to achieve. Here was to be the making of a strategic imbroglio which would set Führer and army at loggerheads for the next five months. Historically the German army, and the Prussian army before it, had always deferred to the fiction that the head of state was warlord – *Feldherr*. However, not since Frederick the Great had led his soldiers in person against those of the tsar and the Holy Roman Emperor had a head of state actually interfered in his generals' planning. Kaisers Wilhelm I and II, at the onset of war with France in 1870 and 1914, had transferred their courts to the army's headquarters; but they had both then surrendered detailed control of operations to their chiefs of staff – the Moltkes, Falkenhayn and Hindenburg in sequence. Hitler would have been willing to do the same had the successors of those men shared his vision of what the reborn German army could, with the Luftwaffe, achieve; but the commander-in-chief, Brauchitsch, was a doubter and his chief of staff, General Franz Halder, a quibbler. Halder was a man of brains, a product of the Bavarian General Staff Academy whose graduates were thought intellectually more flexible than those of the Prussian *Kriegsakademie*. His war experience, however, had been as a staff officer employing the step-by-step tactics of the Western Front; his original arm of the service had been the artillery, also dominated by step-by-step thinking; and he was a devout member of the State Lutheran Church and thereby conditioned to recoil from Hitler's brutal philosophy of domination, national and international, yet not to defy constituted authority by opposing it. As a result, he proposed a form for 'Case Yellow' which, as he admitted elsewhere, would postpone the mounting of a decisive offensive against France until 1942. As outlined on 19 October, his plan was designed to separate the British Expeditionary Force from the French army and to win ground in Belgium which would provide airfields and North Sea ports for the German navy's and air force's operations against Britain, but not to achieve outright victory.

Thus he acquiesced in the letter of Führer Directive No. 6 but succeeded in denying its spirit. The expedient temporarily baffled Hitler, who lacked allies among the military establishment able to help him argue against Halder. On 22 October he unsettled his chief of staff by demanding that 'Case Yellow' begin as soon as 12 November; on 25 October he confronted Brauchitsch with the suggestion that the army attack directly into France instead of northern Belgium; and on 30 October he proposed to General Jodl, his personal operations officer, that the Wehrmacht's tanks be flung into the forest of the Ardennes, where the French would least expect them. Without expert military support to endorse these proposals, however, he could not jog 'Case Yellow' forward.

General Staff resistance rested on solid ground. Late autumn was no season for undertaking offensive operations, least of all on the sodden plains of rainy northern Europe. The Ardennes, even if its narrow valleys led directly to the open French countryside north of the fortified zone of the Maginot Line, was not the obvious terrain for the deployment of tanks. Hitler's wishes therefore seemed beggars looking for horses to ride – until Halder's plan came the way of fellow professionals and their rejection of its limitations reached Hitler's ear. The process took time, time which saved any revision of 'Case Yellow' from miscarrying, and it resulted ultimately in a fruitful outcome; for Halder was right to argue that late autumn was the wrong season for the attack on France but mistaken in believing that a bold strategy would not yield large results.

The professionals who took Hitler's side were the commander-in-chief of Army Group A, Gerd von Rundstedt, and his chief of staff, Erich von Manstein. The significance of Rundstedt's defection from the General Staff plan was his degree of influence, as one of the most senior generals in the army and commander of the strongest concentration of force on the Western Front. The significance of Manstein's opposition to Halder's 'Case Yellow' was that he enjoyed Rundstedt's support and possessed one of the best military minds in the Wehrmacht. At the outset he knew nothing of Hitler's dissatisfactions with Halder's plan. It merely affronted him as a half-hearted approach to a problem that instinct told him was susceptible of a full-blooded solution. As autumn weather worsened into winter, however, his instinct led him to advance one criticism after another of the Halder plan, each converging as if by steps in blind-man's-buff with Hitler's desires for the outcome of 'Case Yellow', and each at the same time laying the foundations for what would come to be called 'the Manstein plan'.

On 31 October the first of six memoranda he was to write arrived at OKH. It argued that the aim of 'Yellow' must be to cut off the Allied forces by a thrust along the line of the Somme, thus chiming in with Hitler's idea of 30 October for an attack through the Ardennes. Brauchitsch, the commander-in-chief, rejected it on 3 November but conceded that more armour should be allotted to Rundstedt's Army Group A. Meanwhile, as bad weather forced one postponement of the Halder plan after another, Hitler vented his rage in person against his generals for their half-heartedness. He was determined on victory, he warned at the Reich Chancellery on 23 November, and 'anyone who thinks otherwise is irresponsible'. Manstein called support from other middle-rank professionals, notably Guderian, the tank expert, to endorse his conception of a knockout blow into northern France. Even discounting the possibility that the French and British would do him the favour of throwing too strong forces into Belgium – precisely what they were contemplating, though he could not know that – he was moving over more certainly to the conviction that a drive to divide the enemy forces along the line of the Somme was the correct strategy. Guderian's assurance that a tank force, if made strong enough, could negotiate the Ardennes, cross the Meuse and deliver the knockout blow reinforced him in that view.

Hitler, despite his differences with Halder, was still allowing his urge to victory to overcome his doubts in Halder's plan. 'A-Days', which would have set it in motion, were fixed four times in December and a final one for 17 January 1940. On 10 January, however, two Luftwaffe officers crash-landed in Belgium with parts of the 'Yellow' plan in a briefcase. Enough survived after their attempts to incinerate the documents, the German military attaché to Holland discovered, to compromise the offensive and to oblige the army to make a clean breast of things to Hitler. After his rage subsided – it resulted in the dismissal of the commander of the Second Air Fleet and his replacement by Albert Kesselring, who was to prove one of the most talented German generals of the war – Hitler postponed 'Yellow' indefinitely and demanded a new plan 'to be founded particularly on secrecy and surprise'.

Here was Manstein's opening. However, the last of his six memoranda had so tried Halder's patience that he had arranged in December for Army Group A's chief of staff to be given command of a corps, a theoretical promotion but, since the corps was in East Prussia, an effective dismissal of his troublesome junior from a post of influence. Protocol required, however, that corps commanders on appointment should pay their respects to the head of state. The ceremony ought to have been a formality; but on this occasion chance took Schmundt, Hitler's Wehrmacht adjutant, by Manstein's Coblenz headquarters where he got wind of the Manstein plan. It so uncannily matched the Führer's aspiration, though in a 'significantly more precise form', that he ensured Manstein should have a whole morning with the Führer on 17 February. Hitler was entranced, converted, and thereafter did not rest until Brauchitsch and Halder too had accepted the Manstein plan – which he passed off as his own conception.

OKH then demonstrated its institutional strengths. The direct descendant of the old Prussian Great General Staff, it worked merely as the handmaiden of a strong master. Hitler had thitherto shown the strength of will but not of mind to call forth its talents. Now that it had a clear expression of its master's voice, it concentrated all its efforts on transforming the elements of the Manstein-

Hitler conception – for an attack by strong armoured forces through the Ardennes forest into the rear of the Franco-British field army north of the Somme – into a detailed and watertight operation order. It worked fast. Only a week after Hitler's morning of enlightenment by Manstein, it produced a proposal, codenamed *Sichelschnitt*, 'Sickle Stroke', which was a transformation of their half-formed ideas. The theme of its plan was a reversal of Schlieffen's from 1914. That great chief of staff – already dead by the time his conception was tested on the field of battle – had based his victory plan on the expectation that the French would push into Germany south of the Ardennes, allowing the German armies to outflank them through Belgium. 'Sickle Stroke' was based on the expectation that in 1940 the French, with their British allies, would push into Belgium, allowing the German armies to outflank them through the Ardennes. It was a brilliant exercise in double-bluff, all the more so because it reinsured against the expectation's miscarrying. For, even if the Franco-British army did not push into Belgium, the unexpectedness of the Ardennes thrust and the power and mass of the armoured force with which it was to be mounted promised an excellent chance of catching the enemy in the rear and toppling him off-balance.

'Sickle Stroke' allotted the three German army groups the following missions. Army Group B, the northernmost, commanded by General Fedor von Bock, was to attack into Holland and northern Belgium, with the aim of tempting the Franco-British field army as far eastward as possible and seizing territory from which it could be outflanked from the north. Army Group C (commanded by General Wilhelm Ritter von Leeb), the southernmost, was to engage the garrison of the Maginot Line, penetrating it if possible. Rundstedt's Army Group A, in the centre, was to advance through the Ardennes, seize crossings over the great water obstacle of the Meuse between Sedan and Dinant, then drive north-west, along the line of the River Somme, to Amiens, Abbeville and the Channel coast. It was to command seven of the ten available Panzer divisions to spearhead the advance, leaving none for Leeb and only three for Bock.

Bock, displeased by his secondary role, emphasised to Halder the risks of the plan in a brilliantly withering 'worst case' analysis. 'You will be creeping by, ten miles from the Maginot Line, with the flank of your breakthrough and hope that the French will watch inertly! You are cramming the mass of the tank units together into the sparse roads of the Ardennes mountain country, as if there were no such thing as airpower! And you then hope to be able to lead an operation as far as the coast with an open southern flank 200 miles long, where stands the mass of the French army.' To German officers of their generation, Bock's warning to Halder recalled the German army's last 'open flank' operation into France, in 1914 – the long dusty roads overcrowded with marching troops, the French nowhere to be found, the unprotected lines of communication ever lengthening, the great walled fortress of Paris, bulging with troops and artillery, looming unreduced in the rear until, like a thunderclap, the French counter-stroke was launched, the first Battle of the Marne lost, the German spearheads sent trundling into reverse and the urgent footfalls of manoeuvre warfare drowned by the thud of spades digging the first trenches of the Western Front.

Bock was right to warn that the stagnation of another Western Front awaited the Wehrmacht if 'Sickle Stroke' miscarried; he was wrong to warn that it might miscarry as the Schlieffen Plan had done in 1914. For one thing, the Maginot Line, unlike the fortress of Paris in 1914, was not a *place d'armes* from which a counter-attack force could spring panther-like against the German army's flank. On the contrary, its conformation and structure imprisoned its garrison within it, consigning them to a purely frontal defence against frontal attack, which it was not Rundstedt's role to deliver. For another, the German army would not be 'creeping by' the Maginot Line; its tank spearheads, if they could negotiate the Ardennes and cross the Meuse, would be driving onward at thirty or forty miles a day, as they had in Poland and as the French army, wherever its mass stood, was not organised to do. As to airpower, there was certainly 'such a thing', but the Luftwaffe was superior in quality of aircraft and in tactics of ground-air operations, considerably superior in numbers and far superior in fighting experience to the *Armée de l'Air* and the Advanced Air Striking Force of the RAF combined.

Hermann Goering's Luftwaffe would reveal its deficiencies later in the war; but in 1940 its strengths were paramount. Unlike its British and French equivalents, which had over-diversified in aircraft production and procurement – trying to build too many types at home and then being forced into purchasing from America to replace unsatisfactory models – it had concentrated on procuring a large number of a few types of aircraft, each of which was finely adapted to its specialised function. The Messerschmitt 109 was an excellent example of what today would be called an 'air-superiority fighter', fast, manoeuvrable, heavily armed and with a high rate of climb. The Junkers 87 was a formidable ground-attack dive-bomber, particularly when protected by the Me 109 and as long as ground-air defence depended upon the visually aimed anti-aircraft gun. The Heinkel 111 was an effective medium bomber, at least for daylight operations. Some alternative German types – the Dornier 17 bomber, the Messerschmitt 110 heavy fighter – were to prove misconceptions; but in 1940 the Luftwaffe was burdened with none of the obsolescent or obsolete types which equipped the French and British squadrons. Moreover, its senior officers included a number of first-rate men – Milch, Jeschonnek and Kesselring – whose transfer from the army to the air force was a token of their competence; too many senior officers of the French and British air forces, by comparison, were also-rans who had forsaken the army to restart frustrated careers.

The commonality of training shared by German air force and army officers – Jeschonnek had passed first out of the *Kriegsakademie* – ensured that the

Wehrmacht's tactics of ground-air operations were fine-tuned. The staffs of its ten Panzer divisions knew that when they called for air support it would arrive on time, where and how they required it. This ensured a massive increment to the Panzers' power, which was in any case formidable. German tanks were not, model for model, notably superior to those of the British and French armies. The Mark IV Panzer, the army's future main battle tank, was well armoured but undergunned. The Mark III, its workhorse, was inferior in protection both to the British Infantry Tank Mark I and the French Somua, the latter an advanced design whose all-cast hull would influence that of the American Sherman of 1942-5. However, the German tanks were integrated into 'all-tank' formations, the Panzer divisions, which were not only 'tank-heavy' – that is, unencumbered by unmechanised infantry or artillery – but also trained to maximise the tank's characteristics: speed, manoeuvrability and independence of action. By contrast, the British had only one armoured division, which was still in the process of forming; while the French, with more tanks than the Germans (3000 to 2400), had distributed half (1500) among their slow-moving infantry divisions, allotted others (700) to bastard 'cavalry' and 'mechanised' divisions, and kept only 800 to form five armoured divisions, of which in 1940 three were active and one – commanded by the wayward Charles de Gaulle – was still forming. Germany's ten Panzer divisions were not only homogeneous in composition, as a result of the reorganisation of the 'light' into true tank divisions since the Polish campaign; they were also subordinated to higher Panzer headquarters commanded by Hoepner (XVI Panzer Corps), Hoth (XV), Guderian (XIX) and Reinhardt (XLI). Guderian's and Reinhardt's Corps, with Wietersheim's XIV Mechanised Corps, a formation of motorised infantry divisions which included integral tank battalions, actually composed a separate entity, Panzer Group von Kleist. At the time of its creation it was a revolutionary organisation, the largest armoured force existing in any army and the forerunner of the great tank armies which were to sweep across the battlefields of the world in 1941-5.

The Maginot mentality

It was these dense concentrations of tanks which made the German army so menacing an opponent of those of the Western Allies, as the two sides watched and waited either side of the Franco-German frontier in the spring of 1940. The French army, 101 divisions strong, scarcely differed in character from that of 1914; it wore the same boots, manned the same artillery, the venerated 75 mm, and marched to the same tunes as under 'Papa' Joffre; many of its commanders had been staff officers to the generals who had led it to war in that terrible August twenty-six years earlier. Moreover, it was still a marching army, its pace of manoeuvre determined by the age-old rhythms of soldier's stride and horse's walk. So too was that of the bulk of the German army, whose 120 infantry divisions were as roadbound as those of the enemy. But the ten German Panzer divisions were not roadbound; the Luftwaffe squadrons that supported them were not even earthbound. Together they indeed threatened 'lightning war' against the groundlings of the Western Alliance. How did the Western generals hope to give them check?

The strategy of the West was founded first, of course, on its belief in the inviolability of the Maginot Line, that 'Western Front in concrete' which had consumed the disposable margin of the French defence budget since the first funds for its construction were voted in January 1930. However, the French commitment to an 'impermeable' military frontier long predated that role. As early as 1922 the French army had determined that its soldiers should never again, as in 1914, have to fight a defensive battle in the open field; and every demographic and economic development since – the declining birth-rate, the static industrial base – had only reinforced that resolution. The original vote for the Maginot Line was for 3000 million francs; by 1935, 7000 million had been spent, one-fifth of the year-on-year military budget, but only 87 miles of fortification had been completed. Fortification experts were satisfied (rightly, as the events of 1940 were to demonstrate) that the money had bought effective protection as far as the line ran, which was along the Franco-German border – but there remained 250 miles of totally unfortified frontier, where France abutted Belgium. Not only had the money lacked to extend the line. The maintenance of good relations with Belgium had argued against its being found; for, on Hitler's reoccupation of the Rhineland in 1936, Belgium had revoked its military treaty with France, declaring itself 'independent' – though not neutral – but made clear its resistance to being left on the wrong side of the Maginot Line if it were extended northwards.

In the event of a German offensive, which seemed certain to be based on the exploitation of Belgian weakness (as in 1914), the French high command would therefore have to launch its mobile field army, with the British Expeditionary Force, into Belgian territory, without having been able to co-ordinate plans with the Belgian General Staff beforehand or reconnoitre the ground on which it was to fight. Nevertheless the French were obliged to accept this highly unsatisfactory basis on which to prepare a defensive battle. On 24 October 1939 General Maurice Gamelin, the French commander-in-chief, issued orders for an advance to the line of the river Schelde in Belgium in the event of a German attack. Three weeks later, on 15 November, when the disadvantages of that scheme had been realised, he issued an amended Directive No. 8 which set the line of advance on the river Dyle, a shorter front which connected the two big Belgian water obstacles, the Schelde estuary and the Meuse, from which river-mobile troops would hold the gap between it and the Maginot Line.

Directive No. 8 had the advantage of bringing the Franco-British force closer behind the projected positions of the Belgian army, which was twenty-two divisions strong on mobilisation and had an excellent military reputation; for all the scorn the Allies were later to heap on the Belgians, the Germans had regarded them since 1914 as tenacious opponents and would continue to do so even after the débâcle that was to come. Their front was also protected, as in 1914, by strong fortifications, particularly along the Meuse, on which much money had been spent.

Could the Belgians, even if fighting independently, win a delay on their frontier with Germany? Directive No. 8 promised an effective strategy. Its success would depend on the operational efficiency of the Franco–British forces of manoeuvre. Of these the British Expeditionary Force composed a homogeneous element, though of mixed quality. 'It was no use', the British Ambassador in Paris had written to Lord Halifax, the Foreign Secretary, in January 1940, 'pointing to the size of the British Navy and Air Force . . . French public opinion wanted large numbers of troops in Europe.' The British had done their best; by December 1939 they had sent all five of their excellent home-based regular divisions to France. However, because the British military system was indeed regular, and yielding trained reserves in very small numbers by comparison with the conscripted armies of France and Germany, that almost exhausted its military resources. Extra divisions had to be found from the voluntary reserve, the Territorial Army, 'Saturday night soldiers' as they were known at home, high in enthusiasm but low in experience and skill. The five extra divisions sent to France between January and April were all Territorial; a final three sent in April were so deficient in training and equipment that even the British categorised them as 'labour' formations. Further, all thirteen were infantry divisions; in May 1940 Britain's only tank formation, the 1st Armoured Division, was still not ready for action. Nevertheless there was an impressive consistency of organisation and spirit in the British Expeditionary Force. The regulars had mobilised with Tommy Atkins's traditional and cheerful indifference to the identity of the King's enemies – or allies ('going to fight them bloody Belgiums', a Tommy had explained to Siegfried Sassoon in 1914) – and the Territorials eagerly aped their sang-froid.

By contrast, the French army was a piecemeal collection of divisions and units, good, indifferent and plain bad. The good included the ten 'active' conscript infantry divisions, which were kept at full strength in peacetime, the seven regular divisions of the colonial army and those North African divisions of the *Armée d'Afrique* which had been brought to France. Less good were the category 'A' reserve divisions mobilised from the younger reservists; some of the category 'B' divisions, mobilised from reservists of over thirty-two, were militarily inert and even insubordinate. Lieutenant-General Alan Brooke, the future British chief of staff, recalled a march-past of such men in November 1939 with disgust: 'men unshaven, horses ungroomed, clothes and saddlery that did not fit, vehicles dirty and complete lack of pride in themselves or their units. What shook me most, however, was the look in the men's faces, disgruntled and insubordinate looks . . . although ordered to give the "eyes left", hardly a man bothered to do so.' The French tank and motorised divisions were of better human quality, but organised on no coherent system; the five light cavalry divisions (DLC) included horse and armoured-car units, the three light mechanised divisions (DLM) armoured cars and light tanks, the four armoured divisions (DCR) tanks only and the ten motorised divisions of track-borne infantry. They were distributed haphazardly among the armies, providing none of the commanders with an equivalent of the solid mass of armoured troops which would form the cutting-edge of Rundstedt's Army Group A. Perhaps the only French units logically trained and equipped to perform an allotted function were the fortress divisions in the Maginot Line, which included units of Indo-Chinese and Madagascan machine-gunners; but they, by definition, were prisoners of their positions and unavailable for deployment elsewhere.

The German army which opposed this miscellaneous Allied host impressed above all by the homogeneity of its composition. It maintained only three types of division: armoured (Panzer), motorised and infantry. The parachute divisions formed part of the Luftwaffe. By May 1940 all ten of its Panzer and all six of its motorised divisions were deployed in the west; so too were 118 infantry divisions, which, since the Polish campaign, differed little in fighting efficiency, whether they were pre-war 'active' or wartime reserve. The only oddities in the German order of battle were the 1st Cavalry Division (effectively a motorised formation), the elite mountain infantry divisions and the two motorised divisions drawn from the SS, the Nazi Party militia. The SS had already demonstrated in Poland a tendency to illegal brutality that it was to amplify in France. Otherwise its units differed from those of the army only in an evident determination to excel in courage on the battlefield.

The simplicity of the German army's organisation was reflected in its command arrangements. Authority over its formations ran from Hitler through his personal headquarters, OKW (*Oberkommando der Wehrmacht*, the Supreme Command), as yet an undeveloped instrument of control, to the army high command (OKH) and then directly to the army groups. In practice, as foreshadowed in Poland, Hitler would deal directly with the General Staff, locating his headquarters close to it, but leave direct operational control to its experts. The Luftwaffe's liaison staff at OKH directly co-ordinated air operations with the army's. On the Allied side, by contrast, operational authority rested with the French Supreme Commander, General Maurice Gamelin, but was exercised first through a Commander Land Forces (General Doumenc) and then by the commander for the north-east, General Alphonse Joseph Georges, under whom came not only the French Army Groups 1, 2 and 3 but also the British Expeditionary Force. The BEF's commander, General Lord Gort, answered operationally to Georges but politically to the British cabinet; but by May 1940, because Gamelin answered politically to his own cabinet, he had developed the habit of dealing directly with Gort rather than through Georges, while Gort ultimately looked to London for orders rather than to La Ferté (Georges's HQ), Montry (Doumenc's) or Vincennes (Gamelin's). It was a further structural weakness of the Allied command system that Gamelin's headquarters were near Paris, those of Doumenc halfway to those of Georges in northern France, those of Gort separate from his, and those of both the British and French air forces separate again. The Royal Air Force in France actually answered to two headquarters: Gort directly controlled the RAF component of the BEF, but the much larger Advanced Air Striking Force came under Bomber Command in Britain. The French air force had three levels of command above its operational squadrons, three separate squadron headquarters, and liaison staffs with both elements of the RAF.

Structural deficiencies were compounded by personal failings. Gort was a famously brave officer who had won the Victoria Cross in the First World War but identified over closely with his fighting battalion commanders. Georges had never properly recovered from a wound suffered during the assassination of the King of Yugoslavia at Marseilles in 1934. Gamelin, once operations officer to Joffre, was simply old – sixty-eight – and, what was worse, tired by age. De Gaulle, who visited him in his remote 'convent-like' headquarters at Vincennes during the phoney war, brought away the impression of a researcher testing the chemical creations of his strategy in a laboratory. Air Marshal Arthur Barratt, commander of the British Air Forces in France, had a more caustic judgement: 'a button-eyed, button-booted, pot-bellied little grocer'. Gamelin's operational directives read like philosophical tracts. No word, written or spoken, that issued from Vincennes carried fire to the men at the front.

Perhaps only a Prometheus could have done that – and there was nothing Promethean about Gamelin. Even the British army, a brotherhood of professional warriors and eager amateurs, approached the war with a sense of *déjà vu;* 'as we have beaten the Germans once, why do we have to do it again?' might have encapsulated their attitude. The French army, drawn from the whole of the nation, scarred by its terrible sufferings of 1914–18 and divided by the extremism of its politics, was touched by a similar sense of pointless repetition, but still more acutely. Albert Lebrun, the French President, noted after his visit to the front a 'slackened resolve, relaxed discipline. There one no longer breathed the pure and enlivening air of the trenches.' Winston Churchill, First Lord of the Admiralty, was 'struck by the prevailing atmosphere of calm aloofness, by the seemingly poor quality of the work in hand, by the lack of visible activity of any kind' on the French front. General Edouard Ruby, of the Second Army, found that 'every exercise was considered as a vexation, all work as a fatigue. After several months of stagnation, nobody believed in the war any more.'

In part the French did not believe because the war was foreseen, not only by common soldiers but also by the generals, as a repetition of the trenches, long-drawn-out and indecisive. The common soldiers and generals of the German army had been given in Poland a vision of a different outcome; if they as yet lacked the faith to believe it could be repeated in the west, Hitler had no doubts. 'Gentlemen,' he told his staff on the eve of 'Case Yellow', 'you are about to witness the most famous victory in history!' On 27 April, persuaded by his reading of captured Allied documents relating to their intervention in Norway that he could not be condemned for his imminent violation of Dutch and Belgian neutrality, he announced to Halder that the attack in the west would begin in the first week of May. Weather forecasts enforced postponement of the date from 5 to 6 May, then 8 May. Finally on 7 May he postponed it again to 10 May 'but not one day after that'. He held to his resolve.

'Late in the evening of Friday 9 May from the Dutch frontier to Luxembourg,' wrote Professor Guy Chapman, 'outposts facing Germany became aware of a vast murmuring on the German side as of the gathering of a host.' A warning of impending attack from the Belgian military attaché in Berlin, delayed in deciphering, was received in Brussels just before midnight. The Belgian high command at once put its army on alert; but by then the German vanguards were already moving to the attack. At 4.30 on the morning of 10 May, airborne units began landing near The Hague and Leyden in Holland and on the crossings of the Meuse in Belgium. The most daring of the airborne attacks was against the Belgian fort of Eben Emael, guarding the junction of the Meuse with the Albert Canal, both key obstacles in the Belgian defence plan. German glider-borne infantry crash-landed on the roof of the fort, penned the defenders inside and, using concrete-piercing charges, overwhelmed them by the sheer surprise of their descent.

Surprise afflicted no one worse than the Dutch, who were genuine neutrals. They had taken no part in the First World War, wanted no part of the Second and commended themselves as an enemy only because parts of their territory, notably the strip known as the 'Maastricht appendix', offered an easy way round the Belgian water obstacles. The ability of the Dutch to defend their territory was minimal. Their army, only ten divisions strong, had not fought a war since 1830. Their air force had only 125 aircraft, half of which were immediately destroyed on the ground by surprise attack. Their best hope of delaying defeat, as they had learnt in the Eighty Years War against the Spanish three centuries earlier, was to retreat inside their waterlogged zone around Amsterdam and Rotterdam and trust to the network of its canals and rivers to delay the invader. The strategy which had cost Spain decades of campaigning was unhinged by German airpower in a few hours. By overflying the water defences of 'Fortress Holland' with streams of Junkers 52 transport aircraft on the morning of 10 May the Luftwaffe landed the whole of the 22nd Airborne Division in its heart, there to await the arrival of Army Group B's tanks. Despite the brave resistance of the Dutch army, the blowing of several vital bridges through the miscarriage of German surprise attacks, and the intervention of the French Seventh Army, the German airborne troops did not have long to wait. On the morning of 13 May, as the German armoured spearheads reached out to join hands with them as they were on the point of capturing Rotterdam, the Luftwaffe misunderstood a signal from the ground announcing their success and bombed the city centre flat. It was the first 'area' operation of the Second World War and a raid which killed 814 civilians. But it effectively ended Dutch resistance, prompting the Queen of the Netherlands to embark on a ship of the Royal Navy for a British port – she had asked to be taken to another part of her kingdom – and causing the Dutch high command to capitulate the following day. As Queen Wilhelmina left, she forecast that 'in due course, with God's help, the Netherlands will regain their European territory'. The Dutch people, who were to pass through the cruellest of German occupations in western Europe, were not to foresee that the Dutch empire in the East Indies would also be lost to them before liberation eventually came.

No word of criticism has ever been levelled against the Dutch by either victors or vanquished of 1940. Not so the Belgians. Although the German army found their soldiers stalwart in action – the official historian of the German 18th Division spoke of their 'extraordinary bravery'; the German opponent of Hitler, Ulrich von Hassell, judged that 'among our adversaries the Belgians fought the best'; while Siegfried Westphal, later to be chief of staff of the German armies defending France against invasion in 1944, noted that 'it was astonishing to see that the Belgians fought with increasing tenacity the nearer the end of the war approached' – the British and French, both during the crisis of 1940 and ever afterwards, insisted on laying blame for much of what befell them on the Belgian army, King and government.

King Leopold's chief military adviser, General Robert van Overstraeten, has been characterised as the 'evil genius' of the 1940 campaign, resisting liaison with the French and British before the German attack and succumbing to defeatism as soon as it began. There is certainly something in both charges; but the truth was that Belgium found itself in an impossible position. Short of allowing France and Britain to garrison its territory from the onset – which would have compromised the neutrality it still believed to be its best hope of averting invasion – it had no option but to keep its military distance from the Allies, while fortifying its eastern frontier as best it could against the Wehrmacht. Even so, van Overstraeten did allow British and French officers wearing civilian clothes to reconnoitre the positions they intended to take up if Germany attacked; and, though he refused to co-ordinate defence plans with the Allies, he did transmit to them Belgian intelligence of German intentions, including details of the original 'Case Yellow' captured at Mechelen on 9 January, and subsequent indications of their scheme to envelop and destroy the Franco-British army on the Western Front.

Van Overstraeten's professional objection to closer co-operation with the Allies lay in his belief that nothing would induce them to defend the whole of Belgium. His (correct) judgement was that they intended to advance no further than the centre of the kingdom; his equally correct but harsher judgement was that they would allow the Belgian army to 'sacrifice' itself in its forward positions on the Albert Canal while they consolidated their own behind it on the

Dyle Line. In the event, they did not even win the time to consolidate. The French Seventh Army, though commanded by Henri Giraud, a genuine fighting general and future rival of de Gaulle for leadership of Free France, made poor time along the North Sea coast on its mission to bring support to the Dutch and the Belgian left flank. It had further to advance than the Germans of Army Group B coming from the opposite direction, who proved more adept than it in negotiating water obstacles even when defended. Its motorised reconnaissance elements also came under German air attack. By 12 May its advance was blunted near Breda, its objective, and on the following day it was ordered to fall back to guard the left flank of the Dyle Line near Antwerp. It did so pursued by the advance guard of the 9th Panzer Division.

The Allied deployment on to the Dyle was already going wrong. A 'domino effect' was in train. As the Dutch army fell back from its forward positions into Fortress Holland around Amsterdam and Rotterdam, it uncovered the left flank of the Belgians on the Albert Canal, where they were outflanked by the 9th Panzer Division. On the right they were outflanked by the 3rd and 4th Panzer Divisions, which were about to be let across the precipitous defile of the Meuse – the most formidable of military obstacles in north-west Europe – by the German airborne troops' descent on Eben Emael. While the Royal Air Force tried vainly in a series of suicidal bombing missions to destroy the Meuse bridges in the face of the German advance, the Belgians began to fall back, hoping to feel behind them the support of the French First Army and the BEF advancing to the Dyle.

'Steps in a dream'

Both these forces were in forward motion, the BEF passing by Brussels, the French First Army by Maubeuge, with the Ninth Army of General André Corap on its right. For the British their line of advance was familiar country. It ran through Marlborough's campaigning ground, past Waterloo and across more recent battlefields of their military history, Ypres and Mons. 'It was almost', wrote the American war correspondent, Drew Middleton, 'as if they were retracing steps taken in a dream. They saw again faces of friends long dead and heard the half-remembered names of towns and villages.' Dream was shortly to become nightmare for them, and for their French allies too. The Dyle, to which they were advancing, was scarcely a natural obstacle at all; the artificial obstacles they had been led to believe the Belgians had erected along it were scattered or absent altogether (the British would encounter those that they had emplaced a few years later; collected and transported to Normandy, they would form a principal element in the German fortifications of the D-Day beaches). The French had two 'cavalry' and one mechanised division with them; the British had almost no armour at all. Opposite were the 3rd and 4th Divisions of Hoepner's armoured corps, with over 600 tanks, their crews battle-hardened and trained for rapid advance by the experience of the Polish campaign. No wonder that an eerie cynicism suffused Hitler's reminiscence of this stage of the campaign: 'It was wonderful the way everything turned out according to plan. When the news came through that the enemy were moving forward along the whole front, I could have wept for joy; they had fallen into the trap . . . they *had* believed . . . that we were striking to the old Schlieffen Plan.' Hitler's own first experience of battle had occurred only fifty miles from the Dyle, in the dying stage of the 'old Schlieffen Plan' in October 1914. It had been a bitter and bloody baptism. Now: 'how lovely Felsennest [Crag's Nest, his 'Sickle Stroke' headquarters] was! The birds in the morning, the view of the road up which the columns were advancing, the squadrons of planes overhead. There I was sure everything would go right for me . . . I could have wept for joy.'

There were soon to be tears of anguish in his adversaries' headquarters – but not from the hard-boiled Major-General Bernard Montgomery commanding the British 3rd Division whose troops on 11 May were digging in cheerfully on the Dyle Line; nor from Sir Edmund Ironside, the British chief of staff, whose diary tells that he judged 'on the whole the advantage is with us' and looked forward to 'a really hard fight all this summer'; nor from Gamelin, who remained 'above all preoccupied with Holland' and had the previous day delegated his powers of command in Belgium to Georges; not even from General Gaston Billotte, to whom Georges had in turn delegated authority on the northern front and who, with thirty divisions to cover fifty-five miles of line, had more than adequate force to fulfil his mission. On the 'line of engagement' along the Dyle, the Allies, despite the disturbing developments on their flanks and the softening of Belgian resistance in front of them, had reason to believe that they outnumbered the approaching Germans – as they did – and would be able to check their advance.

The Allied appreciation of the situation in Belgium, however, rested on the misapprehension (in which Hitler was then exulting) that there they faced the main axis of the German offensive and confronted their main concentration of force. As in 1914, their intelligence resources had failed to establish where the German *Schwerpunkt* lay. In 1914 it was the French cavalry, beating the thickets of the Ardennes when it should have been roaming Flanders, which missed the German spearheads; in 1940 it was the Allied air forces, flailing vainly at the German spearheads in Belgian Flanders when they should have been overflying the Ardennes, which had lost touch with essentials. From 10 to 14 May, the seven Panzer divisions of Army Group A nudged forward nose to tail along the Ardennes defiles in a traffic concentration so dense that General Günther Blumentritt calculated that if deployed on a single-tank 'front', the tail of the column would have been in East Prussia; they breasted up towards the weakest spot on the Allied front to form an irresistible force. These seven Panzer divisions – 1, 2, 5, 6, 7, 8, 10 – deployed between them 1800 tanks. In front of them they found in first line the two Belgian divisions of Chasseurs Ardennais, an old-fashioned elite of forest riflemen whose bravery counted for nothing against armour. When they had been brushed aside, the Panzers found themselves opposed by Corap's Ninth Army and part of Huntziger's Second. Although neither formed an elite by any estimation, with the Meuse at their front their reservists should nevertheless have been able to hold, at least in normal times; but May 1940 was not normal times. Almost as soon as the German vanguards of Army Group A made touch with the Meuse defences, they were able to find a way across. Corap's and Huntziger's outpost guards took fright, the banks of the river were abandoned and the breach in the Allied defensive dyke was opened.

General André Beaufre, then a junior staff officer at French general headquarters, described the impression the news made on General Georges at his command post at La Ferté early in the morning of 14 May:

The atmosphere was that of a family in which there had been a death. Georges . . . was terribly pale. 'Our front has been broken at Sedan. There has been a collapse. . . .' He flung himself into a chair and burst into tears. He was the first man I had seen weep in this campaign. Alas, there were to be others. It made a terrible impression on me. Doumenc [Georges's subordinate] – taken aback – reacted immediately. 'General, this is war and in war things like this are bound to happen!' Then Georges, still pale, explained: following terrible bombardment from the air the two inferior divisions [55 and 71] had taken to their heels. X Corps signalled that the position was penetrated and that German tanks had arrived in Bulson [two miles west of the Meuse, and so inside the French-defended area] about midnight. Here there was another flood of tears. Everyone else remained silent, shattered by what had happened. 'Well, General,' said Doumenc, 'all wars have their costs. Let's look at the map and see what can be done.'

There is much in Beaufre's description of this scene that yields to exegesis. First, Sedan: the name of the town where Napoleon III had surrendered to the Prussians in September 1870 was in French ears synonymous with disaster. Second, the 'two inferior divisions': the 55th and 71st Divisions of Huntziger's Second Army were both composed of older reservists, and both had indeed taken to their heels at the approach of the German tanks. Third, what the map suggested might be done: the German penetration of the French line had occurred at a point so sensitive – as Manstein had intended – that any counter-measure adopted would have to be massive and almost instantaneous if it were to stop the rot. The story of Allied strategic decision in the next week would be one of seeking the telling blow.

The details of the story from the German side, however, boded even worse for Georges than he had grasped. For the Meuse had first been crossed not, as he believed, on the day before he had his nervous collapse, but the day before that, 12 May. As darkness fell, patrols of the motorcycle reconnaissance battalion of the 7th Panzer Division commanded by Erwin Rommel had found an unguarded weir across the Meuse at Houx, north of Sedan. Creeping across it, they reached an island in midstream from which a lock-gate led to the west bank. During the night reinforcements joined them there, so that by 13 May 'Sickle Stroke' had already struck at the foundations of the Gamelin plan. The next morning Rommel's engineers began to lay pontoon bridges across the river, while his tanks, waiting to cross, destroyed French bunkers on the other side with gunfire. By evening the bridges were completed and the first of his tanks had crossed the river – only 120 yards wide at this point.

The French might have dealt successfully with this bridgehead. It was as yet precarious. They tried a counter-attack, with a force that included a tank battalion, and Gamelin was told, 'the incident at Houx is in hand'. However, the tanks withdrew after taking a few prisoners, leaving Rommel's bridgehead intact, if not yet a burgeoning threat. Meanwhile French attention was diverted southwards by the assault of Army Group A's main Panzer formation at Sedan. They had been deploying in the open flood plain of the river, after three days of nose-to-tail driving through the defiles of the Ardennes, all through the morning of 13 May. General P. P. J. Gransard observed 'the enemy emerging from the forest . . . an almost uninterrupted descent of infantrymen, of vehicles either armoured or motorised'. The French artillery brought them under fire; but it was answered by German bombing, first by high-level Dornier 17s, then by diving Stukas. The effect on the French infantry regiments was shattering. 'The noise, the horrible noise', repeated the wounded brought to a field ambulance; better troops were to feel the same terror under air attack through-out this and subsequent wars. 'Five hours of this nightmare', wrote General Edouard Ruby, deputy chief of staff of Second Army, 'were enough to shatter [the troops'] nerves.' By three in the afternoon the Stukas drew off. As soon as they did so, the assault pioneers of the 1st, 2nd and 10th Panzer Divisions began dragging their inflatable boats to the water's edge. Setting off under a suddenly amplified hail of enemy fire – the French manned their weapons as they saw the danger they faced – the boat crews suffered heavy casualties and were here and there driven back, but along the whole line of assault, from Donchery to Bazeilles, established a series of footholds on the far bank. Bazeilles was a place of legend in French military history; it was there in 1870 that the elite *coloniales* had fought to the death against the Germans in 'the house of the last cartridge'. In 1940 it was the Germans who were ready to do or die at Bazeilles. Hans Rubarth, a pioneer sergeant of the 10th Panzer Division, ordered his men to throw their entrenching tools out of their overloaded boat in midstream: 'No digging for us – either we get through or that's the end.' Before the day was out, nine of his eleven men had become casualties but the group had taken its objective. Rubarth was promoted lieutenant in the field and awarded the Knight's Cross of the Iron Cross, Germany's highest decoration for bravery.

Such exploits, many times repeated, carried the storming parties of all three Panzer divisions across the Meuse during the afternoon of 13 May. In front of them isolated outposts of French infantry held their ground with great courage; but others ran at the sight of tanks – sometimes at the sight of French tanks, often merely at the rumour of tanks. French tanks did appear towards evening; they belonged to the 3rd Armoured and 3rd Motorised Divisions, but the counter-attack they had been sent to deliver was not driven home. As they withdrew from the river's edge, the Germans reinforced their own tank units, which by pontoon bridges had been transported to the French bank and prepared for the coming breakout.

That evening Gamelin, still at Vincennes, 120 miles from the crisis-point, issued an order of the day: 'The onslaught of the mechanical and motorised forces of the enemy must now be faced. The hour has come to fight in depth on the positions appointed by the high command. One is no longer entitled to retire. If the enemy makes a local breach, it must not only be sealed off but counter-attacked and retaken.'

During 14 May Gamelin's troops – who were far too widely dispersed to 'fight in depth' – did attempt counter-attacks against the German bridgeheads. None was successful, in part because the target was diffuse. The blade with which 'Sickle Stroke' would be delivered had not yet formed. Its component elements were still struggling out of their bridgeheads: the 6th and 8th Panzer Divisions north of Sedan; the 2nd, 1st and 10th to the south. The danger posed by the 5th and Rommel's 7th at Dinant had not yet impressed itself on the French high command's consciousness. In a strict military sense, it would have been best to wait until the Panzer divisions had coalesced and started inland, before their supporting infantry had crossed the river to join them. Then the armoured column might have been caught 'in flank' and decapitated. As it was, the French 3rd Armoured Division wandered about the battlefield on 14 May seeking ineffectively whom it might devour. While the Panzer bridgeheads were enlarged, the German tanks refuelled and reammunitioned and the start-lines were drawn for a plunge into the French heartland.

Which of the German spearheads would be first away? The Panzer concentration around Sedan was the stronger, but that further north at Dinant faced the poorer troops of Corap's Ninth Army. André Corap, a fat and jovial colonial soldier with a talent for making his men like him, was opposed, moreover, by the wiry and ascetic Erwin Rommel, whose soldiers idolised him because he clearly cared only for beating the enemy. Rommel had won the *Pour le Mérite*, Germany's highest military decoration, as a captain by a brilliant stroke of personal initiative during the First World War, and destroyed much of an Italian division in the process. On 15 May 1940, by a similar initiative, he broke through Corap's tentative 'stop-line' before it could be manned and advanced seventeen miles for the loss of fifteen German dead. During the afternoon the 6th Panzer Division, crossing at Monthermé, north of Sedan, joined in the Ninth Army's destruction. The Indo-Chinese machine-gunners who had defended the crossings with devoted bravery for three days were bypassed (their soldierly qualities portended of the bitterness with which Vietnam would be contested by Ho Chi Minh's followers in the post-war years). Their French comrades-in-arms, whom the 6th Panzer Division met as it drove forward, showed no such tenacity – nothing, indeed, but pitiful demoralisation. Karl von Stackelberg, a war correspondent accompanying the German tanks, was astonished to encounter formed bodies of French troops marching to meet them:

> There were finally 20,000 men, who here . . . in this one sector and on this one day, were heading backward as prisoners. Unwillingly one had to think of Poland and the scenes there. It was inexplicable. How was it possible that, after this first major battle on French territory, after this victory on the Meuse, this gigantic consequence should follow? How was it possible these French soldiers with their officers, so completely downcast, so completely demoralised, would allow themselves to go more or less voluntarily into imprisonment?

Not all French soldiers would give up the fight so easily. In the north the First Army was still resisting steadily, as it would do until its remnants were completely surrounded at Lille. And on 15 May Charles de Gaulle, who had been appointed to command the 4th Armoured Division four days previously, received orders from General Georges to attack at Laon, which lay in the German Panzers' path, and 'gain time' for a new front to be established north of Paris. Although the 4th Armoured Division was still in the process of forming, de Gaulle, long an enthusiast for armoured warfare and a patriot whose love of country was fortified, not diminished, by its army's current demoralisation, accepted the challenge with ardour. 'I felt myself borne up by a limitless fury,' he wrote later. ' "Ah! it's too stupid! The war is beginning as badly as it could. Therefore it must go on. For this the world is wide. If I live, I will fight wherever I must as long as I must until the enemy is defeated and the national stain washed clean." All I have managed to do since was resolved upon that day.'

De Gaulle managed to do little when he finally brought his division into action on 17 May. His tanks made inroads into the positions of the 1st Panzer Division, one of whose staff officers, Captain Graf von Kielsmansegg, who thirty-five years later would command the NATO forces in Germany, decided on showing them that 'discretion was the better part of valour'. However, they were too few to do more than frighten the Germans and by evening they turned about and withdrew to refuel.

The Germans had grown collectively nervous that day – although Guderian, commanding the 2nd and 10th Panzer Divisions, champed at the bit and sought by every means to get forward. But Hitler, recorded Halder, 'is anxious about our own success, doesn't want to risk anything and would therefore be happiest to have us halt.' Halder himself was concerned to line the 'walls' of the developing 'Panzer corridor' with his infantry, which was lagging behind the tanks; Brauchitsch, the commander-in-chief, was adamant that he should. The Panzers had advanced forty miles since the crossing of the Meuse four days earlier, were converging into a solid armoured mass of seven divisions and had the clear evidence of the collapse of the French Ninth and Second Armies everywhere before their eyes. The French First Army, the BEF and the Belgians were giving ground to the north, while the French to the south, immobilised in the Maginot Line and unable to manoeuvre for lack of transport, were clearly unable to intervene against the Panzers. Nevertheless the German high command, prompted by Hitler's anxieties, on 17 May imposed a halt on the advance.

German anxieties paled by comparison with those of the Allies. The Belgians, for the second time in the century, faced the prospect of defeat and occupation. The British were confronted by the fear of losing their only army – and large parts of their air force – if they continued to stand by their allies on a collapsing battle line. The French foresaw their army breaking into two, the better part falling victim to encirclement in Belgium and the northern departments, while the remnants struggled to form a new and doubtfully defensible front on the approaches to Paris. The potential for disaster loomed as large as in 1914 but the crisis was actually more acute. Then the French army had suffered defeat in the Battle of the Frontiers but retreated in good order under an imperturbable commander; in 1940 it was retreating in disorder, a disorder which grew worse daily under the nominal orders but not the effective command of a general who was surrendering to events. On 16 May Paul Reynaud, the French Prime Minister, sent for new men: to the Madrid embassy for Philippe Pétain, hero of Verdun, to join him as his deputy; to Syria for Maxime Weygand, chief of staff to Foch in the victory campaign of 1918, as a replacement for Gamelin. Both were very old – Weygand, at seventy-three, five years older than Gamelin, Pétain older still – but at this moment of agony their heroic reputations seemed a reassurance that something might yet be snatched from the yawning jaws of defeat.

Gamelin was now discredited. In Paris on 16 May he conferred with Reynaud and Winston Churchill – Prime Minister since 10 May, when the House of Commons had withdrawn its confidence from Neville Chamberlain – and admitted that he had no troops available to stem the German onrush. 'I then asked, "Where is the strategic reserve?" ' Churchill recorded. 'General Gamelin turned to me and, with a shake of his head and a shrug, said "Aucune" ["There is none"]. There was a long pause. Outside in the garden of the Quai d'Orsay clouds of smoke arose from large bonfires and I saw from the window venerable officials pushing wheelbarrows of archives on to them.' (The burning of official papers was to be a token of apprehended defeat at capitals and headquarters throughout the Second World War.) 'I was dumbfounded. . . . It had never occurred to me that army commanders having to defend five hundred miles of engaged front would have left themselves unprovided with a [strategic reserve]. . . . What was the Maginot Line for?'

Churchill left for England promising to send six additional squadrons of British fighters to join the few already in France. However, so complete was German air superiority that fighter reinforcements could make no difference at this stage of the battle. What was needed was the strategic reserve he had discovered did not exist. Weygand, who assumed command from Gamelin on 20 May, attempted to improvise one by proposing ('the Weygand Plan') on 21 May that the encircled Allied forces north of the German break-in should co-ordinate convergent attacks against the Panzer corridor with the French armies still operating to its south. This reflected a correct appreciation of how to deal with *Blitzkrieg* and had in fact been proposed by Gamelin two days before, but the authority to execute it was lacking. Georges was now a broken man, while Billotte, to whom he had delegated authority, was killed in a motor accident on 21 May. The troops were lacking too. De Gaulle had attempted another vain counter-attack with his depleted 4th Armoured Division on 19 May; and on 21 May two British divisions, supported by two tank battalions, succeeded in denting the flank of the Panzer corridor at Arras, so alarming Rommel, commanding on the spot, that he estimated he had been attacked by five enemy divisions. However, these formations represented almost the whole Allied force available to Weygand for manoeuvre. The Ninth Army had disintegrated. The First Army and the BEF were constricted between the North Sea and the advancing Germans. The as yet unengaged French armies south of the Panzer corridor lacked transport, tanks and artillery. Meanwhile, after the German high command's hesitation of 17 May, the Panzers had driven on. By 18 May they were driving across the battlefields of the First World War, skirting the river Sambre on their northern flank and the river Somme on the south. On 20 May Guderian's divisions reached Abbeville at the mouth of the Somme, thus effectively dividing the Allied armies into two.

These were heady days for Heinz Guderian. He was dedicated to the development of the Panzer arm and even before Hitler's rise to power was an advocate of what would be called *Blitzkrieg*. Frustrated by the timidity of his own high command – Brauchitsch, abetted by Halder, represented the fainthearts – he had had to restort to subterfuge in evading its orders to proceed with caution after crossing the Meuse. His creative disobedience had not yet won a great victory; he and the whole of the German Panzer force would have difficulty in achieving it. On 20 May Hitler reviewed plans for 'Case Red', the advance into the French heartland which would complete 'Sickle Stroke' and also complete the destruction of the French army – as long as the Panzer arm was kept intact. So it was that the British counter-attack at Arras, which had so alarmed Rommel, now alarmed Hitler once again. Rundstedt, commanding Army Group A, agreed with him that the Panzers had advanced too far for safety and should not proceed until the slower-moving infantry had lined the 'walls' of the Panzer corridor against a repetition of the Arras surprise. Brauchitsch, supported by Halder, now abandoned his earlier caution, urged that the Panzers should press home their attacks against the encircled Allies in the north, and even tried to transfer command of part of the striking force from Rundstedt to Bock, the situation of whose Army Group B, advancing on a front through Belgium, had now aroused Hitler's anxiety. When Hitler learned of the attempt on 24 May, however, he cancelled it and reiterated his refusal to allow the Panzers to press into the coastal lowlands which he claimed, from his own trench experience of the First World War, were quite unsuitable for armoured operations.

Hitler's 'stop order' would keep the Panzers halted for two whole days, until the afternoon of 26 May – two days which in retrospect have been deemed strategically decisive for the outcome of the Second World War. Unbeknown to the Germans, the British government had on 20 May decided that part of the BEF might have to be evacuated from the Channel ports and had instructed the Admiralty to begin assembling small ships on the British south coast to take them off. The operation would be codenamed 'Dynamo'. It was not yet to comprehend a full-scale evacuation; the government still hoped that the BEF, with the French First Army, would be able to break through the Panzer corridor to join the surviving bulk of the French armies on and south of the Somme – which was the point of the Weygand Plan. However, the BEF was itself becoming wearied by its battle in Belgium which had entailed a fighting withdrawal from the Dyle to the Schelde, and Gort was increasingly concerned by his responsibility for safeguarding Britain's only army. On 23 May he had received an assurance from Anthony Eden, who was serving as War Minister, that the government would make naval and air arrangements to assist them should they have to withdraw on the northern coast. On the same day he concluded that the Weygand Plan could not be realised for lack of troops, tanks and aircraft, and withdrew from Arras the two divisions which had attacked Rommel with such effect on 21 May. 'Nothing but a miracle can save the BEF now,' Alan Brooke, commanding Gort's II Corps, wrote on 23 May; but on that day Gort's decision to disengage and draw the BEF back towards the coast in effect laid the basis for its salvation.

For Hitler had anticipated events. He was right to fear that the Panzers would get bogged down in the canals and rivers around Dunkirk, to which port Gort now directed the BEF. He was wrong to give the 'stop order' when he did, two days before the British – and a substantial portion of the French First Army – reached the watery sanctuary of the 'Canal Line'. When the stop order was revoked on 26 May that part of the Allied army he most wanted and needed to destroy was – temporarily – safe. Protected by the Aa Canal and the Colme

Canal, the fugitive enemy could start embarking in the flotillas of destroyers and small boats which Admiral Bertram Ramsay began to send cross-Channel from the headquarters of Operation Dynamo that same day. Hitler had been assured by Goering that the Luftwaffe would prevent any evacuation from the Dunkirk pocket. During 24-26 May its aircraft did indeed raise havoc inside it and would continue to do so as long as the evacuation lasted, until 4 June. But it could not stop the evacuation ships closing the shore – the total it sunk was six British and two French destroyers in nine days of air attack – nor could it reduce the resistance of the Dunkirk defenders, many of them French, many of those French colonials, who gave ground with extreme reluctance against concentric German attack.

The Belgian army was forced into a capitulation north of the Dunkirk pocket at midnight on 27 May. It surrendered in almost exactly the same area where, in 1914, it had been able to consolidate a defensive position and continue the fight until 1918. Then, however, it had been supported by French and British armies which remained intact and combatant. Now, unfairly condemned for deserting them by allies who were themselves on the point of collapse, it had no option but to ask for terms. So, too, shortly would the divisions of the French First Army which were encircled at Lille and running out of ammunition. So bravely had they fought that, when they marched out to surrender on 30 May, the Germans rendered them the honours of war, playing them into captivity with the music of a military band. It was significant evidence of the fighting spirit of the French army of 1940 that a large proportion of these stalwarts were not French at all, but North African subjects of the French empire.

The evacuation of the BEF – and the French troops in the Dunkirk pocket who could be got to the beaches – was now in full swing. Only 8000 were got off on 26-27 May; but on 28 May, as the fleet of naval ships and civilian small craft standing in to the shore grew, 19,000 were embarked. On 29 May 47,000 were rescued; on 31 May, the day Gort himself left for England, 68,000. By 4 June, when the last ship drew away, 338,000 Allied soldiers had been saved from capture. The number included almost the whole manpower of the BEF less its temporarily irreplaceable equipment, and 110,000 French soldiers, the majority of whom on arriving in England were immediately transhipped and returned to French ports in Normandy and Brittany to rejoin the rest of the French army still in the field.

This now consisted of sixty divisions, some survivors of the battle on the Meuse, some withdrawn from the Maginot Line; only three were armoured, all much depleted, particularly de Gaulle's 4th Armoured, which on 28-30 May had once again tried but failed to dent the flank of the Panzer corridor, this time near Abbeville. Two British divisions remained in France: the 1st (and only) Armoured Division and the 51st Highland Division, defending the coast west of Dunkirk (the British rifle regiments committed to the defence of Calais had already been overwhelmed). Against them the Germans deployed eighty-nine infantry divisions and fifteen Panzer and motorised divisions, the latter organised into five groups, each of two Panzer and one motorised divisions. These Panzer-motorised combinations formed powerful offensive instruments, which provided the model for the tank-infantry formations with which offensive operations would be conducted throughout the Second World War and, indeed, ever since. The Luftwaffe continued to deploy about 2500 strike aircraft, fighters and bombers, which they could now operate from captured airfields close to the battle line. The French air force, though reinforced by machines hastily purchased from the United States, and supported by 350 aircraft of the RAF, could only operate some 980.

The Weygand Line

Weygand, his 'plan' having collapsed, now pinned his hopes for resistance on the defence of a position which would be called the 'Weygand Line'. The resilient old general had not yet abandoned hope; he had even outlined a defensive scheme which mirrored that of the German offensive plan in its modernity. The 'Weygand Line', running from the Channel coast along the line of the rivers Somme and Aisne to join the Maginot Line at Montmédy, was to be held as a 'chequerboard' of 'hedgehogs' (NATO would adopt a similar scheme for the defence of the Central Front in Germany in the 1970s). The 'hedgehogs' – villages and woods – were to be filled with troops and anti-tank weapons and to continue resistance even if bypassed by enemy spearheads.

The theory was excellent, the practice lamentable. The Weygand Line broke almost as soon as it was attacked by the right wing of the Panzer array between Amiens and the coast on 5 June. The fault lay not with the fighting spirit of the French troops, which had greatly revived, but with their material weakness. They were outnumbered and lacked tanks, effective anti-tank weapons and air cover. Colonials and reservists fought with equal valour. 'In those ruined villages', wrote Karl von Stackelberg, 'the French resisted to the last man. Some hedgehogs carried on when our infantry was twenty miles behind them.' On 5 and 6 June the Germans were stopped dead at several points and even suffered crippling tank losses. If the Weygand Line had had 'depth', the German advance might have been held by its outposts, but, once its crust was broken, there were no troops behind it to seal off the breach or counter-attack. Rommel, leading the 7th Panzer Division across country when it was checked by hedgehogs commanding the roads, quickly found a way into the rear and was directed by the headquarters of Bock's Army Group B, to which he was now subordinate, to turn towards the coast and encircle the defenders of the Weygand Line's left wing from the rear; in the process he would force the surrender of the last British infantry division left in France, the 51st Highland.

On 9 June Rundstedt's Army Group A moved to the attack on the Aisne. Led by Guderian's Panzer group of four armoured and two motorised divisions, it was briefly checked, notably by the resistance of the French 14th Division under General Jean de Lattre de Tassigny, a future Marshal of France, whose reputation for defiance in the teeth of defeat was established that day. Yet on the Aisne, as on the Somme, the Germans were now too strong for any display of French courage to hold them in check. The previous evening Pétain, the Deputy Prime Minister, told his former chief of staff, Bernard Serrigny, that Weygand foresaw the possibility of holding the line for three days at most and that he himself intended to 'push the government to request an armistice. There is a meeting of the Central Committee tomorrow. I shall draft a proposal.' Serrigny warned that the next day would be too late. 'Action should be taken while France still has the façade of an army and Italy has not yet come in. Get a neutral to intervene in the approach. Roosevelt seems the obvious choice. He can bring his power to bear on Hitler.'

This was a counsel of desperation. Roosevelt had already declared to Reynaud his inability to influence the course of events in Europe, by the dispatch either of more new material or of the United States fleet, while Mussolini, who had told the British ambassador on 28 May that not even the bribe of French North Africa would keep him neutral, was now bent on declaring war before its termination deprived him of a share of the glory and its rewards. On 10 June, the day Pétain had indicated he intended to press the government to treat for terms, Reynaud evacuated it from Paris. As it headed for Tours on the Loire, whither Churchill would fly for his fourth meeting with his ally on 11 June, the German Panzer groups were outflanking the city to the west and east. The day after the government left, it was declared an 'open city', to spare it destruction. However, Hitler did not choose to attack it – perhaps, quite unnecessarily, fearing another Commune. Churchill, on his flying visit to Reynaud in Tours, reminded him of 'the absorbing power of the hour-to-hour defence of a great city'. However, those Parisians with cars were already streaming southwards in tens of thousands, while those who remained behind opened for business as usual when the first German soldiers arrived on 14 June; three days later they were thronging the terrace of the Café de la Paix, happy to be sightseers in the tourist capital of the world.

Frenchmen in uniform were still fighting, often to the death. Like the Belgians, they found at the approach of defeat an outraged capacity for self-sacrifice. At Toul, behind the Maginot Line, the 227th Regiment of Infantry fought on long after it had been surrounded. At Saumur the students of the Cavalry School held the bridges over the Loire from 19 to 20 June until their ammunition ran out. The garrison of the Maginot Line itself, 400,000 strong, refused all calls to surrender; only one section of blockhouses was ever to be captured by German attack. South of the Loire an officer of the Fifth Army watched a 'small group of Chasseurs Alpins from the 28th Division' cross on 17 June. They were 'led by a sergeant covered with dust, their uniforms in rags, marching in order and in step, the men bent forward, pulling with both hands at the straps of their equipment. Some were wounded, the dressings stained with dirt and blood. Some slept as they marched, ghosts bowed under the weight of their packs and rifles. They passed in silence, with an air of fierce determination.'

Comrades-in-arms of these mountain troops were meanwhile confronting Mussolini's attack on the Riviera across the Alpes Maritimes, Italy having declared war on 10 June. Four French divisions stood in the path of twenty-eight Italian divisions. They held their ground without difficulty, yielding nowhere more than two kilometres of front, losing only eight men killed against Italian casualties of nearly 5000. Eventually, in desperation, the Italian high command asked for German transport aircraft to land a battalion behind the French lines, as a token of success. 'The whole thing is the usual kind of fraud,' wrote Halder. 'I have made it clear I won't have my name mixed up in this business.'

The humiliation of France

Resistance in the Alps and the Maginot Line could avail not at all against German triumph in the heartland. The British landed the 52nd Lowland and the Canadian Division at Cherbourg on 12 June, to assist returning French troops to open a new front in the west; both had to be evacuated almost immediately to avoid capture. The day before, Churchill had seen for himself the hopeless pass to which France had been brought. At Tours, Weygand, all fight gone, told him and the French ministers: 'I am helpless, I cannot intervene, for I have no reserves... C'est la dislocation.' De Gaulle, determined on some 'dramatic move' to keep the war going, proposed to Churchill in London on 14 June that 'a proclamation of the indissoluble union of the French and British peoples would serve the purpose', and Churchill offered such a Declaration of Union to Reynaud on 16 June. His ministers rejected it categorically. Jean Ybarnegaray no doubt spoke for many in saying that he 'did not want France to become a dominion' (of the British Empire). Pétain was now chiefly concerned that France should not fall into disorder; even more than defeat and continuing casualties he feared a takeover by the left. His determination to seek an armistice which would allow the conservatives to continue in office was at least a policy; Reynaud had none. On the evening that Churchill's offer of union was rejected, President Lebrun decided that the old marshal should be asked to form a government. General Edward Spears, Churchill's personal emissary to France, left for England at once, taking with him Charles de Gaulle, who, promoted general on 25 May and appointed Under-Secretary for Defence on 10 June, was almost the only member of the government determined to carry on resistance. Next day, 18 June, de Gaulle broadcast from London to the French people: 'This war has not been settled by the Battle of France. This war is a world war ... whatever happens the flame of resistance must not and will not be extinguished.' He called on all Frenchmen who could join him on British soil to continue the fight. For this defiance de Gaulle would shortly be court-martialled and condemned as a traitor by the Pétain regime.

Pétain had himself broadcast to the French people the day before de Gaulle: 'Frenchmen, at the appeal of the President of the Republic, I have today assumed the direction of the government of France. . . . I give myself to France to assuage her misfortune. . . . It is with a heavy heart that I say we must end the fight. Last night I applied to our adversary to ask if he is prepared to seek with me, soldier to soldier, after the battle, honourably the means whereby hostilities may cease.' Hitler, the insistent 'front fighter', would treat 'soldier to soldier' but without the 'honour' his defeated enemy craved; Versailles had eaten too deep into his psyche for that. When the emissaries sent by Pétain met their German counterparts near Tours on 20 June they found themselves transported first to Paris and then eastward. On 21 June, at Réthondes, near Compi'gne, General Charles Huntziger, whose Second Army had been one of the first victims of the Panzer onslaught, alighted from a German military convoy outside the railway coach in which the German delegates had signed the armistice of November 1918. An exultant Hitler observed his arrival; General Wilhelm Keitel, head of OKW, presented the armistice terms. They did not allow for negotiation: Pétain's government was to remain sovereign, but Paris, northern France and its borders with Belgium, Switzerland and the Atlantic were to become a zone of German occupation; Italy, on terms to be discussed with Mussolini, was to occupy south-eastern France. The French army was to be reduced to 100,000 men and 'occupation costs', set at an exorbitant franc:mark exchange rate, were to be met from the French budget. The French empire – in North and West Africa and Indo-China – was to remain under the control of the French government (which was shortly to establish its capital at Vichy), as was the French navy, which was to be demilitarised. All prisoners taken in the campaign, including the garrison of the Maginot Line, though it had not surrendered, were to remain in German hands. France, in short, was to be emasculated and humiliated, as Hitler believed Germany had been in 1918. The terms, in truth, were far more severe than those imposed at Réthondes twenty-two years previously. Then Germany had been left the bulk of its national territory and its soldiers their freedom to return to civilian life. Now the most productive part of French territory was to be occupied and two million Frenchmen, 5 per cent of the population but perhaps a quarter of France's active manhood, were to go into German captivity with no term fixed for an alteration of these penalties. The delegation argued, but, as Léon Noœl, the former ambassador to Poland, observed, while it did so 'fighting was still going on, the invasion was spreading and fugitives were being machine-gunned on the roads'. Huntziger applied for instructions to Pétain at Bordeaux, where the French government had withdrawn. He was instructed to sign forthwith and did so on the evening of 22 June. Meanwhile a delegation led by Noœl signed terms at Rome with the Italian government, which provided for the occupation of the Franco-Italian border up to fifty kilometres' depth on the French side. The armistice with both Germany and Italy was then timed to come into force at 25 minutes past midnight on the morning of 25 June.

By then some German spearheads had penetrated deep into the 'free zone' which the armistice left to the new Vichy government. There were German tanks south of Lyon, German tanks outside Bordeaux; for a time there were even German tanks in Vichy. As the armistice terms took effect, they withdrew, without heel-dragging; the campaign of 10 May to 25 June 1940 had not cost the German army dear. The French counted some 90,000 dead in what many of their village war memorials, incongruously to British and American eyes, call 'the war of 1939-40'; the Germans had lost only 27,000. Theirs had been, in its last weeks, almost a war of flowers. 'Reached here without difficulty,' Rommel wrote to his wife from Rennes in Brittany on 21 June. 'The war has become practically a lightning Tour de France. Within a few days it will be over for good. The local people are relieved to see everything happening so peacefully.' The German army, imbued with the magnanimity of victory, behaved with all the 'correctness' to their beaten enemy that army orders prescribed. The French, as if shell-shocked by the catastrophe they had undergone, responded with an almost grateful meekness. Virtually no part of France had been spared the sight of beaten French soldiers – young conscripts, older recruits, black Senegalese, Arab light infantrymen, Polish and Czech volunteers, infantry, cavalry, artillery, tanks – falling back, dirty, hungry, tired, directionless, sometimes leaderless, on through fields and orchards ripening for harvest under a sun and skies whose daily brilliance remain, for victors and vanquished alike, inseparable from their memories of the 'summer of '40'. Amid the persisting normalities of life – Sunday lunch, first communions, *jours de fête* - the sensation of a predestined national doom, averted in 1918 by the tenacity of their British allies and the miraculous intervention of the Americans, overwhelmed the nation. This was how it had been in 1815, when the enemies of France had beaten the first Napoleon in Belgium; this was how it had been in 1870 when the Germans had beaten the second Napoleon in Lorraine. The victory of 1918 now seemed merely an intermission. The decline of *la grande nation*, set about by philistines and barbarians, might seem irreversibly charted. Pétain, hero of Verdun, embodied the spirit of his countrymen in June 1940 above all because they saw in him a being inured to loss and suffering.

The Germans, by contrast, were in lightened spirits. 'The great battle of France is over,' wrote Karl Heinz Mende, a young engineer officer who had fought the campaign from start to finish. 'It lasted twenty-six years.' The British, too, were in lightened spirits, perhaps perversely so. 'Personally,' wrote King George VI to his mother, 'I feel happier now that we have no allies to be polite to and to pamper.' Winston Churchill, face to face with realities, confronted the future in starker terms. 'The Battle of France is over,' he told the House of Commons on 18 June. 'I expect that the Battle of Britain is about to begin.'

FOUR

Air Battle: The Battle of Britain

The Battle of France, though sensational by reason of its brevity and decisiveness, had been an otherwise conventional military operation. In their support of the German armoured spearheads, aircraft had played a major part in bringing victory; but neither they nor indeed the tanks they had overflown had wrought the Allied defeat. That defeat was the outcome of defects in strategy, military structure and readiness for war, psychological as well as material, which were buried deep in the Western democracies' reaction to the agony they had undergone in the First World War.

The Battle of Britain, by contrast, was to be a truly revolutionary conflict. For the first time since man had taken to the skies, aircraft were to be used as the instrument of a campaign designed to break the enemy's will and capacity to resist without the intervention or support of armies and navies. This development had long been foreseen. Aircraft had been used as weapon platforms – by the Italians in Libya in 1911 – almost as soon as they had become viable as vehicles. For much of the First World War they had served as auxiliaries to the ground and sea forces, but from 1915 onwards airships had been used intermittently as bomb-carriers against Britain by the Germans and, later, bombing aircraft were used by both Germany and Britain against each other's cities. By the 1930s bombers, drawing on the technology of the increasingly dependable and long-range civil airliner, had become instruments of strategic outreach; it was that development that in 1932 drew from Stanley Baldwin, then a member of the coalition government, the incautious (and inaccurate) forecast that 'the bomber will always get through'. The terror inflicted by German and Italian bombers on the Republican population of Spain in 1936-8 seemed to endorse his warning. As the air historian Dr Richard Overy writes:

> By 1939 it was widely believed that the air weapon was coming of age. The experience of the First World War . . . persuaded many, politicians and generals among them, that the next war would be an air war. This was founded partly on the uncritical expectation that Science was now harnessed sufficiently closely to military life to produce a stream of new weapons; of secret devices from the air whose nature could only be guessed at. It was founded too on the more critical scrutiny of what aircraft had actually done

in the First World War. In reconnaissance work, in the support of troops on the ground, in co-operation with the navy on the first clumsy aircraft carriers and primarily in the carrying out of bombing campaigns independent of surface forces, the aircraft threatened to dwarf the contribution of the other services or even to supplant it altogether.

The belief that air forces might supplant armies and navies as war-winning instruments of power took root earliest and deepest in three countries with widely disparate strategic needs: the United States, Britain and Italy. In the United States, isolationist after 1918 and vulnerable only to transoceanic attack, it was the ability of the aircraft to destroy battle fleets which commanded attention. Successful experiments in the aerial bombing of captured German battleships prompted the foremost American exponent of independent airpower, General William Mitchell, to agitate for the creation of an independent air force, with such insubordinate vigour that he was obliged in 1925 to defend his stand at court martial. Britain, committed to the defence of both the Empire and the home base, and experienced in 'strategic' bombing against Germany at the end of the First World War, had created an autonomous air force in 1918 which thereafter formulated its own empirical concept of broad deterrence of attack by independent air operations. Curiously it was in Italy that a comprehensive theory of air strategy emerged in its most developed form. Giulio Douhet, universally recognised as the Mahan (if not the Clausewitz) of airpower, seems to have arrived at his vision of 'victory through airpower' by a recognition of the futility of First World War artillery tactics. In his book *Command of the Air* (1921) he argued that, rather than bombarding the periphery of enemy territory with high explosive, where it could destroy only such war material as an adversary deployed there, the logic of the air age required that it be flown to the centres of enemy war production and targeted against the factories, and workmen, that made the guns. Douhet's perception was conditioned by Italy's experience of the First World War, which it had fought on narrow fronts dominated by artillery that was supplied from factories located chiefly in modern Czechoslovakia, at no great distance from its own airfields.

Douhet's theory extended to a belief that the bomber would prove immune to defensive counter-measures, whether mounted by fighters or guns, and that a bomber offensive would achieve its effect so quickly that the outcome of a future war would be decided before the mobilisation of the combatants' armies and navies was complete. In that respect, he was a true visionary, since he foresaw the logic of the nuclear 'first strike'. However, he insisted that the long-range bomber, carrying free-fall high explosive, could deliver the disabling blow – and there few would follow him. The United States Army Air Force, when it entered the Second World War *en masse* in 1942, believed that its advanced Flying Fortress bombers, built to embody the Douhet ideal, were instruments of 'victory through airpower'; the unlearning of that mis-conception in its deep-penetration raids of 1943 was to be painful. The Royal Air Force, whose commitment to strategic bombing was pragmatic rather than doctrinaire, expected less of its early offensive against Germany (and achieved even less than it expected). The Luftwaffe of 1939-40 did not espouse any strategic bombing theory at all; in 1933 it had examined Germany's capacity to build and operate a long-range bomber fleet and concluded that the effort required exceeded its industrial capacity even in the medium term. Its chiefs, most of whom were ex-army officers, therefore devoted themselves to building the Luftwaffe into a ground-support arm, and this was still its role at the end of the Battle of France, despite the reputation it had won as an instrument of mass destruction in the attacks on Warsaw and Rotterdam.

When on 16 July 1940, therefore, Hitler issued his next Führer Directive (No. 16) on 'Preparations for a landing operation against England', the Luftwaffe's professional chiefs were perturbed by the scope of the tasks allotted to them: to 'prevent all air attacks', engage 'approaching naval vessels' and 'destroy coastal defences . . . break the initial resistance of the enemy land forces and annihilate reserves behind the front.' Here was a demand for nothing less than the achievement of the preconditions of victory before the army and navy had been committed. Hermann Goering, Air Minister and chief of the Luftwaffe, who at heart was still the fighter ace he had made himself in the First World War, made light of the difficulties. On 1 August, when the preliminaries of the Battle of Britain were already in progress, he predicted to his generals: 'The Führer has ordered me to crush Britain with my Luftwaffe. By means of hard blows I plan to have this enemy, who has already suffered a crushing moral defeat, down on his knees in the nearest future, so that an occupation of the island by our troops can proceed without any risk!' To Milch, Kesselring and Sperrle, professional commanders respectively of the Luftwaffe itself and the two air fleets (2 and 3) committed to support Operation Sealion (as the plan for the invasion of Britain was codenamed), the difficulties and risks of the air offensive Goering had so lightly agreed to undertake loomed larger than he had given any hint of perceiving.

First among the difficulties was the improvised nature of the Luftwaffe's operational base. Air Fleets 2 and 3, hastily redeployed to the coasts of Belgium, northern France and Normandy in the weeks after the French armistice, were making use of captured enemy airfields; every local facility – of supply, repair, signals – had to be adjusted to their needs. The Royal Air Force, by contrast, was operating from home bases it had occupied for decades. Another advantage RAF's Fighter Command enjoyed was that of defending its own territory. While the Luftwaffe at the very least would have twenty and more generally fifty or a hundred miles to fly before coming to grips with its enemy, Fighter Command could engage as soon as its aircraft reached operational height. That conserved not only fuel – crucial when a Messerschmitt 109's operational range was a mere 125 miles – but also ensured that the pilots of damaged aircraft could bale out over friendly soil or, on occasion, bring them to earth. The Luftwaffe's parachuting pilots or crash-landed aircraft would, by contrast, be lost for good; many German pilots, parachuting into the Channel, would be doomed to drown.

Fighter Command, besides operating close to its own bases, had the use of a highly trained and integrated control and warning system. Its four groups, 13 (Northern), 12 (Midlands), 11 (South-Eastern), and 10 (South-Western), were under the control of a central headquarters located at Uxbridge, west of London, by which the hardest-pressed groups (usually No. 11 Group protecting London and nearest northern France) could be reinforced from those temporarily unengaged. Fighter Command headquarters, moreover, could draw on information from a wide variety of sources – the ground Observer Corps and its own pilots – to 'scramble' and 'vector' (direct) squadrons against a developing threat; but it depended most of all on the 'Chain Home' line of fifty radar warning stations with which the Air Ministry had lined the coast from the Orkneys to Land's End since 1937. Radar worked by transmitting a radio beam and measuring the delay and direction of the pulse reflected from the approaching target aircraft – a sequence which established distance, bearing, height and speed. It was a British invention, credit for which belongs to Robert Watson-Watt of the National Physical Laboratory. By 1940 the Germans had also produced radar devices of their own, but their Würzburg and Freya stations were few in number, inferior to their British counterparts and no help to the Luftwaffe in conducting offensive operations. Radar conferred on Fighter Command a most critical advantage.

Fighter Command enjoyed one more advantage over the Luftwaffe: higher output of fighter aircraft from the factories. In the summer of 1940 Vickers and Hawker were producing 500 Spitfires and Hurricanes each month, while Messerschmitt was producing only 140 Me 109s and 90 Me 110s. The Germans had a larger force of trained pilots on which to call, with an overall military figure of 10,000 in 1939, while Fighter Command could add only 50 each week to its complement of 1450. This was to confront the RAF, at the height of the battle, with the paradoxical crisis of a lack of pilots to man aircraft; but at no stage of the coming battle was it to lack aircraft themselves. Indeed, despite Churchill's magnificent rhetoric, Fighter Command fought the Battle of Britain on something like equal terms. It would manage throughout to keep 600 Spitfires and Hurricanes serviceable daily; the Luftwaffe would never succeed in concentrating more than 800 Messerschmitt 109s against them. These fighters, evenly matched in speed (about 350 mph) and firepower, were the cardinal weapons of the battle by which victory was to be decided.

Nevertheless the Luftwaffe might have established the air superiority by which its powerful force of bombers – 1000 Dornier 17s, Heinkel 111s and Junkers 88s and 300 Junkers 87s – could have devastated Britain's defences, had it operated from the outset to the same sort of coldly logical plan by which the German army had attacked France in 1940. On the contrary, it had no considered strategy, no equivalent of 'Sickle Stroke', and fought Fighter Command instead by a series of improvisations, all posited on Goering's arrogant belief that Britain could be brought 'down on its knees' by any simulacrum of a 'hard blow' that he directed against it.

Aerial stalemate

The Battle of Britain, historians would agree in retrospect, was to fall into five phases of German improvisation: first the 'Channel Battle' (*Kanalkampf*) from 10 July to early August; then 'Operation Eagle', beginning on 'Eagle Day' (*Adlertag*), 13 August, the 'classic' phase of aerial combat between the Luftwaffe and the Royal Air Force, which lasted until 18 August; next the Luftwaffe's switch of offensive effort against Fighter Command's airfields from 24 August to 6 September; then the Battle of London, from 7 to 30 September, when the Luftwaffe's fighters escorted its bombers in daily, daylight and increasingly costly raids against the British capital, and finally a series of minor raids until the Battle's 'official' end on 30 October. Thereafter the badly mauled German bomber squadrons were switched to destructive but strategically ineffective night operations, a phase that Londoners would come to call 'the Blitz', in a homely adaptation of the term coined by the world's press to denote Hitler's overwhelming ground offensives against Poland and France.

The *Kanalkampf*, which opened on 10 July, began with German bomber raids, in a strength of twenty to thirty aircraft, against English south coast towns – Plymouth, Weymouth, Falmouth, Portsmouth and Dover – and on convoys when intercepted. Later it was extended to the mouth of the Thames. Some material damage was inflicted and about 40,000 tons of shipping sunk; but the Royal Navy, which had to be beaten if the fleet of tugs and barges Hitler was having assembled in the Dutch and Belgian estuaries for the Channel crossing was to be passed safely across the Narrows, remained untouched. During the period 10–31 July about 180 German aircraft were shot down, for the loss of 70 British fighters; a hundred of the German aircraft destroyed were bombers, so the 'exchange rate' in fighters, on which decision the Battle of Britain would turn, stood even.

Hitler was becoming impatient with the aerial stalemate. He had persuaded himself that the British were already beaten, if only they would recognise it, and shrank from unleashing the invasion – because of both its risks and his own hope that they would shortly concede defeat – but was now determined that, since no other means were available, the Luftwaffe must force Britain to accept the necessity of treating with Germany. He still insisted to his generals that he had no desire to humiliate Britain (as he had France), let alone to destroy her (as he had Poland). He clung to the illusion that his new European empire and Britain's old empire of the oceans might not merely coexist but even co-operate, to each other's advantage. 'After making one proposal after another to the British on the reorganisation of Europe', he told Vidkun Quisling, his Norwegian puppet, on 18 August, 'I now find myself forced against my will to fight this war against Britain. I find myself in the same position as Martin Luther, who had just as little desire to fight Rome but was left with no alternative.'

On 1 August he issued Führer Directive No. 17, ordering the Luftwaffe to 'overpower the English air force with all the forces at its command in the shortest possible time'. The objectives were to be 'flying units, their ground installations and their supply organisations, but also . . . the aircraft industry including that manufacturing anti-aircraft equipment'. On the same day Goering assembled his subordinates at The Hague to harangue them on the outcome he expected from *Adler* (Operation Eagle). Theo Osterkamp, a Great War ace already made cautious by his experience in the Channel Battle, expressed reservations: 'I explained to him that during the time when I alone was in combat over England with my Wing I counted . . . about 500 to 700 British fighters . . . concentrated in the area around London. Their numbers had increased considerably [since] the beginning of the battle. All new units were equipped with Spitfires, which I considered of a quality equal to our own fighters.' Goering was angry and dismissive. He claimed that the British were cowardly, that their numbers were much depleted and that the Luftwaffe's superiority in bombers made the British defences of no consequence. *Adlertag* (Eagle Day) was shortly afterwards fixed for 7 August.

In fact Operation Eagle, beset by bad weather, stuttered into life on 8 August; eventually 13 August was declared Eagle Day. By then, however, the Luftwaffe had already experienced setbacks, largely through spreading its effort too wide. On 12 August, a typical day of the operation, it attacked RAF airfields, Portsmouth harbour, shipping in the Thames and – inexplicably for the only time throughout the Battle of Britain – the 'Chain Home' radars. It lost 31 aircraft, the RAF 22. On Eagle Day itself it also attacked, in darkness, a Spitfire factory near Birmingham, losing 45 aircraft to the RAF's 13 (from which six pilots, the key element in Britain's air defences, were saved). On 15 August it lost 75, the RAF 34. Throughout the week it persuaded itself that the 'exchange ratio' was in its favour (in fairness it must be said that the RAF grossly exaggerated its estimate of German aircraft destroyed). On 14 August the Luftwaffe reported to Halder: 'Ratio of fighter losses 1:5 in our favour. . . . We have no difficulty in making good our losses. British will probably not be able to replace theirs.'

German losses, particularly in dive-bombers, were running so high, however, that on 15 August Goering was already beginning to institute a change of plan, and of commanders. Doubters like Osterkampf were promoted out of front-line responsibility, and aggressive young leaders (like Adolf Galland, who would shortly be decorated with the Knight's Cross by Hitler – reluctantly, since he 'looked Jewish') were promoted into their place. Goering outlined to them the objectives of the third stage in the Battle of Britain: the RAF fighter airfields. Bad weather averted the inception of this effort, the first genuine concentration of force that OKL (the Luftwaffe high command) had ordered since the battle was undertaken. Not until 24 August did the RAF feel its effect, but then with alarm. Manston, the most forward of its fighter stations, was put out of action by a determined strike, and North Weald, in the north-east London suburbs, was badly damaged. At Manston the ground staff were demoralised, taking to the air-raid shelters and refusing to emerge. The Luftwaffe flew 1000 sorties that day and destroyed 22 RAF fighters for the loss of 38 of its own aircraft. There was worse to come: on 30 August and 4 September serious damage was inflicted on aircraft factories, while Biggin Hill, a main fighter station covering London, was attacked six times in three days, the operations room destroyed and seventy ground staff killed or wounded. Between 24 August and 6 September Fighter Command lost 290 aircraft in constant defensive engagements; the Luftwaffe lost 380 aircraft, but only half of those were fighters.

The crunch

The Luftwaffe had begun to win the battle – but not fast enough for Hitler's and Goering's patience. The autumnal gales threatened. If the invasion barges were to be got across the Channel Narrows in 1940 Britain's resistance would have to be broken in the next few weeks: Fighter Command would have to be beaten in the air so that the Royal Navy could be bombed out of the Channel. On 31 August OKL decided that on 7 September the *Schwerpunkt* (focus of attack) would be shifted from the airfields to London. Thitherto it had been spared; as an OKW order of 24 August stated: 'Attacks against the London area and terror attacks are reserved for the Führer's decision.' He had withheld it because he had still hoped to bring Churchill to the conference table – and also to avert retaliation against German cities. Now he was driven to the calculation that only by an attack on London would 'the English fighters leave their dens and be forced to give us open battle', as the ace Adolf Galland put it.

Thus the Battle of Britain reached its climax: the assault by dense formations of Heinkel, Dornier and Junkers bombers protected by phalanxes of Messerschmitt 109s and 110s, against (Galland's description) 'the seven-million-people city on the Thames . . . brain and nerve centre of the British High Command'. It was an assault that had to brave a ring of 1500 barrage balloons, 2000 heavy and light anti-aircraft guns and the constantly maintained ranks of 'the Few', 750 Spitfires and Hurricanes. For ten days in mid-September, days of blue sky and brilliant sun remembered by all witnesses, the skies of south-eastern England were filled each morning by German raiders in hundreds proceeding towards London to be intercepted by British fighters rising to meet them and to disperse, sometimes to re-form, sometimes not, as the battle engulfed them. Desmond Flower, a young conscript of the Middlesex Regiment, recalls the spectacle:

> Sunday in Sevenoaks was the same as Sunday throughout Kent, Surrey, Sussex and Essex. The hot summer air throbbed with the steady beat of the engines of bombers which one could not see in the dazzling blue. Then the RAF would arrive; the monotonous drone would be broken by the snarl of a fighter turning at speed, and the vapour trails would start to form in huge circles. I lay on my back in the rose garden and watched the trails forming; as they broadened and dispersed a fresh set would be superimposed upon them. Then, no bigger than a pin's head, a white parachute would open and come down, growing slowly larger; I counted eight in the air at one time.

Some of the parachutes may have been British, for on 9, 11 and 14 September Fighter Command lost heavily in repelling the German formations. However, its success in sparing London damage – it was not true, as the German military attaché was reporting from Washington, that 'the effect in the heart of London resembles an earthquake' – now prompted the Luftwaffe to maximise its effort. On 15 September the largest bomber force yet dispatched, 200 aircraft with a heavy fighter escort, approached London. Fighter Command had early warning: its forward airfields had been repaired since the opening of the assault on the capital, and Air Chief Marshal Sir Hugh Dowding, commander-in-chief of Fighter Command, gave permission for the Midlands Group, No. 12, to lend its squadrons to the defence. Visiting No. 11 Group's headquarters at Uxbridge that morning, Churchill asked Air Vice-Marshal Keith Park, the group commander, 'What other reserves have we?', and got the answer he had heard from Gamelin in Paris three months before: 'There are none.' But Dowding's plunge was a considered decision, not a miscalculation. It measured means against ends with discrimination, and the decision was justified by results. Some 250 Spitfires and Hurricanes intercepted the German bombers well east of London and by the end of the day when a second wave had been met and turned back, had shot down nearly sixty. It was the Luftwaffe's most spectacular defeat in the battle (though not the costliest) and decisive in its deterrent effect. Hopes that Britain's resistance could be broken while the invading season held collapsed: on 17 September Hitler announced the postponement of Operation Sealion until further notice.

Postponement of Sealion did not evoke a termination of Eagle. For one thing, Goering had always regarded the two operations as quite separate and clung to the hope that this personal offensive against Britain could achieve a strategic result independent of the efforts of the army and navy. For another, Hitler wished to sustain the pressure on Churchill's government, which he had persuaded himself must perceive the inevitability of an accommodation as clearly he did himself. Daylight attacks on London and other targets were therefore maintained throughout September and on some days inflicted heavy damage; on 26 September, for example, a surprise raid on the Spitfire factory in Southampton stopped production for some time. The equation of aerial effort, however, was speaking for itself. As Galland explained to a resistant Goering at the Reichmarschall's hunting lodge, whither Galland had been bidden to shoot a stag in reward for his fortieth victory, on 27 September, 'British plane wastage was far lower and production far higher than the German intelligence staff estimated and now events were exposing the error so plainly that it had to be acknowledged.'

Acknowledgement was conceded slowly: daylight raids continued, at mounting cost, into October; but night raids – inaccurate though they were, besides inviting both retaliation and the accusation of 'terror tactics' which Hitler eschewed – began to become the norm. During October six times the tonnage of bombs was dropped by night as by day; and after November, in 'the Blitz' proper, night bombing supplanted daylight operations altogether. By then the Battle of Britain could be said to be over. It had been a heroic episode. 'The Few' deserved their epitaph: some 2500 young pilots had alone been responsible for preserving Britain from invasion. The majority were citizens; but significant numbers were Canadians, Australians, New Zealanders and South Africans (including the icy 'Sailor' Malan, who tried to send German bomber pilots home with a dead crew, as a warning to the rest). A few were aliens, Irishmen and Americans, impatient at their countries' neutrality, and a vital minority were refugees, Czechs and Poles; the latter, who formed 5 per cent of 'the Few', were responsible for 15 per cent of the losses claimed to have been inflicted on the Luftwaffe.

The victory of 'the Few' was narrow. During the critical months of August and September, when the Battle of Britain was at its height, Fighter Command lost 832 fighters, the Luftwaffe only 668. It was the loss of nearly 600 German bombers which made the balance sheet read so disfavourably to the attacker. Had Hitler and Goering been privy to the extent of their success during the height of the battle, when a quarter of Fighter Command's pilots became casualties and fighter losses for a period (11 August to 7 September) exceeded production, they would undoubtedly have surpassed their effort. Had they done so, the Luftwaffe might then have made itself the first air force to achieve a decisive victory in combat as an independent strategic arm, thus fulfilling the vision that Douhet and Mitchell had glimpsed in the dawn of military aviation. As it was, the pragmatism of Dowding and his Fighter Command staff, the self-sacrifice of their pilots and the innovation of radar inflicted on Nazi Germany its first defeat. The legacy of that defeat would be long delayed in its effects; but the survival of an independent Britain which it assured was the event that most certainly determined the downfall of Hitler's Germany.

FIVE

War Supply and the Battle of the Atlantic

Supply of food, of raw materials, of finished products, of weapons themselves, lies at the root of war. From the earliest times man has gone to war to take possession of resources he lacks and, when at war, has fought to secure his means of livelihood and self-protection from his enemy. The Second World War was no exception to this rule. In the view of Professor Alan Milward, principal economic historian of the conflict, its origins 'lay in the deliberate choice of warfare as an instrument of policy by two of the world's most economically developed states. Far from having economic reservations about warfare as a policy, both the German and Japanese governments were influenced in their decisions for war by the conviction that war might be an instrument of economic gain.'

Milward's judgement that economic impulsion drove Japan to war is incontestable. It was Japan's belief that her swelling population, overflowing an island homeland deficient in almost every resource, could be supported only by taking possession of the productive regions of neighbouring China which had brought her into direct diplomatic conflict with the United States in 1937-41; it was America's reactive trade embargoes, designed to hamstring Japan's strategic adventurism, that in 1941 drove the Tokyo government to choose war rather than circumscribed peace as its national way forward. In the year of Pearl Harbor 40 per cent of Japan's requirements of steel had to be imported to the home islands, together with 60 per cent of her aluminium, 80 per cent of her oil, 85 per cent of her iron ore and 100 per cent of her nickel. America's threat to deny her oil and metals, against a guarantee of good behaviour as Washington should judge it, was therefore tantamount to strangulation. The 'southern offensive' was an almost predictable outcome.

Hitler could not argue economic insufficiency to justify his strategic adventurism. In 1939, when a quarter of the population was still employed on the land, Germany was almost completely self-sufficient in food, needing to import only a proportion of her consumption of eggs, fruit, vegetables and fats. She also produced all the coal she consumed and a high proportion of her iron ore, except for armaments-grade ore which was supplied from Sweden. For rubber and oil – commodities for which coal-based substitutes would be found during the war – she was wholly dependent on imports, as she was also for most non-ferrous metals. However, through peaceful trade, her high level of exports (particularly of manufactures such as chemicals and machine tools) easily earned the surplus to fund and make good those deficiencies. Had it not been for Hitler's social-Darwinian obsession with autarky – total national economic autonomy – Germany would have had no reason to prefer military to commercial intercourse with her neighbours.

Paradoxically it was Germany's adversaries, Britain and France, and her half-hearted ally, Italy, which had the better economic reasons for going to war. Italy was a major energy importer, while her industry, particularly her war industry, was rooted in a tradition of craftsmanship quite inconsistent with the ruthless mass consumption of the modern battlefield. Italian aero-engines were works of art – no consolation to the *Regia Aeronautica* pilots when replacement aircraft failed to come off the production-line at a rate to match attrition in the skies over Malta and Benghazi. France, too, maintained military arsenals run on artisan principles; and, though the country fed itself with ease and exported luxuries in plenty, it depended on its empire and its trading partners for many raw materials and some manufactured goods – for advanced aircraft, for example, from the United States in the crisis of 1940, and for rubber from its colonies in Indo-China.

Britain's case was the most paradoxical of all. In high gear, its industry could produce all the weapons, ships, aircraft, guns and tanks that its mobilised military population could man on the battlefield. As it had demonstrated in the First and events would prove in the Second World War, moreover, it could continue to find a surplus of armaments to export (to Russia) or to re-equip exile forces (Poles, Czechs, Free French), even at the nadir of its military fortunes. However, it could do so only by importing much of its non-ferrous metal and some of its machine-tool requirements to supply its factories, all its oil, and – most critically of all for an overpopulated island – half its food. At a pinch the Japanese, by living on unhusked rice, could survive at near-starvation level. The British, if deprived of North American wheat, would in the few months it would take to exhaust the national strategic reserve of flour and powdered milk have undergone a truly Malthusian decline and halved in numbers.

Hence Winston Churchill's heartfelt admission, once victory came, that 'the only thing that really frightened me during the war was the U-boat peril. . . . It did not take the form of flaring battles and glittering achievements, it manifested itself through statistics, diagrams and curves unknown to the nation, incomprehensible to the public.' The most important statistics were easily set out. In 1939 Britain needed to import 55 million tons of goods by sea to support its way of life. To do so it maintained the largest merchant fleet in the world, comprising 3000 ocean-going vessels and 1000 large coastal ships of 21 million gross-register (total capacity) tons. Some 2500 ships were at sea at any one time: the manpower of the merchant service, a resource almost as important as the ships themselves, totalled 160,000. To protect this fleet the Royal Navy deployed 220 vessels equipped with Asdic – the echo-sounding equipment developed by the Allied Submarine Detection Investigation Committee in 1917 – consisting of 165 destroyers, 35 sloops and corvettes and 20 trawlers. The ratio between merchant ships and escorts was thus about 14:1. Convoy, the practice of assembling merchant ships in organised formations under naval escort, was no longer a controversial procedure as it had been in the First World War; the Admiralty was committed to it before war broke out and it was introduced on the oceanic routes immediately and in coastal waters as soon as practicable.

U-boats and surface raiders

The principal enemy of convoy was the submarine, or U-boat (*Unterseeboot*). As in 1914 the Germans also deployed a number of surface commerce raiders, including both orthodox warships and converted merchantmen, but their number was small; between September 1939 and October 1942 less than a dozen auxiliary raiders gained great waters, of which the most successful, *Atlantis*, sank twenty-two vessels before interception and destruction by HMS *Devonshire* in November 1941. Germany's battleships, battlecruisers, pocket battleships and cruisers occasionally raided the sea lanes, but they too were few in number, and judged too valuable, to be risked often, particularly after the humiliating defeat of the pocket battleship *Graf Spee* off Montevideo by three British cruisers in December 1939. German aircraft achieved some success as ship destroyers – in May 1941, a peak month, they sank 150,000 tons (the average displacement of a Second World War merchant ship was 5000 tons) – and mines, whether laid by aircraft, surface ship or submarine, were a constant menace. German fast coastal craft, known to the British as E-boats, were prolific minelayers in British coastal waters in 1941-4, and constituted a relentless threat to coastal convoys; in April 1944 a raid on an American troop convoy practising disembarkation for D-Day at Slapton Sands in Devon drowned more GIs than were lost off Normandy on 6 June itself. However, the attacks of aircraft and surface ships, large and small, on merchant shipping were extraneous to the real battle at sea in European waters in the Second World War. That was one, as Winston Churchill rightly denoted, between the convoy escort and the U-boats.

In September 1939 Karl Dönitz, the German U-boat admiral, had fifty-seven U-boats under command, of which thirty were short-range coastal types and twenty-seven ocean-going. The German navy's pre-war expansion programme, the 'Z-plan', called for the construction of a fleet of 300, with which Dönitz claimed he could certainly strangle Britain to death. He was to achieve that total in July 1942, allowing him to maintain 140 boats on operations and sink shipping at an annual rate of 7 million tons, a figure which exceeded British building of replacement shipping more than five times. By then, however, thanks to the inescapable dynamic of warfare, almost every term in the equation by which he had calculated the inevitability of Britain's strangulation by U-boat tactics had changed to his disadvantage. Requisition and chartering of foreign ships had added 7 million tons to the British merchant fleet, the equivalent of a year's torpedoing. American shipyard capacity, enormously expanded by an emergency mobilisation, had been added to the British, promising an output of 1500 new ships in 1943 (including many of 10,000–15,000 tons), more than three times as many as the U-boats were sinking. Naval construction in the United States would add 200 escorts a year to the fleet between 1941 and 1945. Over 500 of these would go to join the Royal Navy's escort fleet in the North Atlantic which, having reached a strength of 374 in March 1941, had almost doubled since the outbreak. Long-range aircraft based in North America, Iceland and Britain were progressively reducing the 'air gap' in which U-boats could safely operate on the surface, their preferred mode because of their low submerged speed; and integral aircraft protection for convoys, provided by 'escort carriers', was soon to level a direct threat against attacking U-boats. Only in facilities for basing his boats had Dönitz's position improved; in electronic and cryptographic warfare the conflict hung in the balance; the promise of secret underwater weapons, which favoured Germany, could not be realised for some years. Nevertheless the U-boats had already inflicted severe material and psychological damage on the Allied, particularly the British, war effort; and in mid-1942 the eventual outcome of the Battle of the Atlantic was evident to no one. The 'statistics, diagrams and curves' were pregnant with menace.

Thus far the Battle of the Atlantic (Churchill had coined the term) had passed through four distinct phases. From the outbreak of the war until the fall of France, the U-boat fleet had been confined by geographical constraints and Hitler's concern for the sensitivities of neutrals to operating within the immediate vicinity of the British Isles. After June 1940, when Germany gained possession of the French Atlantic ports (where in January 1941, with remarkable prescience, Hitler ordered the construction of bomb-proof U-boat 'pens' to begin), the fleet began to operate in the eastern Atlantic, concentrating in particular on the 'Cape route' to West and South Africa and occasionally penetrating into the Mediterranean, since the Italians were proving themselves inept submariners. From April to December 1941, thanks to their increasing expertise in anti-convoy tactics, and despite the delineation of an American 'Neutrality Zone' in which the United States Navy gave notice of its intention to attack marauding submarines, the U-boat captains began to extend their operations into the central and western Atlantic; after June 1941, when Britain began to run convoys to North Russian ports with war supplies, the U-boats, frequently supported by German warships and shore-based aircraft, also began to operate in Arctic latitudes. Finally, after December 1941, Dönitz's men carried the submarine war to the Atlantic coast of the United States and into the Gulf of Mexico, where, during a gruesomely named 'Happy Time' of several months caused by the US Navy's temporary inability to organise convoy on coastal routes, they sank coastwise shipping by hundreds of thousands of tons.

Until June 1940 the U-boats had been confined by the same facts of geography as had kept the German High Seas Fleet close to their home bases during the First World War. Using the Baltic as their training ground (as they were to do throughout the war), they attacked British shipping in the North Sea but were denied egress via the Channel by the mine barrier in the Dover Straits and could reach the Atlantic only by making the long passage round the north of Scotland – if, that is, they had the range to do so. Few had that range. Only eight of the Type IX were truly oceanic, with a range of 12,000 miles; eighteen could cruise as far as Gibraltar; the remaining thirty could not leave the North Sea. Despite these limitations the U-boats had some notable successes, including the sinking of the battleship *Royal Oak* in the Royal Navy's main base at Scapa Flow in October 1939 and the aircraft carrier *Courageous* which was sunk while flagrantly neglecting anti-submarine precautions in home waters. From the outbreak of war to the fall of France, total merchant sinkings in the North Atlantic did not exceed 750,000 tons and 141 ships.

The capture of the French Atlantic ports in June 1940, however, transformed the basis of U-boat operations. Possession of Brest, Saint-Nazaire, La Rochelle and Lorient put Dönitz's boats on the doorstep of Britain's trade routes, thus ensuring that the pattern of sinkings, thitherto arbitrary and sporadic, should become regular and consistent. As soon as his crews cleared the Bay of Biscay they found themselves astride the route from Britain to the Cape along which travelled Nigerian oil and South African non-ferrous ores; and by reaching out only a little further into the Atlantic they could attack convoys carrying meat from the Argentine and grain from the United States.

Ships sailing individually were desperately vulnerable to interception. As the Royal Navy's experience in the First World War had proved, independent sailing presented U-boats with a succession of targets: a captain who missed one lone ship on a well-used trade route could still count upon the appearance of another and thus achieve a respectable success rate by the operation of the probability factor alone. Convoy upset probability. Because the submerged speed of a submarine was at best equal to and often lower than that of a merchant ship, a U-boat captain who was wrongly positioned for an attack when a convoy hove into view would miss all the ships in it and might have to wait days before another appeared, with no greater certainty than before of finding himself correctly positioned to attack it.

Dönitz, a First World War submarine captain, had recognised the mathematical disadvantage at which his naval arm operated and conceived a method to overcome it. By experimentation with surface torpedo-boats, during the period when Germany was denied U-boats by the Versailles Treaty, he demonstrated that 'packs' of submarines, if disposed in a chain on the surface where their speed exceeded that of merchant ships, could identify the approach of convoys across a wide band of ocean, be concentrated against one by radio command from shore and inflict mass sinkings by a concerted raid in numbers that would overwhelm the escorts. Once Germany acquired the French Atlantic ports, it was these 'wolf pack' tactics that were to make the Battle of the Atlantic the knife-edge struggle for advantage which overshadowed Churchill's conduct of the British war effort from mid-1940 to mid-1943.

Convoy, which was the Royal Navy's defence against the wolf packs, offered only partial protection to the Atlantic lifeline. The naval escorts themselves – in the early days perhaps only two or three destroyers and a corvette were available to shepherd forty freighters and tankers across 3000 miles of ocean – were little direct threat to a determined U-boat formation. Asdic, the echo-sounder used to detect submerged U-boats, was ineffective beyond 1000 yards and reflected only range and bearing, not (until 1944) depth. The depth charges used to attack U-boats, triggered by water-pressure fuses, had to be set by guess and fractured the U-boat hull only if detonated close by. Most U-boat attacks, moreover, were delivered from the surface at night, when radar was more useful than Asdic, but until 1943 radars were too primitive to give early warning or accurate ranging.

The radio intelligence war

It was the measures taken to route convoys away from known or suspected U-boat patrol lines which best assured their safety, together with ancillary measures – particularly aerial patrol – to force U-boats to submerge while convoys passed by. Until May 1943 a shortage of aircraft and their shortness of range left an 'air gap' between North America, Iceland (available as a base to Britain after the German invasion of Denmark in April 1940) and Britain itself in which U-boats operated without fear of surveillance; the gap was closed when the very-long-range Liberator (B-24) with an endurance of eighteen hours came into service. Rerouteing, on the other hand, was a stratagem employed from the very beginning of the Atlantic war, and there was always a strong sense of direct conflict between the two sides. On the German side, the officers of the B-Dienst (Observer Service) used wireless intercepts and decrypts of cipher transmissions to establish convoy positions and read their orders; on the British (later Anglo-American) side, the cryptographers of the Government Code and Cipher School at Bletchley and the staff of the Admiralty Operational Intelligence Centre monitored the signals sent between U-boats and from Dönitz's headquarters at Kernevel, Lorient (after March 1942 Berlin), to detect the formation of patrol lines and the vectoring of wolf packs against their targets. Rerouteing was by far the most successful of convoy protection measures. Between July 1942 and May 1943, for example, the British Admiralty and US Navy Department intelligence centres managed to reroute 105 out of 174 threatened North Atlantic convoys clear out of danger, and minimised attacks on another 53 by rerouteing; only 16 ran directly into wolf-pack traps and suffered heavy loss.

The success achieved by Captain Rodger Winn, RNVR, and later by his American counterpart, Commander Kenneth Knowles, depended ultimately upon the skill of the Bletchley Park cryptographers in decrypting the Kernevel U-boat traffic fast enough for its significance to be applied to convoy operations. That traffic was, of course, enciphered on the Enigma machine, and the 'Shark' key used by the U-boat service proved particularly resistant to Bletchley's efforts; it was not broken until December 1942 and then not regularly until 1943. Much of the vital radio intelligence used by the Operational Intelligence Centre up to that time was of lower-grade, position-fixing quality. High Frequency Direction Finding (HF/DF or 'Huff Duff') enabled ships to detect and locate shadowing U-boats from the transmissions they sent back to U-boat headquarters, and so for convoys to be rerouted or protecting aircraft summoned. Meanwhile, because of the Admiralty's ill-advised persistence in the use of a book code instead of a machine cipher, the B-Dienst was able to read convoy traffic and direct wolf packs on to chosen routes with sometimes disastrous effect.

The crux of this radio intelligence war began with the move of the U-boats from the eastern to the central Atlantic after April 1941. Substitution and rationing in Britain had allowed the import requirement to be reduced from 55 to 43 million tons, but the minimum level of subsistence was approaching and had to be measured against a rate of sinking which threatened to outstrip replacement building. In February 1941 the United States had enacted a Lend-Lease law which, in effect, allowed Britain to borrow war supplies against the promise to repay after victory; and from April 1941 the United States was operating a Neutrality Patrol which effectively excluded U-boats from the Atlantic west of Bermuda, under the terms of the Pan-American Neutrality Act of 1939. However, the U-boat fleet now had over 2000 miles of ocean in which to intercept convoys and was adding to its numbers at a considerably higher rate than it was losing U-boats on operations; during 1941 the building rate exceeded 200, while the total lost since September 1939 was less than fifty.

The eight months of extended U-boat warfare in the Atlantic in 1941 therefore proved extremely successful to the German navy. In May it suffered the loss of the great battleship *Bismarck*, unwisely unleashed as a commerce raider, at the end of a great chase by most of the British Home Fleet; but that defeat was offset by the sinking of 328 merchant ships of 1,500,000 tons at a time when the British yards were launching less than a million tons of new construction annually. The casualties took down with them almost every category of material of which the home islands stood in dire need – wheat, beef, butter, copper, rubber, explosives and oil, as well as military equipment.

Apologists for the British effort could show, however, that two-thirds of the ships lost had been sunk out of convoy and that U-boat losses for the year totalled 28, suggesting that the escorts' success was increasing. Dönitz was certainly prepared to draw that conclusion: as soon as the United States Navy became an overt combatant rather than a hostile neutral (as it had been since September 1941), he transferred the weight of his effort to the United States coast. From January 1942 onwards, up to twelve U-boats were cruising off the American east coast and in the Gulf of Mexico at any one time; between January and March they sank 1.25 millon tons of shipping, equivalent to an annual rate four times higher than that achieved in the North Atlantic during 1941.

By May, however, convoy had been introduced on America's Eastern Sea Frontier and sinkings at once declined in those waters. Moreover, the rate of new building, both of merchantmen and escorts, began to accelerate spectacularly, as the American shipyards revived and were mobilised for new construction. Of particular importance was the appearance of a standardised tanker, the T10, and a freighter, the Liberty ship, both of which were larger (14,000 and 10,000 tons) and faster than their pre-war equivalents, besides – most important of all – being quick to build. Three months was the average construction time; by October 1942 American yards were launching three Liberty ships a day and in November the *Robert E. Peary* was built from the keel up in four days and fifteen hours – a public-relations stunt, but grim evidence to Dönitz of the challenge that prefabrication techniques posed to the efforts of his U-boat captains.

The critical point

By July 1942, though neither side yet perceived it, the Battle of the Atlantic was approaching its crux. There were diversions from the central issue. In March 1942 the British destroyed the dock at Saint-Nazaire which offered an Atlantic coast home to the *Tirpitz*, Germany's largest battleship. It was then harboured in north Norway, where it had been joined in February by the *Scharnhorst* and *Gneisenau* after their daring dash through the Channel – an incident that provoked much recrimination between the Admiralty and the RAF over who bore responsibility for the failure to intercept them. The three heavy ships levelled a menace against the Arctic convoys for months to come and caused lethal disruption to convoy PQ17 in July. In December 1943 *Scharnhorst* made a brief sortie against an Arctic convoy, provoking much anxiety at the Admiralty, but suffered an identical fate to that of *Bismarck*; and the *Tirpitz* menace did not end until November 1944 when she was sunk by bombing at her moorings in Troms| Fiord. But these episodes largely summarise the contribution of the German surface fleet to the *Kriegsmarine*'s war. The need to mount the North African 'Torch' landings in November 1942 temporarily drained away Allied merchantmen and warships from the Atlantic lifeline. There was a serious interruption of Ultra decrypts of U-boat radio traffic in February 1942 which lasted for most of the year, at a time when the B-Dienst was enjoying increased success against the Royal Navy's no. 3 book code. The British, American and Canadian convoy control systems were simultaneously experiencing problems of 'shaking down' into a routine of co-operation; the

Royal Canadian Navy, which was undergoing an expansion from six to nearly 400 warships in service, the largest of any armed force in any country during the Second World War, found particular difficulty in matching the expertise of its larger partners. Since the outbreak of the war the Admiralty and the Royal Air Force had been locked in a quarrel over the deployment of long-range aircraft, the Admiralty rightly but vainly arguing that convoy protection produced a better return of effort than spectacular but often ineffective offensive bombing of German cities. Against this background Dönitz was working to extend the range of his U-boats by experimentation with refuelling at sea from submarine 'milch cows', and to equip his boats for the increasingly dangerous surface transit of the Bay of Biscay against attack by such long-range aircraft as the RAF did allocate to the Admiralty via Coastal Command. In the first half of 1942 Coastal Command aircraft mounting the new and powerful Leigh searchlight began to surprise U-boats in the Bay at night and attack them with depth charges. Two were destroyed in July, although Dönitz had already ordered that they must make the passage submerged, despite the delay entailed in reaching the Atlantic cruising grounds. The significance of the Leigh Light was that it gave aircraft 'eyes' in the final 2000 yards of their approach when radar did not work. In a see-saw of technical duelling, its usefulness was to be reduced when the Germans learned to develop passive radar detectors which indicated danger before the Leigh Light could be activated. Over the course of the Bay of Biscay battle, however, which lasted into 1944, the advantage consistently returned to the Allies. It was not until the deployment of the first schnorkel-equipped U-boats (which could recharge batteries while cruising submerged) in early 1944 that the danger levelled by anti-submarine aircraft began to be offset.

By then, however, the climactic phase of the Atlantic battle had come and gone. From July 1942 onward, when Dönitz at last achieved his target figure of 300 U-boats, he redeployed his effort to the central Atlantic, where the Allied escort fleet had been weakened by the transfer of British ships to help the Americans introduce convoy on the Eastern Sea Frontier. He was also now becoming more adept at organising patrol lines and concentrating wolf packs against convoys, and he was having greater success in locating convoys because of the advantage in cryptography that the B-Dienst currently enjoyed over Bletchley. The British responded with two experimental measures which would bear fruit later: the creation of a 'support group' of escorts to go to the rescue of a convoy under attack and the adaptation of merchant ships to fly off aircraft, the merchant aircraft carriers. The MACs however, proved clumsy forerunners of the true escort carriers, the first of which, USS *Bogue*, would not appear until the following March; while the persisting shortage of escorts would compel the 20th Support Group to be disbanded after two months.

As a result, U-boat sinkings in the North Atlantic in November 1942 reached the total of 509,000 tons, a figure exceeded only once before in the previous May, during the 'Happy Time' off the American coast. Vicious Atlantic weather halved sinkings in December and January; but in February 1943, despite continuing bad weather, 120 U-boats sank nearly 300,000 tons in the North Atlantic and the toll seemed set to rise. During March, in a running battle against two convoys eastbound from North America to Britain, codenamed HX229 and SC122, forty U-boats sank twenty-two out of ninety merchantmen and one escort of the twenty which were defending them. The tonnage sunk, 146,000 was the highest in any convoy battle and led Dönitz and his crews to believe that they had victory in their grasp. Altogether they sank 108 ships in the North Atlantic in March, totalling 476,000 tons, the majority lost in convoy. Wolf-pack tactics, supported by the position-finding and decrypting successes of the B–Dienst, appeared to have achieved mastery over convoy protection.

The appearance was illusory. Not only was the shipping replacement rate rising to meet losses (by October 1943 new construction had actually made good the amount of shipping lost since 1939, and built a superior merchant fleet into the bargain); U-boat losses were also beginning to equal launchings, at a monthly rate of about fifteen. Those statistics spelt doom to Dönitz's effort. The explanation of the shift lay in several directions: Bletchley recovered its edge over the B–Dienst in May 1943, making rerouteing even more successful; escorts were becoming more plentiful, allowing the creation of permanent 'support groups', five in number in April; to the existing escorts were added two escort carriers embarking twenty anti-submarine aircraft which could force all U-boats in the vicinity of a convoy to submerge, thus effectively negating their offensive potential; improved radar, Asdic and depth-charge launchers (Hedgehog and Squid) levelled a direct tactical threat against the U-boat in close combat; but, above all, the increased availability of long-range patrol aircraft for the Atlantic battle wrought a strategic shift of advantage to the British-US-Canadian side. The long-range aircraft, particularly the Liberator bomber, equipped with radar, Leigh Light, machine-guns and depth charges, was flying death to a surfaced U-boat. In the Bay of Biscay – after a short and disastrous episode of 'fighting it out', ordered by Dönitz – the aircraft forced all U-boats to make their passage to the North Atlantic hunting-grounds submerged, at a fourfold time penalty; in great waters they completely disrupted Dönitz's wolf-pack tactics by dispersing his patrol lines and savaging his concentrations wherever they appeared. In May 1943 U-boat losses, inflicted at a ratio of about 3:2 between aircraft and escorts, reached forty-three, which exceeded replacement more than twice. On 24 May Dönitz, accepting the inevitable, withdrew his fleet from the ocean, conceding later in his memoirs: 'We had lost the Battle of the Atlantic.'

That did not mark the end of the U-boat war. In May 1944 the first schnorkel-equipped U-boat made its trial cruise. The schnorkel, a retractable air-breathing tube, allowed a submarine to cruise submerged while using its diesel engines. The device, invented by a Dutch naval officer in 1927, anticipated the development both of the closed hydrogen-peroxide system, which the Germans would bring into service in 1945, and of nuclear propulsion, in that it transformed the submersible U-boat into a true submarine, capable of operating below the surface throughout an operational mission. Misused, it could kill a crew by suffocation, and two U-boat crews are believed to have died in that way; properly employed, it would have revived the U-boat threat. Had the Germans not lost their main Atlantic ports to the American army in August 1944, schnorkel U-boats would have reopened the Atlantic battle, to the Allies' very great cost.

Measured across the space of the Atlantic struggle, from September 1939 to May 1943, the cost can be seen to have fallen most heavily on Dönitz's submarine arm. Although the Allies lost 2452 merchant ships in the Atlantic, of nearly 13 million gross-register tons, and 175 warships, mostly British (a term which included Canadian, Polish, Belgian, Norwegian and Free French escorts also), the *Kriegsmarine* lost 696 out of 830 U-boats dispatched on operations, almost all in the Atlantic, and 25,870 killed out of 40,900 crewmen who sailed; another 5000, plucked from the wrecks of their depth-charged boats, became prisoners. This casualty rate – 63 per cent fatal, 75 per cent overall – far exceeded that suffered by any other arm of service in the navy, army or air force of any combatant country.

The cost was certainly not in vain. Given that the economic odds disfavoured Germany from the outset, that its industry was organised 'in breadth' for a short war rather than 'in depth' for a long war, and that Hitler's campaign of conquest was notably unsuccessful in adding either productive capacity or raw material resources to the Reich's war-making capacity – it failed, for example, to acquire any large source of oil or non-ferrous metal ores for the German war machine – the delaying effect of the U-boat war on the transformation of Britain into an Anglo-American *place d'armes* for the eventual liberation of Europe may be seen as crucial. Moreover, while Germany fed itself easily between 1940 and 1944 from its own agricultural output and the requisitions made on farming in the occupied lands to the east and west, Britain was constantly held close to the level of minimum subsistence by U-boat depredations on its food imports. Rationing, though fairly applied and beneficial to the classes nutritionally deprived before the war, created a climate of latent crisis among the British which distorted and diminished their capacity to strike back at the enemy. Britain's military threat to Germany during the Second World War was as intense as that levelled during the First though in relative terms Britain was not much weaker in 1940-4 than in 1914-18. It was the U-boats, marginally assisted by the Luftwaffe, which made the difference.

The U-boats were also to prove of crucial significance in diffusing and diminishing the support brought by British and especially American industry to their allies and their own ancillary theatres of war. Russian industry was devastated by the German invasion of White Russia and the Ukraine in 1941, and the Soviet Union's capacity to sustain resistance was only salved by the almost incredibly swift transfer of factories from the western provinces to the trans-Ural regions in the terrible winter of 1941-2. Between July and October, for example, 496 factories were transported by train from Moscow to the east, leaving only 21,000 out of 75,000 metal-cutting lathes in the capital; overall the Russian railways moved 1523 factories from west to east between June and August, and between August and October it was calculated that 80 per cent of Russian war industry was 'on wheels', moving from the threatened zones to areas of safety in western or eastern Siberia. The disruption of production entailed by this unprecedented industrial migration could only be made good by substitutions from Western sources, of weapons and munitions but above all of the elements of war's infrastructure – vehicles, locomotives and rolling stock, fuel, rations and even such simple but vital supplies as boots, the felt winter boots for lack of which tens of thousands of German soldiers lost toes in the winter of 1941-2. Between March 1941 and October 1945 the United States supplied the Soviet Union with 2000 locomotives, 11,000 rail wagons, nearly 3 million tons of gasoline, 540,000 tons of rails, 51,000 jeeps, 375,000 trucks and 15,000,000 pairs of boots. It was in American boots and trucks that the Red Army advanced to Berlin. Without them its campaign would have foundered to a halt in western Russia in 1944.

Boots and trucks proved far more important items of war supply than the 15,000 aircraft, 7000 tanks and 350,000 tons of explosives which Lend-Lease also consigned to the Soviet Union; far more important than all the aid sent by Britain during the course of the war – 5000 tanks, 7000 aircraft, even the 114,000 tons of rubber. Vital though these war supplies were, however, they reached Russia between 1941 and 1944 by the most circuitously inconvenient routes, thanks to Dönitz's U-boat campaigns. The 'North Russia run' from Britain to Murmansk and Archangel had to be routed almost as far west as Greenland and as far north as Spitsbergen (on which a strange little sub-war for possession of weather stations was fought in 1941-2) during the summer months of 1941-4, to avoid air and sea attack by German units based in Norway; when the winter ice drove the convoys eastward, losses rose grievously, forcing Churchill to interrupt sailings on several occasions – to Stalin's woundingly expressed scorn. The alternative route through the Persian Gulf was roundabout and terminated at the railhead of a long and inadequate railway system. The Pacific route, to Vladivostok, was also affected by ice and the danger of enemy attack, and it connected with the wrong end of the longest railway line in the world, the Trans-Siberian.

Hitler's investment in his U-boat fleet thus more than justified the cost. It exerted a partial strangulation on the offensive effort of the immediate enemies, the Russians most of all but also the British, delayed the build-up of a large American expeditionary force on his doorstep and hampered the development of a hostile 'peripheral' strategy in the Mediterranean; the closing of that sea to regular British convoying in 1940-2, largely through aerial but partly through submarine threat, forced the desert army to depend on supply via the Cape route, 12,000 miles long, at very great cost to its efficiency.

Had Hitler achieved the creation of a 300-boat fleet before 1942, added significantly to its size thereafter, or managed to introduce his advanced schnorkel and revolutionary hydrogen-peroxide types before 1944, partial might have become total strangulation. None the less, Germany considerably maximised the advantage of being able to operate from the centre of its strategic area – the advantage of 'interior lines' that continental powers had traditionally exercised against oceanic enemies in the European world. Dönitz proved by far the most useful of all the military subordinates who lent their services to Hitler's campaign of conquest – far more useful than Goering, the Luftwaffe chief, or even von Braun, the father of his pilotless missiles – and it was entirely appropriate that he should have been nominated to succeed him as Führer in the last days of the Reich. In pitiless and self-immolatory dedication to the creed of total war, Nazism found no equal within the Wehrmacht to the U-boat arm. Its 'aces' – Günther Prien, Otto Kretschmer, Manfred Kinzel, Joachim Schepke – whether believing Nazis or not, personified its ethos of the superman and even succeeded, for all the cruelty they inflicted, in winning the respect of their enemies for their warrior prowess. The British officer who interrogated Kinzel, 'ace of aces' with a slate of 270,000 tons of shipping sunk, ruefully expressed the hope that 'there were not too many like him'.

The Atlantic was not the only *via dolorosa* of war supply. The Burma Road and 'the Hump', over which supplies were driven or flown (at 14,000 feet) to Chiang Kai-shek's army in south China, were others. The Takeradi route, from West to East Africa, provided the desert air force with aircraft disembarked from Atlantic convoys and assembled ashore. The Lake Ladoga 'ice road' saved Leningrad from total starvation in the winters of 1941-3. And ultimately the Japanese, who had succeeded in turning the archipelagos within their Pacific 'island perimeter' into the foundation of a watery 'continental' strategy during 1942, achieved extraordinary feats of maritime supply in keeping their far-flung garrisons combatant during 1943-4, when MacArthur's 'island-hopping' plan cracked the carapace of their oceanic fortress. In 1945, at the end of their own 'Battle of the Pacific', when they lost almost their whole merchant fleet to America's submarines, the home islands were on the brink of starvation by midsummer – a whimpering end to Japan's campaign of conquest which was stopped short by the cosmic bang of Hiroshima and Nagasaki.

However, none of these logistic efforts compared in duration, magnitude and importance with the Battle of the Atlantic. It was truly both a battle and a war-winning enterprise. Had it been lost, had those 'statistics, diagrams and curves' which blighted Winston Churchill's days and nights in 1940-2 turned wrong, had each U-boat on its patrol line succeeded in sinking only one more merchant ship in the summer of 1942, when losses already exceeded launchings by 10 per cent, the course, perhaps even the outcome, of the Second World War would have been entirely otherwise. The 30,000 men of the British Merchant Navy (one-fifth of its pre-war strength) who fell victim to the U-boats between 1939 and 1945, the majority drowned or killed by exposure on the cruel North Atlantic sea, were quite as certainly front-line warriors as the guardsmen and fighter pilots to whom they ferried the necessities of combat. Neither they nor their American, Dutch, Norwegian or Greek fellow mariners wore uniform and few have any memorial. They stood nevertheless between the Wehrmacht and the domination of the world.

PART II
THE WAR IN
THE EAST
1941–1943

EUROPE ON THE EVE OF HITLER'S INVASION OF RUSSIA, JUNE 1941

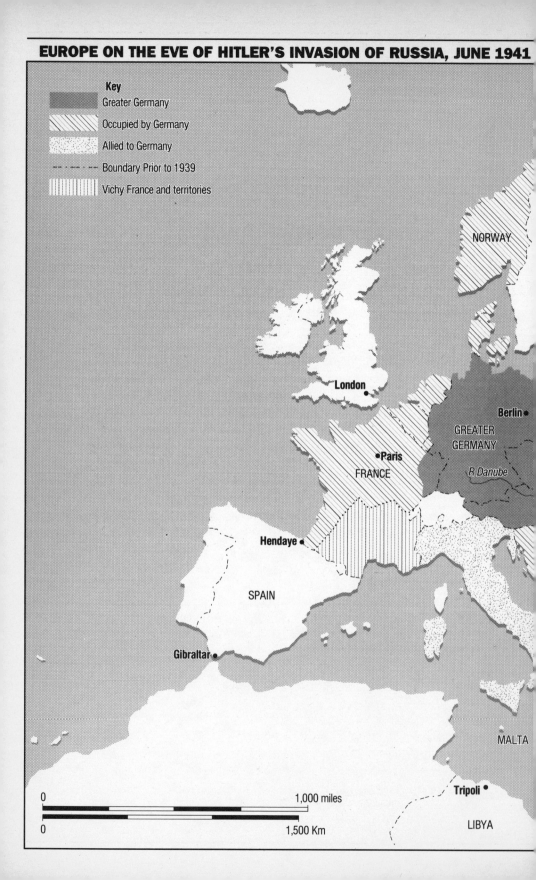

Key

Greater Germany

Occupied by Germany

Allied to Germany

Boundary Prior to 1939

Vichy France and territories

NORWAY

London

Berlin•

GREATER
GERMANY

•Paris

R Danube

FRANCE

Hendaye•

SPAIN

Gibraltar•

MALTA

Tripoli•

0   1,000 miles

0   1,500 Km

LIBYA

SIX

Hitler's Strategic Dilemma

On 19 July 1940 Hitler convened the Reichstag in the Kroll Opera House in Berlin to witness his mass creation of the new German marshalate. It was a consciously Napoleonic gesture and, like Napoleon's elevation of eighteen of his generals to be Marshals of the Empire on 9 May 1804, was designed to glorify the head of state rather than honour his military servants. His three army group commanders, Bock, Leeb and Rundstedt, his personal chief of staff, Keitel, the army commander-in-chief, Brauchitsch, four of the most successful field commanders, Kluge, Witzleben, Reichenau and List, and three Luftwaffe chiefs, Milch, Sperrle and Kesselring, were on the roll; Goering was appointed to the novel rank of Reichsmarschall, a distinction he decided entitled him to yet another splendid uniform, and was decorated with the Great Cross of the Iron Cross, the fifth – and last – award of an honour previously conferred by the Prussian kings on Blücher, Moltke and Hindenburg.

Although the creation of the marshalate was the sensational event of that day, the actual point of the occasion was to review for the puppet deputies the course of the Second World War thus far and to state the terms on which it might be concluded. Hitler's speech was intended as an appeal, via world opinion, to Britain, exposing the hopelessness of her position and inviting her government to make peace. William Shirer, the American journalist, who witnessed it and was a connoisseur of Hitler's speeches, thought it his finest performance: 'The Hitler we saw in the Reichstag tonight was the Conqueror, and conscious of it, and yet so wonderful an actor, so magnificent a handler of the German mind, that he mixed superbly the full confidence of the conqueror with the humbleness which always goes down so well with the masses when they know a man is on top.' His appeal came at the very end of his long oration: 'In this hour, I feel it to be my duty before my own conscience to appeal once more to reason and common sense in Great Britain as much as elsewhere. I consider myself in a position to make this appeal since I am not the vanquished begging favours, but the victor speaking in the name of reason. I see no reason why this war must go on.'

He did not, however, disclose, or even apparently harbour, any view of how it might be ended. Since the arrangement of the armistice with France, intellectually and emotionally Hitler had given himself a vacation from responsibility from which he was loath to return. In the company of two old comrades of the trenches he had toured the First World War battlefields of the

Western Front where he had fought with great bravery as a common soldier. He had visited the sights of Paris, to muse at Napoleon's tomb and view the Opéra, the supreme expression of his taste in architecture. He had wandered through his favourite South German landscapes, breathing the mountain air and the adulation of simple people. He had waited for a week in one of his many purpose-built headquarters, at Freudenstadt in the Black Forest, for word that Churchill was recognising the reality of defeat. It was with reluctance that he had returned to the burdens of leadership, made all the heavier by the need to decide the future. Britain or Russia? That was the choice of enemy he confronted at the crossroads to which his decision for war ten months earlier had now brought him.

Either choice was disagreeable and dangerous. He could not be defeated by Britain but he could be humiliated in the attempt to invade her; moreover, he clung to his dream of winning Britain's co-operation rather than beating her into subjection. On the other hand, he had long and ardently desired the defeat and subjection of Russia; but he recognised the dangers of the attempt. Russia was strong, her centres of power remote; only the fear that time would make her stronger and the urge to incorporate her fertile and productive western territories – those Germany had briefly possessed in 1918 – drove him to seek for ways through the risk of an eastern offensive.

During the days after his Reichstag speech, Hitler addressed himself to debating these dilemmas with his commanders. Erich Raeder, his Grand Admiral, warned that, 'if the preparations for Sealion' (by which he meant the defeat of the RAF) 'cannot definitely be completed by the beginning of September, it will be necessary to consider other plans.' In fact Hitler, even during his 'vacation' after the French armistice, had told Schmundt, his chief Wehrmacht adjutant, that he was considering an attack on Russia – which was not what Raeder meant by 'other plans' – and had set Colonel Bernhard von Lossberg, one of OKW's operations officers, to draft a study (which Lossberg codenamed 'Fritz', after his son). He now set OKH to the same task.

At the end of July he reconvened discussions with his commanders at the Berghof, his Bavarian retreat. On 31 July he told Brauchitsch and Halder that he was reversing his decision, taken in mid-June, to demobilise thirty-five divisions to provide manpower for the economic war against Britain, would in fact increase the strength of the army to 180 divisions (he had already ordered a doubling of the number of Panzer divisions from ten to twenty) and would accelerate the transfer, already begun, of forces to the east, so that by the spring of 1941 he would have 120 divisions close to Russia's border.

This decision could be interpreted as a precautionary move. He had been alarmed by Russia's occupation of Latvia, Lithuania and Estonia in mid-June and by its annexation of Bessarabia and North Bukovina from Romania on 28 June – an annexation in which he was bound to acquiesce, since Russia's claim to those provinces had been agreed in the Molotov-Ribbentrop pact of 22

August 1939. These acquisitions of territory could be seen as threatening. They consolidated a move westward of Russia's strategic boundary, which since September of the previous year had engulfed 286,000 square miles inhabited by 20 million people. Hitler did not, however, believe that Russia intended to attack. It was rather that the boundary changes enlarged Russia's opportunities for further strategic expansion while narrowing Germany's. The occupation of the Baltic states threatened Finland, effectively a German protectorate, and extended Russia's area of control in Baltic waters (where Germany, among other things, trained its U-boat crews). The annexation of Romania's Danubian provinces threatened Bulgaria, a German client state, and improved Russia's opportunity of seizing the Mediterranean entrance to the Black Sea.

It was the evidence these 'forward' moves gave of Russia's determination to pursue its own advantage in the teeth of Germany's proven military power that persuaded Hitler he could not defer a test of strength with her for ever – and, if so, it must be sooner rather than later. Foreign Armies East, the OKH intelligence branch which monitored Soviet capabilities and intentions, had reported in May, from the military attaché in Moscow, that the Red Army, though capable of raising 200 infantry divisions for war, remained so disorganised by the great military purge of 1938 that it would take twenty years 'until it reached its former heights'. Its information on Russian arms production, particularly of tanks, which would have warned otherwise, was defective: the size of the Russian tank fleet was reckoned at 10,000 (against Germany's 3500), when in fact it was 24,000. Hitler was prepared to pit his tank fleet against the Russian, even at odds of three to one; and he had no doubt that 120 German divisions could defeat 200 Russian, if Stalin succeeded in mobilising such a number.

When the twelve new marshals came to collect their batons at the Chancellery on 14 August, therefore, Hitler's talk was of the emerging need to fight the Soviet Union. Field Marshal von Leeb's record of Hitler's remarks reveals the trend of his calculations:

> Probably two reasons why Britain won't make peace. Firstly, she hopes for US aid; but the US can't start major arms deliveries until 1941. Secondly she hopes to play off Russia against Germany. But Germany is militarily far superior to Russia. . . . There are two danger areas which could set off a clash with Russia: number one, Russia pockets Finland; this would cost Germany her dominance of the Baltic and impede a German attack on Russia. Number two, further encroachments by Russia on Romania. We cannot permit this, because of Romania's gasoline supplies to Germany. Therefore Germany must be kept fully armed. By the spring there will be 180 divisions. . . . Germany is not striving to smash Britain because the beneficiaries will not be Germany but Japan in the east, Russia in India, Italy in the Mediterranean and America in world trade. That is why peace is possible with Britain.

A pattern of evasion and delay

On 27 August Hitler sent Schmundt and Dr Fritz Todt, his chief of war construction, to East Prussia to search for a suitable site for another new headquarters from which an eastern campaign might be conducted. On 6 September he approved the transfer of Bock's Army Group B from west to east, where thirty-five divisions, including six Panzer, were now deployed. And on 14 September, when his commanders again convened at the Chancellery for a war conference, he reviewed further reasons for postponing Operation Sealion against Britain; three days later he announced that it was postponed again.

However, he could not yet commit himself to a firm decision for the attack on 'the Bolshevik enemy'. On 15 September Lossberg submitted his 'Fritz' plan to Jodl; ultimately 'Fritz' was to be the plan according to which the Wehrmacht would march eastward, but, as a communication between subordinate and superior within Hitler's personal staff, it remained meanwhile a contingency document. The transfer of German divisions into Poland continued, camouflaged as a move to validate Germany's guarantee of Romania's new frontiers announced at the time of the 'Vienna Award' of 30 August which transferred half of Transylvania from it to Hungary. Hitler also sent a 'military mission', in the unusually great strength of a whole army division, into Romania itself, together with a Luftwaffe air defence force of a thousand men. His diplomats were simultaneously beginning the discussions with Romania, Hungary and the puppet state of Slovakia which would lead to their joining the new Tripartite Pact, signed on 2 September between Germany, Italy and Japan, binding any two to come to the assistance of the third if it were attacked. All these were necessary and useful preliminaries to the mounting of an eastern offensive. Yet they did not amount to a direct provocation of the Soviet Union – though its leaders conceived dire suspicions of what the Tripartite Pact (in fact designed to support Japan in its burgeoning conflict with the United States) portended – nor did they commit Hitler to the decision for such an offensive itself.

As the need to accept or reject such a decision sharpened, Hitler fell into a characteristic behaviour pattern of evasion and delay. It had overcome him for weeks after the Polish triumph, while he had fenced with his generals over the strategy for an attack on the Western Allies. It had seized him in an acute form twice during the Battle of France, once before and once during the attack on the Dunkirk perimeter. Now it was manifested in a search for means of winning the war by broadening its base. If he could not talk the British round, or defeat them by invasion – Sealion was cancelled for good on 12 October – he would achieve the same effect by multiplying the enemies they had to face and the fronts on which they had to fight. Mussolini had opened an offensive into British-garrisoned Egypt from Libya on 13 September. On 4 October, while the offensive still seemed to promise success, Hitler met Mussolini at the Brenner Pass, on their joint frontier, to discuss how the war in the Mediterranean, for two hundred years Britain's principal foothold outside its island base, might be turned to her decisive disadvantage. He suggested to his fellow dictator that Spain might be coaxed on to the Axis side – thus giving Germany free use of the British Rock of Gibraltar – by offering Franco part of French North Africa, and that France might be persuaded to accept that concession by compensation with parts of British West Africa. Mussolini proved enthusiastic – and understandably so, since the scheme included his acquisition of Tunis, Corsica and Nice (annexed by Napoleon III in 1860) from France. Hitler accordingly hurried home to Berlin to arrange visits to Franco and Pétain. Back in the capital, he constructed with Ribbentrop a letter to Stalin inviting Molotov, the Soviet Foreign Minister, to visit at an early date, when Germany and the Soviet Union might agree between themselves how to profit from Britain's current defencelessness.

A week later, on 20 October, he left in his command train, *Amerika*, to meet Pétain and Franco. The meeting with Franco took place on 23 October at Hendaye on the Franco-Spanish frontier. It has become famous in the diplomatic history of the Second World War for Hitler's furious parting shot that he would 'rather have three or four teeth extracted than go through that again'. Franco, supported by his Foreign Minister, Serrano Suñer ('Jesuit Swine', in Hitler's characterisation – he preserved a Benedictine catechumen's defensive antipathy for the Society of Jesus), stonewalled throughout the hours of negotiation. When his train left at two in the morning, Hitler had not advanced an inch towards co-belligerency with Franco. Pétain, whom he met on 24 October, proved equally unresponsive, but nevertheless succeeded in convincing Hitler that they had had a meeting of minds. The marshal's reputation, antiquity, soldierly bearing and evident patriotism were all to Hitler's taste. Though Pétain had conceded nothing more than a promise to consult his government, which obeyed him automatically, Hitler decided to believe that they were united in a productive hostility to Britain.

Hitler now had the outlines – despite Franco's heel-dragging – of a larger coalition war to present to Molotov at his forthcoming visit. While he waited for the Soviet Foreign Minister to arrive, he was distracted by the errant behaviour of Mussolini, who chose this moment to mount an attack from Albania (occupied by the Italian army in April 1939) into Greece. Mussolini claimed to be motivated by the fear that the British would establish positions in Greece if he did not, and he certainly had legitimate strategic reasons for wishing to deny them naval and air bases any closer to his own along the Adriatic than those they already possessed in Egypt and Malta. However, his purpose in striking into Greece on 28 October was an egocentric wish to emulate Hitler. Once fulsome in praise of his political 'genius', Hitler, whose rise to power had trailed in the wake of his own, and who had sought domestic plaudits for the remilitarisation of the German Rhineland while Mussolini was conquering an overseas empire in Ethiopia, had cast him into the shadows by the triumphs of *Blitzkrieg* in

Poland and France. Mussolini's own abortive participation in the Battle of France (and the Battle of Britain, in which the *Regia Aeronautica* had briefly and ingloriously joined) had aroused the derision of neutrals and enemies alike. He was accordingly determined to win in Greece his share of the laurels which had fallen in disproportionate number to the Wehrmacht.

The failure of his invasion of Greece – the tale of its miscarriage belongs in the next chapter – confounded and outraged Hitler as he awaited Molotov's arrival. It not only upset his scheme to transform the Balkans into a satellite zone by peaceful diplomacy; it was also a provocation to the Soviet Union at a moment and in an area when and where he sought to lull its suspicions. Moreover, it had the immediately undesirable effect of furnishing the British with a pretext for returning to the continent. On 31 October Britain occupied Crete and the Aegean island of Lemnos with troops sent from Egypt, and in the next few days transferred air units to southern Greece, thus putting Romania's Ploesti oilfields, his main source of supply, in danger of bombing attack.

These developments provoked him to an outburst of contingency planning. He ordered OKH to prepare plans for capturing Gibraltar and occupying, if necessary, the French *zone libre*, and to prepare another plan for the invasion of Greece. These orders would result in the appearance of Führer Directives 18 (Felix), 19 (Attila) and 20 (Marita) on 12 November and 10 and 13 December. He also curtailed active consideration of Mussolini's request for German assistance in his offensive against the British in Egypt. 'Not one man and not one pfennig will I send to North Africa,' he told – ironically – General Erwin Rommel. The Panzer units Mussolini wanted would instead be earmarked for intervention in Greece from positions inside Bulgaria, Germany's First World War ally, which Hitler was now trying to coax into the Tripartite Pact, while Mussolini's army was left to manage its desert campaign against the British as best it could.

Even though distracted by unwelcome developments on the margin of his empire and thrashing apparently between strategic options, nevertheless throughout October and November Hitler remained fundamentally pre-occupied by the decision for an eastern campaign. 'What will transpire in the east', he told Bock, his army group commander in Poland, in early November, 'is still an open question; circumstances may force us to step in to forestall any more dangerous developments.' However, he was sustaining his transfer of divisions from west to east, while both OKW and OKH proceeded with the drafting of plans. 'Political discussions have been initiated', he minuted to his commanders on the eve of Molotov's visit, now arranged for 12 November, 'with the aim of establishing what Russia's position will be. . . . Irrespective of the outcome of these discussions, all the preparations orally ordered for the east are to continue.' By 11 November, therefore, it was clear that only if Molotov came bearing guarantees of Russia's acquiescence in Hitler's mastery of the continent could Hitler be deterred from mobilising for the eastern offensive.

Molotov came in no acquiescent mood. Despite the extent of Hitler's military victory and the power of his armed forces, the Soviet Union, he quickly made clear, was determined to hold Germany strictly to the terms of the Molotov-Ribbentrop Pact (which defined their respective spheres of influence in eastern and southern Europe), to pursue its own interests as a great power and to demand knowledge of Germany's intentions in its relationship with third parties. Ribbentrop, at a preliminary meeting with Molotov, disclosed the German side of the bargain on offer: Russia was to share in the despoiling of the British Empire in return for siding with the Tripartite Pact powers. The Soviet Union would be free to expand southwards towards the Indian Ocean while Japan completed its conquests in Asia and Germany extended its area of control into Africa.

Molotov showed himself uninterested. At his subsequent meetings with Hitler, he insisted on the letter of the Molotov-Ribbentrop Pact and on Russia's freedom to pursue its traditional interest in the Black Sea region. The Soviet Union wanted to annex Finland, which had been assigned to its sphere by the pact. It wanted to guarantee Bulgaria's frontiers (apparently whether or not Bulgaria asked for such a guarantee), thereby challenging Germany over control of that country. It also wanted a revision of the Montreux Treaty of 1936 to improve its rights of passage between the Black Sea and the Mediterranean via the Turkish Straits. Molotov demanded to know what spheres of interest the Tripartite Pact delimited between Germany, Italy and Japan, particularly Japan, its old enemy in Asia. In a final exchange with Ribbentrop, conducted in the German Foreign Minister's air-raid shelter during an RAF night attack, he revealed that Russia's interest in the Baltic did not stop with the annexation of Finland (Russian, of course, between 1809 and 1918) but included the question of Sweden's continuing neutrality and control of the Baltic exit to the North Sea, most sensitive of all Germany's home waters. As a parting shot, when Ribbentrop tried to remind him of how greatly Russia would profit by assisting in the dismemberment of the British Empire, whose defeat was at hand, Molotov asked, 'If that is so, then why are we in this shelter and whose are those bombs which are falling?'

Next morning Molotov left for Moscow. Although he had been in Berlin only forty-eight hours, his visit had lasted long enough to convince Hitler that 'the final struggle with Bolshevism', which had been a leitmotiv of his political creed since the earliest days of his 'struggle', could not now be deferred. In the last week of his life, he still recalled the outrage Molotov's intransigence aroused in him: 'He demanded that we give him military bases on Danish soil on the outlets to the North Sea. He had already staked a claim to them. He demanded Constantinople, Romania, Bulgaria and Finland – and *we* were supposed to be the victors.' Memory only marginally exaggerated the reality. When the draft of a proposed treaty written by Molotov reached Berlin on 25 November, it contained clauses requiring the withdrawal of German troops from Finland (an agreement allowing them to use Finnish territory had been signed on 12 September) and allowing the Soviet Union to acquire bases in Bulgaria. Hitler instructed Ribbentrop to make no reply.

A blueprint for 'cauldron' battles

The documents to which he devoted himself in the first weeks of December were military, not diplomatic. On 5 December the plans for a Russian campaign which OKW and OKH had been preparing separately since June and August respectively were brought together for joint staff discussion under his auspices at the Chancellery. OKW's plan, prepared by Lossberg and still codenamed 'Fritz', agreed with that submitted by OKH (it had been completed by General Friedrich von Paulus, the future defender of Stalingrad) in accepting that the encirclement of the Red Army close to Russia's borders was the precondition for success. The danger of engulfment by the vast spaces of the Russian interior had dominated German General Staff thinking since the previous century. That danger had prompted Schlieffen, the author of Germany's war plan for 1914, to eschew the option of striking eastward against the tsar's army – believed though it then was to be as inferior to the German army as Hitler held the Red Army to the Wehrmacht – in favour of attacking France. Schlieffen had recalled 1812, when Napoleon's failure to defeat the Russians in their borderlands had first drawn him to Moscow and then condemned him to drag the Grand Army back again through the winter snows. Hitler too recalled the retreat from Moscow, which had destroyed the Grand Army, but he believed that the Red Army could itself be destroyed by deep armoured thrusts through and behind its frontier positions, creating 'cauldrons' in which its fighting units would be rendered down to inert pulp. OKH's plan was a blueprint for such cauldron battles: the three army groups of the western triumph (to be entitled North, Centre and South) would direct themselves respectively on Leningrad, Moscow and Kiev; but, on their march to the Baltic, the capital and the Ukraine, their Panzer spearheads would encircle the Red Army in three great pockets, which the follow-up infantry would then reduce piecemeal.

Lossberg's OKW plan was even more insistent on this point, and, though it was considered on 5 December apparently only in the form of verbal comments from Jodl, the OKW operations officer, it greatly influenced the trend of the discussions. Halder's advocacy of the OKH plan laid great emphasis on the need to strike for and capture Moscow at an early stage. There was a degree of traditionalism in this accordance of priority, but Halder justified it, with considerable reason, by reference to the centralism of the Soviet system. Under Stalin, all authority was concentrated in Moscow; moreover, the Russian transport system, which in that largely roadless land meant the railways, was also centred on the capital. So, too, by German intelligence estimates, was much of the country's industry. Halder's war diary reveals that the General Staff believed 44 per cent of Soviet war production facilities to be located in the

Moscow–Leningrad region, 32 per cent in the Ukraine and only 24 per cent east of the Ural mountains. This industrial intelligence was faulty; but the rest of Halder's analysis was correct. It was disquieting, therefore, that even on 5 December 1940 Hitler showed himself already more drawn to Lossberg's 'Fritz' proposals which argued for postponing a final drive on Moscow until Army Group North had encircled the Russians in its sector against the Baltic coast and Army Group South had created a great 'cauldron' in the Ukraine. 'In terms of the weapons,' Hitler remarked, 'the Russian soldier is as inferior to us as the French. He has a few modern field batteries, everything else is old, reconditioned material . . . the bulk of the Russian tank forces is poorly armoured. The Russian human material is inferior. The armies are leaderless.' Hitler was well informed about the damage Stalin's monstrous purge of experienced generals had done to the Red Army's high command; the *Sicherheitsdienst* (SD) (the Nazi security service) had, indeed, supplied the NKVD (as the KGB was then known) with much of the evidence to incriminate them. By contrast, the Abwehr (the intelligence branch of the German armed forces) had failed altogether to identify the progress Soviet military industry had made in the development of new and advanced armoured vehicles, particularly the T–34 tank, which would shortly establish itself as the best tank in any army.

In the two weeks that followed the Chancellery meeting, OKH laboured to transform its draft plan into a Führer Directive. Jodl co-operated in the task, lending to it some of OKW's thinking derived from Lossberg's 'Fritz'. Nevertheless the emphasis on Moscow persisted until Hitler ordered a redrafting which directed Army Group Centre (which had Moscow as its objective) to lend armour to Army Group North for its encirclement of the Russian armies in the Baltic region. 'Only after this, the most urgent task, has been accomplished, followed by the capture of Leningrad . . . are the offensive operations to be continued with the object of seizing the vital transport and armaments centre, Moscow.' Führer Directive 21, when issued on 18 December, actually included an instruction for Army Group Centre to 'swing strong units of its mobile forces to the north, in order to destroy the enemy forces fighting in the Baltic area, acting in conjunction with Army Group North . . . in the general direction of Leningrad'. The directive also included a codename for the Russian operation. It was to be known, after the medieval emperor who legend held, lay sleeping in a Thuringian mountain ready to come to Germany's aid in her hour of need, as Barbarossa.

The starting date for Barbarossa lay in June 1941, many months in the future; all that Führer Directive 21 prescribed by way of timing was a stipulation that preparations preliminary to the attack deployment were to be 'concluded by 15 May 1941'. After December, however, Hitler amended the Barbarossa plan little, if at all. On 7–9 January 1941 he assembled his commanders at the Berghof to hear his justification, in detail, for a switch of strategic effort to the east.

There he indicated that his objectives lay as far away as Baku, on the Caspian, the centre of the Russian oil industry which German forces had penetrated in 1918. Early in March (before 3 March) he issued instructions to Jodl which assigned all but the immediate operational zone of the Wehrmacht to the responsibility of the SS and 'Reich Commissioners' appointed by himself; the implication, as he made clear in a speech to 250 senior Wehrmacht commanders at the Chancellery on 30 March, was that 'special measures' (execution, or deportation) were to be taken against Communist Party functionaries and 'hostile inhabitants'. Otherwise – although, in the words of Walter Warlimont, deputy chief of OKW's operations staff, 'during January and February the forthcoming Russian campaign gradually absorbed the efforts of the entire Wehrmacht', in redeployment, creation of military infrastructure and detailed offensive planning by army group, army, corps, divisional, regimental and battalion staffs – the objects and objectives of Barbarossa were altered not at all. The decision to which Hitler had set his hand in December 1940, which had been in the forefront of his mind since his overthrow of France in June, and which had in truth dominated his 'world outlook' since the day he had set out to take power in Germany nearly twenty years earlier, was to remain the fixed point of all he thought and did throughout the first half of 1941, however much supervening events might work to alter it.

The '1812 factor'

Hitler's certainty of purpose was not matched among his entourage. Numbers of his senior commanders and staff officers were intimidated by the '1812 factor'. Halder and Brauchitsch, when first discussing the project on 30 July, concluded: 'The question whether, if a decision cannot be enforced against England and the danger exists that England allies herself with Russia, we should first wage against Russia in the ensuing two-front war, must be met with the answer that we should do better to keep friendship with Russia. A visit to Stalin would be advisable . . . we could hit the English decisively in the Mediterranean, drive them out of Asia.' However, though Halder continued to utter warnings of the dangers throughout the autumn, he did not carry opposition to the sticking-point; Brauchitsch, who had been terrorised by his one open difference of opinion with Hitler after the Polish campaign, altogether lacked the nerve to do so. Jodl, who early had his own doubts, suppressed them when he detected the inflexibility of Hitler's intention, and on 29 July browbeat Warlimont, his deputy, and the three section chiefs of OKW's operations staff into quelling their own. Manstein and Guderian, rising commanders who were to shine in Russia, were disquieted by the '1812 factor' of space swallowing numbers, and Bock, as a very senior officer, expressed something of this to Hitler when the Führer visited him in hospital on 3 December: Russia, he suggested, was 'an enormous country whose military strength was unknown' and 'such a war might be difficult even for the Wehrmacht', thus offending his leader without deflecting him. Ewald von Kleist, the senior Panzer general, claimed (but after the war): 'Most of us generals realised beforehand that if the Russians chose to fall back there was very little chance of achieving a final victory without the help of [a political] upheaval.' Although that may have been their outlook, nevertheless they collectively kept it to themselves. The army may have been intimidated by the technical difficulties of an advance to the White Sea, the shores of the Caspian and the banks of the Volga – Hitler's 'AA' (Archangel-Astrakhan) line, 1600 miles east of Warsaw, nearly 2000 from Berlin, marked the area of conquest he believed would bring about Russia's collapse – but they did not differ fundamentally from him in perceiving the Russian war as inevitable, nor (unless in intensity of feeling) in welcoming a confrontation with the Bolshevik and Slav enemies of Germany.

Reasoned opposition came not from the ground commanders but from the representatives of their sister and (to some degree) competing services, the navy and air force. Goering, as head not only of the Luftwaffe but also, however improbably, of the economic planning authority, was concerned by the economic effort a war with Russia would entail. He continued to believe, moreover, in the benefits to be won by sustaining an air offensive against Britain. Goering had confronted Hitler with his arguments on 13 November, immediately after Molotov's visit to Berlin, forecasting that the course on which Russia seemed bent would draw it into war with Britain, an outcome from which Germany was bound to benefit. Meanwhile, he advocated, Germany should maintain its current strategy. When Hitler turned the economic argument against him, however, claiming that Russian conquests would supply the food and oil needed to beat Britain down, he withdrew his objections and thereafter largely co-operated in the Barbarossa preparations.

Raeder, Hitler's Grand Admiral, was a more persistent opponent. He saw Hitler the day after Goering, raised the danger of fighting a two-front war, rightly emphasising that Germany's leaders had always sought to avoid such a strategic predicament, and urged that no new enterprise should be undertaken until Britain was beaten. Raeder had influence with Hitler. It was he who had advocated the attack on Norway, the success of which had reinforced his prestige. It was also he who had persuaded Hitler to prepare invasion plans against Britain, and who had then deflected the Führer from undertaking Sealion by warning of the likelihood of its miscarriage. He had already produced alternatives to Barbarossa – notably Felix, the plan to hamstring Britain in the Mediterranean by capturing Gibraltar – and he was also proposing initiatives in the Balkans and towards Turkey, which would put pressure on Britain at the Mediterranean's eastern end. Goering shared his strategic outlook. They were both attracted by the opportunities presented by seizing French North Africa, so that Italy could be supported in Libya and Britain outflanked in Egypt. Raeder went further: he wanted to take the Atlantic islands – the Azores, Canaries and Cape Verde islands, Spanish and Portuguese possessions – which would give Germany control of the western mid-Atlantic, particularly since he was outraged at what he called 'the glaring proof of [America's] non-neutrality'. However, while Hitler was excited by the prospect of bringing the Atlantic islands under German control, he continued to set his face inflexibly against the idea of adding the United States to the list of his enemies. Within a year, his curious concept of honour between allies would prompt him to follow Japan into war with America. In the autumn of 1940, however, even as he withdrew from the thought of risking thirty-six of the Wehrmacht's best divisions on the turbulent tides of the Channel, he clung as if by the force of dogma to the principle of placating Britain's natural co-belligerent in the face of almost any provocation she might offer. Russia he would brave in its lion's den; the United States he would not confront at all.

There was more than strategic calculation to this diversity of policy. He had no admiration for the American people, as he did for the British, nor did he fear their military power in the immediate term. He did not, indeed, view the United States as a military power at all. It was its commercial and productive capacity which figured in his 'correlation of forces', and he did not believe that that capacity could be brought to bear against Germany until the war had run its course much further. However, it was precisely because his attitude to America was devoid of ideological content that he chose to disregard all provocation she might offer him in the months while Barbarossa was in the making. The maintenance of diplomatic, if not friendly, relations with the United States was a necessary simplification of the strategic balance sheet that would allow the preordained struggle with the Soviet Union to be brought on and carried through with the least possible diversion of effort.

Hitler's attitude towards Russia, by contrast, was suffused by ideology, drawn from many sources – racial, economic, historical – and fermented by his own rancours and ambitions into a self-intoxicating potency. He was obsessed, perhaps most of all, by the 'story' of German history: how the Teutonic tribes, alone among the peoples on Rome's western borders, had resisted the power of the empire, beaten it down, raised warrior kingdoms of their own and then turned eastward to carry their standards into the Slav lands. The epics of the Teutons, as Varangian bodyguards of the Byzantine emperor, as Viking venturers on the northern seas and founders of princedoms along the Russian rivers, first outposts of 'civilisation' in the east, as Norman conquerors of England and Sicily, as knights of the Baltic shore, formed a theme to which he returned night after night in the monologues which passed for his 'table talk'. The survivals and implantations of German settlers east of consolidated *Deutschtum*'s central European front – in Poland, Hungary, Romania, Czechoslovakia, even in Russia proper, outside the Baltic states, where 1.8 million German colonists were living as late as 1914 – evoked in him feelings of the 'manifest destiny' of the German race akin to those of the British, as they contemplated the diaspora of the English-speaking peoples about the oceanic world, in Victoria's heyday. Yet while the British saw the bounds of their world destined to grow wider and wider still, as if by the operation of some beneficently divine hand, Hitler was conditioned, by his obsession with the tribulations of the Germans, to see them as a people under threat, from which they were to be preserved only by unrelenting struggle.

The threat was manifold and amorphous, but it lay in the east, its instruments were the 'motley of Czechs, Poles, Hungarians, Serbs and Croats, etc.' (the 'etc.' included all the diverse Slav and non-Slav peoples of Russia), 'and always the bacillus which is the solvent of human society, the Jew', and its permanent trend was towards the fragmentation and subjection of the German nation. Bolshevism, which he was determined to see as directed by Jewry, in his lifetime invested that threat with unifying and aggressive force. 'Cosmopolitan' Judaism denied the principles of racial singularity and purity which stood at the pinnacle of his value system; Bolshevism, by its espousal of the cause of the 'masses', itself a term of contempt, and its substitution of faith in economic forces for trust in the warrior's strong arm, repudiated the creed of aristocratic populism on which Hitler had founded his appeal to his folk. 'Jewish Bolshevism' had therefore to be confronted head on, its dominions wrested from its leaders by brute force, and the 'life space' (*Lebensraum*) thus liberated settled with the 'higher peoples' – Germans of the Reich proper, Germans of the eastern settlements, associated 'Germanics' of northern Europe – who, if they did not win supremacy in war, were fated to subjection and enslavement by the myriad hordes of their inferiors.

'Irrevocable and terrible in its finality', as David Irving, Hitler's biographer, has characterised his Barbarossa decision, it was therefore 'one he never regretted, even in the jaws of ultimate defeat'. However, though the decision was certainly fixed by December 1940, six months were to elapse before the forces necessary to implement it were set in motion. In the meantime a sequence of events centred on the Balkans, where German and Soviet power politics were most directly engaged against each other, was to distract his attention from the inception of the coming campaign. For all its appalling risk, Barbarossa was characterised by a certain 'stark simplicity': which would prove the stronger on the field of battle, the Wehrmacht or the Red Army? In the Balkans, during the months while the German army's divisions completed their redeployment to the start-lines from which Barbarossa would be launched, Hitler found himself embroiled in the complexities of an ancient strategic quandary: which way to throw his power among small states, militarily insignificant in themselves, which might nevertheless, by invoking the help of stronger protectors, disrupt the smooth unrolling of his chosen strategy?

SEVEN

Securing the Eastern Springboard

'Crossroads of Europe' is a catchphrase designation for the Balkans, conveying little more than unfamiliarity with the region by those who use it. The Balkans, spined and herringboned by some of the highest mountains on the continent, offer few highways, and none deserving to be called a path of conquest. No single power, not even the Roman Empire at its height, has dominated the whole region: cautious generals have consistently declined to campaign there if they could. It has been a graveyard of military operations ever since the Emperor Valens succumbed to the Goths at Adrianople in 378.

Yet, though the Balkans do not offer easy passage to conquerors, it is the fate of the peoples who inhabit them to be campaigned over. For, precisely because the region is a jumble of mountain chains and blind valleys, where even the rivers must negotiate defiles and gorges impassable by man or beast, it marks a natural barrier between European and Asian empires. In the sixteenth and seventeenth centuries, when Islam was on the march, the Balkans were the battleground where Turk fought Habsburg. In the nineteenth, when Turkey had fallen sick, they offered the fronts on which her enemies – Austria, Russia and their satellites – drove the Ottomans back upon their Anatolian fastnesses. And possession of the coasts of the Balkans and their archipelagos – the Ionian islands, the Dodecanese, the Cyclades – have been contested by power-seekers even longer and more consistently; for, as Sicily does in miniature, and Malta on yet a smaller scale, the Balkans dominate the sea-passages and seas by which they are washed. Venice, greatest of Italian city-states, made herself mistress of the Adriatic by control not of her own lagoon but of the fortress harbours which run the length of the Adriatic's Balkan shore – Zara, Cattaro, Valona – and the Ionian islands at its mouth. In her heyday, Venice also extended powerful tentacles into the eastern Mediterranean by her occupation of the Greek Peloponnese and its satellite islands of Naxos, Crete and Cyprus. The Turks, whatever the ebb and flow of their military fortunes, always assured themselves of an ultimate base of Balkan power by clinging to possession of the Bosphorus, channel of communication between the Black Sea and the Mediterranean. In the face of bribes, threats and direct attack – by the Russians in the nineteenth century, the emergent Balkan states in the early twentieth, the British and French in the First World War – Turkey clung limpet-like to Istanbul (the Constantinople and Byzantium of old) in the sure knowledge that it was control of the 'the Straits' which in European eyes made her a power to be reckoned with and not, as she would become if she relinquished it, merely a Levantine appendage.

Because the Balkans form both a land barrier and a maritime base, or cluster of bases, at the point where Asia meets Europe and the Mediterranean the Black Sea, the strategy of any commander drawn into the area will tend to be both 'continental' and 'maritime', and the one will run at cross-purposes with the other. This, as Professor Martin van Creveld, the closest student of German war-making in the months between the fall of France and the inception of Barbarossa, has pointed out, is precisely the complication into which Hitler fell at the end of 1940. His Balkan policy thitherto had been to allow Italy to play the great power in its relations with the maritime and historically 'Italian' sphere of influence – Albania, Greece, Yugoslavia – while drawing the inland zone – Hungary, Bulgaria and Romania – into Germany's. Hungary and Romania had fallen willingly under his sway, signing the Tripartite Pact and allowing German troops to be stationed on their territory; Bulgaria had proved more resistant, but for reasons of understandable caution, not hostility. Yugoslavia had successfully trodden a middle path, insisting on its neutrality but averting a breach with the Axis. Then Britain's persistence in belligerence had upset his Balkan scheme. Having failed in his efforts to beat down her air defences in the Battle of Britain, as a preliminary to an invasion in which he did not fully believe, Hitler subsequently acquiesced in the Italian attack on Greece (of which he was probably forewarned at his meeting with Mussolini at the Brenner Pass on 4 October), because Britain, whose sole remaining continental ally was Greece, thereby came under increased strategic pressure from another direction. He had calculated that the offensive should diminish Britain's capacity to prosecute its war in Egypt with the Italian Libyan army, and thereby strengthen the 'pincers' he was seeking to construct by drawing Spain and Vichy France into his anti-British alliance.

This complex, but also tentative, strategic design was compromised by the humiliating failure of the Italian offensive. Before the invasion of 28 October, Hitler was considering the dispatch of a German intervention force to North Africa and had actually sent a senior officer (Ritter von Thoma, whom the British would later know well as an opponent) to study the problem of deploying an 'Afrikakorps'. Once the miscarriage of Mussolini's invasion of Greece became apparent, however, Hitler felt constrained to rescue his ally – who had anyhow refused the help of an Afrikakorps – from humiliation, even though direct German intervention against Greece, which required the acquisition of bases in Bulgaria, would alarm the Russians at precisely the moment he was keenest to allay their anxieties (or even, had Molotov brought assurance of acquiescence in German continental hegemony to Berlin on 12 November, agree binding non-aggression terms with them). Mussolini's Greek adventure thus had the direct effect of driving Hitler into heightening his war effort against Britain, though in her Mediterranean empire rather than against her coasts; it also had the indirect effect of committing him to a seizure of territory – useful but not essential to the launching of Barbarossa – which made any agreement of 'spheres of influence' between him and Stalin impossible. In that respect the Greek campaign was to be decisive in determining the future course of the Second World War.

Mussolini's Greek venture

Mussolini's venture into Greece was an operation Hitler was justified in believing ought to have succeeded. The Greek army was greatly outnumbered and was obliged to divide its forces so as to defend Thrace – the coastal strip at the head of the Aegean – against Bulgaria. On paper it should have been overwhelmed in the opening stage of the invasion; but Italy's forces were also divided, by the garrisoning of Ethiopia and Libya, and it could therefore deploy only a fraction of its much larger army on the Albanian-Greek frontier. The Italian army of 1940 was not, moreover, what it had been in 1915. Then, committed to war on a single, equally mountainous front against Austria, it had fought courageously in one offensive after another, and not without effect. By October 1917 its efforts had impelled the Austrians to appeal for help to the Germans lest its twelfth offensive on the Isonzo succeed in breaking through. Under Mussolini, however, Italian formations had been reduced in size in order to increase their number, a typical demagogic act of window-dressing. The divisions which Mussolini launched into Greece on 28 October 1940 were therefore weaker in all arms, but particularly in infantry, than their Greek equivalents; they were also weaker in motivation. Mussolini's reasons for seeking war with Greece went no further than a desire to emulate his German ally's triumphs, settle trifling old scores with Greece, reassert Italy's interest in the Balkans (he was piqued that Romania, an Italian client, had accepted German protection for its Ploesti oilfields earlier in October) and secure bases from which his British enemy's eastern Mediterranean outposts might be attacked. None of these reasons counted for much with his soldiers. They began their assault through the Epirus mountains without enthusiasm; even the Alpini regiments, Italy's best troops, appeared in poor heart. Their Greek opponents, by contrast, defended with a will. General John Metaxas, head of government, was enabled early in the campaign to transfer forces from Thrace to the Albanian front, thanks to Turkey's warning to the Bulgarians that its thirty-seven divisions concentrated in Turkey-in-Europe would be used if Bulgaria tried to profit from Greece's difficulty. In the meantime the Greeks allowed the Italian attackers to wear themselves out in frontal attacks on their mountain positions. When their own reinforcements arrived, they counter-attacked, on 14 November, and drove the invaders back in confusion. Mussolini summoned reserves from all over Italy, some of which were flown to Albania in German aircraft, but by 30 November the Greeks opposed fifteen of his divisions with eleven of their own, his whole invading force had been thrown back inside Albania and the Greek counter-offensive was still gathering strength.

Hitler, who had already ordered OKW on 4 November to prepare an operational plan for a German offensive against Greece, was by then committed to its launching. For all the diplomatic difficulties it would cause – affront to Yugoslavia, Greece's neutralist neighbour, anxiety to Turkey, which was even more strongly determined to remain neutral, alarm to Bulgaria, which shrank from offending Russia by granting Germany the bases the Greek operation required – and for all the military difficulties the operation entailed, particularly those of committing mechanised formations to the least 'tankable' terrain in Europe, he now saw no means of avoiding the initiative, except at the price of conceding his British enemies strategic and propaganda advantages he could not allow them. Mussolini, for better or worse – and Hitler was never to waver in his loyalty to the founder of fascism – was seen by the world as his political confederate as well as military ally. Hitler was determined to rescue him from humiliation at the hands of the Greeks, all the more so because he rightly held the Greeks in high esteem as soldiers; he was also determined to deny the British long-term possession of bases on Greek soil, from which they could menace his extraction of Balkan resources – foodstuffs, ores, above all oil – essential to his war effort.

Thus far the Greeks had been careful not to grant the British anything more than short-range tactical facilities. The bases the RAF had set up since 3 November were located in the Peloponnese, on the Gulf of Corinth and near Athens, from which its aircraft could just support the battlefront in Albania. Greece had resisted requests for larger bases near Salonika which would have brought the Ploesti oilfields in Romania within range of its bombers. Hitler had good reason to fear the worst, therefore, from a consolidation of the Greek victory over Mussolini. South-eastern Europe provided half of Germany's cereal and livestock requirements. Greece, with Yugoslavia, was the source of 45 per cent of the bauxite (aluminium ore) used by German industry, while Yugoslavia supplied 90 per cent of its tin, 40 per cent of its lead and 10 per cent of its copper. Romania and, to a marginal extent, Hungary provided the only supply of oil which lay within the radius of German strategic control; the rest came from Russia under the terms of the Molotov-Ribbentrop Pact. If those oilfields, and the railways which carried ores and agricultural produce out of the Balkans to Germany, were brought under British bomber attack, his ability to prosecute the war would be seriously compromised. Moreover, he recognised the depth and antiquity of Britain's penetration of the Mediterranean strategic zone. British admirals and generals had campaigned in the eastern Mediterranean for 150 years; Nelson's reputation had been made by his victory at the Nile in 1798. The British had ruled the Ionian islands from 1809 to 1863, had possessed Malta since 1800, Cyprus since 1878, and maintained a fleet and an army in Egypt since 1882. In 1915 a British army had almost captured the Black Sea straits and between 1916 and 1918 sustained an offensive front against Bulgaria on Greek soil (the Salonika campaign). Moreover, the intimacy of their relationship with the Greeks was assured by their title as 'lovers of liberty', won by the help the British had given them in their war of independence against Turkey in the 1820s. Byron's reputation as a romantic hero in both countries was a touchstone of their peoples' common antipathy to tyranny.

However, Britain's tentacles reached further than that. Although she had fought Turkey in the First World War and established a homeland for the Jews in Palestine after 1918 in the teeth of Muslim antipathy, she was also a historic protector of the Turks against Russia, in which cause she had fought the Crimean War of 1854–6, and a sponsor of Islamic nationalism by her foundation of the states of Iraq and Trans-Jordan. Her reputation as an exponent of self-determination for small nationalities also stood high in central and south-eastern Europe, where Yugoslavia in particular owed her existence partly to British support for the cause of Slav independence at the post-1918 peace conferences. Britain's only clear-cut enmity in the Balkans was with Bulgaria, her opponent in the First World War, and that was offset by King Boris's concern to placate Russia, which he could not afford to offend unless assured of full-blooded German support.

The ambiguity of a Balkan entanglement – which automatically involves an intrusive land power not only in the conflict between central Europe's vital interests and those of Russia but at the same time in the maritime complexities of Mediterranean politics – therefore worked to divert and fragment Hitler's strategic purpose in the winter and spring of 1940-1. His overriding aim – to attack and destroy Russia's fighting power in an early *Blitzkrieg* campaign – was fixed by December 1940; his desire to rescue his toppling Italian ally from public humiliation and to circumscribe the activity of his irrepressible British enemy before he embarked on the Russian war – both in some sense residues of his vacillation of the autumn – drew him into a series of initiatives, some calculated, some adventitious, which were to end in his fighting a larger Balkan-Mediterranean campaign than he had ever intended when he first contemplated venturing southward.

'Fox killed in the open'

In early January 1941, when he met with his commanders at the Berghof (7–9 January) and exposed to them the Barbarossa strategy in its entirety, the southern difficulty seemed to centre less on the Greeks than on the British. Though planning for Operation Marita (the invasion of the Balkans) was in full flow, he was still not contemplating the outright occupation of Greece. A mere seizure of bases in Greece from which the Luftwaffe might dominate the eastern Mediterranean seemed an adequate strategic solution of the situation in that sector. He was even optimistic that the Greeks, whose promised defeat by Mussolini in a spring offensive he was treating with sceptical (and as it turned

Hitler in Poland, September 1939. On his right is Rommel, then the commander of the Führer's escort battalion, on his left is Keitel, his chief of staff.

Refugees from the war zone arriving in an area not yet overrun by the Germans, France, May 1940. Columns of refugees seriously hampered Allied troop movements.

Above: Men of the Royal Tank Regiment, who had taken part in the Arras counterattack, return from Dunkirk. *Below*: Troops of the BEF waiting to be evacuated from Dunkirk. In all, 338,000 British and French troops were taken from the beaches.

Facing page: General Heinz Guderian, commanding XIX Panzer Corps, in his command vehicle during the campaign in France. The picture clearly shows two signallers (front) sitting at an Enigma machine.

Right: The human face of the Blitz. A Heavy Rescue Squad gently pulls a survivor from the rubble in November 1940. Some 40,000 civilians were killed in the Blitz.

Facing page: An He111 over the London docks, September 1940. The He111's success against second-rate opposition during the Spanish Civil War encouraged the Luftwaffe to underestimate the threat posed by the RAF's eight-gun fighters.

Left: Architect of victory, Air Chief Marshal Sir Hugh Dowding, AOC-in-C Fighter Command, 1936–40. His support for the development of radar in the late 1930s, and his careful husbanding of scant resources during the Battle of France, when he was under great pressure to commit vital squadrons to a lost cause, ensured that Fighter Command enjoyed the narrowest of margins over the enemy in the summer of 1940. *Right*: An RAF armourer prepares a Spitfire.

Left: HM destroyer *Skate* testing stern-dropped depth charges on exercise during the Battle of the Atlantic. Stern dropping gave way to firing ahead as the battle progressed.

Below: HM destroyers *Wishart* and *Anthony* depth-charging U-6761 in February 1941. The U-boat, first detected by an American aircraft, was destroyed and all but seven of her crew picked up by the destroyers.

Below: The German battleship *Bismarck* engages HM battle-cruiser *Hood* during its North Atlantic sortie, 24 May 1941; the photograph was taken from the *Prinz Eugen*.

Hitler during his one-day 'victory' visit to Paris, 23 June 1940.
Albert Speer is on his right and Martin Bormann is in profile on his left.

Victorious German infantrymen prepare to raise the Swastika over the Acropolis in Athens, 27 April 1941. The Greek campaign has lasted only three weeks.

Left: Junkers 52 aircraft parachutists over Crete, 20 May 1941. Heavy German casualties sustained in Crete led Hitler to suspend further major airborne operations.

out justified) caution, might bring the Italians to accept a bilateral peace treaty. The British, on the other hand, were demonstrating a determination to persist in defiance of Axis military superiority. Not only had they deployed air units to mainland Greece, and troops to Crete and some of the Aegean islands; they had also inflicted direct defeats on the Italians. On the night of 11-12 November a Royal Navy task force, centred on the aircraft carrier *Illustrious*, surprised the Italian fleet in its Taranto base in the heel of Italy and sank three battleships at their moorings by aerial torpedo attack. This success, following earlier surface engagements in July, confirmed the Royal Navy's dominance over the Italian fleet, despite the latter's superiority of numbers in the inland sea. Worse was to follow: on 9 December the British army in Egypt, commanded by General Sir Archibald Wavell, launched a counter-offensive against the Italian army which Marshal Rodolfo Graziani had led sixty miles inside the frontier from Libya in September. Conceived as a 'five-day raid', it achieved such success that Wavell decided to sustain his advance. In three days Lieutenant-General Richard O'Connor, his tactical commander, had captured 38,000 Italians, for a total loss of 624 British and Indians killed and wounded, overrun a large fortified enemy position and found nothing beyond it to bar his advance into Libya. At Bardia, the first town inside the Italian colony, General 'Electric Whiskers' Bergonzoli signalled to Mussolini in the aftermath of the British counter-attack, 'We are in Bardia and here we stay'; but by 5 January Bardia had fallen to the Army of the Nile, as the 4th Indian and 7th Armoured Divisions had been grandiloquently designated by Churchill, and their spearheads were pressing on along the coast road towards the port of Tobruk. On 21 January Tobruk fell yielding another 25,000 prisoners; the port was to provide O'Connor's army with logistical support for its continued advance. O'Connor now divided his forces: the remnants of the Italian invaders of Egypt were falling back on Tripoli, capital of Libya, along the Mediterranean coast road which veered north around the bulge of Cyrenaica; a direct route through the desert offered the prospect of cutting them off by a fast mobile thrust. O'Connor accordingly launched the 7th Armoured Division into the desert behind them and on 5 February it arrived out of the sands ahead of the fleeing Italians at Beda Fomm. 'Fox killed in the open,' O'Connor signalled in clear – to pique Mussolini – to Wavell; the hunting metaphor described a victory which brought the British 130,000 prisoners in the course of an advance of 400 miles in two months.

Churchill exulted in Wavell's triumph. 'We are delighted that you have got this prize,' he wrote to Wavell. But the victory, though spectacular, was not really one of modern warfare. The Army of the Nile was little more than the sort of colonial 'movable column' with which the British Empire's native enemies had been defeated in the campaigns of the nineteenth century. Its success was due not to its superiority over the Italian troops, who had fought bravely in defence, but to the incompetence of their leadership and, as in Greece, the attenuation of their means of making war, the result of Mussolini's appetite for campaigning over a wider front than Italy's resources could support.

Hitler's efforts to check a British offensive had earlier been frustrated by Mussolini's reluctance to accept help; now he would not brook refusal. 'The crazy feature is', he complained to his staff, 'that on the one hand the Italians are shrieking for help and cannot find drastic enough language to describe their poor guns and equipment, but on the other hand they are so jealous and childish they won't stand for being helped by German soldiers.' On 3 February, rather than Manstein, he chose Rommel to lead an Afrikakorps willy-nilly to Graziani's assistance, because of his proven ability to inspire soldiers; on 12 February the vanguards of the Afrikakorps, to consist of the 15th Panzer and 5th Light Divisions, began to arrive at Tripoli; by 21 February Rommel had his forces in position to begin preparing a counter-offensive.

Nevertheless Hitler's determination to restore Axis prestige and consolidate Germany's strategic position in the Balkans could not wait on a future desert victory. The British were profiting from their superiority in arms in the one strategic region where they still enjoyed freedom of action to puncture the imperial pretensions of Mussolini in humiliating detail. On 9 February their Mediterranean fleet had appeared off the Italian port of Genoa, and bombarded the harbour without suffering riposte; it was a foretaste of the defeat they were to inflict on the Italian fleet at the Battle of Cape Matapan (Tainaron) in Greek waters on 28 March. In East Africa, where Italian forces had seized undefended British Somaliland in August 1940 and made incursions into the Sudan and Kenya, the British counter-attacked. A British force based in the Sudan had entered northern Ethiopia and the colony of Eritrea, Italy's oldest possession in East Africa, on 19 January; and on 11 February another British army, based in Kenya, began an offensive into southern Ethiopia and Italian Somaliland. British Somaliland was retaken without a fight on 16 March. Worse was to follow. During February the British had been in continuous conclave with the Greek government on the nature of the assistance it would be willing to accept as a guarantee against German intervention. Metaxas, the Greek dictator, had died on 19 January; General Alexandros Papagos, the army commander-in-chief, was less cautious in negotiating measures which might provoke Germany to action. A figure of four British divisions was eventually agreed as an acceptable contribution to reinforce the eighteen Greek divisions deployed on the northern frontier. Their advance guards – withdrawn from the desert army, which was thereby dangerously depleted – began to disembark on 4 March. It was the start of an ill-fated venture.

This initiative made up Hitler's mind. Bulgaria, which on 17 February had secured a non-aggression pact with Turkey (overawed by Germany's military might in a way Greece was not), acceded to the Tripartite Pact on 1 March. As a result the Wehrmacht's 'army of observation' in Romania, which by 15 February had reached a strength of seven divisions, was free to begin bridging the Danube into Bulgaria and construct its attack positions for Operation

Marita. In view of Britain's deployment of the four divisions to Greece, Hitler now decided that Marita's objects would not be limited to securing a strategic position in Greece from which the Luftwaffe might dominate the Aegean and eastern Mediterranean; they were to comprehend the occupation of Greece outright.

He was not prepared to risk the reopening of another 'Salonika front' from which Britain (with France) had harried Germany's southern flank of operations in 1916-18. Here, as so often elsewhere in his conduct of the Second World War, Hitler's strategic calculations were influenced by his experience and memories of the First, in which he had fought as a common soldier. Then the British had profited from their maritime mobility to sustain campaigns which diverted Germany's armies from their war-winning task in the great theatres; he was not prepared to concede them the opportunity a second time.

During the spring of 1941 he was, indeed, attempting to play their own game back at them. His failure to persuade Franco and to pressure Pétain – who had dismissed the pro-German Laval from his government on 13 December – to join the anti-British alliance had closed the western Mediterranean to him as a forum of opportunity. In the eastern Mediterranean and its hinterland, however, he detected openings for the same sort of subsidiary campaigning and subversion as Germany, with and through its then Turkish ally, had conducted against British interests in 1915-18. For example, he had hopes of persuading the French administration of Syria and Lebanon to accept German military assistance, and so eventually the basing there of Luftwaffe units with which the Suez Canal and the oilfields of Iraq might be brought under attack. In Iraq itself, a former British mandate, the nationalist party was pro-German; his contacts with it were indirect, passing through the Mufti of Jerusalem, leader of another anti-British Arab party, but he could calculate on its dissidence to complicate Britain's efforts to sustain control of the Middle East. Indeed, throughout the region Churchill's difficulties resembled Mussolini's in his African empire – those of straining to make over-stretched resources meet over-large responsibilities.

The threat that German interference in the Levant and Iraq offered to the British bulked so large in their assessment of risk in the spring of 1941 that it would prompt them to take possession of both areas later in the year. For Hitler, by contrast, any advantage he might win in either was likely to prove ephemeral and therefore did not merit any major investment of force. That was not the case with Greece, where Britain's involvement had produced a direct and provocative challenge to his military control of the continent and, though they did not guess it, threatened the unhampered development of his campaign against Russia. It had in consequence to be crushed outright; he could not, for example, count upon any eventual success from Rommel's counter-offensive in Libya (to be delivered in late March) which might oblige the British to re-embark the divisions they had just deployed from Egypt to the Greek mainland, even if that were its probable result. Operation Marita had to produce a direct and clear-cut victory.

During the first weeks of March he was working to complete the preliminaries essential to its launching, the last of which required concessions by Yugoslavia. For military reasons, which OKH had made clear to him in relentless detail, neither Albania nor Bulgaria provided suitable terrain or adequate logistical bases from which the Marita forces could operate. Albania was crowded with beaten Italian troops and could be reinforced only from the sea or by air. Bulgaria's roads, bridges and railways were few and primitive. The Wehrmacht therefore needed to deploy troops along the southern Yugoslav railway system in order to open a third front at Monastir and on the Vardar river – traditional invasion routes – if the Greek army and its British confederates were to be overwhelmed with dispatch.

Yugoslav resistance

German pressure on Yugoslavia to accede to the Tripartite Pact, as Romania, Hungary and now Bulgaria had done, had been unrelenting since the previous October. With great courage the Yugoslavs had resisted. In their negotiations with Berlin they insisted that the Balkans would be best designated a neutral zone in the ongoing European war; in private, Prince Paul, the regent, an Anglophile who had been educated at Oxford and said he 'felt like an Englishman', did not conceal his sympathies for Britain's cause. Moreover, as the husband of a Greek princess, he had no desire to co-operate in the defeat of his southern neighbour. During the winter and spring of 1940-1, as Hungary, Romania and finally Bulgaria began to fill with German troops, his ground for resisting German pressure shrank under his feet. His government nevertheless contested every demand that the Germans thrust upon them; eventually, on 17 March, in return for what must almost certainly have been a worthless assurance that Yugoslav territory would not be used for military movements, it terminated diplomatic resistance and agreed to join the pact. The signatures were entered at Vienna on 25 March.

Hitler exulted in the result – but too soon; incautiously as a former citizen of the Habsburg Empire with which the Serbs had played such havoc, he had failed to allow for the impetuosity of the Serb character. On the night of 26-27 March a group of Serb officers, led by the air force general Bora Mirković, denounced the treaty, seized the capital, Belgrade, next day, obliged Paul to resign as regent and then had the uncrowned king Peter, installed as monarch. Paul, who might have rallied support among the kingdom's Croat population, which differed automatically with the Serbs in politics and was heavily penetrated by pro-Axis sympathies, accepted the coup as a *fait accompli* and went into exile. A government was set up under the leadership of the air force chief of staff, General Dušan Simović, who later headed the Yugoslav government in exile.

The Mirković coup still appears in retrospect one of the most unrealistic, if romantic, acts of defiance in modern European history. Not only did it threaten to divide a precariously unified country; it was also bound to provoke the Germans to hostile reaction, against which the Serbs could call on no external assistance whatsoever to support them. They were surrounded by states that were wholly inert, like Albania, or as threatened as themselves, like Greece, or actively hostile, like Italy, Hungary, Romania and Bulgaria, with all of which they had bitter and long-standing territorial disputes. If Croatia, which would shortly take its own independence under Italian tutelage, is added to the roll of the Serbs' enemies, the behaviour of General Mirković and his fellow-conspirators of 27 March appears the collective equivalent of Gavrilo Princip's firebrand assault on the Austro-Hungarian monarchy personified by Archduke Franz Ferdinand in June 1914. It ensured the extinction of the Serb national cause as if by reflex; it would also doom Serbia, as in 1914, to invasion, defeat and occupation and with it the peoples of Yugoslavia, of whom the Serbs had assumed the leadership in 1918, to an agony of protracted civil and guerrilla warfare for the next four years. Of none of this do Mirković, Simović or any of the other Serb patriots – reserve officers, cultural stalwarts and the like – who staged the 27 March coup seem to have taken the least reckoning. There is no doubt that they had been encouraged in their foolhardiness by the British and the Americans. Colonel William 'Wild Bill' Donovan, future head of the Office of Strategic Services and in 1941 President Roosevelt's personal emissary to Belgrade, had arrived in the capital on 23 January bearing an exhortation about the preservation of national honour; Winston Churchill was meanwhile pressing his ambassador to 'pester, nag and bite' the Yugoslav government to stay outside the Tripartite Pact. But Western warnings and encouragement were ultimately beside the point. The 27 March coup was an autonomous Serb initiative, to be seen with hindsight as the last outright expression of sovereign defiance made by any of the small peoples who lay between the millstones of German and Russian power since Poland's rejection of Hitler's ultimatum in August 1939 and their subjection to Stalinism.

It was to be punished with vehemence and without delay. Hitler judged that the Serbs' defiance simplified his strategic options in the approach to Marita. Diplomatically it put Yugoslavia in the wrong; for all the popular enthusiasm displayed for the coup – crowds cheering the Allied cause in Belgrade, whose streets were bedecked with British and French flags – the new government could with some reason be denounced as illegitimate. Militarily, it provided OKH with a solution of its logistic difficulties: the Yugoslav railway system, inherited from the Habsburg Empire, connected with those of Austria, Hungary, Romania and Greece (as Bulgaria's did not), and thereby provided the Wehrmacht with a direct approach to its chosen battlefront in Macedonia. Hitler did not pause to seize the advantage he had been offered. 'I have decided to destroy Yugoslavia,' he told Goering, Brauchitsch and Ribbentrop, summoned post-haste to the Chancellery on 26 March. 'How much military force do you need? How much time?' The answers to these questions already lay in the files of contingency plans in army and Luftwaffe headquarters. In early afternoon he met the Hungarian minister to offer him a port on the Adriatic for his country's part in the coming campaign, and then the Bulgarian minister, to promise him the Greek province of Macedonia. 'The eternal uncertainty is over,' he told him, 'the tornado is going to burst upon Yugoslavia with breathtaking suddenness.' Next day in more pensive mood he told the Hungarian minister (whose head of state, Admiral Horthy, had decided to decline the bribe of an Adriatic port), 'Now that I reflect on all this, I cannot help believing in a Higher Justice. I am awestruck at the powers of Providence.'

The Yugoslav conspirators persisted in blissful ignorance of the opportunity Hitler felt they had offered him. They believed that they could placate Germany by declining to accept a British mission and that their coup could not be regarded as a repudiation of Yugoslav accession to the Tripartite Pact because the signature had never been ratified. In fact the terms stipulated that ratification was assured by signature, while in Hitler's eyes the coup put them in the enemy camp in any case. On the day of the coup itself he issued Führer Directive No. 25: 'The military revolt in Yugoslavia has changed the political position in the Balkans. Yugoslavia, even if it makes initial professions of loyalty, must be regarded as an enemy and beaten down as quickly as possible. . . . Internal tensions in Yugoslavia will be encouraged by giving political assurances to the Croats. . . . It is my intention to break into Yugoslavia [from north and south] and to deal an annihilating blow to the Yugoslav forces.'

Halder had directed OKH's planning staff to prepare plans for such an offensive the previous October. The forces positioned for Marita easily sufficed for an invasion of Yugoslavia as well: the Second Army, stationed in Austria, would simply advance directly on Belgrade, while the Twelfth, positioned to attack Greece through Bulgaria, would now move into southern Yugoslavia before doing so; an Italian army would also attack from Italy towards Zagreb, capital city of the Croats, who were Italy's clients, while the Hungarian Third Army would seize the trans-Danubian province of Vojvodina, where Hungary claimed rights.

Yugoslavia's fate

The Yugoslav army, a million strong, was organised into twenty-eight infantry and three cavalry divisions; but it contained only two battalions of 100 tanks, and those antiquated. The whole army belonged, indeed, to the era of the Balkan wars of 1911-12 rather than to the modern world – its movements depended on the mobilisation of 900,000 horses, oxen and mules – and, moreover, it was not mobilised. Its General Staff – which General Sir John Dill, the Chief of the (British) Imperial General Staff, visited secretly immediately after the coup on 1 April – behaved, by his report, 'as if it had months in which to make decisions and more months in which to put them into effect'. Though its deputy chief conferred with Papagos, the Greek commander, in Athens on 3-4 April, it refused to co-ordinate a joint strategy of concentrating its forces in the south to support the Greeks (and the British contingent arriving to join them) but insisted on lining the whole frontier (with Italy, Germany, Hungary and Bulgaria, a sector 1000 miles long) against the threat of invasion – as the Russians were also currently doing in their border zone.

'He who defends everything', in Frederick the Great's chilling military aphorism, 'defends nothing.' The attempt to defend everything was the mistake the Poles had made in 1939, though with some excuse, since the economically valuable parts of their country lay in frontier regions. It was also the mistake towards which the Greeks were tending, divided as they were by their urge to protect the exposed salient of Thrace as well as the traditional invasion routes in Macedonia. But no country has perhaps ever as irrationally dispersed its forces as the Yugoslavs did in April 1941, seeking to defend with ancient rifles and mule-borne mountain artillery one of the longest land frontiers in Europe against Panzer divisions and 2000 modern aircraft.

The Yugoslav air force, which had masterminded the coup of 27 March, was overwhelmed in the opening hours of the German attack on 6 April; of its 450 aircraft 200 were obsolete and most were destroyed outright in an initial air offensive which also caused 3000 civilian deaths by a terror raid on Belgrade. The German army's plan, with which those of the Italian Second and Hungarian Third Armies were integrated, nullified Yugoslav strategy from the start. It turned on throwing armoured columns down the valleys of the rivers – the Danube, the Sava, the Drava, the Morava – which penetrate the mountain chains on which the Yugoslavs had counted to protect their country's heartland; the columns would then turn to converge and so envelop the Yugoslav formations they had outflanked. It proved brilliantly successful. As the official Yugoslav history of the war subsequently conceded:

Three initial attacks determined the fate of the Yugoslav army, on April 6 in Macedonia, April 8 in Serbia, and April 10 in Croatia. On all three occasions the Hitlerites breached the frontier defences, pushed deep into the interior and dislodged the Yugoslav defences from their moorings. After the breakthrough of the frontier defences, the Yugoslav troops were soon outmanoeuvred, broken up, surrounded, without contact with each other, without supplies, without leadership.

What the official history seeks to conceal is the active responsibility of much of the 'Yugoslav leadership' for the débâcle. Yugoslavia – originally, by the designation of the Allied peace treaties with Austria and Hungary in 1919, 'the Kingdom of the Serbs, Croats and Slovenes' – was in no sense a nationally unified state. It had inherited all the tendencies that had racked the Habsburg monarchy's Slav dominions before 1914 and sought to check them merely by imposing Serb dominance over those minorities which had always preferred Vienna to Belgrade. The invasion of 6 April was seized by the Croat and Slovene nationalists as an opportunity for secession; on 10 April the Croatian Ustashi, a group of extreme right-wing nationalists, proclaimed an independent state, and on 11 April the Slovenes did likewise: both would shortly accept Axis tutelage. Some of the Croat formations of the Yugoslav army mutinied and went over to the enemy in the opening stages of the campaign; the chief of staff of the (Croatian) First Army Group actually conspired with the Ustashi leadership in opening talks with the Germans on 10 April. These were the preliminaries of a collaboration which were to result in the cruellest of all the internecine wars that would torment occupied Europe during the Hitler years. Nevertheless, Yugoslavia's Serb majority cannot escape its share of responsibility for the suddenness of their country's defeat. All but one of the army's divisions was under Serb command, and most of those divisional generals surrendered to the panic which the rapidity of the Wehrmacht's onslaught induced. So feeble was the army's resistance that the German invaders suffered only 151 fatal casualties in the course of the campaign; the XLI Panzer Corps lost a single soldier dead, though it was in the forefront of the advance to Belgrade. The only senior Serbian officer who resisted the disabling spirit of collapse was Draza Mihailovicü, deputy chief of staff of the Second Army, who took to the hills at the signing of the armistice with Germany on 17 April. There, with a band of fifty faithfuls, he founded the nucleus of the Chetnik movement, which consisted of Serbian freedom-fighters loyal to the crown. Until Tito's communist Partisans emerged as a major force in 1942, the Chetniks sustained the principal guerrilla resistance against the regimes of occupation – German, Italian, Bulgarian, Hungarian, puppet Croatian – imposed on Yugoslavia.

In Greece, which the Germans also invaded on 6 April, the Wehrmacht met stiffer resistance. The Greek army was already mobilised, had fought a successful offensive against the Italians and was commanded by generals whose campaigning experience stretched back to the Graeco-Turkish war of 1919-22. Moreover, it was supported by a British expeditionary force of three divisions which had brought with it modern tanks and aircraft. Hitler regarded the Greek soldiers as the valorous descendants of Alexander's hoplites and the Theban Sacred Band – a unique mitigation of his disdain for non-Teutons – and he so admired the bravery they had shown in their war with Mussolini that he instructed OKW, before the campaign began, to release from captivity all Greeks taken prisoner as soon as an armistice should be signed.

Neither Greek valour nor British arms would avail to postpone an armistice. The Greek plan was flawed, and neither advice nor deployments from Britain could avoid defeat. Papagos, the Greek commander, insisted on keeping four of his eighteen divisions on the Metaxas Line, along the Bulgarian frontier, and disposed three with the British formations – the 6th Australian and 1st New Zealand Divisions, with the 1st British Armoured Brigade – a hundred miles to the rear on the Aliakhmon Line hinged on Mount Olympus. He counted on the Yugoslavs to protect the left flank of both positions and had even arranged a scheme with the Yugoslavs to react to an Axis attack by opening an offensive into Albania against the Italians – who on 20 March had once again tried and failed to revive their own Balkan offensive – with the bulk of the Greek army, fourteen divisions. Professor Martin van Creveld describes the dispositions – without exaggeration – as 'suicidal'. The defending forces were aligned in three separate positions which depended on their security on a fourth, entirely extraneous Yugoslav force protecting their flanks. 'Should the Germans', van Creveld observes, 'succeed in breaking [the Yugoslavs] rapid and total disaster was inevitable. Yugoslavia and Greece would be cut off from each other, the Metaxas and Aliakhmon lines outflanked, and the Greek army in Albania attacked from the rear. After that it would be a small matter to mop up the rest of the Allied and Yugoslav forces separately.'

Collapse in Greece

The course of the campaign developed exactly as thus predicated. In two days of fighting, 6-7 April, the Germans broke the resistance of the Yugoslavs in Macedonia and forced the Greek defenders of the Metaxas Line, who had stoutly resisted frontal assault, to surrender on 9 April. They were thus freed to turn the left flank of the Aliakhmon Line, defended by New Zealanders, and press on down the ancient invasion route which leads from the Vardar Valley in Macedonia into central Greece. A detached force meanwhile unhinged the main body of the Greek army which in Albania was confronting the Italians, who were thus granted the opportunity to begin the decisive advance they had been unable to win by their own efforts in six months of fighting.

General George Tsolakoglu, commanding the Greek First Army on the Albanian front, was so determined, however, to deny the Italians the satisfaction of a victory they had not earned that, once the hopelessness of his position became apparent to him, he opened quite unauthorised parley with the commander of the German SS division opposite him, Sepp Dietrich, to arrange a surrender to the Germans alone. It took a personal representation from Mussolini to Hitler to bring about an armistice in which Italy was included on 23 April.

Elsewhere the Graeco-British front was collapsing concertina-like as one position after another was outflanked by the invaders. The Greek Prime Minister, Alexander Koryzis, committed suicide on 18 April, leaving the rest of the Greek government unable to agree with General Sir Henry Wilson, commanding the British expeditionary force, how best to sustain resistance. In fact the British had been in full retreat from the Aliakhmon Line since 16 April. Though they lacked the numbers and equipment to resist the Germans, they had the motorised transport in which to withdraw; the Greek army, like the

Yugoslav, belonged to an earlier age of warfare and 20,000 of its soldiers fell into German hands in the wake of the British retreat.

The British made a stand at Thermopylae, where the Spartans had fallen defying the Persians 2500 years before, but were quickly hustled southward by German tanks. That day and every day they were harried by the Luftwaffe, which, by the report of *The Times* correspondent, was 'bombing every nook and cranny, hamlet, village and town in its path'. It had destroyed Piraeus, the port of Athens, on the first day of the war with Greece, so that the fugitives had to head for the Peloponnese to find harbours for their return flight to Crete and Egypt. A German parachute drop on the Isthmus of Corinth on 26 April was timed just too late to cut them off. By then the British – most of them Australians and New Zealanders forming the Anzac Corps, whose predecessor had established the Antipodean military legend at Gallipoli only twenty-six years earlier – had passed through Athens and reached haven. Retreating though they were, 'no one who passed through the city', wrote a Royal Artilleryman, Lieutenant-Colonel R. P. Waller, would ever forget the warmth of the Athenians' farewell. 'We were nearly the last British troops they would see and the Germans might be on our heels; yet cheering, clapping crowds lined the streets and pressed about our cars, so as to almost hold us up. Girls and men leapt on the running boards to kiss or shake hands with the grimy, weary gunners. They threw flowers to us and ran beside us crying, "Come back – You must come back again – Goodbye – Good luck".'

It would be three and a half years before British soldiers returned to Athens, then to participate in a grim and bloody civil war between the parties of left and right which had learned the politics of violence as guerrilla fighters against the German occupation. In April 1941, sunny and flower-scented in the memory of the soldiers who were leaving Greece with the taste of defeat in their teeth, that cold and bitter December would have seemed an unimaginable legacy of the whirlwind campaign they had fought against the Germans. The three British divisions which together with the six Greek divisions spared from the Albanian front had battled against eighteen of the enemy had, rightly, the sensation of having fought the good fight. The Greek campaign had been an old-fashioned gentlemen's war, with honour given and accepted by brave adversaries on each side. In the aftermath, historians would measure its significance in terms of the delay Marita had or had not imposed on the unleashing of Barbarossa, an exercise ultimately to be judged profitless, since it was the Russian weather, not the contingencies of subsidiary campaigns, which determined Barbarossa's launch date. The combatants had not felt they were participating in wider events. The Greeks, with British help, had fought to defend their homeland from conquest. The Germans had battled to overcome them and had triumphed, but in token of respect to the courage of the enemy had insisted that the Greek officers should keep their swords. That was to be almost the last gesture of chivalry between warriors in a war imminently fated to descend into barbarism.

EIGHT

Airborne Battle: Crete

The Balkan campaign, save for its brevity, had been a conventional operation of war in every respect. Even the breakneck speed of the German advance, now that *Blitzkrieg* in Poland and France had accustomed the world to the Wehrmacht's methods, seemed rather a revelation of the developing pattern of modern warfare than a further instalment of the military revolution that Hitler's generals had instituted. Indeed, it had been less revolutionary than the victory of 1940. The sheer disparity in quality between the Wehrmacht and its Balkan opponents, who had furthermore brought defeat upon themselves by the perverse ineptitude of their defensive arrangements, was all the explanation necessary for the catastrophe which had overcome them.

The Balkan campaign might have ended on that note, with the hoisting of the swastika flag over the Acropolis in Athens on 27 April as a fitting symbol of a triumph of the strong over the weak. But it did not: even as the cost of the campaign was counted – 12,000 British casualties (of whom 9000 were prisoners), uncounted Yugoslav and Greek dead, against a mere 5000 German killed, wounded or missing – and its spoils were divided – Yugoslav Bosnia, Dalmatia and Montenegro given to Italy, South Serbia and Greek Thrace to Bulgaria, the Vojvodina to Hungary, Croatia to the puppet Croatians of the Ustashi movement – Hitler was lending an ear to those in his circle who argued that the Balkan campaign was incomplete and urged that Germany's victory should be crowned by a descent upon Crete by the one largely untried instrument of *Blitzkrieg*, Germany's airborne army.

Germany was not the first advanced state to have created an airborne force. That cachet belonged to Italy, where the idea of strategic bombing had also been born. As early as 1927 the Italians had experimented with the delivery of infantrymen directly to the battlefield by parachute. The technique had then been taken up by the Red Army, which by 1936 had sufficiently perfected it to demonstrate at large-scale manoeuvres held in the presence of Western military observers the dropping of an entire regiment of parachutists and the subsequent airlanding of a whole brigade; this spectacular operation was made possible by the Red Air Force's development of transport aircraft large enough to hold complete units of fully equipped soldiers.

The Red Army's primacy in airborne tactics was to be severely retarded, however, by Stalin's great military purge of 1937-8, of which forward-looking officers were the principal victims. Its airborne units survived, and were to mount a number of operations in the Second World War, notably on the river

Dnieper in the autumn of 1943, but they were never accorded the independent and decisive role their advocates had hoped for them. In Germany, however, the concept of airborne operations was taken up enthusiastically by the Wehrmacht's new generation of military pioneers. As in France, where military parachute training was deemed an airforce activity, it was the Luftwaffe which was constituted the directing authority. In 1938 General Kurt Student, a flying veteran of the First World War, was appointed Inspector of Parachute Troops and shortly afterwards was given command of the first parachute division, designated 7 Flieger. It was this division which provided the units used in Norway and Holland in 1940. By 1941 its associated units, constituting Student's XI Air Corps, stood ready to extend the German conquest of the Balkans deeper into the Mediterranean area.

Hitler's closest military advisers, the operations officers of OKW, were anxious that XI Air Corps should be used to capture Malta. When asked to advise whether Crete or Malta was the more important objective in the Mediterranean, 'All officers of the Section,' General Walter Warlimont recalled, 'whether from the Army, Navy or Air Force, voted unanimously for the capture of Malta, since this seemed to be the only way to secure permanently the sea route to North Africa.' Keitel and Jodl, their chiefs, accepted their conclusions; but when on 15 April they confronted Student with this opinion he overcame them. He had already decided that Malta was too strongly garrisoned and defended to yield to an airborne assault. Crete, on the other hand, with its 'sausage-like form and single main road', offered an ideal target to his parachutists; moreover, he argued, they would be able to reach out towards the other Mediterranean islands required by German strategy-makers – not only Malta but also Cyprus – and thereby consolidate an impregnable land-sea position intermediate between Fortress Europe and Britain's increasingly tenuous foothold in the Middle East.

Goering, who saw in Student's plan an opportunity to rehabilitate the reputation of the Luftwaffe after its failure to overcome the Royal Air Force in the Battle of Britain, warmly endorsed his subordinate's conception and on 21 April presented it to Hitler. Since the capture of Crete had not figured in his original plans, Hitler was initially resistant but eventually agreed to lend it his support and on 25 April issued Führer Directive No. 28, codenamed *Merkur* (Mercury) for the Crete operation. Student, who was to remain the driving force of the operation throughout its inception and course, at once arranged for 7th Airborne Division to be brought to Greece from its training centre at Brunswick; he also persuaded OKH to let him use one of the divisions earmarked to garrison Greece, the elite 5th Mountain, and to lend him some of the light tanks from the 5th Panzer, which were not needed for Barbarossa. The mountain division was to provide a follow-up force, transported in local craft under Italian naval protection. The airborne division, consisting of three parachute and one airlanding regiment, was to storm the island by direct assault, flying in a fleet of 600 Junkers 52 aircraft, some of which would also tow eighty gliders carrying light tanks and the manpower of the 7th Division's spearhead, I Battalion of the 1st Assault Regiment. An air force of 280 bombers, 150 Stukas and 200 fighters would cover and support the operation. In all, 22,000 soldiers were to be committed; command of the whole campaign was to be under General Alexander Löhr's Fourth Air Fleet.

Student's plan was straightforward. He intended to use each of his three parachute regiments against the three towns on the north coast of the island, from west to east Maleme, Retimo and Heraklion, where airstrips were located. Once captured, these would be used for the landing of heavy equipment and as bases to 'roll up' the British defences along the single road which ran along the island's 170-mile length. At Maleme, which he had decided should be his *Schwerpünkte*, he intended to commit the 1st Assault Regiment, which would crash-land in gliders directly on to the airfield. Although he expected to be outnumbered by the defenders, he was sure that surprise, the high quality of his troops and the air superiority assured by the Luftwaffe's overwhelming strength would subdue them in a few days of brutal action.

His judgement that his force was superior in quality to the British garrison was correct. Major-General Bernard Freyberg, its commander, was a fire-eater, a legendary hero of the First World War in which he had won the VC commanding a battalion of the Royal Naval Division on the Somme, after an equally gallant – and romantic – passage of arms at Gallipoli. There he had been among the party which had buried the poet Rupert Brooke on the island of Skyros and later, on lone reconnaissance, he had swum the Hellespont, as Leander had done in legend and Lord Byron in reality a hundred years before him. Winston Churchill had christened him 'the Salamander' in tribute to his fire-resisting qualities.

Few of Freyberg's troops on Crete in the summer of 1941, however, matched his robustness. One brigade of regular British infantry had been brought direct from Egypt to garrison the island and was what the Germans called *kampffähig* – 'combat fit'. The rest were fugitives from the Greek fiasco. Two brigades of the 2nd New Zealand Division – with which Freyberg, who had spent his youth in New Zealand, had a special affinity – were intact and also one Australian brigade. The rest of the 40,000 troops on the island were remnants, disorganised and many of them disheartened. All, moreover, lacked essential equipment. 'Crete', wrote the New Zealander Charles Upham (who was to end the war as a double VC), 'was a pauper's campaign, mortars without base plates, Vickers guns without tripods.' A handful of tanks and a regiment's worth of artillery had been brought to the island; but the defenders lacked most essential heavy equipment and, above all, aircraft. On 1 May there were only seventeen Hurricanes and obsolete biplane Gladiators on Crete, and all were to be withdrawn before the Germans arrived. Worst of all, the British defenders could not count on local assistance. Since the 5th Cretan Division had been mobilised for war against the Italians and had been captured on the mainland, the only Cretan soldiers left on the island were recruits and reservists, with one rifle between six and five rounds per rifle.

The role of Ultra

Crete nevertheless might and perhaps ought to have been held; for, unbeknown to the Germans, their intentions were betrayed to the British well before the first parachutists had emplaned. The whole logic of an airborne operation was thereby compromised from the start. Like the proponents of armoured warfare and strategic bombing, the military parachutist pioneers had conceived their operational theory in reaction to the trench warfare they had witnessed in the First World War. It was the self-betrayal of the effort needed to mount a trench-breaking offensive which had affronted them: the laborious assembly of men and material, the ponderous and protracted process of preliminary bombardment, the agonised inching forward across no-man's land through barbed-wire barriers and earthwork zones. The bombing enthusiasts had reacted to that spectacle with the argument that high explosive was better delivered against the centres of production from which the enemy's artillery and machine-gun defences were supplied. The apostles of armoured warfare had argued – and in 1939-40 demonstrated – that deep defences were best overcome by launching against them a weapon impervious to the firepower the defenders deployed. The military parachutists proposed an intermediate but even more arresting alternative: to overarch ground defences by airpower which would deliver aggressive infantrymen at the soft spots immediately behind the enemy's front, his headquarters, communication centres and supply points. It was a brilliantly daring leap of strategic imagination; but its success rested on the precondition that the enemy remain unaware of the stroke poised against him – otherwise the parachutists committed to deliver it would suffer the same (if not worse) fate as the infantrymen of the trenches going over the top against the enemy alerted by the preliminary bombardment. Their helplessness during descent, the necessary lightness of the equipment they would use to fight if they survived, doomed them to undergo appalling losses against defenders who had been warned of their approach.

The British defenders of Crete had been warned. Ultra, the intelligence source derived from the interception and decryption of enemy ciphers by the Government Code and Cipher School at Bletchley, had hitherto yielded little information of value to the conduct of ground operations between the British and the Germans. Until the end of the campaign in France, Bletchley had had great difficulty in breaking the cipher 'keys' used on the German Enigma ciphering machine through which the different Wehrmacht headquarters communicated. The difficulties were in part intrinsic – the Enigma machine was designed to confront an eavesdropper with several million possible solutions to an intercept – and in part those of any experimental enterprise:

Bletchley was accumulating procedures which hastened the process of breaking but had not yet systematised them. There was another difficulty: Bletchley's success depended chiefly upon the exploitation of mistakes made by German Enigma machine operators in encipherment procedure. German army and navy operators, perhaps because they were drawn from old-established signals services, made few mistakes. It was the younger Luftwaffe which provided Bletchley's listeners with the bulk of their opportunities; but, though 'breaks' into the Luftwaffe key considerably assisted the Air Defence of Great Britain to resist and deflect bombing attacks during the winter blitz of 1940-1, they were of less use in opposing the Germans in the Battle of the Atlantic or in the ground campaigns in Greece and North Africa.

Crete, however, was to be a Luftwaffe campaign. Thus the vulnerability of its 'Red' key, as Bletchley denominated it, to British decryption on a regular, day-to-day basis and in 'real time' – at a speed, that is, equivalent to that at which German recipients of Enigma messages deciphered them themselves – was to compromise the security of the parachute operation from the outset. On 26 April, for example, the day after Hitler issued his *Merkur* directive, two intercepted 'Red' messages were found to refer to Crete: the Fourth Air Fleet mentioned the selection of bases for 'Operation Crete', while its subordinate VIII Air Corps asked for maps and photographs of the island. Thereafter the warnings accumulated almost daily. On 6 May Ultra revealed that German headquarters expected preparations to be completed by 17 May and outlined the exact stages and targets of the German attack. On 15 May it detected that D-Day had been postponed from 17 May to 19 May. And on 19 May it warned that 20 May was to be the new attack date and that the German parachute commanders were to assemble immediately with maps and photographs of Maleme, Retimo and Heraklion. All this information, disguised as intelligence collected by a British agent in Athens, was transmitted in 'real time' to Freyberg, who thus, on the morning of 20 May, knew exactly when, where and in what strength Student's parachutists and glider infantry were going to land.

Between foreknowledge and forestalment, however, there always yawns the gap of capability. That predicament was Freyberg's. Against an attacking force whose defining characteristics were mobility and flexibility he opposed a defending force almost totally bereft of the means of movement. Its units were in the right place, but, should one be driven from any of the vital airstrips, it could not be replaced; the Germans would be enabled to land reinforcements and heavy equipment and the battle for the island would probably be lost in consequence.

Defending Maleme airstrip were the 21st, 22nd and 23rd New Zealand Battalions. New Zealanders were to be reckoned by Rommel, on his experience in the desert campaign, the best soldiers he met in the Second World War: resilient, hardy, self-confident, they had little opinion of any soldiers but themselves. When on the early morning of 20 May they brushed themselves clean of the dust thrown up by the Luftwaffe's preparatory bombardment and cocked their weapons to resist the parachute assault they knew must follow, they harboured no sense of the harshness of the battle to come. Lieutenant W. B. Thomas of the 23rd Battalion found his first sight of the German parachutists 'unreal, difficult to comprehend as anything at all dangerous':

> Seen against the deep blue of the early morning Cretan sky, through a frame of grey-green olive branches, they looked like little jerking dolls whose billowy frocks of green, yellow, red and white had somehow blown up and become entangled in the wires that controlled them. . . . I struggled to grasp the meaning behind this colourful fantasy, to realise that those beautiful kicking dolls meant the repetition of all the horrors we had known so recently in Greece.

Lieutenant Thomas's sense of unreality was understandable. He was witnessing the first purposeful parachute operation in history. The Germans' earlier jumps, in Norway and Holland, had been small-scale, lightly opposed and strongly supported by conventional ground forces. The *Sprung nach Kreta* was a true leap into the unknown, the pitting of pioneers in a military revolution against forces they could overcome only by their own unaided effort. Student's men were in a sense primitives: the British and American equivalents, already training for future parachute operations of their own, would regard their equipment and technique with horrified incredulity. The Germans had no control over their descent; they jumped from their Junkers 52s, in groups of twelve, their parachutes opened by static line, but were then suspended by a single strap attached to the harness in the middle of the back. Slipstream and wind carried them indeed 'like dolls' to their landing, from the shock of which their padding, helmets and rubber boots were supposed to protect them. Those not injured by the impact – and jump injuries were numerous – then collected their weapons from parachuted containers, assembled in squads and moved to the assault. The glider infantry of the 1st Assault Regiment, crash-landed in groups of fifteen, reinforced them with heavier equipment.

Student's theory of airborne assault took no account of Cretan terrain or New Zealand tenacity. The harsh and broken ground around Maleme injured many of his parachutists as they landed and pulverised a high proportion of the gliders; the New Zealanders dealt pitilessly with the survivors. They shot the enemy in the air: 'You'd see one go limp and kind of straighten up with a jerk and then go limp again, and you knew he "was done for".' They shot them as they landed, so that next day a visiting staff officer to 23rd Battalion found 'bodies everywhere, every ten-twelve yards. One stepped over them as one went through the olive groves.' Sixty New Zealanders released from a field punishment centre in Maleme, where they were serving time for minor military offences, killed 110 Germans in the first hour of the assault.

The losses suffered by the German parachute battalions around Maleme in the first hours of 20 May were truly appalling. One company of III Battalion, 1st Assault Regiment, lost 112 killed out of 126; 400 of the battalion's 600 men were dead before the day was out. Only a hundred men of the glider-borne I Battalion survived its landing unwounded; II Battalion also suffered heavily. IV Battalion, led by Captain Walter Gericke (who would survive to command the West German army's parachute division as a NATO general), alone preserved the bulk of its strength. It and the survivors of the other three struggled throughout the day of 20 May to assemble their remaining strength, fight off the remorseless New Zealanders and move towards their objective, the Maleme airstrip. They made no progress; in the 21st New Zealand Battalion's area the parachutists who fell in the streets of the village of Modhion were attacked 'by the entire population of the district, including women and children, using any weapon, flintlock rifles captured from the Turks a hundred years ago, axes and even spades.' They helped to add to the 1st Assault Regiment's casualties, which by the end of the day included two battalion commanders killed and two wounded, together with the regimental commander. The 1st Assault Regiment, which regarded itself as the Wehrmacht's elite, had by nightfall suffered much – perhaps 50 per cent losses – and achieved nothing.

Its sister regiments, 1st, 2nd and 3rd Parachute, directed against Heraklion, Retimo and Suda respectively, all on the north coast, also suffered heavily on 20 May. In one or two places, airborne assault achieved its intended suprise: near Suda, Crete's main port, ten glider infantrymen who landed close to an artillery regiment killed 180 gunners who lacked small arms to defend themselves. Elsewhere, though, it was generally the Germans who were slaughtered. The 3rd Parachute Regiment, landing just east of the 1st Assault Regiment around Canea and Suda, arrived directionless; their commander, Süssmann (who also commanded the division), had died in a glider crash on take-off. Its I Battalion, led by Baron von der Heydte, an untypical parachutist by reason both of his undisguisedly aristocratic disdain for Nazism and of his marked intellectuality – he was to write a remarkable memoir of the Cretan campaign and end his career as a professor of economics – got down relatively unscathed. Its III Battalion, however, was almost wiped out during the day's fighting, justifiably in the view of the New Zealanders, whose senior medical officer had been shot, with many of his patients, by members of this battalion during their initial assault. Its II Battalion attacked a feature defended by the New Zealand Division's logistic troops; Company Sergeant-Major Neuhoff describes the results of his encounter with the petrol company of its Composite Battalion: 'We advanced to attack the hill . . . we proceeded, without opposition, about half way up . . . suddenly we ran into heavy and very accurate rifle and machine-gun fire. The enemy had held their fire with great discipline and allowed us to approach well within effective range before opening up. Our casualties were extremely heavy, and we were forced to retire leaving many dead behind us.'

Yet their opponents, as the New Zealand official history records, were 'for the most part drivers and technicians and so ill-trained for infantry fighting'.

Student, who had not yet left his rear headquarters in the Hotel Grande Bretagne in Athens, remained all day in ignorance of the fate his cherished division had suffered. Far into the night of 20/21 May he sat at his map table, as von der Heydte recalled, 'waiting and waiting for the news which would bring him confirmation that he had been right in proposing the attack on the island to Goering a month previously. Everything had seemed so simple in prospect, so feasible and so certain. He had thought that he had taken every possibility into consideration – and then everything had turned out contrary to plans and expectations.' The truth – as I. M. D. Stewart, the medical officer of the 1st Welch Regiment, a veteran and the most meticulous historian of the campaign, later recorded – was that he had 'dissipated' his airborne division 'in scattered attacks about the island':

> Thousands of its young men now lay dead in the olive groves and among the buttercups and the barley. His glider troops and four of his parachute battalions . . . had been shattered, reduced within the space of fifteen minutes to a few dozen fugitive survivors. Other battalions had suffered little less severely. Yet he still had not captured an airfield. Now he had left only his tiny airlanding reserve. If these few hundred men should fail on the morrow [21 May] the only possible relief for the Division would have to come by sea.

On the evening of the first day of the first great parachute operation in history, therefore, the advantage appeared to have passed decisively to the opposition – an ill-organised force of under-equipped troops almost bereft of air cover and supporting arms. Yet, despite all the agony Student's men had suffered and all the mistakes he had made, on 21 May he would succeed in recovering the initiative and turning the battle to his advantage. How so? The explanation, one of Freyberg's staff officers was to reflect ruefully in the aftermath, was the absence of 'a hundred extra wireless sets'; for the defenders had failed to recognise the extent of their own success and had failed to report it to Freyberg's headquarters, which in turn had failed to radio the orders to recoup and regroup. Next morning Winston Churchill reported to the House of Commons that the 'most stern and resolute resistance' would be offered to the enemy. Meanwhile Freyberg lacked that clear picture of his battle which would allow him to react as commander. He communicated with the New Zealanders defending the Maleme airfield – Student's *Schwerpunkt* – through the headquarters of 5 Brigade; the brigade in turn communicated indirectly with its battalion commanders; and Lieutenant-Colonel L. W. Andrews, the commander of the crucial battalion, 22nd, mistakenly believed that his brigade commander planned to support him. A brave man – he had won the Victoria Cross in the First World War – he decided on the evening of 20 May, after an initial counter-attack supported by two of the only six heavy tanks on Crete had failed, to regroup on high ground overlooking the airfield for a concerted push the next day, and this regrouping inadvertently conceded the vital spot to the Germans and so rescued them from the inevitability of disaster.

While Andrews took the wrong decision for good reasons, Student was arriving at the right decision for bad reasons: he had no ground for thinking that fresh troops would fare any better at Maleme than those already dead. Indeed, the universal military maxim, 'never reinforce failure', should have warned him against committing his reserve at that point. He nevertheless decided to do so. On the afternoon of 21 May his last two companies of parachutists fell among the New Zealand division's Maori battalion and were slaughtered – 'not cricket, I know,' wrote one of their officers, 'but there it is.' At the same time Student's airlanding reserve, the spearhead of 100th Mountain Rifle Regiment of the 5th Mountain Division, began to crash-land in Junkers 52s on the Maleme airstrip from which Andrews had withdrawn his defending 22nd Battalion the previous evening. 'Machine-gun bullets tear through the right wing,' wrote a war correspondent aboard. 'The pilot grits his teeth. Cost what it may he has to get down. Suddenly there leaps up below us a vineyard. We strike the ground. Then one wing grinds into the sand and tears the back of the machine half round to the left. Men, packs, boxes, ammunition are flung forward . . . we lose the power over our own bodies. At last we come to a standstill, the machine standing half on its head.'

Nearly forty Junkers 52s succeeded in landing on the Maleme airstrip in this way, bringing 650 men of II Battalion, 100th Mountain Rifle Regiment. The mountain riflemen, like Student's parachutists, also regarded themselves as an elite, and with justification. While the New Zealanders struggled to come to terms with the new threat, the mountaineers were moving to consolidate the German position at Maleme airfield, with the intention of extending their foothold next day.

Some of the mountaineers' reinforcements were meanwhile approaching Crete by ship. They were to suffer an unhappy fate; but so too were the ships of the Royal Navy which intercepted them. The Alexandria squadron easily overcame the Italian escort to the fleet of caiques and barges carrying the remainder of 100th Mountain Rifle Regiment towards Crete, causing 300 of them to be drowned; but during 22 May the Luftwaffe inflicted a far more grievous penalty on the British ships and crews. The battleship *Warspite* was damaged, the cruisers *Gloucester* and *Fiji* sunk, together with the destroyers *Kashmir* and *Kelly* – the latter commanded by the future Earl Mountbatten of Burma. This was not the end of the navy's losses; before 2 June it also lost the cruisers *Juno* and *Calcutta* and the destroyers *Imperial* and *Greyhound*, which were sunk, and suffered damage to the battleship *Valiant*, the aircraft carrier *Formidable*, the cruisers *Perth*, *Orion*, *Ajax* and *Naiad* and the destroyers *Kelvin*, *Napier* and *Hereward*. When the tally was taken, the Battle of Crete, though less shocking in its effect on British morale than the future loss of the *Prince of Wales* and *Repulse* was to prove, was reckoned the costliest of any British naval engagement of the Second World War.

Student gains the upper hand

Ashore, meanwhile, the battle had begun to run irreversibly the Germans' way. The New Zealand counter-attack to recapture Maleme airfield failed in the early hours of 22 May; throughout the day Student, with brutal recklessness, directed a stream of Junkers 52s at the airfield. Those that crashed on impact, as many did, were pushed off the runway for the next arrival. Meanwhile the Luftwaffe operated overland in overwhelming strength, shooting and bombing anything that moved. 'It is a most strange and grim battle that is being fought,' Churchill told the Commons that afternoon. 'Our side have no air . . . and the other side have very little or no tanks. Neither side has any means of retreat.' The truth was that the British had no tanks that counted and no means of moving, while the Germans were accumulating growing numbers of fresh, first-class soldiers to manoeuvre against the defenders.

Freyberg now decided to withdraw eastward and regroup for a counter-attack. However, this regrouping was composed not of a single unit but of the bulk of his best troops, the New Zealanders and the regular British battalions. The withdrawal conceded yet more vital ground to the parachutists and mountain riflemen around Maleme, who were growing steadily in numbers. On 24 May they were repulsed from the village of Galatas, then took it, then lost it again to the New Zealand counter-attack Freyberg had planned as his decisive riposte; but it could not reach as far as Maleme, into which the Germans had now crowded almost the whole of the mountain division. When the Germans resumed their attack the British were driven relentlessly eastward, abandoning one position after another.

On 26 May, Freyberg told Wavell, commanding in the Middle East, that the loss of Crete could only be a matter of time. Next day Wavell decided on evacuation before the dominance of the Luftwaffe made that impossible. The garrison of Heraklion, against which the parachutists had made no impression, was taken off on the night of 28 May. The garrison of Retimo, which had also resisted all attacks, could not be reached by the navy and had to be abandoned. During 28-31 May the main force left its positions east of Maleme and began a long and agonising trek southwards across the mountains to the little port of Sphakia on the south coast. It was a shaming culmination to a benighted battle. The minority of troops which actually fought kept together as best they could; those who had left Greece disorganised now lost all semblance of unity. 'Never shall I forget the disorganisation and almost complete lack of control of the masses on the move,' wrote Freyberg, 'as we made our way slowly through the endless stream of trudging men.' When he and the rest of his broken army reached Sphakia they sheltered under the cliffs waiting for the navy to rescue them under cover of darkness. The navy suffered heavily in the attempt but by 1 June had succeeded in taking off 18,000 troops; 12,000 remained to fall prisoner to the Germans and nearly 2000 had been killed in the fighting.

These figures confirmed, if the evidence were needed, that Crete had been a catastrophe. It had entailed the loss of two formed divisions of troops, New Zealand, Australian and British, urgently needed to fight the burgeoning war in the desert against Rommel's expeditionary force. It had also added unnecessarily to the roll of humiliation which Hitler had inflicted on the British Empire, most of all because both he and his enemies knew by what a narrow margin his parachutists had been rescued from defeat. Had Maleme not been abandoned on the second day, had Freyberg's counter-attack been launched two days earlier, the parachutists would have been destroyed in their foothold, the island saved and the first definitive check to Hitler's campaign of conquest imposed in a blaze of spectacular publicity. As it was, the German war machine had been seen once again to triumph, in a new and revolutionary form, in the very centre of Britain's traditional strategic zone, and against a principal instrument of its overseas power, the Mediterranean Fleet.

Yet, not only with hindsight, Crete could also be seen as a highly ambiguous victory. 'Hitler', Student recorded, was 'most displeased with the whole affair.' On 20 July he told his parachute general, 'Crete proves that the days of the paratroopers are over. The paratroop weapon depends upon surprise – the surprise factor has now gone.' He had refused to allow the German propaganda machine to publicise the operation while it was in progress and he now set his face against mounting operations of the same type in the future. Crete had killed 4000 German soldiers, most from the 7th Parachute Division; nearly half the 1st Assault Regiment had died in action. Gericke, who had come across the dropping zone of its III Battalion on 23 May, was appalled by the evidence of what had befallen it. 'Frightful was the sight that met our eyes. . . . Dead parachutists, still in their full equipment, hung suspended from the branches [of the olive trees] swinging gently in the light breeze – everywhere were the dead. Those who had succeeded in getting free from their harness had been shot down within a few strides and slain by the Cretan volunteers. From these corpses could be seen all too clearly what had happened within the first few minutes of the battle of Crete.' Not only the men but the whole structure of the airborne force had suffered disastrously; 220 out of 600 transport aircraft had been destroyed, a material loss quite disproportionate to the material advantage gained. The seizure of Crete had not been and would not prove essential to German strategy; a successful attempt on Malta, desired by OKW, would by contrast have justified any loss suffered by its mounting. The occupation of Crete, moreover, would involve the Germans in a bitter anti-partisan campaign, their conduct of which would blacken their name and lay the foundations of a bitter hatred of them not erased in the island to this day.

The British and Americans, both energetically raising parachute divisions, drew from Crete a conclusion different from Hitler's: that it was that particular form rather than the underlying principle of airborne operations which had proved unsound. In their great descents on Sicily, Normandy and Holland, they would eschew Student's practice of launching parachutists directly on to an enemy position in favour of landing at a distance from the objective and then concentrating against it. In Sicily and Normandy they would also risk large-scale airborne offensives only in co-ordination with a major amphibious assault from the sea, thus distracting the enemy from a concerted response against the fragile military instruments of parachute and glider. In Sicily and Normandy this careful reinsurance was to justify itself. In Holland, in September 1944, when they abandoned caution and essayed a Crete-style assault of their own, the disaster which overtook Montgomery's parachutists was to prove even more complete than that suffered by Student's. In the broad if not the narrow sense, therefore, Hitler's appreciation of Operation *Merkur* was correct: parachuting to war is essentially a dicing with death, in which the odds are loaded against the soldier who entrusts his life to silk and static line. There is a possibility that a combination of luck and judgement will deposit him and his comrades beyond the jaws of danger, enable them to assemble and allow formed airborne units to go forward to battle; but the probability is otherwise. Of the four great parachute endeavours of the Second World War, two – Sicily and Normandy – managed to evade the probabilities, two – Crete and Arnhem – did not. The demise of independent parachute forces since 1945 is the inevitable outcome of that unfavourable reckoning.

NINE

Barbarossa

Even while the news of his flawed victory in Crete was reaching him, Hitler's thoughts were engaged with events far away. Indeed, at the height of the battle he had been preoccupied with two quite unrelated matters: the abortive sortie of his 'wonder' battleship *Bismarck* into the North Atlantic and the flight on 10 May of his deputy, Rudolf Hess, bearing an unauthorised offer of peace to the British. *Bismarck*'s destruction on 27 May could be represented by his propaganda machine as a sort of epic; Hess's crazed initiative – which puzzled the British quite as much as it mystified his fellow Nazis – continued to enrage Hitler for weeks and months afterwards. He had Goebbels describe it as the result of 'hallucination'; but the defection struck him a personal blow. Hess was not only an 'old fighter' but his amanuensis, who had taken down *Mein Kampf* from dictation during their incarceration at Landsberg after the Munich *Putsch* of 1923; he was also a comrade-in-arms from the List Regiment, a society of 'young Germans' whose brotherhood during the First World War had brought Hitler the one truly fulfilling experience of his lonely and confused youth.

Memories of the sacrifice offered by the List Regiment in the *Kindermord bei Ypern*, perhaps aroused by Hess's flight, must have put the devastation of 7th Parachute Division and the destruction of the 1st Assault Regiment into perspective. Hitler was himself the survivor of a massacre in 1914 even more extensive than the parachutists had suffered in May 1941. No other division of the Wehrmacht, in Poland, Norway, the Low Countries, France and the Balkans, had suffered losses approaching those of Student's elite. However, not only were such losses commonplace by First World War standards; they also counted for little beside the strength which had accrued to the Wehrmacht since the war had begun. The Wehrmacht's losses thus far in twenty-one months of war had, by the standards of twentieth-century bloodletting, been inconsiderable: in Poland 17,000 dead and missing; in Scandinavia 3600; in France and the Low Countries 45,000; in Yugoslavia 151; in Greece and Crete less than 5000. On the other hand, the strength of the German army had risen since September 1939 from 3,750,000 men to 5,000,000; the Luftwaffe numbered 1,700,000, including anti-aircraft (flak) and parachute troops; and the navy 400,000. The Nazi Party's army, the Waffen-SS, had increased from 50,000 to 150,000 men. The most striking assertion of strength was in the army. On mobilisation for war in September 1939, the *Feldheer* (field army) had included 106 divisions, of which ten were armoured and six motorised; by June 1941, on the eve of Barbarossa, the roll had increased to 180 infantry, 12 motorised and 20 Panzer divisions. The multiplication of armoured formations had been achieved by halving the number of tanks each contained. For all that, the German army, with the airfleets which supported it, was not only larger but disproportionately stronger in every way – in weapons, in reserves, above all in operational skill – than in 1939. Hitler's rearmament programme of 1935-9 had merely lent military weight to his adventures in foreign policy; his war-making had permeated the whole of German society. One German male in four was now in uniform; most had directly experienced victory, had trodden the soil of occupied territory and had seen soldiers of the victor nations of 1918 taken into captivity. The red-white-black and swastika flag had been raised 'from the Meuse to the Memel, from the Belt to the Adige', as the national anthem proclaimed it should fly, and German soldiers now stood ready to carry it even deeper into the zone of conquest Hitler had worked out as his own: into Stalin's Russia.

The Balkan campaign, often depicted by historians as an unwelcome diversion from Hitler's long-laid plan to attack the Soviet Union and as a disabling interruption of the timetable he had marked out for its inception, had been in fact no such thing. It had been successfully concluded even more rapidly than his professional military advisers could have anticipated; while the choice of D-Day for Barbarossa had always depended not on the sequence of contingent events but on the weather and objective military factors. The German army found it more difficult than expected to position the units allocated for Barbarossa in Poland; while the lateness of the spring thaw, which left the eastern European rivers in spate beyond the predicted date, meant that Barbarossa could not have been begun very much earlier than the third week of June, whatever Hitler's intentions.

The outlook for Barbarossa nevertheless rested on German over-optimism. 'Massive frontier battles to be expected; duration up to four weeks,' Brauchitsch had written at the end of April 1941, 'but in further development only minor resistance is then still to be reckoned with.' Hitler was more emphatic. 'You have only to kick in the door,' he told Rundstedt, commanding Army Group South, on the eve of Barbarossa, 'and the whole rotten structure will come crashing down.' Hitler's prognosis was in part ideologically determined; he was committed to a view of Soviet Russia which represented its citizens as the crushed and brutish creatures of a Bolshevik tyrant, 200 million Calibans quailing under the eye of a Prospero corrupted by absolute power. There was irony in this mirror image. But Hitler's belief that Soviet communism had a hollow centre was supported not only by prejudice but also by realities: in 1939 the giant Red Army had performed lamentably against minuscule Finland; and that humiliation was explained in turn by the massacre of its senior officers – far more complete than any war could have inflicted –

instituted by Stalin in 1937-8.

In extension of his purge of the party and secret police (NKVD), which had assured his political primacy, Stalin had then accused, tried, convicted and executed for treachery over half the senior commanders of the Red Army. The first to go was Tukhachevsky, his chief of staff, the leading representative of that group of former tsarist officers who had convincingly turned coat during the Civil War of 1918-20 and thereafter provided the army with the professional leadership of which it was in such dire need in the era of post-war reconstruction. He had given firm evidence of his commitment to the new Russia. It was he who had led the offensive against Warsaw in 1920 and put down the Kronstadt rising of 1921. He had also pioneered the creation of the army's tank arm and the organisation of the large mechanised corps whose existence had by 1935 put the Red Army at the very forefront of modern military development. Perhaps because of his very strategic radicalism, he was singled out for extinction at the outset of the military purge and shot, with seven other generals, on 11 June 1937.

The shootings thereafter were to proceed apace. By the autumn of 1938 three out of five of the Red Army's marshals were dead, thirteen out of fifteen army commanders, 110 out of 195 divisional commanders and 186 out of 406 brigadiers. The massacre of those in administrative and politico-military appointments was even more extensive: all eleven deputy commissars for defence were shot, seventy-five out of eighty members of the Military Soviet, and all military district commanders, together with most of their chiefs of political administration – those party commissars whose function was to ensure that soldiers should not take decisions or commitments which might attract the disfavour of the party.

The effects of the purge

It was difficult to detect any pattern in Stalin's bloodlust. The purge certainly exterminated many of the ex-tsarists who had thrown in their lot with the Bolsheviks after 1917; yet it did not spare the chief of staff, Yegorov, whose proletarian credentials were impeccable, who was replaced by Shaposhnikov, a graduate of the Imperial General Staff College. Nor was it the case that commanders suffered worse than commissars, since the 'politicals' were executed in even greater numbers than the 'soldiers'. If there was a clue to Stalin's murderous motives, it seemed to lie in the history of personal rancours and alliances formed during the Civil War. Just as the principal victims of the party purge were those who had opposed or not assisted Stalin's aggrandisement of his role as First Secretary after Lenin's death, so the principal victims of the military purge were those who had been identified with Leon Trotsky's command of the Red Army in its struggle with the Whites. The anti-Trotsky faction had been centred on the First Cavalry Army, commanded by S. M. Budenny and K. E. Voroshilov, which had recalcitrantly conducted its own strategy during the struggle against the Whites in South Russia where Stalin was political commissar, in 1918-19, and signally failed to assist Tukhachevsky in the abortive advance on Warsaw in 1920. The First Cavalry Army had been an aberrant element in Trotsky's – and Tukhachevsky's – anti-White war; but it had been the military instrument associated with Stalin's early struggle against Trotsky, and its old comrades, in the years of his triumph, therefore stood high in his favour. The four men whom the purge elevated to high place in the Soviet military command after the purge, Budenny, S. K. Timoshenko, L. Z. Mekhlis, G. I. Kulik, were all First Cavalry Army officers and veterans of the southern Russian campaign. Voroshilov, already Defence Commissar before 1937 but much enhanced in power by Tukhachevsky's fall, was also a First Cavalry Army man.

Their advancement was not to Russia's military advantage. Budenny had a fine moustache but no military brain. Mekhlis, the chief commissar, seemed, in Professor John Erickson's words, to have 'combined a monumental incompetence with a fierce detestation of the officer corps'. Timoshenko was at least competent, but a political rather than a military commander. Kulik, head of ordnance, was a technological reactionary who opposed the distribution of automatic weapons to soldiers on the grounds that they were incapable of handling them and halted the production of anti-tank and anti-aircraft guns. Voroshilov was worst of all; in 1934, apparently for no better reason than Tukhachevsky's advocacy of the independent armoured force, he argued: 'It is almost axiomatic that such a powerful force as the tank corps is a very far-fetched idea and we should therefore have nothing to do with it.' Immediately after Tukhachevsky's removal, he abolished all tank formations larger than a brigade.

Voroshilov's obscurantism – and that of other First Cavalry Army veterans – was exposed by the Finnish war. The humiliation inflicted on the Russians by the Finns, whom they outnumbered two-hundredfold, demanded hasty reforms. Voroshilov, nominated to the comparatively harmless posts of Deputy Prime Minister and Chairman of the Defence Committee on 8 May 1940, was replaced by Timoshenko as Commissar for Defence. Though his conduct of operations against Finland had been less than masterly, Timoshenko at least grasped that the Red Army stood in urgent need of reorganisation. Under his aegis, steps were taken to re-establish Tukhachevsky's large armoured formations, consisting of two armoured and one mechanised division forming an armoured corps; to begin the construction of fixed defences on Russia's new military frontier, which stood some 200 miles further westward than before the annexation of eastern Poland in 1939; to demote commissars to a consultative status in the command structure and to bring forward soldiers of proven ability to high command. First among those was G. K. Zhukov, who in 1939 had won the Battle of Khalkin-Gol in Russia's undeclared war against Japan on the disputed border of Mongolia. Zhukov, according to a colleague, Lieutenant-

General P. L. Romanenko, describing his presentation to a high-level staff conference held in the Kremlin in December 1940, had not altogether grasped the dynamism with which Germany's Panzer formations operated, the large scale on which they were organised or the closeness with which they combined their attacks with those of their supporting Luftwaffe squadrons. According to Romanenko, he expected 'a relatively weak saturation level of equipment in formations' – an old-fashioned battlefield, in short, where infantry predominated and tanks merely leavened the mass, rather than the dense concentrations of armour with which, as in France, infantry formations were tossed about by the flail of armour like sheaves on the threshing floor. Nevertheless, his was clearly a modern military mind.

The fighting potential of a Russian army – Red or tsarist – was never in doubt. Russian soldiers had proved brave, hardy and patriotic fighters against the enemies who had taken their measure in the past – Turkish, Austrian, French and British as well as German. As artillerymen they stuck to their guns – and the quality of Russian artillery material had always been excellent. As infantrymen they were tenacious in defence and aggressive in attack. Russian armies, when they had failed, had done so not because their soldiers were poor but because their generals were bad. It had been the fate of too many Russian armies to be cursed by incompetent leadership, in the Crimea, in Manchuria, and never more so than in the First World War. The Revolution had swept away the likes of Rennenkampf and Samsonov who had succeeded in leading overwhelmingly superior armies to defeat by the outnumbered Germans in East Prussia in 1914. It had replaced them during and after the Civil War with young and dynamic leaders who had learned the art of victory in the face of the enemy. The question now was whether those who had survived the purge – and they were, by definition, junior officers of conformist quality – retained the self-confidence to act with decision and energy on the battlefield.

The prospect for the 479 officers newly appointed major-general in June 1940 (the largest mass promotion in the history of any army) was not wholly discouraging. One by-product of the purge was a tightening of the Red Army's disciplinary code, which subjected the Soviet conscript to a positively Prussian standard of military obedience. Another, paradoxically, was a demotion of the commissars; these political officials, whom the Revolution had originally imposed on the army to forestall treachery by ex-tsarist commanders, had been empowered with the right to veto military orders until 1934. That right was reimposed during the purges but withdrawn again after the débâcle of the Finnish campaign. The 'political deputies' of the new divisional commanders were therefore restricted in their responsibilities to the political education of the soldiers and the maintenance of party orthodoxy among the officers. There lay an important alleviation of professional military anxieties. Another encouraging factor was the improvement of equipment. For all Kulik's efforts to retard the modernisation of the Red Army, the material reaching its formations was of good quality. One effect of Stalin's industrialisation programme had been to encourage the development of modern tanks, based on designs purchased outright from the American tank pioneer, Walter Christie. His revolutionary propulsion and suspension systems had resulted in the models that would evolve eventually into the T-34, which was to prove itself the best all-round tank of the Second World War. Soviet industry was also producing useful military radio sets and a prototype radar; while the aircraft industry, with an annual output of 5000 machines, was busily accumulating a fleet which, like the tank park, would by 1941 be the largest in the world.

Through his arbitrary terrorisation of inventive scientists and technologists in the armaments industry during the era of the purges, Stalin did much to interrupt the transformation of the Soviet forces into an advanced instrument of war; and, when he was not directly persecuting the innovators, he often spoke with two tongues in their support. Thus, at the important Kremlin conference of December 1940, he ridiculed his creature Kulik's advocacy of a return to large marching divisions of infantry supplied by animal transport, comparing him to the peasant who preferred the wooden plough to the tractor; nevertheless he permitted the disbandment of the mechanised transport department and starved the army of trucks (which, when the time came, would have to be supplied by Lend-Lease from Detroit). Yet his influence was not wholly malign. Having wrought his spite in the purge, he thereafter accepted the need for reform which the army's disastrous showing in the Winter War with Finland had revealed, accepted the common-sense advice of men like Timoshenko, recognised the talents of others who had made their name in the Mongolian campaign against Japan, like Zhukov and Rokossovsky, and in Finland, like Kirponos, and promoted other younger men of good military standing, like Konev, Vatutin, Yeremenko, Sokolovsky and Chuikov. Above all he sustained the growth of the Red Army itself. It was a source of pride to him and the Russian people that the army was the largest in the world, and from its very size he took confidence in Russia's ability to defend itself and exercise influence beyond its borders. By the spring of 1941 its war strength numbered between 230 and 240 'rifle' divisions (about 110 in the west) – formations 14,000 strong, though largely dependent on horsed transport for supply – as well as 50 tank divisions and 25 mechanised divisions which were fully equipped. The Soviet tank park numbered 24,000 and, if of mixed quality, could draw on an annual output of 2000, of which an increasing number were T-34s; by the end of 1941 tank production targets would stand at between 20,000 and 25,000, while Germany would never succeed in producing more than 18,000 tanks in any year. The Red Air Force, drawing on an annual output of 10,000 machines in 1941, stood at a strength of at least 10,000 in 1940; lacking as yet equivalents to the best German aircraft, and wholly subordinate to the army though it was, it was nevertheless the largest air force in the world.

In crudely material terms, therefore, Stalin as warlord stood on equal, perhaps superior, footing to Hitler. As strategist, however, he was as yet in no way his match. Hitler's decision to provoke war in 1939 was to prove a catastrophic miscalculation; in its prosecution, however, he displayed exactly the same cynical estimation of motive and ruthless exploitation of weakness as had won him such spectacular diplomatic victories in 1936-9. Stalin also operated with ruthlessness and cynicism; but his estimation of motive and assessment of reality were clouded by a coarse and over-cunning solipsism. He ascribed to adversaries a pattern of calculation as brutal and grasping as his own. Thus, because he took such satisfaction in the quantity of territory he had added to the Soviet Union since 1939, he appears to have thought that it would be a primary German aim, in the event of war, to take it from him. He certainly made it his primary object to hold what he had. Consequently much of Soviet military effort in the spring of 1941 was dedicated to the construction of new frontier defences, to replace those abandoned by the advance from the 1939 frontier in the previous two years. At the same time the Red Army was deployed so as to defend the frontier's every kink and twist, in defiance of all traditional military wisdom about 'defence in depth' and the maintenance of counter-attack reserves. The defences of the 1939 frontier were actually stripped to provide weapons for the new ones; while the armoured formations, which might have been held in support behind the zone under fortification, were dispersed piecemeal throughout the five western military districts, concentrated neither for a counter-stroke nor for deep blocking operations.

Stalin's dissipation of his own forces was matched by his disregard for others' warnings of the danger in which they stood. Entirely cold-hearted though he was in his dealings with Hitler, he insisted on regarding reports of his fellow dictator's aggressive intentions as 'provocation'. Such reports came to him in profusion after March 1941, from his own ambassadors and military attachés, from Soviet agents, Russian and non-Russian, from foreign governments already at war with Germany, particularly the British, even from neutrals, including the United States. Primary indications of German intentions were provided by the systematic flights of German reconnaissance aircraft over Soviet territory – by the same squadron commanded by Theodor Rowehl that had overflown Britain in 1939 – and the penetration of the Soviet border zone by German patrols dressed in Russian uniforms. These were supplemented as early as April by reports from Richard Sorge, the Comintern spy in Tokyo who was privy to the dispatches of the German ambassador (which he helped him compose), that preparations for war were complete. On 3 April Winston Churchill, whose source (unrevealed, of course, to the Russians) was Ultra intelligence, sent Stalin word that the Germans had deployed armour released by Yugoslavia's accession to the Tripartite Pact directly to southern Poland. By a strange dereliction of duty, Stafford Cripps, the warmly pro-Soviet British ambassador to Moscow, delayed transmission of the message until 19 April; but Stalin had had equally good Western intelligence of German intentions weeks before that. Early in March, Sumner Welles, the American Under-Secretary of State, had passed to the Russian ambassador in Washington the gist of an official inter-governmental communication which, in its original form, read: 'The government of the United States, while endeavouring to estimate the developing world situation, has come into possession of information which it regards as authentic, clearly indicating that it is the intention of Germany to attack the Soviet Union.' The imprecision of this weighty warning was dissipated throughout the spring by messages from the Comintern agent Alexander Foote and the mysterious but quickly authenticated 'Lucy' network, both based in Switzerland. 'Lucy', whose identity remained obscure – perhaps a member of the exiled Czech secret service, the most effective run by any government during the war; perhaps a cell of the Swiss intelligence agency; perhaps an outpost of Bletchley – signalled Moscow in mid-June with a list of German objectives, a Wehrmacht order of battle for Barbarossa and even the current date for D-Day, 22 June. At about the same time (13 June), the British Foreign Minister told the Soviet ambassador that evidence of an imminent German attack was mounting and offered to send a military mission to Moscow.

Stalin's wishful thinking

Stalin by then had a plethora of evidence that German (with Romanian and Finnish) forces stood ready in millions to attack Russia's western frontier. Yet in the face of it all he clung to his belief and hope that every unwelcome interpretation of the facts was the fruit of Western ill-will. Cripps appears to have been so baffled by Stalin's wishful thinking that he consistently represented it to London as proof of Russia's intention to yield to a German ultimatum. In that judgement he was, of course, half right. Since the failure of Russia's diplomatic offensive against Bulgaria, Yugoslavia and Turkey four months earlier, Stalin was in chastened mood. Badly frightened also by the Hess mission, which he insisted on seeing as an attempt by Hitler to make peace with Britain so that he could attack Russia, Stalin had reverted to his earlier policy, fixed in August 1939, of dealing with Germany by granting her concessions, and he was now concerned most of all to placate Hitler by a meticulous fulfilment of the economic terms of the Molotov-Ribbentrop Pact. Train-loads of oil, grain and metals, as by strict quota, continued to pour across the frontier throughout June; what was to prove the final delivery crossed in the early hours of 22 June itself.

In this climate of appeasement, the Red Army's commanders, denied access to reliable intelligence ('reliable' intelligence, in Stalin's topsy-turvy world, was automatically deemed 'unreliable', as Professor John Erickson has demonstrated), and fearful of offending their timorous warlord, were prevented from taking any precautionary measures. M. P. Kirponos, commander of the Kiev military district, who was to show himself the most independent-minded of Stalin's generals in the weeks before Barbarossa (and who died in the great encirclement battle at Kiev in September which was the worst outcome of Stalin's blindness), deployed some of his units to the frontier in early June, was reported by the local NKVD to Beria, Stalin's secret policeman, for 'provocation' and required to countermand the order. In mid-June, when he tried to man his defensive positions again, he was flatly told, 'There will be no war.' This was not merely a statement for private consumption. On 14 June, eight days before the launching of Barbarossa, the Soviet national newspapers printed a government statement to the effect that 'rumours of the intention of Germany to break the [Molotov-Ribbentrop] Pact are completely without foundation, while the recent movements of German troops which have completed their operations in the Balkans to the eastern and northern parts of Germany are connected, it must be supposed, with other motives which have nothing to do with Soviet-German relations.' On 14 June in 'eastern and northern parts of Germany', which meant those parts of Poland and Czechoslovakia captured or annexed between 1938 and 1939, nearly 4 million German troops, organised in 180 divisions, with 3350 tanks and 7200 guns supported by 2000 aircraft, stood ready to march to war. They were to be accompanied by fourteen Romanian divisions and shortly to be joined by the Finnish, Hungarians and puppet Slovak armies, together with a volunteer Spanish (the 'Blue') and several Italian divisions. To those Russian commanders in the front line who sensed the massing of this mighty host and asked for orders, advice, even reassurance from above, the answer, as Kilch, the chief of artillery at Minsk (a primary Wehrmacht objective), later bitterly recalled, was 'always the same – "Don't panic. Take it easy. 'The boss' knows all about it." '

In reality, Stalin was as surprised as any subordinate by the unleasing of Barbarossa, and persisted in refusing to face facts even as the German attack units moved to their start-lines. When Timoshenko, Defence Commissar, and Zhukov, chief of staff, arrived at the Kremlin on the evening of Saturday, 21 June, with news that the Germans had cut the telephone lines into Russia and that a German deserter had brought news that the offensive would begin at four o'clock next morning, Stalin replied that it was too early to issue a warning order. He mused, 'Perhaps the question can still be settled peacefully. . . . The troops of the border districts must not be incited by any provocation, in order to avoid complications.' Implacably Zhukov confronted him with a draft directive for preparatory measures and, after insisting on some minor amendments, Stalin signed.

However, the directive did not order mobilisation nor fully alert the border troops to the danger in which they stood. In any case, it reached them too late. Even as the Leningrad, Baltic, Western, Kiev and Odessa military districts began to man their defences, the German offensive was upon them. Mass air raids and a gigantic artillery bombardment fell upon airfields and fortified zones. Behind this wall of fire the German army in the east, the *Ostheer*, moved to the attack.

Its three masses – Army Groups North (Leeb), Centre (Bock) and South (Rundstedt) – were each aligned on one of the historic invasion routes which led into European Russia, towards Leningrad, Moscow and Kiev respectively. The first followed the coast of the Baltic, through territory Germanised by Teutonic knights and Hanseatic traders for 500 years; from it came many of the families which had officered the Prussian and German armies throughout their history. Manstein and Guderian, who were to win Hitler his greatest eastern victories, descended from landowners of those parts; Stauffenberg, who was to fail by the narrowest of margins to kill Hitler on 20 July 1944, was married to a woman born at Kovno on the river Neman. The second route was that followed by Napoleon in 1812, running through the ancient, formerly Polish cities of Minsk and Smolensk. The third, demarcated by the crest of the Carpathian mountains in the south and separated from the northern and central routes by the huge freshwater swamp of the Pripet, 'the Wehrmacht hole', as the Germany army would call it, since no military operations were possible within the 40,000 square miles it covered, led into the black-earth country of the Ukraine, Russia's wheatbowl and the gateway to the great industrial, mining and oil-bearing regions of the Donetz, the Volga and the Caucasus.

The Pripet Marshes apart, no natural barrier stood between the Germans and their objectives on any of these routes. They are crossed, it is true, by several of Russia's enormous rivers, notably the Dvina and the Dnieper, but rivers, as the French had discovered in their much more defensible country the previous year, offer little obstacle to aggressively led armies, less still if the armies are mechanised and supported by airpower. In the vast spaces of the steppe, the Russian rivers were mere interruptions in country ideally suited for armoured advance. The sparsity of the road and rail network, and the spring and autumn floods which liquefy Russia's dirt roads, were better protection. However, the Germans had deliberately chosen high, dry summer for their onslaught, while the Russians, by crowding the bulk of their standing army into the narrow frontier zone behind the thin and incomplete belt of fortifications called the Stalin Line, liberated the Wehrmacht from dependence on a road network to make rapid ground into their rear. The shallowest of penetrations would suffice to put the Russian 'fronts' (as their army groups were designated) between the Panzers' jaws; thereafter they could be devoured almost at leisure by the columns of marching infantry following in the tanks' wake.

Strongest in armour and committed to the largest encirclement mission was Army Group Centre, whose spearheads were Panzer Groups 3 and 2, commanded by Hoth and Guderian. Its orders were to encircle as much of the Russian army in White Russia as possible, hack or hug it to death, and then press forward to secure the 'land bridge' between the headwaters of the Dvina and Dnieper rivers by which the Minsk-Smolensk road passes to Moscow. Its attack was prepared by the aircraft of Second Air Fleet, which on the morning of 22 June destroyed 528 Russian aircraft on the ground and 210 in the air; by the end of the day, across the whole front of the attack, the Red Air Force had lost 1200 machines, a quarter of its front-line strength.

Hoth's and Guderian's Panzers were simultaneously pressing through the Stalin Line. Brest-Litovsk, the frontier fortress city in which the Germans had dictated peace twenty-three years earlier, was isolated on the first day; V. S. Popov, commanding the XXVIII Rifle Corps, described its fortress as 'literally covered all over with uninterrupted artillery and mortar fire'. For a week it was to be defended heroically by the survivors of its garrison; but their sacrifice was irrelevant. By the time it fell it had been by-passed and the German spearheads were ranging far to its east.

Bock, commanding Army Group Centre, was nevertheless misled by the tenacity of the Brest-Litovsk defenders to believe that they were covering the withdrawal of neighbouring Russian defenders towards the Dnieper-Dvina 'land bridge'. On 24 June he accordingly put it to OKH that his Panzer groups should abandon their mission to close their first set of pincers around Minsk, 200 miles from their start-line, and proceed immediately towards Smolensk. Halder, not yet accustomed to the headless-chicken behaviour of Soviet troops at this – for them – almost leaderless stage of the war, and fearing that Hoth's Panzer Group 3 might press too deep and get cut off, refused. Hoth therefore turned inwards on 24 June. As he did so, Guderian's Panzer Group 2 began to feel the pressure of Russian troops deflected southwards by Hoth against his flank, apparently seeking to escape into the Pripet Marshes where, by Halder's estimation, they might form a 'stay behind' army and menace the German follow-up forces as they advanced to consolidate the ground the Panzers had won. Accordingly he ordered Fourth and Ninth Armies to destroy these fugitives trapped between Hoth's and Guderian's extending pincers as fast as their infantry formations could be brought forward.

By 25 June, therefore, Army Group Centre was fighting no less than three encirclement battles: one, the smallest in scale, around Brest-Litovsk; one in the salient of Bialystok, the most senseless of the frontier meanders in which Stalin had marooned the Red Army; and one at Volkovysk. Twelve divisions were surrounded at Bialystok and Volkovysk; by 29 June, when Army Group Centre's infantry had released the Panzer groups for a further advance, a fourth encirclement battle – threatening the destruction of another fifteen divisions – was in progress around Minsk.

These battles, moreover, were being fought with a brutality and ruthlessness not yet displayed in the Second World War, perhaps not seen in Europe since the struggle between Christians and Muslims in the Ottoman wars of the sixteenth century. Not only did many encircled Russians, unlike all but the most intransigent Frenchmen, fight with the tenacity of despair; they were attacked by the Germans with a pitiless ferocity that no Norwegian, Belgian, Greek or even Yugoslav soldier had yet had to face. Hitler had set the tone of the campaign. In an address to his generals on 30 March 1941 he had warned:

The war against Russia will be such that it cannot be conducted in a knightly fashion; the struggle is one of ideologies and racial differences and will have to be conducted with unprecedented, unmerciful and unrelenting harshness. All officers will have to rid themselves of obsolete ideologies. I know that the necessity for such means of making war is beyond the comprehension of you generals but . . . I insist that my orders be executed without contradiction. The commissars are the bearers of ideologies directly opposed to National Socialism. Therefore the commissars will be liquidated. German soldiers guilty of breaking international law . . . will be excused. Russia has not participated in the Hague Convention and therefore has no rights under it.

The Soviet Union had indicated on 20 August 1940 its desire to accede to the Hague Convention – which since 1907 had regulated the treatment of prisoners and non-combatants in war – but the approach was tentative: after 22 June 1941, therefore, its soldiers were protected by none of the Hague or Geneva provisions which spared those of signatory powers from mistreatment. As a result, it was not only commissars who were subjected to 'special treatment'; as Professor Omar Bartov has shown, the anti-Bolshevik indoctrination of the Wehrmacht's members, many of whom in 1941 had grown up under Nazism, resulted in the arbitrary massacre of prisoners from the start of the campaign. The commander of XLVIII Panzer Corps, for example, was obliged to protest to his soldiers only three days after the campaign had begun that 'senseless shootings of both prisoners and civilians have taken place. A Russian soldier who has been taken prisoner while wearing a uniform, and after he put up a brave fight, has the right to decent treatment.' Five days later he was forced to circulate the corps again: 'Still more shootings of prisoners and deserters have been observed, conducted in an irresponsible, senseless and criminal manner. This is murder.' But his strictures were fruitless. So common did the mistreatment of Russian prisoners become at the very outset of Barbarossa that by early 1942 another German formation, the 12th Infantry Division, was warning its soldiers that Red Army men were 'more afraid of falling prisoner than of a possible death on the battlefield. . . . Since November last year . . . only a few deserters have come over to us and during battles fierce resistance was put up and only a few prisoners taken.' This was not surprising; word of how the enemy treats prisoners circulates with lightning rapidity inside any army. It is news equalled in importance only by that of the survival rate for wounded in the army's own hospitals – but with this difference: poor prognosis for the wounded discourages soldiers from fighting hard, while bad treatment of captives has the opposite effect. During the course of the Second World War, the Wehrmacht took 5,700,000 Russians prisoner; of these 3,300,000 died in captivity, the majority in the first year of the campaign, victims above all of the lack of provision the Wehrmacht had made for feeding, housing and transporting such myriads. The result, succinctly summarised in a document circulated inside the *Grossdeutschland* Division in April 1943, was 'a stiffening of the enemy's resistance because every Red Army soldier fears German captivity'.

Systematic maltreatment was, however, a secret to which Germans alone were privy in June and July 1941. While their Russian opponents fought doggedly, they made little effort to beat a fighting retreat out of encirclement, in part because their commanders feared the consequences of ordering any withdrawal – their conditioned fear of Stalin would shortly be validated by the institution of summary executions for dereliction – in part because they lacked the means of escape. The German infantry divisions were themselves having great difficulty in catching up with the Panzer spearheads once they launched themselves into the blue; at this stage Barbarossa was following a pattern whereby armoured divisions lunged forward at fifty miles a day, pausing only to deal with resistance or take in supplies, while the plodding infantry laboured behind across the steppe at twenty miles a day or less. Between 22 June and 28 July, for example, the 12th Infantry Division marched 560 miles, an average of fifteen miles a day, all under broiling sun and the weight of 50 lb of equipment, ammunition and rations per man. This marathon greatly exceeded in distance the march on Paris made by von Kluck's infantry in August 1914; it seems probable that the exhausted *Landser* were sustained in their agonising trek, which bloodied feet and wore shoulders raw, only by the knowledge that the Panzers were winning the battle ahead of them. The encircled soldiers of the Red Army had no such spur. Commanded by generals paralysed by fear of Stalin's disfavour and the NKVD's execution squads, denied any prospect of living to fight another day, they commonly hunkered down in the pockets the Panzers drew around them and awaited the end which would follow when their last rounds were expended.

By 9 July those in the Minsk pocket had capitulated to Army Group Centre, but its two armoured groups, now reorganised as the Fourth Panzer Army under the dynamic (and strongly Nazi) General Günther von Kluge, were already pressing far beyond Minsk to complete a fourth encirclement at Smolensk. This pocket, in which the Dnieper-Dvina 'land bridge' lay, contained by 17 July some twenty-five Russian divisions, centred on Vitebsk, Mogilev and Smolensk itself, the largest concentration of Russian numbers the Germans had yet corralled. Since Army Group Centre's infantry formations on the Minsk-Smolensk axis lay anything up to 200 miles behind its Panzer spearheads at this date, Bock, who was determined to 'clean up' his front in the shortest possible space of time, was now obliged to commit his valuable Panzer and motorised (soon to be renamed 'Panzergrenadier') divisions to close combat. A cordon of tanks, half-tracks and dismounted truck-borne infantry, diverted from the Panzer drive down the Moscow road, was therefore strung round Smolensk between 17 and 25 July and drawn ever tighter around the trapped Russians until, on 5 August, all resistance came to an end.

By then, however, Bock had grasped that his difficulty in closing the ring was due not simply to the resistance of the Russians within but also to a determined effort from without to reinforce and resupply the trapped divisions. The Dnieper-Dvina gap, while it had remained open, had been used as a 'land bridge' in reverse, to carry troops and ammunition westward as fast as it could be sent by the high command. On 10 July the high command had been reorganised, as Stalin, recovering from the initial paralysis imposed by *Blitzkrieg*, recognised how inappropriate to war was the existing machinery. He had recently assumed the formal title of head of government; on 10 July he created the post of Supreme Commander, to which he had the Supreme Soviet appoint him on 7 August. The State Defence Committee (GKO), consisting of Stalin, Voroshilov, Beria, Molotov (Commissar for Foreign Affairs) and Malenkov (Stalin's deputy in the party), had been set up on 30 June; directly subordinate to it was the Stavka (Operations Staff), which when reorganised on 10 July included Stalin, Molotov and Voroshilov from the party and Timoshenko, Budenny, Shaposhnikov and Zhukov from the army. The General Staff, extended to oversee all branches of the armed forces, was subordinated to the Stavka on 8 August. By then Stalin occupied all the highest appointments in the Soviet state – Chairman of the GKO, Defence Commissar and Supreme Commander – and directly controlled all the rest. This self-elevation entailed risk. The odium of defeat now attached immediately to his person. However, so desperate was Russia's situation, after less than two months of war, that Stalin must have accepted he could not survive the consequences of further disaster. Victory alone could save him.

The carrot and the stick

There was almost no shift or expedient at which Stalin would not grasp in this supreme crisis of the Soviet state to assure its survival – as well as his own. In September he decreed the creation of new units of 'Guards', quintessential symbols of the *ancien régime*; in 1917 Guards officers had had the skin stripped from their hands in revolutionary hatred of the white gloves they had traditionally worn. Now Stalin decreed that regiments, divisions, even armies which resisted the Germans most stoutly should add 'Guards' to their titles. New distinctions were meanwhile created for heroes and victors, named after the generals who had fought Napoleon: the Orders of Kutuzov and Suvorov. Old distinctions of rank were soon to be revived, including the 'shoulder boards' which had been torn from officers' uniforms in 1917. Even the hierarchy of the Orthodox Church, persecuted and vilified for two decades, was suddenly restored to esteem as the servant of 'Mother Russia', a matriarch resurrected by the autocrat who had violated her children without pity in the era of collectivisation and the purges.

But with the carrot went the stick. The 'dual authority' of the commissars was restored on 16 July; on 27 July an order sentencing nine senior officers to death was read out to all officers and men. The condemned included the signals officer of the Western Front and the commanders of the Third and Fourth Armies and of the 30th and 60th Rifle Divisions. Others were shot in secret, or simply committed suicide rather than face the executioners of the NKVD; its 'Special Sections' (how terrible a meaning did 'special' acquire in the Second World War – 'special leader', 'special command', 'special treatment', all spelt death to the defenceless and disfavoured) were deployed in the rear of the fighting units to shoot deserters and menace with machine-guns those who even thought of quitting their posts.

Yet the difficulty of sustaining resistance grew greater with every day of combat. On 10 July three fronts had been set up – North-Western, nominally commanded by Voroshilov, Western, under Timoshenko, and South-Western, under Budenny – to correspond with the three German army groups attacking them. This was a rational means of bringing under command the reinforcements and supplies which the GKO was mobilising for the defence. In July 1941, however, new men and equipment were scarcely to be found; while existing units and weapons were being consumed like chaff in the furnace of battle. By 8 July, OKH reckoned it had destroyed 89 out of 164 Russian divisions identified; as a running check on that estimate, Army Group Centre was able to show that it had captured 300,000 prisoners, 2500 tanks and 1400 guns (the majority with their crews dead about them, so tenaciously did the Russian gunners fight). Stalin himself counted 180 divisions committed to battle, out of 240 mobilised; he hoped eventually, if Hitler would allow him the time, to raise 350. At present, however, replacements were being used up as soon as found: during the Smolensk encirclement battle (4–19 July), Army Group Centre's fourth, another 310,000 prisoners were taken, along with 3200 tanks and 3100 guns. Russian industry, suddenly thrown into high gear, was producing 1000 tanks a month (and 1800 aircraft) but losses exceeded these figures.

As Army Group Centre completed its destruction of the Soviet Sixteenth, Nineteenth and Twentieth Armies in the Smolensk pocket, Army Group North was accelerating its rate of advance along the Baltic coast towards Leningrad. Lakes, forests and rivers had impeded Leeb's spearheads at the outset. Although he had only three Panzer divisions at his disposal and he achieved no encirclement as spectacular as Bock's, by 30 June Army Group North had occupied Lithuania and secured bridgeheads across the lower reaches of the Dvina where the Stalin Line was supposed to run. Racing through it, Panzer Group 4 arrived at Ostrov, on Latvia's pre-1940 frontier with Russia, and ten days later stood on the Luga, only sixty miles from Leningrad and the last major water obstacle outside the city.

Army Group South had initially made slower progress than Centre and North. Commanded by Rundstedt, who had directed the great breakthrough across the Meuse thirteen months earlier, it consisted of two disparate blocs, a northern *masse de manoeuvre* of German infantry led by the five armoured divisions of Panzer Group 1, and, to the south, the allied contingent formed of Romanian and Hungarian divisions, equipped with inferior French weapons supplied during the years of the Little Entente. The satellite divisions' mission was to cross the rivers Dniester and Bug and capture Odessa and the Black Sea ports, while the German infantry and tanks marched deep into the steppe towards Kiev, capital of the Ukraine and founding city of Russian civilisation. Rundstedt's vanguards passed easily through the Soviet frontier defences, swamping the fortifications of Przemysl, which had sustained siege for 194 days in 1914-15. It then ran up against a major concentration of Soviet force belonging to the South-Western Front, under the direction of one of Stalin's best generals, Kirponos, whose political commissar was Nikita Khrushchev, the future First Secretary of the Soviet Communist Party, with the outstanding General K. K. Rokossovsky as one of his tank commanders. The South-Western Front was particularly strong in armoured formations – it contained six mechanised corps – and had a high proportion of the T-34s in service. Kirponos determined to deal with Rundstedt's *Blitzkrieg* in absolutely correct fashion, by pinching the spearheads of Kleist's Panzer Group 1 between concentric attacks mounted by the Fifth and Sixth Armies; Fifth, operating out of the impenetrable marshes of the Pripet, had a firm base for its thrust; Sixth, whose positions were in the open steppe, did not. Although both armies pressed their attacks, their pincers never met and Kleist pushed between to capture Lvov (as Lemberg the Austrian capital of Galicia until 1918, then a Polish city until 1939) on 30 June. The commander of the garrison was General A. A. Vlasov, who managed to fight his way out on this occasion; a year later he would fall into German hands near Leningrad and defect, to set up the 'Vlasov Army' of anti-Stalinists. His loyalty to the regime may have been shaken during the evacuation of Lvov, when the local NKVD massacred its Ukrainian political prisoners rather than let them be liberated by the Germans.

Kirponos persisted in his efforts to mount 'pinching' operations against Kleist's Panzers on 29 June and 9 July; but the power of the Panzers and the flail of the Luftwaffe kept Rundstedt's spearhead moving forward, increasingly constricted within a narrow axis of advance, to be known as the 'Zhitomir Corridor', but reaching inexorably towards Kiev. On 11 July Kirponos held a command conference at Brovary, only ten miles east of the city. It was there decided that the Fifth and Sixth Armies – shadows of their former selves, despite constant reinforcement and re-equipment – should continue to hack at the approaching Germans. He was counting on the arrival of two new corps, LXIV and XXVII, to lend weight to their efforts, though, according to Professor John Erickson, what he had heard disturbed him: 'short of weapons, horse-drawn guns, disorganised staffs, no wireless sets; [in XXVII Corps] only one division had a commander.' As the military soviet of the South-Western Front dispersed from the conference, in gloom approaching despair, the headquarters came under heavy German air attack. Kirponos had already glimpsed the danger to which his failure to pinch off Kleist's penetration of his front exposed Kiev, and indeed his whole command: Panzer Group 1's advance now constituted one arm of a counter-pincers; should the Germans bring down tanks from the north, from Bock's Army Group Centre, a second pincer arm would be created and he, his men and the whole of the Ukraine would be enveloped within it.

The question of Moscow

The same thought was simultaneously exercising Hitler. He and the army high command had differed in their view of how the Russian campaign should be fought from the moment of initial planning a year earlier. Their differences had been significantly reconciled in the Barbarossa directive of December 1940. But OKH, and particularly Halder, still believed that the Russians' fighting power could best be overcome by driving headlong at Moscow, while Hitler was above all anxious to seize as much Russian territory as possible in one gulp, devouring the Russians defending it in giant encirclements on the way. His confidence as a commander, however, was rapidly increasing. He had left the conduct of the Polish campaign to his generals, and had largely been talked into the Scandinavian invasions by Admiral Raeder. Before and during the campaign in the west he had given his generals orders but had also suffered from severe attacks of indecision and second thoughts, notably outside Dunkirk. Since the inception of Barbarossa, however, he had found an increasing certitude. It was in the fullest sense his war, it had started triumphantly, and as its course developed he grew overbearing in its direction. 'The Führer's interference is becoming a regular nuisance,' wrote Halder on 14 July; a little later, he enlarged on this theme:

> He's playing warlord again and bothering us with such absurd ideas that he's risking everything our wonderful operations so far have won. Unlike the French the Russians won't just run away when they've been tactically defeated; they have to be defeated in a terrain that's half forest and marsh. . . . Every other day now I have to go over to him [Hitler's headquarters and those of OKH, though close to each other at Rastenburg in East Prussia, were separate entities]. Hours of gibberish, and the outcome is there's only one man who understands how to wage wars . . . if I didn't have faith . . . I'd go under like Brauchitsch [the army C-in-C] who's at the end of his tether and hides behind an iron mask of manliness so as not to betray his complete helplessness.

Hitler's differences with Halder and OKH emerged into the open on 19 July when he issued a new Führer Directive, No. 33, outlining his conception of the next stage of operations. It laid down that Army Group Centre's two Panzer groups, 3 (Hoth) and 2 (Guderian), were to be diverted from the drive on Moscow to co-operate respectively with Leeb and Rundstedt in their advances on Leningrad and Kiev. A supplement, issued on 23 July, rammed the point home. The drive on Moscow was postponed until mopping-up operations around Smolensk had been completed. In amplification of this order, Brauchitsch issued orders to Army Group Centre which Guderian was called to hear at a conference at Novi Borisov on 26 July. There he was directed to take his tanks off the Moscow road and lead them southwards to destroy the Soviet Fifth Army on the fringe of the Pripet Marshes.

Guderian was outraged. His divisions had been reduced by heavy fighting and long traverses of roadless country to 50 per cent of their tank strength. On the other hand, his leading elements, which had already advanced 440 miles in six weeks, stood only 220 miles from Moscow and, in the period of dry weather that could be guaranteed before the coming of the autumn rains, might certainly be led to reach the capital. As he had been promoted to the status of army commander at Novi Borisov, he was also now independent of Kluge (for whom he nursed a reciprocated antipathy) and so answerable directly to Bock, whose views coincided with his own. With Bock's acquiescence, in which OKH tacitly joined, he therefore embarked on a delaying operation to frustrate Hitler's reordering of the Barbarossa strategy. It took the form of involving his Panzer group (renamed Panzer Army Guderian) in a battle for the town of Roslavl, seventy miles south-east of Smolensk, where the roads to Moscow, Kiev and Leningrad met. His purpose was to entangle his forces so deeply with the Russian defenders that the justification for their diversion to assist Rundstedt would be overtaken by events and so allow him to proceed towards Moscow as originally ordered.

Guderian's disguised insubordination almost worked. His argument for heightening the pressure at Roslavl was validated by the appearance of Russian reserves in that sector, sent to Timoshenko by Stalin from the training units and hastily embodied militias which were now his only source of fresh troops. Moreover, Hitler had had second thoughts. In Führer Directive No. 34, issued on 30 July, he postponed the diversion of Army Group Centre's Panzer groups to assist their tank-poor neighbours and arranged to visit Army Group Centre on 4 August to assess its situation for himself (a dangerous excursion, did he but know it, for its headquarters was the focus of the 'military resistance' which would strike against him in July 1944). Hoth, commanding Panzer Group 3, accepted the Führer's arguments for going to the assistance of Leeb on the Leningrad axis. Bock and Guderian resisted his arguments for joining Rundstedt. There followed what has been called a 'nineteen-day interregnum' during which Guderian edged southwards but attempted to retain the bulk of his striking force on the Moscow road.

The 'nineteen-day interregnum' (4–24 August), which may well have spared Stalin defeat in 1941, was characterised not only by slow German progress on all fronts but also by a succession of changes of mind. On 7 August, OKW and OKH conferred, and Jodl and Halder were able to persuade Hitler of the need to resume the advance on Moscow, which resulted in Führer Directive No. 34A. Three days later he took fright at renewed resistance on the Leningrad front and insisted that Hoth's tanks depart immediately to Leeb's assistance. The Führer, Jodl told Colonel Adolf Heusinger, the OKW operations officer, 'has an intuitive aversion from treading the same path as Napoleon; Moscow gives him a sinister feeling.' When the whole chain of command – Brauchitsch, Halder and Heusinger at OKW, Bock at Army Group Centre, Guderian as Bock's principal field commander – demonstrated that it was continuing to prevaricate, Hitler, who had recovered his sense of how the campaign was unfolding, lost patience. He repeated his orders that Army Groups North and South should proceed to their objectives and dictated a letter to Brauchitsch accusing him of a lack of 'the necessary grip'. Brauchitsch suffered a mild heart attack. Halder, who had urged him to resign when the letter arrived, did so himself 'to stave off an act of folly'. It was refused; Hitler, now as later, treated offers of resignation as acts of insubordination. Halder nevertheless felt that 'history will level at us the gravest accusation that can be made of a high command, namely that for fear of undue risk we did not exploit the attacking impetus of our troops.' Bock, in his diary, echoed his frustration: 'I don't want to "capture Moscow". I want to destroy the enemy's army and the bulk of that army is *in front* of me.' Both left it nevertheless to their subordinate Guderian to confront the Führer with the boldest statement of their anxiety. Overcome by Guderian's exposition of what he believed to be the strategically correct path, when Halder visited Bock's headquarters on 23 August Bock telephoned Schmundt, Hitler's Wehrmacht adjutant, with a request for Guderian 'to be granted audience', while Halder agreed to take him back to OKW in his liaison aircraft.

Arriving in time to make the onward journey to the Rastenburg evening conference (Hitler had recently instituted a timetable for meeting his staff officers at noon and midnight), Guderian was greeted by Brauchitsch with the news: 'I forbid you to mention the question of Moscow to the Führer. The operation to the south [the Kiev attack] has been ordered. The problem now is simply how it is to be carried out. Discussion is pointless.' Guderian grudgingly obeyed, but in the course of the confrontation dropped so many hints about the 'major objective' on Army Group Centre's front that Hitler eventually raised it himself. Given his chance, Guderian launched into an impassioned plea for sustaining the drive on Moscow. He was heard out; Hitler had a special regard for the Panzer pioneer, which had recently been reinforced by his acceptance of Guderian's warnings of Russia's unanticipated tank strength. However, when the general had spoken, the Führer went on to the offensive. His commanders, he said, 'know nothing about the economic aspects of war'; he explained the necessity of seizing Russia's southern economic zone from Kiev to Kharkov, and emphasised the importance of capturing the Crimea, from which the Soviet air force menaced Romania's Ploesti region, still the main source of Germany's natural oil supply. Since the other officers present made it clear that they supported the leader, and Brauchitsch and Halder had pointedly not accompanied him, Guderian felt obliged to desist from opposition. The only concession he extracted was that his Panzer group should be committed to support Rundstedt in its entirety and allowed to return to the Moscow axis as soon as the battle for Kiev was won. Halder and Brauchitsch were loud in recriminations to his face when he returned to OKH from OKW, and Halder vilified him to Bock on the telephone during his homeward flight to Novi Borisov. But the die was now cast. After nearly three weeks of inertia, the *Ostheer* was to resume the attack with a full-blooded offensive into the black-earth region of the south. Whether it could then complete its thrust towards Moscow would depend on the seasons. The descent of the cold weather was only two and a half months distant and then Generals January and February would be fighting on Stalin's side.

Stalin, however, was already planning a counter-offensive. On 16 August he had created the Bryansk Front, under A. I. Yeremenko, to close the gap which had appeared between the Central and South-Western Commands (temporary headquarters superior to fronts). To this new front he consigned as much of the new Soviet equipment as could be spared, several T-34 tank battalions and some batteries of Katyusha rockets ('Stalin organs' the Germans called them) which fired eight fin-stabilised projectiles with very large warheads. With these weapons and two new armies, the Thirteenth and Twenty-First, Yeremenko attempted to counter-attack into the gap which yawned between Rundstedt's armoured spearhead, supplied by Kleist's Panzer Group 1, and Guderian's Panzer Army, approaching from the north. He was simply putting his head into a trap. Kleist had already pulled off a successful encirclement of 100,000 Russians at Uman on 8 August. The converging Panzer groups now stretched out their pincers to enclose the much larger Russian concentration around Kiev. Guderian, who offered an exposed flank eventually 150 miles long as he beat his way southward from the Moscow axis, was vulnerable to a Russian slicing stroke; but his 3rd and 17th Panzer Divisions, led by thrusting young generals, Walter Model, a future army group commander, and Ritter von Thoma, who was to make his name in the desert against the British, brooked no opposition. They drove forward and on 16 September joined hands with Kleist's tank force at Lokhvitsa, a hundred miles east of Kiev. It would take another ten days, during which the Second and Fourth Air Fleets saturated the pocket with bombs, to close all the gaps in its walls through which escaping handfuls of Russians managed to filter. However, on 26 September it had been securely enclosed and 665,000 Russian soldiers were prisoners within it – the largest single mass ever taken in an operation of war before or since. Five Russian armies and fifty divisions had been destroyed, uncounted thousands killed; they included Kirponos, mortally wounded in an ambush close to his final command post at Lokhvitsa on 20 September.

The aftermath of the Kiev encirclement yielded the worst of the spectacles which horrified even the hardest-hearted among the German conquerors, as the captives were marched back across the steppe to the wholly inadequate prisoner cages in the rear. 'We suddenly saw a broad, earth-brown crocodile slowly shuffling down the road towards us,' recorded an eyewitness. 'From it came a subdued hum, like that from a beehive. Prisoners of war, Russians, six deep. . . . We made haste out of the way of the foul cloud which surrounded them, then what we saw transfixed us where we stood and we forgot our nausea. Were these really human beings, these grey-brown figures, these shadows lurching towards us, stumbling and staggering, moving shapes at their last gasp, creatures which only some last flicker of will to live enabled to obey the order to march? All the misery of the world seemed to be concentrated there.' Nearly 3 million Russians had now been taken prisoner and of these half a million would die, of lack of shelter or food, in the first three months of the approaching winter.

'General Winter'

The sense of the approaching winter had already started to touch the whole of the *Ostheer* in late September, with its threat first of liquefied roads and swollen rivers, then of blizzards and snowdrifts which its men and equipment were equally unprepared to meet. Guderian was hastening his Panzer army back to the central front, burning with anxiety to open the final drive on Moscow before the weather broke. To the south, the Romanians were laying siege to Odessa, which was defended by a hastily constituted Special Maritime Army of 100,000 men and would not fall until 16 October, and the Eleventh Army, commanded by Erich von Manstein, was pushing on across the estuary of the Dnieper to reach the neck of the Crimea on 29 September. That thrust largely settled Hitler's fear that the Crimea might be turned into an unsinkable aircraft carrier for the bombardment of the Romanian oilfields. Manstein's advance also brought the coastal industrial region of the Donetz and the Don under threat. Nevertheless, the unloosening of the Red Army's grip on Russia's southern provinces and the reassembly of Bock's striking force on the central front for a renewed drive on Moscow did not constitute a comprehensive solution to the development of the Barbarossa strategy. The unlocking of the northern front and the investment and eventual capture of Leningrad were also a necessary stage in the conquest.

Army Group North's concerted effort to take Leningrad had begun on 8 August with a determined assault on the line of the river Luga, the outermost line of the city's defences, which was to be co-ordinated with a Finnish-German offensive across the isthmus of Karelia – annexed by Stalin after the defeat of Finland in 1940 – and extending far northward towards the Arctic Circle. Leeb's offensive was complicated by three factors. The first was that Leningrad was protected from the rear by Lake Ladoga, an enormous body of water interposing between the city and any encirclement from the north. The second was that the Leningrad command had mobilised the city's population to construct concentric defence lines around the city, including 620 miles of earthworks, 400 miles of anti-tank ditch, 370 miles of barbed-wire entanglement and 5000 pillboxes – an extraordinary labouring effort to which 300,000 members of the Young Communist League and 200,000 civilian inhabitants, including women in equal numbers to men, were committed. The third factor was that Marshal Carl Gustav Mannerheim, the Finnish leader, was determined, even at this low point in Soviet fortunes, to give no hostages by capturing more territory than that to which he had title. While Leeb laboured forward along the Baltic coast, therefore, Mannerheim's Finnish units hung fire above Lake Ladoga after 5 September, when the tanks of Hoth's Panzer group, detached from Army Group Centre following the Hitler-Guderian conclave of 23 August, were returned to Army Group Centre. Hoepner's Panzer Group 4 was left by itself to breach Leningrad's fortifications and take the city.

A fourth impediment to Leeb's Leningrad *Blitzkrieg* emerged in mid-September. Zhukov, who had advised Stalin to abandon Kiev before it was encircled and for his pains had been dismissed as chief of staff, arrived at the North-Western Front on 13 September to energise the defences. He found the Germans on the outskirts of the old tsarist capital; Tsarskoe Selo (now called Pushkin), the Russian Versailles, had fallen on 10 September (its enchanting follies and pavilions, like those of the Peking Summer Palace designed by imported Western architects, were to perish in a conflagration caused by the invaders). Shortly afterwards Leeb's vanguards reached the Gulf of Finland at Strelna. Leningrad, isolated from the rest of Russia by the Finnish advance to the 1939 frontier and by Leeb's occupation of the Baltic littoral, now connected with the interior only by the water route across Lake Ladoga. The lifeline was tenuous and erratic; Leningraders quickly felt the constriction and would shortly begin to experience the pangs of starvation which would kill a million of them before the lifting of the siege in the spring of 1944. In the immediate term, however, Zhukov's arrival had achieved a decisive effect. His first order was 'to smother the enemy with artillery and mortar fire and air support, permitting no penetration of the defences'. Under his resolute command, the energy of Hoepner's Panzer assaults was broken in the lines of trenches and concrete that the citizens of Leningrad had constructed. The situation has 'worsened considerably', Leeb reported to the Führer's headquarters on 24 September; Finnish pressure in Karelia had 'quite stopped'; the city, with its 3 million inhabitants, was intact. German bombardment was inflicting a toll of 4000 civilian casualties a day and starting 200 fires; but the great *enceinte* of canals and classical palaces remained impervious to the Panzer thrust. Only twenty tanks took part in the final assault. Hitler had already decided that the bulk of Hoepner's Panzer Group 4 must be diverted to the climacteric Operation *Taifun* (Typhoon), to take Moscow.

Führer Directive No. 35, which resolved the ambiguity of the Barbarossa strategy inherent in its direction since OKH and OKW had each presented their conception of the campaign's conduct a year earlier, was issued on 6 September. It laid down that, following the encirclement and destruction of the Red Army on the front of Army Group Centre, Bock was 'to begin the advance on Moscow with [his] right flank on the Oka and [his] left on the Upper Volga'. Panzer Groups 2 and 3 were to be reinforced by Hoepner's Panzer Group 4 brought from the Leningrad front to assure the largest possible breakthrough effort on the Moscow axis. The principal aim of the operation was the defeat and annihilation of the Russian forces blocking the road to Moscow 'in the limited time which remains available before the onset of the winter weather'.

The army which set off on the last stage of the road to Moscow in late September was greatly different from that which had crossed the frontier ten weeks earlier. Battle deaths, wounds and sickness had reduced its strength by half a million, casualties not to be compared to the ghastly loss suffered by the Red Army but clearly enough to depress morale at the front and cast a pall of misery and apprehension over family life in the Reich. The war diarist of the 98th Infantry Division, diverted northward from Kiev to the Moscow front, recorded the ordeal of its 400-mile march.

> The modern general-service carts with their rubber tyres and ball-bearing mounted wheels had long since broken up under the stress of the appalling tracks, and been replaced by Russian farm carts. Good-quality German horses [600,000 had begun the campaign] foundered daily through exhaustion and poor food but the scrubby Russian ponies, although in reality too light for the heavy draught work they were doing, lived on eating birch twigs and the thatched roofs of cottages. Equipment, including many tons of the divisional reserve of ammunition, had to be abandoned at the roadside for lack of transport to carry it. Gradually the most simple necessities of life disappeared, razor blades, soap, toothpaste, shoe-repairing materials, needles and thread. Even in September and before the advent of winter, there was incessant rain and a cold north-east wind, so that every night there was the scramble for shelter, squalid and bug-ridden though it usually was. When this could not be found the troops plumbed the very depths of wretchedness. The rain, cold and lack of rest increased sickness that, in normal circumstances, would have warranted admission to hospital; but the sick had to march with the column over distances of up to twenty-five miles a day, since there was no transport to carry them and they could not be left behind in the bandit-infested forest. The regulation boots, the *Kommisstiefel*, were falling to pieces [in the coming winter their iron-nailed soles would accelerate the onset of frostbite]. All ranks were filthy and bearded, with dirty, rotting and verminous underclothing; typhus was shortly to follow.

The realities of conquest can rarely have been much different. Alexander's hoplites entered Persepolis almost barefoot, Wellington's redcoats came to Paris in rags. Neither of those great victors' armies, however, stood at risk from the Arctic winter. Both, moreover, had already defeated the enemy's main force before they entered his capital. The *Ostheer* had a great battle ahead of it before it could be certain of finding shelter in Moscow. The opening stages promised well. In an encirclement rivalling that of Army Group South's at Kiev, Centre's Panzer groups, Hoth's and Hoepner's (detached from the Leningrad front), surrounded 650,000 Russians between Smolensk and Vyazma. Many gave up without a fight; they were the hastily embodied militiamen of the *Osoviakhim*, the pre-war citizen defence force on which Stalin drew for his reserves. Others fought more doggedly. Guderian, visiting the 4th Panzer Division, found 'descriptions . . . of the tactical handlings of the Russian tanks very worrying'. (It had recently encountered T-34s for the first time.)

> Our defensive weapons available at that period were only successful against the T-34 when the conditions were unusually favourable. The short-barrelled 75-mm gun of the Panzer IV was only effective if the T-34 was attacked from the rear; even then a hit had to be scored on the grating above the engine to knock it out. It required very great skill to manoeuvre into a position from which such a shot was possible. The Russians attacked us frontally with infantry, while they sent in their tanks in mass formations against our flanks. They were learning.

Even more ominously, he reported in his war diary on 6 October the first snowfall of the approaching winter. It melted quickly, leaving the roads, as before, liquid mud. It was difficult to know, at that stage, which was preferable – a prolonged autumn, with all the difficulties of movement that that rainy season brought, or an early winter, which made for firm, frozen going on the roads but threatened blizzards before the final objective was reached.

Stalin could repose no solid hopes in the turn of the seasons. Winter might save Moscow; it might not. There was the gravest doubt whether the remains of the Red Army in European Russia could do so. It had now been reduced to a strength of 800,000 men, divided into ninety divisions, with 770 tanks and 364 aircraft; nine of the divisions were of cavalry, only one – and thirteen independent brigades – of tanks. A large army was stationed in the Russian Far East, but it could not be moved while there was still the danger of war with the Japanese, whom the Russians had fought in Mongolia only two years previously. Hitler, by contrast, had now increased the strength of Army Group Centre itself to eighty divisions, including fourteen Panzer and eight motorised, supported by 1400 aircraft; the other two army groups, though depleted by tank transfers to the Moscow front, retained the bulk of their infantry and were sustaining their pressure against Leningrad and into the southern steppe.

In this supreme crisis Stalin turned again to Zhukov. Though their disagreement in the summer had led to Zhukov's removal as chief of staff, Stalin recognised his superlative talents which had recently if only temporarily saved Leningrad. Now, on 10 October, he was called south to energise the defence of Moscow. Rumours of panic were already touching the city; Red Air Force pilots who on 5 October reported German columns fifteen miles away driving towards it were threatened with arrest by the NKVD as 'panic-mongers'; on 15 October, however, fear took hold in earnest. The 'Moscow panic' began with a warning by Molotov to the British and American embassies to prepare for an evacuation to Kuibyshev, a city 500 miles eastward on the Volga. According to Professor John Erickson, 'The real crisis, however, spilled on to the streets and into plants and offices; a spontaneous popular flight added itself to the hurried and limited evacuation, accompanied by a breakdown in public and Party disciplines. There was a rush to the railway stations; officials used their cars to get east; offices and factories were disabled by desertions.' The panic was not merely unofficial. 'Railway troops were told to mine their tracks and junctions . . . sixteen bridges deep within the city were mined and crews at other mined objectives issued orders to blow their charges "at the first sight of the enemy".'

Zhukov, however, kept his nerve. As at Leningrad, he mobilised citizens, 250,000 Muscovites (75 per cent women), to dig anti-tank ditches outside the city. He brought proven commanders, including Rokossovsky and Vatutin, to the threatened front and he concentrated every reserve that Stalin could send him on the Moscow approaches. Stalin, too, found a public resolution he had not always shown in the closed meetings of the Politburo and the Stavka earlier in the campaign. At the traditional Red Square parade held to commemorate the October Revolution on 7 November, even though Bock's Panzers were only forty miles from the Kremlin, he denounced those who thought 'the Germans could not be beaten', declared that the Soviet state had been in greater danger in 1918 and invoked the name of every Russian hero – pre-, post- and even anti-Revolutionary – to stiffen the sinew of his audience. Inspired by those 'great figures', and fighting under 'great Lenin's victorious banner', with or without the opening of the 'Second Front' promised by the British, he forecast the Red Army's eventual triumph.

The first frosts of winter were now hardening the ground for the enemy, and the Panzer groups were making faster progress towards Moscow than they had done in October. Their tank strength, however, was reduced to 65 per cent, and Guderian, Hoth and Hoepner were all concerned about their ability to push their spearheads to the final objective. Accordingly, on 13 November Halder arrived from OKW at Army Group Centre's headquarters at Orsha to canvass opinion from the army group chiefs of staff – Sodenstern of South, Griffenberg of North, Brennecke of Centre – as to the further conduct of the campaign. Should the *Ostheer*, he asked, make a final dash or instead dig in for the winter to await fairer weather for a culminating victory next year? Sodenstern and

Griffenberg, respectively overstretched and blocked on their fronts, answered that they wished to halt. Brennecke replied that 'the danger we might not succeed must be taken into account, but it would be even worse to be left lying in the snow and the cold on open ground only thirty miles from the tempting objective' – Moscow. Since Hitler had already told Halder (who was himself already looking beyond Moscow) that this was the answer he wanted, the issue was decided on the spot.

'The flight to the front'

The final stage of Operation Typhoon began on 16 November. It was organised as a double envelopment of the Moscow defences, Panzer Groups 3 and 4 moving towards Kalinin north of the city, Panzer Group 2 towards Tula in the south. It was to become known in the *Ostheer* as the *Flucht nach vorn*, 'the flight to the front', a desperate attempt, like Napoleon's in 1812, to get to Moscow for shelter from the snows. But between the *Ostheer* and the city stood Zhukov's last line of defence, the Mozhaisk position, which included the man-made Sea of Moscow to the north of the city and the river Oka to the south.

Despite the arrival of some reinforcements, the Mozhaisk position at first did not hold. Guderian, blocked at Tula, merely swung his Panzer group around the town and chose a new axis for his advance on Moscow. To the north, the German Ninth Army broke through to the Sea of Moscow and the Volga Canal on 27 November, linking up with Panzer Group 3; the 7th Panzer Division, Rommel's old command, actually got across the canal on 28 November.

The German effort was now at crisis-point. At Krasnaya Polyana, Panzer Group 3 stood only eighteen miles from Moscow. The Fourth Army, with its outposts at Burtsevo, was only twenty-five miles from the city. Guderian's Panzer Group 2 was sixty miles away to the south. There is a legend that in the following days an advance German unit saw the golden domes of the Kremlin illuminated by a burst of evening sunshine and that a patrol even penetrated an outlying suburb. If it did, that was the last flicker of energy from an army expiring on its feet. The Russian winter in all its cruelty, unknown and unimaginable to a Westerner, was now beginning to bite; the season was approaching when temperatures would fall below minus 20 degrees Centigrade, inflicting on the *Ostheer* 100,000 frostbite casualties, of whom 2000 would suffer amputations. After 25 November, Guderian's Panzer Group 2 made no further advance, having failed to take Kashira on the main southern rail line; on 27 November he ordered a halt. After 29 November there was no movement by the northern pincer either, both Ninth Army and Panzer Group 3 having lost the ability to drive forward. Bock, writing to Halder at OKH on 1 December, explained Army Group Centre's predicament:

After further bloody struggles the offensive will bring a restricted gain of ground and it will destroy part of the enemy's forces but it is most unlikely to bring about strategical success. The idea that the enemy facing the army group was on the point of collapse was, as the fighting of the last fortnight shows, a pipe-dream. To remain outside the gates of Moscow, where the road and rail systems connect with almost the whole of eastern Russia, means heavy defensive fighting. . . . Further offensive action therefore seems to be senseless and aimless, especially as the time is coming very near when the physical strength of the troops will be completely exhausted.

By the first week of December the ordinary German soldiers of the fighting divisions were almost incapable of movement. Jodl had refused to allow the collection or supply of winter clothing, lest its appearance cast doubt on his assurances that Russia would collapse before the coming of the snows. The men in the firing line stuffed torn newspaper inside their uniforms to repel the cold. Such expedients worked scarcely at all. The Russians, by contrast, were accustomed to and equipped against the temperature; every Russian, military or civilian, possessed a pair of felt boots which experience proved best protected feet against frostbite (America was to supply 13 million pairs during the course of the war) and the Red Army accordingly continued to manoeuvre while the *Ostheer* froze fast.

In the meantime Army Group South had occupied the Crimea (except for Sevastopol) during November. Late in that month, Timoshenko's front met Rundstedt's Panzers head on at Rostov-on-Don (the 'gateway to the Caucasus' and so to Russia's oil), recaptured the city on 28 November after it had been in German hands for only a week, and then forced them back to the line of the river Mius, fifty miles behind Rostov, where they dug in for the winter. Army Group North was meanwhile halted outside Leningrad and after 6 December driven back from Tikhvin, its furthest point of advance along the southern shore of Lake Ladoga. There it established a winter line which subjected the city to slow starvation – a million were to die in the three-year siege, the majority in the first winter – but did not quite cut it off from supply across the lake, by ice-road in winter, later by boat.

The Red Army's great manoeuvre, however, began outside Moscow on 5 December. Reinforcements had been found from new waves of conscripts and from the output of her mobilised factories; a few tanks had even arrived by Arctic convoy from Britain, heralds of a source of supply which would become a flood – of trucks, food and fuel – as Western aid developed. Yet the most important source of reinforcement to Zhukov was already in existence: the Siberian force, from which Stalin – who had made only small withdrawals from it previously – had brought ten divisions, 1000 tanks and 1000 aircraft in October and November. That he felt free to do so was chiefly the result of reassurances transmitted by one of the most remarkable espionage agents in history, Richard Sorge, a German but also a Comintern operative who, as a confidant of the German ambassador in Tokyo, was privy to top-secret German-Japanese confidences and so able to assure Moscow (perhaps as early as 3 October) that Japan was committed to war against the United States and therefore would not use its Manchurian army to attack the Soviet Union in Siberia.

Had Japan decided otherwise – and its historic quarrel rather than its focus of strategic ambition lay against Russia, not America – the Battle of Moscow of December 1941 must have been fought as a Russian defensive, instead of as an offensive, and would almost certainly have resulted in German victory. As it was, Stalin's reinforcements had raised Zhukov's Western Front to a strength equal in numbers, if not equipment, to Army Group Centre's, and in consequence the outcome was the first Russian victory of the war. On the morning of 5 December, the Stavka plan mirrored those by which Hitler's marshals had inflicted such bloody wounds on the Red Army in the summer. Zhukov was to drive headlong at the Germans opposite Moscow, while Konev's Kalinin Front and Timoshenko's South-Western Front drove up from the south. Kluge and Hoepner, commanding the Fourth Army and the Fourth Panzer Army (as Panzer Group 4 had been renamed), decided that they could force their troops no further forward and had gone over to the defensive. Accordingly they were inert when the Russians struck. In the north, the deepest advance was made by Lelyushenko's Thirtieth Army, which advanced as far as the Moscow-Leningrad highway, threatening Panzer Group 3's link with the Fourth Army. By 9 December it had reached Klin and with the neighbouring First Shock Army seemed poised to bring off an encirclement. The Sixteenth and Twentieth Armies, commanded by Rokossovsky and Vlasov, operating closer to Moscow, matched their progress and on 13 December retook Istra, close to the Moscow-Smolensk highway up which Army Group Centre's axis of advance from the frontier had lain.

To the city's south, the Thirtieth and Fortieth Armies attacked Guderian's Panzer Group 2 and by 9 December menaced its main line of supply, the Orel-Tula railway. Guderian's Tula position formed a salient; on its opposite face the Soviet Fiftieth and Tenth Armies succeeded in separating Guderian from Kluge's Fourth Army and driving both away from the Moscow approaches – a displacement widened after 16 December when the Soviet Thirty-Third and Forty-Third Armies joined in.

By Christmas Day 1941 the Russian armies had retaken almost all the territory won by the Germans in the culminating stages of their drive on Moscow. Not only had the *Ostheer* lost ground; its leaders had also lost their Führer's confidence. He had dismissed generals in droves. On 30 November Rundstedt insisted on resigning in protest at his treatment by OKH; Hitler, on a visit to his headquarters, recognised the justice of his protest but accepted his resignation nevertheless. (Reichenau, who replaced him, died almost immediately of a heart attack.) On 17 December he replaced Bock with Kluge at Army Group Centre and on 20 December dismissed Guderian for preparing to withdraw his Panzer group from its exposed position. He also dismissed Hoepner from the Fourth

Panzer Army (between October and December 1941 Panzer groups were redesignated Panzer armies) for unauthorised retreat, and the commanders of the Ninth and Seventeenth Armies. Thirty-five corps and divisional commanders were dismissed at the same time. Most dramatically of all, on 19 December, Hitler relieved Brauchitsch as commander-in-chief of the army. Like Bock, Brauchitsch was ailing; but that was not the reason for his removal. Hitler had come to believe that only his own unalterable will could save the *Ostheer* from destruction. He therefore announced that Brauchitsch would not have a successor but that the Führer would himself directly act as the army's leader.

In that role he lectured and terrorised his generals to stand. Over Mannerheim, his Finnish ally, whose soldiers had marched in parallel with Leeb's to the gates of Leningrad (and also made an advance into Arctic Russia further north) he could not prevail; the marshal, once a tsarist officer, was prudently determined to hold no more territory than his country had possessed before the Winter War. Against his own commanders, however, he used the lash of implied incompetence and imputed cowardice. Führer Directive No. 39, issued on 8 December, had announced that the *Ostheer* would go over to the defensive, as several of its formations had already done; where it would defend was the Führer's prerogative. 'Do you plan to drop back thirty miles? Do you think it isn't all that cold there, then?' he recalled asking those he saw as fainthearts. 'Get yourself back to Germany as rapidly as you can – but leave the army in my charge. And the army is staying at the front.'

The spectre of 'Napoleonic retreat' afflicted Hitler in those days of mid-December and early January – the spectre of losing not only the line won but hundreds of thousands of men on the road to the rear and, most disabling of all, the army's heavy equipment. 'Save at least the army, whatever happens to its guns,' he recalled the falterers pleading to him. Convinced that withdrawal would lose both – 'How do you think you're going to fight further back if you haven't got any heavy weapons?' – he bullied and remonstrated down the telephone from Rastenburg (he did not shift his headquarters to Russian territory until the following year), until, by force of personality and threat of professional extinction, he infused the commanders at the front with a determination to hold against the Red Army and the Russian winter as inflexible as his own. By mid-January 1942, the worst was over: Army Group South's front was holding; Army Group Centre's was penetrated by a large salient north of Moscow but stabilised; Army Group North was entrenched on the fringe of Leningrad, which its artillery was slowly battering to pieces. The Red Army's winter reinforcements sustained a sporadic offensive but, advancing too often in human waves, were diminishing in numbers and force. Hitler had already begun to think of the spring and of the battle he would fight to destroy Stalin's Russia for good.

TEN

War Production

The German armoured pincers which encircled and crushed the Soviet armies in western Russia in June, July and August 1941 were instruments of military victory such as the world had never seen; but they were not instruments of total victory. Although they destroyed one of the Soviet Union's principal means of making war, its mobilised front-line defences, they did not succeed in destroying its industrial resources in the European provinces. Even while the Panzers were on the march, an evacuation soviet, directed by the economic expert A. I. Mikoyan, was rapidly uprooting factories from their path, loading machinery, stocks and workforces on to the overstretched railways, and shipping them eastward to new locations beyond the Panzers' reach. The strategic relocation of industry had begun long before the war, with the effort to make the output of the new industrial and raw-material zones beyond the Urals and elsewhere equal that of the traditional centres around Moscow, Leningrad and Kiev and in the Donetz basin. Between 1930 and 1940 new metallurgical plants were opened at Magnitogorsk, Kuznetsk and Novo-Tagil, industrial complexes at Chelyabinsk and Novosibirsk, aluminium works at Volkhov and Dnepropetrovsk, coalfields at Kuznetsk and Karaganda and a 'second Baku' oilfield in the Urals-Volga region, together with no less than thirty trans-Ural chemical plants. The pace of equalisation was slow; but by 1940, when the Donetz basin produced 94.3 million tons of coal, the Urals and Karaganda fields were producing 18.3 million.

However, Barbarossa stimulated nothing less than 'a second industrial revolution in the Soviet Union', in John Erickson's words. In August to October 1941, 80 per cent of Russian war industry was on the move eastward. The German advance had overrun 300 Soviet war-production factories, and would eventually bring the whole immovable extractive resources of western Russia, particularly the rich coal and metal mines of the Donetz basin, into German hands; but it was not rapid enough to prevent the evacuation eastward of the greater part of Soviet engineering industry from Leningrad, Kiev and the regions west of Moscow. In the first three months of the war the Soviet railway system, while transporting two and a half million troops westward, brought back eastward the plant of 1523 factories for relocation in the Urals (455 factories), Western Siberia (210), the Volga region (200) and Kazakhstan and central Asia (more than 250). The effort was both extraordinary and perilous.

On 29 September 1941 the Novo-Kramatorsk heavy-machine-tool works received orders to strip down its workshops; within five days all its machinery, including the only 10,000-ton presses in the Soviet Union, were loaded on to wagons under German bombing, while the 2500 technicians had to march on the last day to the nearest working railhead twenty miles away.

Ninety factories were evacuated from Leningrad, including the heavy tank works, the last shipments being made by barge across Lake Ladoga after the city was cut off by land from the rest of Russia. When similar German advances interrupted evacuations from the Donetz basin, all workable plant, including the gigantic Dnieper dam, was dynamited. In the teeth of this appalling industrial turbulence, Soviet economic managers succeeded in bringing the relocated plants back into production after an almost miraculously brief delay: according to Erickson, on 8 December 'the Kharkov Tank Works turned out its first twenty-five T-34 tanks [at Chelyabinsk in the Urals], just short of ten weeks after the last engineers left Kharkov, trudging along the railway tracks.'

These beginnings of Russia's 'second industrial revolution' were the worst of news for the Wehrmacht, even though it remained unaware of them for months to come. The worm in the apple of Hitler's spectacular campaigns of 1939-41 was that they had been fought from an economic base too fragile to sustain a long war, but with effects on the will of his enemies which ensured that the war would inevitably lengthen into a do-or-die struggle unless he could quickly crown it with a swift and decisive victory. Hitler's Germany, behind the panoply of the Nuremberg rallies and the massed ranks of the Wehrmacht, was a hollow vessel. As a producer of capital goods in 1939 she stood equal to Britain, with about 14 per cent of world output, compared to 5 per cent for France and 42 per cent for the United States, even in her still depressed condition. When British invisible earnings were added into her gross national product, however, Germany's output declined to third place below the British and American (excluding the Soviet Union); and when allowance was made for access to essential raw materials, including non-ferrous metals but particularly oil, the size of the German economy appeared smaller still.

German economic strategy, quite as much as its military one, was therefore geared to the concept of *Blitzkrieg*. The need was for a quick victory to spare German industry the pressure of producing weapons and munitions in quantity; once the war protracted, and Hitler decided upon the necessity of attacking Russia, German economic strategy changed. On the material front the drive was to lay hands on the extractive resources of the enemy, including those of the Balkans but particularly the coal, metal and (above all) oil-bearing regions of south Russia (together with the vast agricultural wealth of the Ukraine). On the industrial front the emphasis shifted in two different directions. Until 1942 Hitler had been adamant that the military effort should not depress civilian living standards or curtail the output of consumer goods; between January and May 1942, at the insistence of his Armaments Minister, Fritz Todt, and then (after Todt's accidental death) Dr Albert Speer, he accepted that military output as a proportion of gross national product would have to rise. However, while Todt and Speer introduced centralised measures of economic control which did indeed begin to raise output at a spectacular rate (for example, armaments, as a proportion of industrial production, increased from 16 per cent in 1941 to 22 per cent in 1942, 31 per cent in 1943 and 40 per cent in 1944), they did not commit Germany to attempt to match the production of their enemies' economies in quantitative terms. German war-economic philosophy rested on the concept that the country's weapons output should and could outdo the enemy's primarily in quality.

The 'quality war'

This concept proved difficult to implement in aircraft production, where both types and individual models of propeller-driven aircraft fell progressively behind their British and American equivalents after 1942. The German aircraft industry never produced a satisfactory strategic bomber; as early as 1934 the Luftwaffe had decided not to make an effort to develop this type of aircraft. Its single-seat fighters had reached the limit of their development in 1943. Its heavy fighters were all failures. By contrast, its first jet-propelled fighter, the Me 262, was a spectacular success, and if Hitler had encouraged its early production in quantity it would have confronted the Allied strategic bombing campaign with a severe challenge. German tanks, designed for serial development from a basic model, were also of the first quality, as were German small arms. For example, the MP-40 sub-machine-gun, known to Allied soldiers as the Schmeisser, was both the best of its kind in use in any army and one of the simplest to produce. German design engineers had simplified its components so that almost all could be produced by repetitive stamping, only the bolt assembly and barrel having to be machined.

German 'secret weapons' were also a testimony to the success of the 'quality' philosophy. Though Germany's electronics industry failed to match the achievements of the British, which consistently produced better radar equipment of every sort and supplied the scientific and technological basis for American industry's advances in that field, and though its nuclear weapons programme was an abject failure, its success with pilotless weapons and advanced submarines was impressive. The perfection of the schnorkel air-breathing system for submarines, though it came too late in the war to revitalise the U-boat campaign, introduced a method of operation which all post-war navies adopted and used until the coming of the nuclear-powered boat; while the development of the hydrogen-peroxide propulsion system, allowing a submarine in theory to cruise submerged indefinitely, in a sense anticipated the principle of the nuclear-powered submarine. Its pilotless weapons, the V-1 'flying-bomb' and the V-2 rocket, were respectively the first operational cruise and ballistic missiles. All modern equivalents of the two types descend from them, largely because their German designers emigrated to one or other of the superpower states.

Germany's limited success in running a 'quality' war economy must, however, be balanced against other factors. In the first place, Hitler's requirement that levels of civilian consumption should be maintained in the teeth of military output could not be observed by mid-1944. Thereafter the German standard of living fell sharply, both in absolute terms – as imports began to decline and the strategic bombing campaign bit deeper – and as a proportion of a declining gross national product. Gross national product, moreover, had risen only slowly during the war, from 129 billion to 150 billion Reichsmarks between 1939 and 1943, and then only as a result of emergency or extraordinary factors. Working hours were increased; imports of raw materials, goods or credit were exacted from the occupied territories by requisition or on terms of trade highly favourable to Germany; and the German labour force was swelled by an import of foreign labour – some induced, some conscripted, some enslaved – equivalent to a quarter of its pre-war size, or nearly 7 million people. Since Germany was not wholly self-supporting before the war, this workforce was at first fed by imported agricultural produce, but when food imports declined sharply after 1944 the swollen labour force became a charge on the war economy rather than a benefit to its structure.

The economic strangulation of Germany in the autumn of 1944, when the whole of its occupation area in the west and the remains of its conquests in the east were lost in the four months of June to September, manifested itself in a sharp and devastating fall in the index of war production. The general index (1941-2 = 100) fell from 330 to 310 between June and November, the index of ammunition production from 330 to 270, the index of explosive from 230 to 180. Oil, without which the army's tanks could not move or the Luftwaffe's aircraft fly, suffered an even more catastrophic decline in output from the synthetic-oil plants. In May, when imports made their last contribution to consumption, supply exceeded consumption for the first time in the war; by September, as a result of the Allied bombers' 'oil offensive', production from the synthetic-oil plants was only 10,000 tons, one-sixth of consumption, which itself had been reduced by stringent economy from 195,000 tons in May. Only the onset of bad weather and disagreements between the bomber chiefs spared the oil plants from further sustained attack and so averted the total interruption of oil supply before Christmas 1944. Germany narrowly escaped defeat by economic effect alone in 1944; as it was, a revival of production at the main Pölitz synthetic refinery released enough oil for its armies to go down fighting on the Rhine and in Berlin the following spring.

Japan was even more vulnerable to economic strangulation than Germany, and in its last weeks of war-making it was brought almost to that point. Its shipping stock had by then been reduced by sinkings, largely the work of American submarines, to 12 per cent of the pre-war stock, a desperate state of affairs for a country which depended not only on food imports to survive but also on inter-island movement to operate as an organised state. Japan had, of course, largely been motivated to war, at least at the objective level, by economic calculation. Its home population (excluding that of Korea and Manchuria) of about 60 million was too large to be supported by domestic agriculture (which supplied only 80 per cent of consumption), and it was the lure of Chinese rice as much as anything else which tempted the army to open its general offensive into mainland China in 1937. After the effective defeat of Chiang Kai-shek in 1938, most Japanese military activity in China took the form of 'rice offensives', forays into the rural districts designed to capture supplies at harvest time, which continued until the institution of the major Ichi-Go operation in 1944. However, since Japan was a rapidly industrialising nation, it needed not only rice but also ferrous and non-ferrous metals (both ores and scrap), rubber, coal and, above all, oil. In 1940 the supply of iron ore from domestic resources was only 16.7 per cent of demand, of steel 62.2 per cent, of aluminium 40.6 per cent, of manganese 66 per cent, of copper 40 per cent. All supplies of nickel, rubber and oil were imported, and although Japan produced 90 per cent of its own coal it had no reserves of coking coal which is essential for steel production. The government might, of course, have resolved to pursue a policy of exchange through trade; but the world slump, and protectionist measures imposed by Western importing nations as a result, so reversed the terms of trade that progressively more militarist Japanese cabinets set their face against the reduction of domestic living standards that a merely commercial approach to the acquisition of essential resources would have entailed. When the United States began to impose embargoes on strategic exports to Japan during 1940, and encouraged the British and Dutch to do likewise, the army-dominated cabinet rapidly decided on making a surprise attack.

In practice the capture of the 'southern area' – Malaya, Burma and the East Indies – yielded a much lower economic return than the Tojo cabinet anticipated. Imports of raw rubber, for example, which stood at 68,000 metric tons in 1941, declined to 31,000 in 1942, reached 42,000 in 1943 but declined again to 31,000 in 1944, largely as a result of the American submarine campaign, which also progressively curtailed imports of coal, iron ore and bauxite. The effect on Japanese industrial production was direct and proportional; although the Japanese aircraft industry sustained a remarkable increase in output between 1941 (= 100) and 1944 (= 465), as did the naval ordnance factories (1941 = 100, 1944 = 512), the output of motor vehicles in the same period declined by two-thirds. There were significant increases in the launching of naval and merchant ships, which respectively doubled and quadrupled; but as sinkings exceeded launchings the effort was more than nullified. The Japanese gross national product overall grew by a quarter between 1940 and 1944; government war expenditure, however, increased fivefold during the same period and eventually represented 50 per cent of GNP, thus stifling non-military production and leading to a harsh curtailment of civilian consumption. The end result of the war effort was to leave Japan with a population on the brink of starvation, though with a significantly enlarged workforce of trained engineering operatives. In the era of post-war economic revival, that workforce would win for Japanese products the overseas markets the denial of which had provoked Japan to aggression in the first place.

Britain's war effort

Britain, the other great island combatant, had also been threatened with economic strangulation by means of an enemy submarine force. It was even more dependent than Japan upon imports for food, since a century-old policy of utilising cheap shipments from America, Canada, Australasia and Argentina had depressed farming to a level where only half of consumption was met from domestic resources. It was wholly self-sufficient in coal and partially so in ferrous ores; but it depended upon foreign supply for all its oil and rubber and most non-ferrous metals. Moreover, though equal to Germany as the second greatest industrial power in the capitalist world, it imported certain vital products such as chemicals and machine-tools. The war effort it had imposed upon itself, moreover, particularly in 1940-1 when it bore the burden of confronting the Axis alone, could not be sustained out of domestic revenue. In order to pay for the fighters which won the Battle of Britain, the escorts which fought the Battle of the Atlantic and the merchant ships sunk in it, and the tanks which contested the issue with Rommel in the western desert, Britain was obliged to liquidate almost the whole of its overseas holdings of capital, an economic sacrifice which would require fifty years of effort to restore.

Had Germany deployed at the outset of the war the force of 300 U-boats which Dönitz had advised Hitler was necessary to win the Battle of the Atlantic, Britain would surely have collapsed as a combatant long before events in the Pacific brought about the United States' entry. Fortunately Dönitz did not achieve the deployment of that number until 1943, when the balance of forces between the Axis and its enemies had already altered to Hitler's fatal disadvantage. In the interim, as the result of the most ruthless imposition of centralised direction attempted by any country other than the Soviet Union, British industry had achieved a remarkable surge in output of war material. For example, the number of tanks produced increased from 969 in 1939 to 8611 in 1942, the number of bombers from 758 in 1939 to 7903 in 1943 and the number of bombs from 51,903 in 1940 to 309,366 in 1944.

Britain also achieved remarkable advances in quality as well as quantity of equipment. Its inventiveness in the field of electronic warfare was unequalled in the world, while its pioneering development of jet-propulsion systems for aircraft led to the deployment of a jet fighter, the Gloster Meteor, to two front-line squadrons in Europe in the closing weeks of the war (although they did not engage their German equivalent, the Me 262). British aero-engine designers produced the power plant which transformed the P-51 Mustang into the most potent long-range fighter of the war. The de Havilland Mosquito proved one of the most elegant and versatile combat aircraft of the war, performing with distinction as a bomber, day and night fighter and in its reconnaissance and intruder roles. The Avro Lancaster night bomber, although approaching obsolescence by 1945, was the supreme instrument of the RAF's strategic bombing campaign. There is little doubt, however, that the heavy emphasis placed on strategic bombing led to a pronounced structural imbalance in the British war economy, absorbing as much as one-third of the nation's war effort and the cream of its high technology. The sheer weight of British industrial effort committed to the bombing offensive meant that Britain had to turn to the United States for all its transport aircraft, many of its landing craft, vast quantities of ammunition and a large proportion of its tanks. Though British industry had produced the first tank in the First World War, British tanks in 1939–45 were notably inferior not only to their German but also their American equivalents; by 1944 all British armoured divisions were equipped with the American Sherman.

The British economy increased in size by over 60 per cent during the war; but civilian consumption declined by only 21 per cent between 1939 and 1943, about the peak of British war production, when military expenditures were consuming 50 per cent of the gross national product. The home population felt the shortfall, notably in the disappearance of all luxuries from the market and the reduction of many essentials such as fats and proteins from the rationed foodstuff allocations, together with a severe shortage of clothing. The effect of the shortfall was nevertheless disguised. Had Britain attempted to sustain its military outlay from domestic resources, its economy would have been broken. The same was true for the Soviet Union. Despite all the sacrifices made, in the extension of working hours, the liquidation of foreign and domestic capital, the reduction of living standards, the utilisation of marginal farming land, the substitution of ersatz for accustomed commodities, the conscription of women to the workforce (and in Britain also to the armed forces, where they formed a higher proportion than in those of any other combatant country) and a dozen other emergency measures, neither the British nor the Soviet economy could have borne the strains of war without external assistance. That outside help came from the United States.

Early in the course of his invasion of Russia, Hitler expressed regret to General Guderian that he had not heeded his warnings of the extent to which Russian exceeded German tank production. 'Had I known they had as many tanks as that,' he conceded, 'I would have thought twice before invading.' Russian tank production, 29,000 in 1944 when German tank production reached its peak at 17,800, was but one index of the degree to which the Allied war economy exceeded Germany's in scale. It was ultimately the United States which dwarfed Germany as an industrial power, at every level, and in each category of available natural resource and manufactured product. The shortfall in British war production had been offset since March 1941 by American provisions under the Lend-Lease legislation, which allowed the recipient to acquire war material against the promise to pay after the war was over. Lend-Lease helped Britain provide military aid to the Soviet Union between June and December 1941. As soon as Germany declared war on the United States, on 11 December 1941, Lend-Lease shipments began to flow to Russia directly from America, via Vladivostok, Murmansk and the Persian Gulf.

These shipments were on an enormous scale. The Soviet Union became the beneficiary of an outpouring of aid; some of the donations, such as tanks, it did not need; some, such as aircraft, were needed – for Soviet aircraft were not of the first quality – but were not properly utilised. Although the Soviet forces preferred their own weapons, the other donations provided the Soviet Union with a high proportion not only of its war-industrial requirements but also of its means to fight. 'Just imagine', Nikita Khrushchev later remarked, 'how we would have advanced from Stalingrad to Berlin without [American transport]'; at the end of the war, the Soviet forces held 665,000 motor vehicles, of which 427,000 were Western, most of them American and a high proportion the magnificent 2½-ton Dodge trucks, which effectively carried everything the Red Army needed in the field. American industry also supplied 13 million Soviet soldiers with their winter boots, American agriculture 5 million tons of food, sufficient to provide each Soviet soldier with half a pound of concentrated rations every day of the war. The American railroad industry supplied 2000 locomotives, 11,000 freight carriages and 540,000 tons of rails, with which the Russians laid a greater length of line than they had built between 1928 and 1939. American supplies of high-grade petroleum were essential to Russian production of aviation fuel, while three-quarters of Soviet consumption of copper in 1941-4 came from American sources.

Wartime Russia survived and fought on American aid. So too did wartime Britain. While British convoys were shipping eastward some £77 million-worth of equipment and raw material (equivalent, at current prices, to the annual defence budget for 1989), other British convoys, which included an increasing proportion of American ships, were bringing from across the Atlantic the means both to sustain the British civil population and armed forces and to equip the American expeditionary armies preparing to invade Hitler's Europe. The percentage of military equipment supplied to the British armed forces from American sources in 1941 was 11.5, in 1942 16.9, in 1943 26.9 and in 1944 28.7; and the percentage of American-supplied food consumed in Britain in 1941 was 29.1, a proportion which continued at that level throughout the war.

This outpouring of aid, combined with the equipment and maintenance of armed forces which increased in size thirtyfold between 1939 and 1945, was achieved at no damage to the United States economy at all. On the contrary: though annual Federal expenditure rose from 13 billion dollars in 1939 to 71 billion in 1944, inflation was easily contained by tax increases and successful war-loan campaigns. The gross national product more than doubled during the same period, and industrial production also nearly doubled.

This achievement had a simple cause. The United States economy had been depressed since the slump and bank collapse of 1929-31, and, despite the application of Roosevelt's New Deal policies of state-financed reflation, it had not recovered to anything like the same extent as the economies of Germany, where Hitler had run a full-blown Keynesian credit programme, or Britain, where more orthodox budgetary policies had nevertheless encouraged a mild boom during the 1930s. As a result, the American economy was both relatively and absolutely still in a depressed state in 1939. There were 8.9 million registered unemployed and the average utilisation of plant was forty hours a week. By 1944 the average utilisation of plant was ninety hours a week, there were 18.7 million more people in work than in 1939 (the 10 million excess over inducted surplus largely representing women), and the value of industrial output represented 38 per cent of national income compared to 29 per cent in 1939.

In absolute terms these figures represented an extraordinary economic surge. Relatively they spelt doom to Germany and Japan, where productivity per man-hour was respectively half and one-fifth of that in the United States. The American economy was, in short, not only much larger than that of either of its enemies. It was also greatly more efficient. As a result, from having been a negligible source of military equipment in 1939, by 1944 it was producing 40 per cent of the world's armaments. In specific categories, output of tanks had increased from 346 in 1940 to 17,565 in 1944, of shipping from 1.5 million tons in 1940 to 16.3 million tons in 1944 and of aircraft – the most spectacular of all America's wartime industrial achievements – from 2141 in 1940 to 96,318 in 1944.

In 1945 the United States was to find itself not only the richest state in the world, as in 1939, but the richest there had ever been, with an economy almost equal in productivity to that of the rest of the world put together. Her people too had benefited. The pathetic 'Okies' described in John Steinbeck's famous novel of protest, *The Grapes of Wrath*, were by 1944 enjoying a middle-class standard of living from their earnings in the aircraft factories of California, whence they had emigrated from their worn-out farms in the dustbowl. Neighbours who had stuck out the depression on better land had also received their reward. If it was American factories which made the weapons which beat Hitler, it was American farmers who grew the crops to feed his enemies. Paul Edwards, before the war a New Deal worker, recalled: 'The war was a hell of a good time. Farmers in South Dakota that I administered relief to, and gave them four dollars a week and bully beef to feed their families, when I came home they were worth a quarter of a million dollars. . . . What was true there was true all over the United States. . . . And the rest of the world was bleeding and in pain. But it's forgotten now. World War Two? It's a war I would still go to.'

In the final enumeration of Hitler's mistakes in waging the Second World War, his decision to contest the issue with the power of the American economy may well come to stand first.

ELEVEN

Crimean Summer, Stalingrad Winter

It is a paradox of campaigning in Russia that, though winter destroys armies, it is the coming of spring that halts operations. The thaw, saturating the suddenly unfrozen topsoil with thirty inches of snow melt, turns the dirt roads liquid and the surface of the steppe to swamp, the *rasputitsa*, 'internal seas' of mud which clog all movement. Motorised transport buries itself above the axles in bog; even the hardy local ponies and the light *panje* wagons they draw flounder in the bottomless mire. In mid-March 1942 both the Red Army and the *Ostheer* accepted defeat by the seasons. An enforced truce descended on the Russian front until the beginning of May.

Both armies made use of it to repair the losses that winter and the fighting had inflicted. The Stavka calculated that there were 16 million men of military age in Russia and that the strength of the Red Army could be raised to 9 million in 1942; allowing for 3 million already taken prisoner and a million dead, there would still be enough men to fill 400 divisions and provide replacements. Many of the divisions were pitifully weak, but a surplus was found to create a central reserve, while the evacuated factories behind the Urals had produced 4500 tanks, 3000 aircraft, 14,000 guns and 50,000 mortars during the winter months.

The Germans were also enlarging their army. In January the *Ersatzheer* (Replacement Army) raised thirteen divisions from new recruits and 'comb-outs'; another nine were created shortly afterwards. For the first time women volunteers (*Stabshelferinnen*) were inducted to release male clerks and drivers to the infantry in January 1942, and volunteer auxiliaries (*Hilfsfreiwillige*) were also found among Russian prisoners, most of whom turned coat as an alternative to starvation. In this way the 900,000 losses suffered during the winter were made good, though a deficiency of 600,000 remained by April. It was concealed by maintaining divisions in existence even when their infantry strength had fallen by as much as a third; tank, artillery and horse strength had also fallen. By April the *Ostheer* was short of 1600 Mark III and IV tanks, 2000 guns and 7000 anti-tank guns. Of the half-million horses the army had brought to Russia, a half had died by the spring of 1942.

Hitler was nevertheless convinced that the force which remained sufficed to finish Russia off and was determined to launch his decisive offensive as soon as the ground hardened. While Stalin had persuaded himself that the Germans would strike again at Moscow – a blow he was certain would be weakened by

Germany's need to deal with a 'Second Front' in the west – Hitler had an entirely contrary intention. The point of the Kaiser's final offensive into Russia in 1918 had been to take possession of its natural wealth. The wheatlands, mines and, now more important than ever, oilfields had always lain in the south. It was in that direction, into the lands beyond the Crimea, on the river Volga and in the Caucasus, that Hitler now planned to send the Panzers for the summer campaign of 1942, to recoup and add to the great economic conquests brought to Germany by the Treaty of Brest-Litovsk twenty-four years earlier.

The front that the *Ostheer* had drawn across western Russia in November, at the moment when its 'final' offensive against Moscow had been launched, had run almost directly north-south from the Gulf of Finland to the Black Sea, bulging eastward between Demyansk, to Moscow's north, and Kursk, to the capital's south. By May it had assumed a much less tidy configuration. Because of the effect of Stalin's winter counter-offensive, it no longer touched Moscow's outskirts and was now also dented in three places. Between Demyansk and Rzhev an enormous bulge protruded westward, reaching almost as far as Smolensk on the Moscow highway, and a reverse loop enclosed a pocket around Demyansk itself which had to be supplied by air. South and west of Moscow another bulge nearly enclosed Rzhev and almost touched Roslavl, on the Smolensk-Stalingrad railway. At Izyum, south of the great industrial city of Kharkov, yet another pocket bulged westward to cut the line of the Kiev railway and impede entry to Rostov, gateway to the Caucasus. The Red Army's sacrificial attacks of January to March had not lacked result.

Hitler briskly dismissed the danger that the two Moscow salients offered to his front. The Demyansk pocket, he calculated, cost the Red Army more to guard than it cost him to maintain; his occupation of the Rzhev re-entrant kept the threat to Moscow alive; and the Roslavl bulge was unimportant. As for the situation at Izyum, it would be resolved automatically by the opening of Army Group South's drive past Rostov into the Caucasus. The outline of that offensive (codenamed 'Blue') was discussed by Hitler with Halder and OKH on 28 March 1942 and issued in greater detail as Führer Directive No. 41 on 5 April. It comprised five separate operations. In the Crimea, Eleventh Army, commanded by Manstein, would destroy the Russian army in the Kerch peninsula and then reduce Sevastopol, still holding out after five months of siege, by bombardment. Bock (who had assumed command of Army Group South after recovering from illness) was to 'pinch out' the Izyum pocket and enclose Voronezh on the Don in armoured pincers; he had nine Panzer and six motorised divisions for the task (as well as fifty-two less reliable Romanian, Hungarian, Italian, Slovak and Spanish divisions). Once that had been accomplished, Army Group Centre would drive down the Don and cross the steppe to Stalingrad on the Volga, joined by a subsidiary force advancing from Kharkov; finally its spearheads would drive into the Caucasus (as the Kaiser's army had done in 1918), penetrate the mountain range between the Black and Caspian Seas and reach Baku, centre of the Soviet oil industry. To protect these conquests, Hitler intended to construct an impermeable East Wall. 'Russia will then be to us', he told Goebbels, 'what India is to the British.'

The economic arguments for the operation were immeasurable. Hitler declared to his generals that a success in the south would release forces to complete the isolation and capture of Leningrad in the north. However, the point of 'Blue' was to capture Russia's oil. Not only did Hitler need it for Germany (he confessed to intimates of nightmares in which he saw the Ploesti fields burning out of control from end to end); he also wanted to deny it to Stalin. The economic damage Barbarossa had already inflicted on the Soviet Union was immense. By mid-October 1941 the *Ostheer* had occupied territory (which it was to retain until the summer of 1944) where 45 per cent of the Soviet population lived, 64 per cent of the Soviet Union's coal was extracted, and 47 per cent of its grain crops, more than two-thirds of its pig-iron, steel and rolled metals and 60 per cent of its aluminium were produced. The frontier evacuation of factories (of which 303 alone produced ammunition) behind the Urals had saved essential industrial capacity from capture, though at the expense of a grave interruption of supply; but the loss, even the impairment, of the oil supply would prove catastrophic, as Hitler well knew. The 'General Plan' of Führer Directive No. 41 stated quite baldly: 'Our aim is to wipe out the entire defence potential remaining to the Soviets, and to cut them off, as far as possible, from their most important centres of war industry. . . . First, therefore, all available forces will be concentrated in the southern sector, with the aim of destroying the enemy before the Don in order to secure the Caucasian oilfields and the passes through the Caucasian mountains themselves.'

'Blue' opened as soon as the ground was hard enough to bear tanks, on 8 May, with Manstein's attack into the Kerch peninsula of the Crimea. A week later it was over and 170,000 Russians had been taken prisoner; only Sevastopol, which would not fall until 2 July, still held out in the Crimea. Meanwhile, however, the main stage of 'Blue', codenamed 'Fridericus', had been compromised. A Russian counter-attack towards Kharkov, a main tank-building centre as well as a key industrial city, began on 12 May, anticipating Bock's 'pinching out' of the Izyum pocket. In a panic he warned Hitler that 'Fridericus' would have to be abandoned for a frontal defence of Kharkov and, when Hitler dismissed the interruption of his plan as a 'minor blemish', retorted, 'This is no "blemish" – it's a matter of life and death.' Hitler was unmoved: he repeated that the situation would resolve itself as soon as 'Fridericus' gathered weight and merely insisted that the launch date be advanced one day. Events proved him right. Kleist, commanding First Panzer Army, easily penetrated the Russians' line north of their Kharkov thrust, joined up with Paulus's Sixth Army south of Kharkov on 22 May and thus achieved yet another of the encirclements which had dismembered the Red Army the previous year. By the beginning of June

239,000 prisoners had been captured and 1240 tanks destroyed on the Kharkov battlefield. Then followed two subsidiary operations codenamed 'Wilhelm' and 'Fridericus II', which set out to destroy the Izyum pocket and the remnants of the Russian forces isolated by the Kharkov battle respectively. Both were over by 28 June.

That was D-Day for 'Blue' proper. It was to be mounted by four armies in line abreast, the Sixth, Fourth Panzer, First Panzer and Seventeenth, the first two armies subordinated to one army group, the second two armies subordinated to another. Bock continued to command Army Group South, and List, who had begun his rise in the Polish campaign, commanded the new Army Group A on the Black Sea sector. They were opposed by four Russian armies, Fortieth, Thirteenth, Twenty-First and Twenty-Eighth, lacking reserves because of Stalin's belief that the principal German threat lay against Moscow. Fortieth Army was destroyed in the first two days; the other three were forced back in confusion. The southern steppe – the treeless, roadless, almost unwatered 'sea of grass' which the Cossack horsemen had made their own in their escape from tsarist autocracy – offered the army no line of obstacles on which to organise a defence. Across it Kleist's and Hoth's armour swept forward. Alan Clark has described the advance:

> The progress of the German columns [was discernible] at thirty or forty miles' distance. An enormous dust cloud towered in the sky, thickened by smoke from burning villages and gunfire. Heavy and dark at the head of the column, the smoke lingered in the still atmosphere of summer long after the tanks had passed on, a hanging barrage of brown haze stretching back to the western horizon. War correspondents with the advance waxed lyrical about the . . . 'Mot Pulk', or motorised square, which the columns represented on the move, with the trucks and artillery enclosed by a frame of Panzers.

However, the unexpected ease with which the Panzers had broken across the Donetz from Kharkov into the great grassland 'corridor' which stretched from that river a hundred miles eastward to the Don and led southward into the Caucasus now prompted Hitler to agree to a change of plan – disastrously, as it would turn out. Bock, worried that Army Group South might be attacked in flank as it proceeded down the Don-Donetz 'corridor', by Russian forces operating out of the interior towards the Don city of Voronezh, directed Hoth's Fourth Panzer Army to attack and capture the city. Paulus's Sixth Army was to be left to march down the corridor alone, unsupported by tanks, and then leap across from the 'great bend' of the Don to the Volga at Stalingrad, which it was to seize and hold as a blocking-point, to prevent further Russian attacks mounted from the interior against the flank of the main body when it passed by to penetrate the Caucasus.

Hitler, still directing the Russian campaign from Rastenburg, now separated by 700 miles from the vanguard of the *Ostheer*, became anxious that in fighting for Voronezh Bock might waste both time and tanks at a moment when time lacked and tanks were precious. Accordingly he flew to see the general on 3 July but was reassured by Bock's apparent promise that he would not embroil his striking force in close combat. By 7 July, however, it was clear that the promise would not be made good. Hoth's tanks had been drawn into the fighting for Voronezh instead of breaking off the battle to join Paulus's infantry in the march on Stalingrad, and looked to be engaged for some time to come. Peremptorily Hitler ordered them away, and on 13 July replaced Bock with Weichs as commander of Army Group South (now renamed B); but, as he would complain for months afterwards, the damage had been done. His generals' hopes of repeating the great captures of the previous year had been reawakened by the success at Kharkov in May. However, in the Donetz-Don corridor the Red Army, commanded by Timoshenko, had grown wilier. At the Stavka, A. M. Vasilevsky had succeeded in persuading Stalin that 'stand fast' orders issued for their own sake were undesirable, since they served the *Ostheer*'s ends, and in extracting permission for threatened Russian formations to slip away out of danger. So they did, assisted between 9 and 11 July by a temporary fuel crisis in Army Group B which halted Hoth's Panzers. Between 8 and 15 July, after three aborted encirclements between the Donetz and the Don, Army Groups A and B had captured only 90,000 prisoners – by the standards of the previous year a mere handful.

The heightening tension of crisis on the steppe front now prompted Hitler to leave Rastenburg for a headquarters nearer the centre of action. On 16 July OKW was transported *en bloc* to Vinnitsa in the Ukraine, still 400 miles from the Don and isolated in a Rastenburg-like pine forest – malarial, as it turned out – but nevertheless handier for direct personal intervention by the Führer in the conduct of operations. From the Vinnitsa headquarters on 23 July he issued Führer Directive No. 45, codenamed 'Brunswick', for the continuation of 'Blue'. It directed the Seventeenth Army and the First Panzer Army of Army Group A to follow the Russians across the great bend of the Don and destroy them beyond Rostov. Meanwhile the Sixth Army, supported by the Fourth Panzer Army, was to thrust forward to Stalingrad, 'to smash the enemy forces concentrated there, to occupy the town and to block land communications between the Don and Volga. . . . Closely connected with this, fast-moving forces will advance along the Volga with the task of thrusting through to Astrakhan.' Astrakhan lay in the far Caucasus, a land fabled even to Russians; to the *Landsers* tramping eastward, already 1000 miles from home in Silesia and 1500 miles from the Rhineland, it was a place almost at the end of the earth. Hitler's imagination leapt effortlessly at such objectives and he remembered that German soldiers had campaigned as far as that in 1918; but in 1942 a vast space and a still unsubdued Red Army interspersed between his infantry columns and the fulfilment of his dream of empire.

'Never contradict the Führer'

List's advance southward with Army Group A at first went even faster and more smoothly than expected. Once across the Don, Kleist's tanks raced over the Kuban steppe to reach Maikop, where the first oil derricks were seen on 9 August. The oilfield was wrecked but the Luftwaffe commander, Wolfram von Richthofen, whose Fourth Air Fleet was supporting the operation, was certain he could drive the Russians out of the Caucasus passes and clear a way through to the main oilfields beyond. A breakthrough was also important to secure possession of Tuapse, the Black Sea port through which the enemy could be supplied from Bulgaria and Romania. On 21 August, Hitler was brought news that Bavarian mountain troops had raised the swastika flag on the peak of Mount Elbrus, the highest point in the Caucasus (and in Europe), but the achievement did not please him. He wanted more tank advances, not feats of mountaineering. As the tanks reached the foothills of the Caucasus, however, the advance began to slow and Hitler vented his impatience against those around him, first Halder, then Jodl. Halder was in disfavour for other reasons: subordinate operations near Moscow and at Leningrad also failed in August, and Halder's defence of soldiers consigned to carry out what he thought 'impossible orders' only inflamed Hitler's rage against what he called 'the last masonic lodge'.

Jodl, though no friend to Halder's OKH, shared its understanding of the difficulties faced by the troops on the ground. When the reports of two emissaries he had sent to the Caucasus front failed to soften the Führer's harshness towards List, Jodl himself went to visit Army Group A. He found the 4th Mountain Division stuck fast in a defile so narrow that it had no hope of breaking through to the Transcaucasus and its oil. The force advancing on Tuapse was equally blocked by Russian resistance and had no prospect of getting to the port before winter closed the passes. When he insisted to Hitler that List's predicament was insoluble and incautiously indicated that the Führer had contrived the impasse, the result was an outburst of fury. Hitler, acutely sensitive to any implied slur on his powers of command and obsessed by the danger of repeating mistakes made during the First World War, declared that Jodl was behaving like Hentsch, the General Staff officer who had sanctioned the retreat from the Marne in 1914. He banished him and Keitel from his headquarters mess, installed stenographers in his command hut to take a verbatim record of his conferences so that his words could not be quoted against him, and dismissed List, assuming command of Army Group A himself on 9 September. He simultaneously sent the black spot to Halder, via Keitel, a man he chiefly valued for his widely remarked qualities of lackey and sycophant. On 23 September the army chief of staff left the Führer's presence in tears, to be replaced forthwith by General Kurt Zeitzler. 'Never contradict the Führer' was Keitel's advice to Zeitzler on the threshold of office. 'Never remind him that once he may have thought differently of something. Never tell him that subsequent events have proved you right and him wrong. Never report on casualties to him – you have to spare the nerves of the man.' Zeitzler owed his promotion mainly to his friendship with Schmundt, Hitler's chief adjutant, but he was also a dogged infantry soldier with an impressive fighting reputation. 'If a man starts a war,' he retorted to Keitel, 'he must have the nerve to bear the consequences.' During the twenty-two months when he served Hitler as army chief of staff, there were to be repeated passages of blunt speaking between them. After Halder, a 'swivel-chair' soldier, in Hitler's dismissive phrase, he brought a down-to-earth directness to the command conferences which Hitler found reassuring, and they were to rub along effectively even in the worst of crises.

Such a crisis was now in the making. Hitler's allusion to the Battle of the Marne was not without point. Then the German army had overextended itself and the high command had taken too little note of the danger levelled by a strongly garrisoned city on its flanks. Now on the Volga a similar danger loomed. Army Group A, reaching southward towards the Caucasus, maintained – with difficulty – lines of communication 300 miles long which it lacked the strength to protect against Russian forces located in the steppe to its east. Army Group B, which had earlier dawdled down the Donetz-Don corridor, was now being drawn into a battle around Stalingrad, and all the signs indicated that Stalin was transforming the city into a formidable centre of resistance. The parallel between 1914 and 1942 was not exact. At the Marne the German army had been beaten because it failed to find the force to capture Paris on its flank. The risk posed in 1942 was that Hitler would overreact and, by concentrating too much force at Stalingrad, deny his armies in the mountains and the open steppe the means to defend themselves against an enemy counter-stroke. Such was precisely the operational outcome towards which Stalin and the Stavka were now groping their way.

The first inkling of the plan for a Russian counter-stroke had been disclosed to Winston Churchill by Stalin when the British Prime Minister visited Moscow on 12-17 August. The moment was a low point in Anglo-Soviet relations. Although the obstacle presented by Russia's treatment of Poland had been partially removed – by Stalin's agreement the previous December to release his 180,000 Polish prisoners and transport them via Iran to form the 'Anders Army' under British command in Egypt – the Russians now had reason to reproach the British. After the massacre of convoy PQ17 in June, Britain had decided to interrupt the convoying of Soviet supplies to the Arctic ports. More critically, in Washington in July, the opening of the 'Second Front' had been definitively postponed from 1942 to 1943. These reproaches Stalin threw in Churchill's teeth. He was also suspicious of Britain's offer to help defend the Caucasus, where the local Muslims had been supported by a British army in an effort to secede from Russia in 1918 and were even now displaying a favour towards the German invaders which had prompted Beria to send secret police troops to the region. On the eve of the leaders' parting, however, Stalin relented. Desperate for the sort of supplies only Western industry could provide – not weapons, which the Urals factories were beginning to produce in plenty (16,000 aircraft and 14,000 tanks in the second half of 1942), but trucks and finished aluminium – he confronted Churchill with his demands. To smooth the transition from accusation to supplication, he 'let the Prime Minister into the immensely secret prospect of a vast counter-offensive.'

The outline of the plan was still vague. It was not to be clearly defined until 13 September, by which time the Battle of Stalingrad had been raging for three weeks. According to Russian calculations, its inception was even earlier than that. On 24 July, Rostov-on-Don, the sentry-box of southern Russia, had fallen to the German Seventeenth Army. Its tank-heavy neighbours, the First and Fourth Panzer Armies, broke eastward across the Don in the next six days and, while the First wheeled south to drive to the Caucasus, the Fourth turned north-eastward to support Paulus's Sixth Army in the assault on Stalingrad. Resistance had been so slight that a sergeant of the 14th Panzer Division recorded that 'many of the soldiers were able to take off their clothes and bathe [in the Don] – as we had in the Dnieper exactly a year earlier'. By 19 August the Sixth Army was positioned to begin the attack on Stalingrad as Hitler's Fourth Panzer Army approached on a converging route. Stalingrad largely comprised a sprawl of wooden buildings surrounding modern factories in a strip twenty miles long on the west bank of the Volga, which was a mile wide at that point. Much of the city was destroyed in a day of bombing by the VIII Air Corps on 23 August. Through the smouldering ruins the Sixth Army pressed forward for the culminating advance to the Volga shore.

In the month that had elapsed since the crossing of the Don, however, Stalin and the Stavka had improvised at Stalingrad a defence as strong as they had found for Leningrad the previous autumn and for Moscow in December. All three cities had a symbolic importance for Hitler; Stalingrad had a particular significance for Stalin. Not only was it the largest of the many Russian cities to be given his name. It was also the place where in 1918 the 'southern clique' – Stalin, Voroshilov, Budenny and Timoshenko – had defied Trotsky over the conduct of the war against the Whites, the episode which launched his rise to power within the party. During August, accordingly, he rushed men and material to the Stalingrad front, created a ring of defences, appointed new and vigorous commanders and made it clear that his order of 28 July, read out to every Soviet soldier – 'Not a step backward!' – must apply most sternly of all there. 'Unitary command', which would once again relegate commissars from an equal to an advisory status beside generals, was to be reintroduced on 9 October. Meanwhile he counted on his Stalingrad generals to resist retreat as if he himself stood at their elbows. V. N. Gordov and Yeremenko were the commanders of Stalingrad and South-Eastern Fronts respectively, V. I. Chuikov commanded the Sixty-Second Army in the city itself, and Zhukov was in overall charge of the theatre.

Zhukov's meeting with Stalin in the Kremlin on 13 September, however, concerned advance rather than retreat. In a dramatic leap of imagination he and Vasilevsky – now Chief of the General Staff, a post inherited from Zhukov on the latter's appointment as First Deputy Defence Commissar – outlined a plan for a wide encirclement of the German forces on the lower Volga and the destruction of Paulus's Sixth Army in the city. Stalin's arguments for a narrow encirclement were dismissed; that would allow the Germans to break out and slip away. So too was his contention that the necessary force did not exist; in forty-five days it could be assembled and equipped. Stalin thereupon withdrew his objections, adding that the 'main business' was to ensure that Stalingrad did not fall.

It had come close to doing so. After the burning of the wooden quarters of the city on 23 August, the German Sixth Army had found itself drawn into a bitter battle for 'the jagged gullies of the Volga hills with their copses and ravines, into the factory area of Stalingrad, spread out over uneven, pitted rugged country, covered with iron, concrete and stone buildings', as one of Paulus's divisional commanders described it. For every house, workshop, water-tower, railway embankment, wall, cellar and every pile of ruins a bitter battle was waged, without equal even in the First World War.

By 13 September, the day after Paulus had returned from a conference with Hitler at Vinnitsa on the battle for the city, the Russian front line was still at least four and in some places ten miles from the Volga. It was held by three divisions of the Sixty-Second Army, which Chuikov had just been appointed to command, and the garrison deployed about sixty tanks. One of the divisional commanders, A. I. Rodimtsev, was experienced in street fighting, which he had learned with the International Brigade at Madrid in 1936. By contrast, Chuikov noted, 'on the pretext of illness three of my deputies had left for the opposite bank of the Volga.' Between 13 and 21 September the Germans, using three infantry divisions in one thrust and four infantry and Panzer divisions in another, drove down the banks of the Volga to surround the core of the defence – the Tractor, Barricades and Red October factories – and brought artillery fire to bear on the central landing stage to which men and supplies were ferried nightly from the east bank.

However, the struggle had exhausted the vanguard of the Sixth Army, and a pause intervened while fresh troops were assembled for the street battle. It began again on 4 October. Chuikov was no longer defending above ground. His strongpoints had become subterranean and his headquarters troglodyte, its staff officers and specialists inhabiting tunnels and bunkers dug into the western bank of the Tsaritsa river near the Volga landing stage. Only the strongest buildings survived, to be fought over for a fractional advantage of dominance that each conferred. An officer of the 24th Panzer Division during the October battle wrote:

We have fought for fifteen days for a single house with mortars, grenades, machine-guns and bayonets. Already by the third day fifty-four German corpses are strewn in the cellars, on the landings, and the staircases. The front is a corridor between burnt-out rooms; it is the thin ceiling between two floors. Help comes from neighbouring houses by fire-escapes and chimneys. There is a ceaseless struggle from noon to night. From storey to storey, faces black with sweat, we bombed each other with grenades in the middle of explosions, clouds of dust and smoke. . . . Ask any soldier what hand-to-hand struggle means in such a fight. And imagine Stalingrad; eighty days and eighty nights of hand-to-hand struggle. . . . Stalingrad is no longer a town. By day it is an enormous cloud of burning, blinding smoke; it is a vast furnace lit by the reflection of the flames. And when night arrives, one of those scorching, howling, bleeding nights, the dogs plunge into the Volga and swim desperately to gain the other bank. The nights of Stalingrad are a terror for them. Animals flee this hell; the hardest storms cannot bear it for long; only men endure.

The Nietzschean-Nazi rhetoric apart, this is not an exaggerated picture of the Stalingrad battle. Chuikov, no sensationalist and a cool-headed newcomer to the war – he had previously been Russian military attaché in China – describes a succeeding stage.

On 14 October the Germans struck out; that day will go down as the bloodiest and most ferocious of the whole battle. Along a narrow front of four or five kilometres, the Germans threw in five infantry divisions and two tank divisions supported by masses of artillery and planes . . . during the day there were over two thousand Luftwaffe sorties. That morning you could not hear the separate shots or explosions, the whole merged into one continuous deafening roar. At five yards you could no longer distinguish anything, so thick were the dust and the smoke. . . . That day sixty-one men in my headquarters were killed. After four or five hours of this stunning barrage, the Germans started to attack with tanks and infantry, and they advanced one and a half kilometres and finally broke through to the Tractor Plant.

This lunge marked the penultimate stage of the German advance. On 18 October a lull fell over the city. 'From then on', noted Chuikov, 'the two armies were left gripping each other in a deadly clutch; the front became virtually stabilised.' In some places it was less than 300 yards from the Volga. The Red October Factory had been lost to the Germans, the Tractor and Barricades factories were only partly in Russian hands, and Chuikov's front was split into two pockets. But the garrisons, inspired by his famous slogan to fight as if 'there is no land across the Volga', held on, the wounded (35,000 in all) being ferried back each night in the boats that brought replacements (65,000 men) and ammunition (24,000 tons) from the far shore. The Germans of the Sixth Army, though supplied and reinforced more easily, were as much gripped by exhaustion as their enemies. Richthofen, the local Luftwaffe commander, noted in a November entry in his diary: 'The commanders and combat troops at Stalingrad are so apathetic that only the injection of a new spirit will get us anywhere.' But no new spirit was forthcoming. Hitler appeared to have forgotten whatever reason he had ever had for committing the Sixth Army to the battle for the city. Its waging had come to overshadow the strategy of capturing the Caucasus or even the consolidation of the 'steppe front' north of the city, along the line of the Don, against a Russian counter-attack. Hitler's dangerous tendency to obsess himself at his twice-daily command conferences with yards instead of miles and platoons instead of armies had robbed his direction of the struggle of all perspective. If his soldiers now succeeded in pushing Chuikov and the remnants of the Sixty-Second Army over the Stalingrad cliffs and into the Volga, at best he would have achieved a local success at catastrophic cost; the twenty divisions of the Sixth Army had already lost half their fighting strength. If they failed, the *Ostheer*'s largest offensive concentration would have been devastated for no result and the initiative given to the Red Army.

Paulus mounted a final effort on 11 November, in weather that already heralded the cold which would freeze the Volga and restore Chuikov's solid passage to the far shore. Next day a thrust by the Fourth Panzer Army succeeded in reaching the Volga south of the city, thus encircling it completely. That was the last success the Germans were to achieve at this easternmost point of their advance into Russia. For six days local and small-scale battles flickered on, killing soldiers on both sides but gaining ground for neither. Then, on 19 November, in Alan Clark's words 'a new and terrible sound overlaid' the rattle of small arms – 'the thunderous barrage of Voronov's two thousand guns to the north'. The Stalin-Zhukov-Vasilevsky counter-stroke had begun.

The fragile shell

In order to concentrate the largest possible German force against Stalingrad itself, Hitler had economised elsewhere by lining the Don over the steppe front north and south of the city with his satellite troops, Romanians, Hungarians and Italians. The kernel of his Stalingrad concentration, in short, was German; the shell was not. All autumn Hitler had blinded himself to this weakness in the *Ostheer*'s deployment. Now the Russians detected that, by breaking the fragile shell, they would surround and overcome the Sixth Army without having to fight it directly. It was about to suffer an encirclement by which Stalin would gain partial revenge on Hitler for those at Minsk, Smolensk and Kiev which had nearly destroyed the Red Army the previous year.

Zhukov's plan disposed two fronts, South-West (Vatutin) and Don (Rokossovsky), west of the city with five infantry and two tank armies, and the

Stalingrad Front (Yeremenko) to the south with one tank and three infantry armies. The South-West and Don Fronts struck on 19 November, the Stalingrad Front the following day. By 23 November their pincers had met at Kalach on the Don west of Stalingrad. The Third and Fourth Romanian Armies had been devastated, the Fourth (German) Panzer Army was in full retreat, and the Sixth Army was entombed in the ruins on the banks of the Volga.

The inception of Operation Uranus (as the Russians codenamed their counter-offensive) found Hitler at his house at Berchtesgaden, in retreat from the strains of fighting the Russian war. He at once took the train to Rastenburg, where he met Zeitzler on 23 November and, in the teeth of his chief of staff's advice that the Sixth Army must withdraw or be destroyed, peremptorily issued the disastrous order: 'We are not budging from the Volga.' During the next week he cobbled together the expedients that would keep the Sixth Army there. The Luftwaffe would supply it: Paulus's statement that he needed '700 tons' of supplies a day was 'realistically' assumed to mean 300, and the figure of 60 tons currently reaching the city in twenty to thirty Junkers 52 was multiplied by the theoretical availability of aircraft to match that figure. Manstein, the armoured-breakthrough magician, would relieve it; the reserves he would need were said to be available for an operation, 'Winter Storm', that would begin in early December. In the meantime Paulus was not to break out. At most, when Manstein's attack developed, he was to reach out towards him (on receipt of the signal 'Thunderclap') so that the Don-Volga bridgeheads could unite to form the same threat to the Red Army they had constituted before the counter-attack of 19-20 November.

Manstein, as commander of the newly formed Army Group Don, disposed of four armies for 'Winter Storm': the Third and Fourth Romanian, the Sixth (German) and the Fourth Panzer. The first two, always defective in equipment and commitment, were now broken reeds; the Sixth Army was imprisoned; the Fourth Panzer Army could still manoeuvre but had only three tank divisons, 6th, 17th and 23rd, to act as a spearhead. The attempted breakthrough began on 12 December. The Panzer divisions had some sixty miles of snow-covered steppe to cross before they could reach Paulus's lines.

Until 14 December they made good progress; a measure of surprise had been achieved and the Russians, now as ever, found it difficult to resist the initial impetus of a German advance. Time might be on their side but they could not match the Wehrmacht in military skill. Against the Wehrmacht's satellites, however, they were on equal if not better terms. On 16 December the Italian Eighth Army north of Stalingrad, which had thus far escaped the Romanians' fate, was struck and penetrated and a new threat was hurled against Manstein's Panzer thrust. On 17 December the 6th Panzer Division lurched to within thirty-five miles of Stalingrad, close enough to hear gunfire from the city; but the pace of advance was slowing, the Italian front was bending and the Sixth

Army showed no sign of reaching out to join hands. On 19 December Manstein flew his chief intelligence officer into the city in an effort to galvanise its commander. He returned with news that Paulus was oppressed by the difficulties and the fear of incurring the Führer's disfavour. On 21 December Manstein tried but failed to persuade Hitler to give Paulus a direct order for a break-out. By 24 December his own relief effort had ground to a halt in the snows of the steppe between the Don and Volga and he could only accept the necessity to retreat.

Retreat, too, was a consequent necessity for Kleist's dangerously over-extended Army Group A in the Caucasus. The previous autumn his motorised patrols had reached the shore of the Caspian Sea, the Eldorado of Hitler's strategy, but had turned back near the mouth of the river Terek for lack of support. In early January the whole of the First Panzer and Seventeenth Armies began to withdraw from the Caucasus mountain line, and retire through the 300-mile salient created by their headlong dash south-eastward the previous summer. As late as 12 January, Hitler still hoped to hold the Maikop oilfields. When Russian pressure north of Stalingrad cast the Hungarian Second Army into disarray, he was obliged to transfer the First Panzer Army to Manstein to augment his armoured strength. Nevertheless he directed Kleist, whose Army Group A was now shrunk to a single army, the Seventeenth, to withdraw it into a bridgehead east of the Crimea from which he hoped offensive operations could be resumed when the Stalingrad crisis ameliorated.

The hope was quite illusory. During January the German defence of Stalingrad was expiring by inches. Daily deliveries by the Luftwaffe to the three airfields within the perimeter averaged 70 tons; on only three days of the siege (7, 21 and 31 December) did deliveries exceed the minimum of 300 tons needed to sustain resistance. In the first week of January 1943 the forward airfield at Morozovskaya was overrun by Russian tanks; thereafter Richthofen's Ju 52s – diminished in number by the transfer of some to fly paratroops to Tunisia – had to operate from Novocherkassk, 220 miles from Stalingrad. After 10 January, when the main airstrip inside the Stalingrad perimeter fell, landing became difficult, most supplies were airdropped and the wounded could no longer be regularly evacuated. By 24 January nearly 20,000 men, one-fifth of the entombed army, were lying in makeshift, often unheated hospitals, with the outside temperature at minus 30 degrees Centigrade.

On 8 January, Voronov and Rokossovsky sent Paulus a summons to surrender, promising medical care and rations. 'The cruel Russian winter has scarcely yet begun,' they warned. Paulus, who had refused Manstein's appeal to break out three weeks earlier for fear of offending the Führer, could not contemplate such an act of disobedience. The terrible struggle continued. On 10 January the Russians opened a bombardment with 7000 guns, the largest concentration of artillery in history, to break the Sixth Army's line of resistance. By 17 January its soldiers had been forced back into the ruins of the city itself, by 24 January the army had been split into two, and the next day the Russian forces on the eastern shore of the river crossed the Volga and joined Chuikov's Sixty-Second Army stalwarts in their pockets around the Barricades and Red October factories.

Hoping for a gesture of honourable defiance, Hitler promoted Paulus to field marshal's rank by signal on 30 January. No German field marshal had ever surrendered to the enemy and he thus 'pressed a [suicide's] pistol into Paulus's hand'. At this final imposition of authority Paulus baulked. On 30 January his headquarters were overrun and he surrendered with his staff to the enemy. The last survivors capitulated on 2 February, leaving 90,000 unwounded to 20,000 wounded soldiers in Russian hands. 'There will be no more field marshals in this war,' Hitler announced to Zeitzler and Jodl on 1 February at Rastenburg. 'I won't go on counting my chickens before they're hatched.' Rightly he predicted that Paulus 'will make confessions, issue proclamations. You'll see.' (Paulus would indeed lend himself to Stalin's Committee of Free German Officers, which would call on the *Ostheer* to cease resistance and work for a Russian victory.) 'In peacetime in Germany about 18,000 to 20,000 people a year choose to commit suicide,' Hitler continued, 'although none of them is in a situation like this, and here's a man who has 45,000 to 60,000 of his soldiers die defending themselves bravely to the end – how can he give himself up to the Bolsheviks?'

The official reaction to the Stalingrad disaster was altogether more measured. German losses between 10 January and 2 February had in fact totalled 100,000, and few of the 110,000 captured survived transport and imprisonment. For three days normal broadcasting by German state radio was suspended and solemn music, Bruckner's Seventh Symphony, transmitted instead. Hitler, advised by Goebbels, saw in the destruction of the Sixth Army and its twenty-two German divisions an opportunity at least to create a national epic. There was no need for the fabrication of epic in Russia. On the news of Paulus's surrender the bells of the Kremlin were rung to celebrate the first undeniable Russian victory of the war. The Sixty-Second Army was redesignated the Eighth Guards Army, and Chuikov, a future Marshal of the Soviet Union, entrained it the following month for the Donetz. 'Goodbye, Volga,' he recalled thinking as he left Stalingrad, 'goodbye the tortured and devastated city. Will we ever see you again and what will you be like? Goodbye, our friends, lie in peace in the land soaked with the blood of our people. We are going west and our duty is to avenge your deaths.' When next Chuikov and his soldiers fought a battle for a city, it would be in the streets of Berlin.

PART III
THE WAR IN
THE PACIFIC
1941–1943

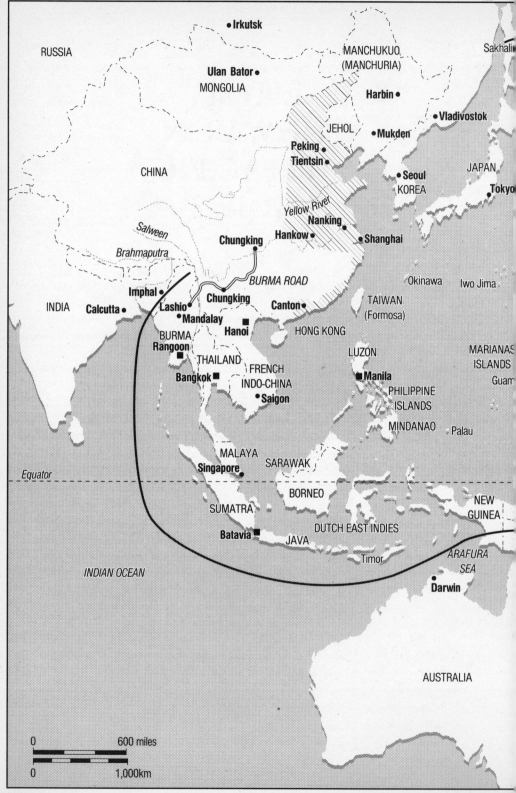

JAPANESE CONQUESTS IN ASIA AND THE PACIFIC, 1941-2

RUSSIA

• Irkutsk

MANCHUKUO
(MANCHURIA)

Sakhali

• Ulan Bator
MONGOLIA

Harbin •

• Vladivostok

JEHOL

• Mukden

JAPAN

Peking •
Tientsin •

• Seoul
KOREA

• Tokyo

CHINA

Yellow River

Nanking •
Hankow •

• Shanghai

Salween

Chungking •

Okinawa   Iwo Jima

Brahmaputra

BURMA ROAD

Imphal •

Chungking •

Canton •

TAIWAN
(Formosa)

INDIA   Calcutta •

Lashio •
• Mandalay

Hanoi ■

HONG KONG

LUZON

MARIANAS
ISLANDS

BURMA
Rangoon ■

THAILAND

FRENCH

■ Manila

Guam

Bangkok ■

INDO-CHINA

• Saigon

PHILIPPINE
ISLANDS

MINDANAO   • Palau

MALAYA

Singapore ■

SARAWAK

Equator

BORNEO

NEW
GUINEA

SUMATRA

DUTCH EAST INDIES

Batavia ■
JAVA

Timor

ARAFURA
SEA

INDIAN OCEAN

• Darwin

AUSTRALIA

0   600 miles

0   1,000km

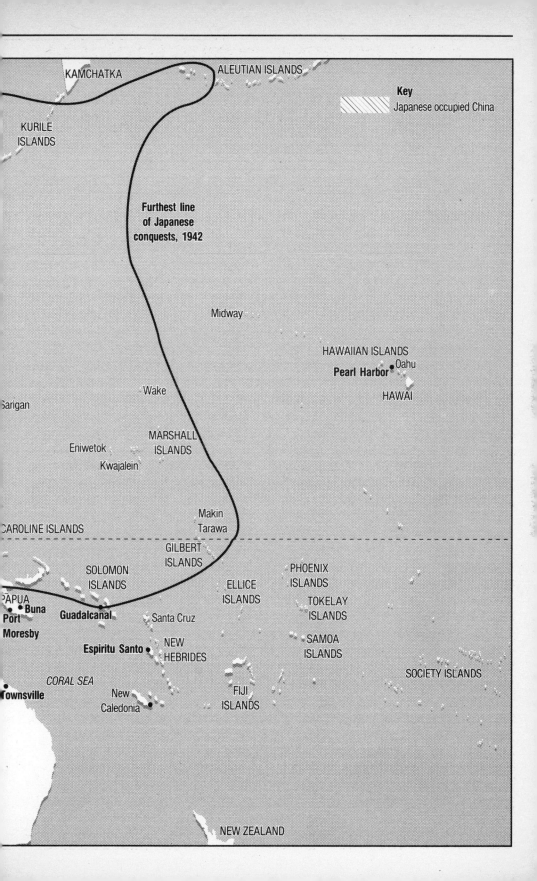

TWELVE

Tojo's Strategic Dilemma

Throughout the second year of his war on Russia, Hitler had laboured under a separate and self-assumed strategic burden: war with America. At two o'clock in the afternoon of 11 December 1941, four days after General Tojo's government in Tokyo had unleashed Japan's surprise attack on Pearl Harbor, Ribbentrop, as Foreign Minister, read out to the American chargé d'affaires in Berlin the text of Germany's declaration of war on the United States. It was an event which Ribbentrop, in perhaps his only truly sagacious contribution to Nazi policy-making, had struggled to avoid. During the period of American neutrality Hitler too had shrunk from acts which might provoke the United States to war against him. Now that Japan had cast the die, he hastened to follow. Ribbentrop emphasised, in vain, that the terms of the Tripartite Pact bound Germany to go to Japan's assistance only if Japan were directly attacked. On hearing the news of Pearl Harbor Hitler raced to tell it to Jodl and Keitel, exulting that 'Now it is impossible for us to lose the war: we now have an ally who has never been vanquished in three thousand years.' (Churchill, on hearing the same news, came to an identical but contrary conclusion: 'So we had won after all.'). On 11 December Hitler convoked the Reichstag and announced to the puppet deputies: 'We will always strike first! [Roosevelt] incites war, then falsifies the causes, then odiously wraps himself in a cloak of Christian hypocrisy and slowly but surely leads mankind to war. . . . The fact that the Japanese government, which has been negotiating for years with this man, has at last become tired of being mocked by him in such an unworthy way, fills all of us, the German people, and, I think, all other decent people in the world with deep satisfaction.' Later that day Germany, Italy and Japan renewed the Tripartite Pact, contracting not to conclude a separate peace nor to 'lay down arms until the joint war against the United States and England reaches a successful conclusion'. Privately Ribbentrop warned Hitler: 'We have just one year to cut Russia off from her military supplies arriving via Murmansk and the Persian Gulf; Japan must take care of Vladivostok. If we don't succeed and the munitions potential of the United States joins up with the manpower potential of the Russians, the war will enter a phase in which we shall only be able to win it with difficulty.'

This view was not only the opinion of a member of Hitler's entourage whose reputation was in eclipse. It was also held by a Japanese commander at the centre of his country's policy-making. In late September 1940 Admiral Isoruku

Yamamoto, Commander of the Combined Fleet, had told the then Prime Minister, Prince Fumimaro Konoye: 'If I am told to fight regardless of the consequences, I shall run wild for the first six months or a year, but I have utterly no confidence for the second or third year. The Tripartite Pact has been concluded and we cannot help it. Now the situation has come to this pass [that the Japanese cabinet was discussing war with the United States], I hope that you will endeavour to avoid a Japanese-American war.' Other Japanese had a fear of war, Konoye among them; the opinion of none of them carried the weight of the admiral at the head of Japan's operational navy. How did it come about not only that his views were overruled but that, less than a year after he expressed his anxieties to Konoye, he should, against his better judgement, have been planning the attack which would commit his country to a life-and-death struggle with a power he knew would defeat it?

The roots of Japan's self-destructive conflict with the West go back far into the country's past, and centre above all on its ruling caste's fear that 'Westernisation' – not that such a term was in use when Portuguese, Dutch and British mariners first appeared off Japan's shores in the sixteenth century – would disrupt the careful social structure on which the country's internal order rested. At the beginning of the seventeenth century, therefore, they closed their coasts to the outside world and succeeded in keeping them shut until the appearance of Western seamen who commanded a new technology, the steamship, in the middle of the nineteenth century forced them to reconsider their remarkable – and remarkably successful – decision. In one of the most radical changes of national policy recorded in history, the Japanese then accepted that, if Japan were to remain Japanese, it must join the modern world, but on terms which guaranteed that the processes of modernisation were retained in Japanese hands. The technology of the Western world would be bought; but the Japanese would not sell themselves or their society to the West in the course of acquiring it.

By the end of the First World War a reformed Japan had made extraordinary progress towards achieving that ideal. A popular children's song of the era of modernisation after the Meiji Restoration of 1867-8, which re-established the power of the central imperial government over the feudal lords, litanised ten desirable Western objects, including steam-engines, cameras, newspapers, schools and steamships. By the 1920s Japan had a universal and highly efficient school system, whose products were working in factories which not only manufactured textiles for sale at highly competitive prices on the world market but also produced heavy and light engineering goods, steel and chemicals, and armaments – ships, aircraft and guns – as modern as any in the world. Japan had already won two important wars, against China in 1894 and Russia in 1905, when she had established rights in the Chinese province of Manchuria; she had also fought on the Western side against Germany in 1914-18, on that occasion with weapons largely manufactured at home.

Japan's designs on China

Japan's emulation of the West did not, however, win her equality of status or esteem with the victor nations in Western eyes. Britain and later America had been grateful for Japan's assistance in the campaign against Germany's Pacific colonies; but, after conceding her a share of those colonies at the peace settlement – a strategically ill-judged concession which British and American admirals would bitterly regret after 1941 – they then combined to deprive her of an equal place among the world's great military powers. It already rankled with the Japanese that they had been compelled to surrender much of the strategically advantageous terrain in China they had wrested from Russia at the end of the Russo-Japanese war of 1904-5. Their relegation by the Washington Naval Treaty of 1922 to a lower place in the world's naval hierarchy by the British and Americans inflicted a wound on the national psyche which its traditional warrior caste, the Samurai, who had made a brilliantly judged leap from feudalism to modernity in order to retain their social dominance in the new Japan, resolved not lightly to forgive. They knew that their navy, in quality of personnel and material and potentially in terms of size, was the equal of the Royal and United States Navies in Pacific waters. They bitterly resented having to accept a treaty which set its numbers of capital ships at three-fifths of those of its wartime allies.

The Japanese army was even more strongly affected by rancour. Less Westernised than the navy, whose officers had been raised in the professional tradition of the Royal Navy, it was very early infected by the spirit of intense racialist nationalism which seized hold of Japanese political life in the inter-war years. Japan, a country of nearly 60 million, had by then ceased to be self-sufficient in food; it had never been, and never could be, self-sufficient in raw materials, least of all those materials on which an industrial revolution, in the throes of which Japan still laboured, most urgently depended – non-ferrous metals, rubber and, above all, oil. The solution that recommended itself to Japanese nationalists was a simple one: Japan would acquire the resources it needed from its neighbours and assure its supply by the most direct of all methods, imperial conquest. China was the obvious source of supply. The Japanese army despised the Chinese both for their economic and political incompetence – the collapse of the imperial system into warlord chaos after 1912 was the most striking evidence for this – and for their inability to resist Western penetration and exploitation. It therefore resolved to found an economic empire in China.

The first step was taken in 1931, when the Japanese garrison of Manchuria, where Japan enjoyed rights of protection over the railway system through which it extracted mineral produce, took possession of the whole of Manchuria from the local warlord, to end his piecemeal efforts at re-establishing Chinese authority in the area. This 'Young Marshal' was an ally of Chiang Kai-shek, commander of the army of the nominally sovereign Nanking government, and his soldiers were swiftly routed. The 'Manchurian Incident' aroused anger both abroad and in Japan, where the civilian government rightly felt its authority had been usurped; but no one moved to chasten the army, despite loud condemnation from the Americans, who had adopted a protective role towards China, based in particular on their missionary connection with the country. One of the humiliations the Chinese had had to accept at foreign hands in the nineteenth century was 'extraterritoriality', the surrender of sovereign power to the Western traders in their commercial settlements. This represented a threat to the Japanese, since it entailed the right to base troops, a right of which they swiftly took advantage themselves. In 1937 the Japanese garrison of the international embassy guard at Peking fell into conflict with Chinese government troops there and initiated a campaign which rapidly spread along the whole Chinese littoral. By 1938 most of fertile China, including the valleys of the Yellow and Yangtse rivers, was under Japanese occupation. Both the new capital of Nanking and the old capital of Peking fell to the invaders, and Chiang Kai-shek, now head of government, withdrew into the interior, to Chunking on the headwaters of the Yangtse.

Japan's army, and less directly the navy, had meanwhile incurred the wrath of foreign powers by causing casualties aboard British and American units of the extraterritorial river flotillas, and by entering into unofficial but bitter hostilities with the Red Army on the Chinese border with Mongolia in 1936. In another clash with the Red Army in 1939, involving armoured forces, the Japanese suffered an undeniable defeat at the hands of the future Marshal Zhukov, among others. Zhukov, who had spent training time in Germany, subsequently volunteered the significant judgement that, while the German army was better equipped than the Japanese, 'taken as a whole' it lacked its 'real fanaticism'. The Japanese army thereafter kept out of harm's way from the Russians – to their very great advantage in 1941, to Japan's catastrophic cost in August 1945. By contrast, the Japanese formed no such respectful opinion of the British or American armed forces, whose governments protested at the attacks on the USS *Panay* and HMS *Ladybird* but took no punitive action.

Foremost among the officers involved in the 'China Incident', as the war which spread from Shanghai in 1937 was called in Japan, was General Hideki Tojo. He was a 'Manchuria Incident' veteran also and in 1938 entered the cabinet as Vice-Minister of War. There he used his position to urge all-out rearmament as a precaution against war breaking out with the Soviet Union as well as continuing with China's Kuomintang government, which had succeeded the abolished empire, and of which Chiang Kai-shek was by then the leading figure. Tojo was a fervent, though not an extreme nationalist; but during the late 1930s extreme nationalists came to play an increasingly malign role in Japanese life. On 26 February 1936 a party of soldiers of the Tokyo garrison, rabidly opposed to what they saw as the appeasing attitude of the old aristocracy which dominated government, attempted to assassinate the Prime

Minister and succeeded in killing two of his predecessors, together with the Grand Chamberlain. This incident temporarily discredited the violent nationalists; paradoxically, however, it strengthened the power of the army because of the speed with which it distanced itself from the mutineers. After a succession of moderate governments, Prince Konoye – a former Prime Minister who commanded wide support – resumed power in July 1940 and accepted Tojo as Minister of War. He also took into his cabinet an ally of the army nationalists, Yosuke Matsuoka, as Foreign Minister. The combined presence of these two strong-headed imperialists at the centre of power was to lead Japan into war.

Matsuoka's first achievement was to commit Japan to the Tripartite Pact of 27 September 1940 with Germany and Italy. It bound the three countries to mutual support in the event of any one being attacked by a power which was not party to the Sino-Japanese dispute or to hostilities in Europe. This was a clear commitment for Japan to fight Russia if it attacked Germany and for Germany to fight the United States if it attacked Japan, but not otherwise a binding co-belligerency. It also recognised Germany's primacy in a European 'New Order' and Japan's in a 'Greater East Asia Co-Prosperity Sphere', under whose guise Japan was to constitute itself ruler of the European empires' Asian colonies after December 1941. Japan had already accepted an alliance with Germany in the 1936 Anti-Comintern Pact against Russia. In April 1941, as a rejoinder to the Ribbentrop-Molotov Pact, Matsuoka negotiated a neutrality treaty with the Soviet Union; but the trend of Japanese diplomacy was now set firmly against accommodation with neutrals and anti-Axis powers and towards ever closer association with the victors of 1940, whose star seemed fixed in the ascendant.

Hitler's attack on Russia in June 1941 – of which the Japanese cabinet was given no forewarning – temporarily shook its confidence and on 16 July it reformed itself to replace Matsuoka with a more moderate Foreign Minister. However, Tojo remained in place as Minister of War, and his was the strongest voice in propelling the cabinet towards confrontation with Hitler's enemies – the British and the Dutch government in exile – and the United States. That policy brought impressive results. In an attempt to appease the Japanese, in July the British closed the 'Burma Road', through which Chiang Kai-shek's armies received aid in southern China. The Dutch, who were even less capable than the British of resisting Japanese pressure, had also been bullied into agreeing to maintain oil, rubber, tin and bauxite supplies from their East Indies, not at the level Japan demanded but in acceptable quantities. In September 1940 the French, beaten in Europe but still a colonial power in the Far East, had been compelled to grant the Japanese basing and transit rights in northern Indo-China, from which the Japanese armed forces could both operate against Chiang Kai-shek and threaten the Dutch army and fleet in the East Indies, the British army and fleet (the latter held in home waters but earmarked for transfer to Singapore in a crisis) in Malaya and Burma, more distantly the British dominion of Australia and eventually also the British possessions around the shores of the Indian Ocean from Ceylon to East Africa.

These were alluring if distant prospects. In the foreground, however, hovered the menace of American power, at nearer hand still the stumbling-block of American disapproval. Not only did Japanese expansion southward threaten the American protectorate of the Philippines, but the United States was also China's protector, almost its guardian angel. Generations of American missionaries and teachers had worked in China's cities and countryside to bring Christianity and Western learning to its people; they had had no more rewarding pupils than China's ruler, Chiang Kai-shek, and his wife. American traders too had received their reward in China; and America's sailors and soldiers had cruised its waters and tramped its plains, under the colours of peacemakers, since the troubled time of the Boxer Rising. The 'China lobby' was the most powerful of foreign-policy interest groups in the United States; it was outraged by the 'China Incident', the cruel war which had now been raging for four years against the Kuomintang government; it was adamant not merely that Japan should be checked from further extension of military power in the Pacific but that it should be forced to draw back from the conquests it had already made.

Preparations for war

In April 1941 Cordell Hull, the American Secretary of State, had laid down four principles of international behaviour to the Japanese, innocuously high-minded by State Department thinking, but requiring a moderation of policy which put a check on Japanese plans of expansion and re-emphasised the humiliatingly inferior role in the Pacific which the United States assigned to their country. The four principles encouraged a party in the Japanese cabinet which thought that the empire's interests were better served by seeking advantages at the expense of the Soviet Union, the so-called 'north' programme. However, the 'south' party, which insisted on the extension of Japanese power, initially by the extraction from Vichy of basing rights in southern Indo-China, remained dominant. The only effect of Cordell Hull's demand was that the cabinet agreed to continue negotiations with the United States while pushing ahead with military preparations. Matsuoka, a lone voice for intransigence, was removed in a cabinet reorganisation on 16 July.

Unknown to the Japanese, the United States had since early 1941 been able to read Japanese diplomatic ciphers, as a result of a remarkable code-breaking operation known in Washington as 'Magic' and equivalent to Britain's Ultra success against the Wehrmacht ciphers. When Roosevelt learned of the Japanese decision at the imperial conference of 24 July to combine a diplomatic with a covert military offensive, he resolved to tighten the screws of economic warfare against Tokyo. On 24 July Japan extracted agreement from Vichy to allow its troops to enter southern Indo-China. On 26 July America, with the agreement of the British and Dutch, imposed further embargoes on Western trade with Japan, thus reducing Japan's foreign trade by three-quarters and cutting off nine-tenths of her oil supply at source.

Japan had by now installed an ambassador in Washington, Admiral Nomura, whose personal relationship with American officials was excellent and whose commitment to the views of the Japanese navy, far more moderate and realistic than those of the army, was genuine. At home, however, the army was pressing for deadlines. On 6 September, at a cabinet conference held in the presence of the emperor, Hirohito, the alternatives were reviewed in their starkest form: to start preparations for war at once; to continue negotiations; or to acquiesce in America's restrictions on Japanese strategic activity, including a withdrawal from Indo-China. Tojo, the Minister of War, had insisted that they be presented in this form. Like the others he was abashed when the emperor reminded his ministers of the awesome consequences of what they were deciding. The conclusion of the conference nevertheless was to continue negotiating while adopting outright preparations for war, the deadline for a successful outcome to be fixed for 10 October.

Delays in negotiation over the next weeks made it obvious that the deadline would have to be set back, and this aroused civilian and naval doubts over the rightness of considering the war option at all. Tojo, as leader of the army party and much influenced by popular impatience with government hesitancy, held out for the aggressive solution. On 5 October a conference convened in his office concluded that diplomacy would settle nothing and that the emperor must be petitioned to approve a military offensive. Over the next week Tojo heightened military pressure on Konoye to choose war and on 14 October made the issue one of the army's confidence in his premiership. Three days later Konoye resigned and Tojo took his place as Prime Minister.

Contrary to Allied wartime propaganda, Tojo was not a fascist, nor ideologically pro-Nazi or pro-Axis; though he was to be executed as a war criminal under the code devised for the Nuremberg trials, his motivation to war and conquest was not the same as that of Hitler and his followers. He did not seek revenge, nor was his racism particular or annihilatory. He was strongly anti-communist and feared the growing power of Mao Zedong in China; but he harboured no scheme to exterminate Japan's Chinese enemies or any other group who might stand in Japan's way in Asia. On the contrary, his chauvinism was exclusively anti-Western. Tojo cultivated the alliance with Germany for wholly expedient reasons and he harboured no illusion that, had Germany rather than America or Britain been a dominant power in the Pacific, it would have behaved any more generously than they to Japan's national ambitions. Tojo's code was simple: he was determined to establish Japanese primacy in its chosen sphere of influence, to defeat the Western nations (eventually and if necessary Russia, the traditional enemy) which would not accept it, to subdue and incorporate China within the Japanese empire, but to offer other Asian states (Indo-China, Thailand, Malaya, Burma, the East Indies) a place within Japan's Asian 'Co-Prosperity Sphere' under Japanese leadership. His vision was of an Asia liberated from the Western presence, in which Japan stood first among peoples who would recognise the extraordinary effort it had made to modernise itself.

On 1 November he chaired a meeting of army, navy and civilian representatives convened to consider the issues of war, peace and the deadline with America. The meeting decided to confront the Americans with one of two new proposals, identified as A and B. By the former the Japanese would offer the Americans a withdrawal of Japanese troops from China to be completed twenty-five years in the future – on the rational supposition that the Americans would reject it. Proposal B would offer a withdrawal of troops from southern Indo-China, where they had just arrived, if the Americans would sell Japan a million tons of aviation fuel. Both moves were to be linked to the establishment of a general peace in the Pacific. Tojo, as Prime Minister but also as representative of the military war party, was momentarily torn; he agreed that a final effort should be made to avoid war by engaging America in discussion of proposal B, however unlikely it was that Washington would accept it. The following day, however, in the presence of the emperor, he expressed his fear that, if Japan did not seize its advantage now, 'I am afraid we would become a third-class nation in two or three years. . . . Also, if we govern occupied areas with justice, the hostile attitude towards us will probably soften. America will be outraged at first but then she will come to understand. Anyway, I will carefully avoid making this a racial war.' The emperor did not on this occasion remind his advisers of the awesomeness of the issues they were discussing. In effect, therefore, in the absence of a subsequent American decision to withdraw from confrontation with Japan, the decision for war was sealed on 5 November; as the generals had brought the admirals to agree the previous day, 30 November was the last date on which American concessions would be accepted. By 25 November the Japanese naval attack force would have sailed from home ports to open the offensive against the United States bases in the Pacific, and Japanese army forces in Indo-China would have begun moves to enter southern Thailand, with the object of invading the British colony of Malaya and, beyond that, Burma and the Dutch East Indies.

Because of their access to Japanese diplomatic traffic through the Magic system, the Americans were aware as early as 7 November 1941 that 25 November marked a key date in the progress of their negotiations with Tokyo. They suspected that it might be the day after which Japan would regard itself as committed to war. However, even though they also had cryptanalytic access to Japanese naval ciphers, they had not identified the preliminary military moves ordered by Tojo and his cabinet, because of the stringent radio silence imposed by Japanese headquarters on the movements of the Combined Fleet and the Twenty-Fifth Army in southern Indo-China. During the second two weeks of November, often in conclave with the British, Dutch and Chinese, the State Department discussed proposal B at length with the Japanese representatives in Washington. The negotiations were fraught with ambiguity. Since Hull knew that the Japanese were proceeding with military preparations while professing to conduct a frank diplomacy, he was disinclined to accord weight to their offers and counter-offers; since they – Nomura and a professional diplomat, Kurusu, sent to assist him – were honourable men, their efforts at negotiation were hamstrung by their personal embarrassment at the double-dealing to which Tojo had made them party.

All ambiguities were resolved on 26 November. Then Cordell Hull bluntly presented them with the United States' ultimate position, which was a firm restatement of the position from which it had begun. Japan was to withdraw its troops not only from Indo-China but also from China, to accept the legitimacy of Chiang Kai-shek's government and, in effect, to abrogate Japan's membership of the Tripartite Pact. The Hull note reached Tokyo on 27 November and provoked consternation. It appeared to go further than any American counter-proposal yet issued. Not only did it link the relaxation of economic embargoes to a humiliating diplomatic surrender. It also demanded, by the Japanese interpretation, a withdrawal from the whole territory which the Chinese emperors had formerly ruled – Manchuria as well as China proper. Since Manchuria was technically not part of ethnic China, and since the Japanese believed they had conquered it by four-square means, this provision of the Hull note confirmed Tojo's belief in the rectitude of his policy. It revealed, as he and his followers had long argued, that the United States did not regard the Japanese empire as its equal in the community of nations, that it expected the emperor and his government to obey the American President when told to do so, and that it altogether discounted the reality of Japanese strategic power. The army and navy at once agreed that the note was unacceptable and, while Tojo instructed his Washington emissaries to persist in the talks, ships and soldiers were meanwhile directed to proceed to their attack stations. A longwinded and misleading restatement of Japanese grievances was transmitted to the Japanese embassy in Washington for presentation to Cordell Hull on the morning of 7 December, intended by Tojo to stand as a declaration of war. Although it was intercepted by Magic, delays in its translation meant that its contents were not formally presented at the State Department until after two o'clock in the afternoon, over an hour after the deadline stipulated by Tokyo. By then Pearl Harbor was under heavy attack, with the result that Tojo as a military leader had the satisfaction of presiding over one of the most shattering surprise attacks in history but as a Japanese traditionalist had the ignominy of inaugurating what Roosevelt denounced as the 'day of infamy'.

THIRTEEN

From Pearl Harbor to Midway

Sunday, 7 December 1941 found the American Pacific Fleet peacefully at anchor in its Hawaiian base of Pearl Harbor. Until April 1940 the fleet's permanent port was at San Diego, California; but the surprise attack on France in May by Germany, Japan's European ally, had caused the Navy Department to decide that the fleet's spring cruise to its forward base in Hawaii should be prolonged pending the return of calm to its western waters. In the Pacific Japan maintained a fleet as strong in battleships as that of the United States and even stronger in aircraft carriers, and was meanwhile prosecuting a great war in China which still left a force of eleven divisions – considerably larger than the American army as then constituted – for operations elsewhere. The Pacific Fleet remained at Pearl Harbor throughout 1940. While its sister formation, the Atlantic Fleet, began to undertake active escort operations off the American eastern seaboard during 1941 in support of Roosevelt's policy of denying those waters to the U-boats, the Pacific Fleet continued its programme of exercises and cruises. Since June 1940 it had undertaken three major alerts and many anti-aircraft and anti-submarine drills; and since October 1941 it had been at a permanent state of readiness. However, the protraction of the warning period had blunted the edge of preparedness. In peacetime the Pacific Fleet always observed Sunday as a holiday. Many officers slept ashore, the crews woke to a late breakfast. So it was on 7 December, the day, Roosevelt was shortly to tell Congress, that would 'live in infamy'.

The Japanese navy was acutely aware that the Pacific Fleet was vulnerable to surprise attack, and its plans were based on the supposition that surprise could be achieved. These plans, devised quite separately from the grand strategic debate in the cabinet, foresaw Japan's war falling into three stages. In the first stage the Combined Fleet would attack the United States Pacific Fleet in Pearl Harbor, while other naval and military forces simultaneously destroyed enemy ships and units and seized essential territory in the so-called 'Southern Area', comprising Malaya, the Dutch East Indies and the Philippines. In an extension of these operations, the army and navy would set up a defensive perimeter in Japan's larger islands and archipelagos in the western Pacific, which would deny America and its allies the opportunity to strike back into Japan's area of strategic dominance. The logic of the Japanese plan rested on the perception that, while the eastern Pacific between California and Hawaii is empty ocean, offering no bases or points of replenishment to a fleet or amphibious force based in the continental United States, the western Pacific is a constellation of islands, whose forward edge might be fortified to make the whole complex impenetrable to an outsider. Moreover, the long eastern flank of this island zone could be so armed with air and naval striking forces that an American fleet sailing to the Western powers' bases in Australia and New Zealand would suffer such damage on passage that no counter-offensive could be mounted there.

The second stage of Japan's plan was to make good the logic of its strategic thinking by constructing fortified bases along a chain running from the Kurile Islands, off Russian Siberia, through Wake (an American possession), the Marshall Islands (ex-German possessions allocated, with the Carolines and Marianas, to Japan at Versailles), the Gilberts (British), the Bismarcks (ex-German, now Australian), northern New Guinea (Australian), the Dutch East Indies and British Malaya. Stage three was to be concerned largely with consolidation: it included the interception and destruction of Allied forces which violated or approached the defensive perimeter, the waging of a war of attrition against the United States with the object of wearing down the American will to fight, and also the extension of the war, if necessary, into the British area of dominance in Burma, the Indian Ocean and perhaps India itself.

The perimeter strategy was rooted deeply in the psyche and history of the Japanese who, as an island people, had long been accustomed to using land and sea forces in concert to preserve the security of the archipelago they inhabit and extend national power into adjoining regions. The key to this strategy was the destruction of the American fleet at Pearl Harbor. Without that, the second and third phases of their war plan would crumble from the outset. The devising of the plan was consigned, paradoxically, to Admiral Isoruku Yamamoto, who was opposed to the Tripartite Pact, admired America and was pessimistic about the outcome of a Japanese-American war; but he conceived it his duty as a patriot and a professional naval officer to construct a feasible scheme.

Yamamoto was by origin a surface fleet officer; he had fought and been wounded in a cruiser at the decisive battle of Tsushima against the Russians in May 1905. Subsequently, however, he accepted that the aircraft carrier was the naval weapon of the future and had learned to fly. However, he still mistrusted his grasp of the essentials of air-sea operations and so, early in 1941, he enlisted the assistance of an outstanding younger naval aviator, Minoru Genda, to help him construct the attack plan. During the spring and summer outline plans were reviewed and criticised within the Combined Fleet and in September were submitted to the Naval General Staff. They comprised five separate but simultaneous operations. On Z-Day (Z was the flag flown by Admiral Togo to signal the opening of battle against the Russians at Tsushima in May 1905) two small amphibious forces would move against the American outposts of Wake and Guam islands, to wipe out those footholds inside the perimeter surrounding the 'Southern Area'. Another amphibious force, concentrated from Japanese bases on Formosa, Okinawa and the Palau Islands, would begin landings in the Philippines, taking the large islands of Mindanao and Luzon as their targets. Land, sea and air forces based in southern Indo-China and south China would invade Malaya (via a lodgement seized in the Kra isthmus of Thailand) and the Molucca Islands in the Dutch East Indies. Meanwhile, in the central act on which the success of all other four depended, the Combined Fleet, with its four large (and later two small) aircraft carriers, would have approached Pearl Harbor by stealth to within a range of 200 miles, launched and recovered its air groups, and departed, leaving the American Pacific Fleet's eight battleships and three carriers burning and sinking at their moorings. Japanese confidence in the viability of the plan was heightened by the Royal Navy's use of carrier aircraft against the Italian fleet in Taranto harbour in November 1940, an operation which Yamamoto's staff officers closely analysed.

There were two impediments to the success of the plan. One was that Japanese torpedoes could not run in the shallow waters of Battleship Row at Pearl Harbor, but modifications were soon made. The other was the danger that the Combined Fleet might be spotted on passage and its security compromised – even though it was to approach Hawaii by the most circuitous of routes, beginning in the stormy waters of the Kurile Islands between Japan and Siberia, and proceeding south-eastward by a route far from commercial shipping lanes. An experimental voyage was sailed by a Japanese liner in October 1941, and when it reported that it had not seen another ship or aeroplane the danger of compromise was discounted.

On 26 November the Carrier Strike Force sailed; the subordinate attack forces followed from their separate ports in the next few days. Nagumo commanded six carriers, two battleships and two heavy cruisers, three submarines, a covey of escorts and an attendant fleet of oilers to support his striking force over its long voyage; the Japanese, with the Americans, were pioneers of replenishment at sea, a technique which enormously extended the range and endurance of an operational fleet. The kernel and justification of his command, however, was his squadron of six carriers which between them embarked over 360 aircraft, including 320 torpedo- and dive-bombers and their fighter escorts for the air strike on Pearl Harbor. If they could be brought to their launch point, 200 miles north of Oahu where Battleship Row lay, the chances of their being deflected from their mission were remote.

Neither American strategic nor tactical intelligence of the planned Japanese strike against Pearl Harbor was adequate. American historians have disputed for years the issue of whether Roosevelt 'knew': those who believed he did imply that he had sought and found in foreknowledge of Japanese 'infamy' the pretext he needed to draw the United States into the war on the side of Britain. It is an extension of the charge that there was a secret understanding between Roosevelt and Churchill, perhaps concluded at their August meeting in Placentia Bay, Newfoundland, to use Japanese perfidy as a means of overcoming American domestic resistance to involvement. Both these charges defy logic. In the second case, Churchill certainly did not want war against Japan, which Britain was pitifully equipped to fight, but only American assistance in the fight against Hitler, which a *casus belli* in the Pacific would not necessarily assure; as we have seen, Hitler's perverse decision to declare war on the United States in the immediate aftermath of Pearl Harbor solved problems of diplomacy which might otherwise have needed months of negotiation between the White House and Congress. In the first case, Roosevelt's foreknowledge can be demonstrated to have been narrowly circumscribed. Although the American cryptanalysts had broken both the Japanese diplomatic cipher Purple and the naval cipher JN 25b, Purple was used only to transmit instructions from the Japanese Foreign Office to its diplomats abroad; in the nature of things, such instructions did not include details of war plans and, though their contents during the last months of peace aroused the suspicions of the American eavesdroppers, suspicion did not amount to proof. War plans, which would have supplied proof, were not entrusted to JN 25b. So stringent was Japanese radio security in the weeks before Pearl Harbor that all orders were distributed between Tokyo, fleet and army by courier, and the striking forces proceeded to their attack positions under strict radio silence. As an added precaution, Nagumo's fleet approached Pearl Harbor inside the forward edge of one of the enormous weather fronts which regularly cross the Pacific at warship speed. This technique, long practised by the Japanese, ensured that the fleet's movements would be protected by cloud and rainstorm from the eyes of any but a very lucky air or sea reconnaissance unit – from any systematic means of surveillance, indeed, except radar.

The strike on Pearl Harbor

Yet Pearl Harbor was protected by radar; in the disregard for the warning it offered lies the principal condemnation of American preparedness for war in the Pacific in December 1941. A British radar set had been installed on the northern coast of Oahu in August and regularly monitored movements in the sea area it covered. Soon after seven o'clock on the morning of 7 December, just as it was about to shut down its morning watch, its operator detected the approach of the largest concentration of aircraft he had ever seen on its screen. However, the naval duty officer at Pearl Harbor, when alerted, instructed him 'not to worry about it' and the radar operator, a private in the Army Signal Corps, did as he was told. The duty officer had wrongly concluded that the echo on the screen represented a flight of Flying Fortresses which were scheduled to land shortly at Hickam Field from California. There was much aerial reinforcement in progress around Hawaii in December 1941; *Lexington* and *Enterprise*, the Pacific Fleet's two carriers (*Saratoga* was Stateside for a refit), were currently delivering aircraft to Wake and Midway islands. The radar blip seemed part of an innocuous pattern.

In fact it represented the first flight of Nagumo's air striking force, released 200 miles from Oahu and detected 137 miles – less than one hour's flying time – from its target at Battleship Row. It totalled 183 torpedo- and dive-bombers, with their escort of Zeros – then and for two years to come the best shipborne fighters in the world – all of whose crews had been relentlessly trained in mock attacks on an exact model of the Pearl Harbor complex for months beforehand. A meticulous espionage programme had established where each battleship and cruiser lay; to each target a group of pilots had been assigned. All that remained was for the attackers to evade the defences and send their bombs and torpedoes home.

There were no defences. Such Sundaying servicemen as were topside when the first Japanese aircraft appeared over Battleship Row and the associated airfield targets at Hickam, Bellows and Wheeler Fields assumed their appearance to be 'part of a routine air-raid drill'. Three-quarters of the 780 anti-aircraft guns on the ships in Pearl Harbor were unmanned, and only four of the army's thirty-one batteries were operational. Many of the guns were without ammunition, which had been returned to store for safekeeping. At 7.49 am the Japanese began their attacks; by 8.12, when the ancient battleship *Utah* was mistakenly sunk, the Pacific Fleet was devastated. *Arizona* had blown up, *Oklahoma* had capsized, *California* was sinking; four other battleships were all heavily damaged. The destruction was completed by the second wave of 168 Japanese aircraft which arrived at nine o'clock. When they left, *West Virginia* had been added to the score of destroyed battleships, *Nevada* was aground – saved by the quick thinking of the junior officer who temporarily commanded her – and *Maryland, Tennessee* and *Pennsylvania* were badly damaged. Another eleven smaller ships had also been hit, and 188 aircraft destroyed, most set ablaze on the ground where they had been parked wing-to-wing as a precaution against sabotage. It was a humiliation without precedent in American history and a Japanese strategic triumph apparently as complete as Tsushima, which had driven Russian naval power from the Pacific in a single morning and established Togo as his country's Nelson.

But Pearl Harbor was no Trafalgar. Even as the Japanese carriers began to recover their aircraft, the first pilots to return confronted the Strike Force commander, Vice-Admiral Chuichi Nagumo, with the demand that they be launched again to complete their devastation of Pearl Harbor. It was a disappointment and anxiety to all of them that they had not found the American carriers at anchor. Failing strikes against them, the next best thing they could achieve was the destruction of the naval dockyards and oil storage tanks, which would at least ensure that the port could not be used as a forward base for a counter-offensive against the Japanese invasions of the Philippines, Malaya and the Dutch East Indies. Genda too lent his weight to their urgings. Nagumo, a doughty warrior but no Nelson, heard them out and then signified his disagreement. 'Operation Z' had succeeded beyond his and Yamamoto's wildest dreams. The rational course now was to withdraw the fleet from danger – who knew where the American carriers might be steering? – and hold it at safety and in readiness for the next stage of the offensive to the south. The rest of the Japanese navy and naval air force, and one-fifth of the Japanese army, was even then risking itself in perilous initiatives against the British, Dutch and American empires in the south-west Pacific. Who could say when and where the Combined Fleet would next be needed?

The tide of Japanese conquest

The 'southern' operation was already in full swing and the Royal Navy was about to feel the weight of Japanese maritime airpower. British plans to defend its scattered possessions in south-east Asia and the Pacific depended on the timely dispatch of capital ships, with carrier support, to the strongly fortified naval base of Singapore, at the tip of the Malayan peninsula between the two largest islands of the Dutch East Indies, Sumatra and Borneo. As a precautionary measure, the new battleship *Prince of Wales* and the old battlecruiser *Repulse* had been sailed to Singapore at the beginning of December. A carrier should have accompanied them, but casualties among those in home waters and the need to keep the only other uncommitted carrier to watch the German battleship *Tirpitz* in its Norwegian fiord meant that they had to sail unescorted. On 8 December, prompted by news that the Japanese had begun to land troops off the Kra isthmus, which joins southern Thailand to Malaya, *Prince of Wales* and *Repulse* with their small escort of destroyers sailed from Singapore to intercept. The Japanese landing troops had already occupied the airfield from which the two capital ships might have been afforded fighter cover, but although their commander, Admiral Sir Tom Phillips, was warned that strong Japanese torpedo-bomber forces were stationed in southern Indo-China he held his course. Early on the morning of 10 December the Japanese bombers found him, and both his capital ships were sunk in two hours of relentless attack. The loss of a brand-new battleship and a famous battlecruiser to Japanese shore-based aircraft was a disaster for which no one in Britain was prepared. Not only did it upset all preconceptions about Britain's ability to command distant waters through naval power; it struck cruelly at the nation's maritime pride. 'In all the war', wrote Winston Churchill, who heard the news by telephone from the Chief of the Naval Staff, 'I never received a more direct shock.'

News quite as bad was on its way; on 8 and 10 December the islands of Wake and Guam, American outposts within the great chain of former German islands on which the Japanese were to base their south-western Pacific defensive perimeter, were attacked. Guam fell at once; Wake, heroically defended by its small Marine garrison, succumbed to a second assault on 23 December, after an American relief sortie had timorously retreated. The British territory of Hong Kong resisted siege, which began on 8 December, but although its Anglo-

Canadian garrison fought to the bitter end it capitulated on Christmas Day. The atolls of Tarawa and Makin in the British Gilbert archipelago were captured in December. And on 10 December the Japanese opened amphibious offensives designed to overrun both Malaya and the Philippines.

The collapse of the British defence of Malaya has rightly come to be regarded as one of the most shameful Allied defeats of the war. The Japanese were outnumbered two to one throughout the campaign, which they initiated with only one division and parts of two others against three British divisions and parts of three others. The British were admittedly outnumbered and outclassed in the air, and had no tanks, whereas the Japanese invasion force included fifty-seven tanks. Superior equipment did not, however, explain the whirlwind Japanese success. That victory resulted from the flexibility and dynamism of their methods, akin to those that had characterised the German *Blitzkrieg* in France in 1940. The British were put off their stroke from the outset. Air Chief Marshal Sir Robert Brooke-Popham, the commander-in-chief, and Percival, his senior general, had intended to forestall a Japanese attack by moving forward across the Thai border to seize the potential landing places in the Kra isthmus, but the same sort of confused warnings that bedevilled American responses to Japan's surprise attacks prevented them from making that move. When the Japanese appeared in their forward defensive zone, they did not contest the advance but fell back to what were deemed better defensive positions further to the rear. The retreat surrendered valuable ground, including the sites of the three northernmost airfields in Malaya, none of which was put out of action and which were soon in use by the Japanese. Much else was left behind which the invaders put to use, including motor vehicles and seagoing vessels. Long columns of Japanese infantrymen with the scent of victory in their nostrils took to the roads in captured cars and trucks, followed by others pedalling southward on commandeered bicycles. Seaborne units embarked in fishing craft began to descend on the coast behind British lines, which were abandoned as rapidly as word of the Japanese appearance in their rear was received. By 14 December northern Malaya had been lost; by 7 January 1942 the Japanese had overrun the Slim river position in central Malaya and were driving the defenders southward to Singapore.

The units which collapsed so easily before the Japanese onrush were mostly Indian. They were not the first-line regiments of the pre-war Indian army which were currently winning victories against the Italians in the Western Desert, but war-raised units manned by recently enlisted recruits and led by inexperienced British officers most of whom had not learned Urdu, the command language by which the Indian army worked. There was therefore a lack of confidence between ranks, and orders for retreat were too often taken as a pretext for pell-mell withdrawal. However, poor morale was not the only explanation of Malaya Command's collapse. Few of its units had been trained in jungle warfare or had made the effort to train themselves. Even the resolute

8th Australian Division was bewildered and disorganised by the appearance of Japanese infiltrators far to the rear of the positions where they were expected. Yet one unit, the British 2nd Argyll and Sutherland Highlanders, showed what might have been achieved in defence. In the months before the war its commanding officer had practised his soldiers in extending their flanks into the jungle beyond the roads running through its defensive positions and demonstrated that the enemy's outflanking tactics might thus be nullified. It fought with great success, though at heavy loss, in central Malaya. Had all its fellow units adopted this practice, the Japanese invasion would certainly have been slowed, perhaps checked, before Singapore was brought under threat.

By 15 January, however, the Japanese Twenty-Fifth Army, having advanced 400 miles in five weeks, was only a hundred miles from the island fortress and in heavy fighting over the next ten days drove the Australians and Indians from Singapore's covering positions. On 31 January their rearguards, piped out of Malaya by the 2nd Argyll and Sutherland's two remaining pipers, crossed the causeway linking Singapore to the mainland and retreated to lines covering the naval base against attack from the northern shore.

The tragedy of the Malaya campaign was now reaching its climax. Singapore had just been reinforced by the British 18th Division, brought from the Middle East, so that despite the toll of units lost in the retreat from the north Percival commanded forty-five battalions to oppose thirty-one in General Tomoyuku Yamashita's Twenty-Fifth Army. General Sir Archibald Wavell, the victor of the war against Italy in the Middle East and now commander-in-chief in India, also counted on the arrival of air and sea reinforcements to support the troops on the ground and believed that the much-bruited strength of the Singapore naval base defences would assure its resistance for several months. As the most casual reader of the history of the Second World War now knows, however, Singapore's defences 'faced the wrong way'. This legend is false. The island's strongpoints and heavy guns had been positioned to repel an attack from the mainland; but the guns had been supplied with the wrong ammunition, unsuitable for engaging troops. Singapore is separated from Johore by a channel less than a mile wide at its narrowest. The northern shore of the island, moreover, was over thirty miles long, requiring Percival to disperse his battalions – when some had been concentrated in central reserve – at one to the mile. 'Who defends everything', Frederick the Great had written, 'defends nothing.' It is a harsh truth of war. Yamashita concentrated his forces (now reinforced by the Imperial Guards Division) against six Australian battalions on the north-west corner of the island and on 8 February launched them across the narrow waters of the Johore Strait. Under this overwhelming attack the Australian 22nd and 27th Brigades rapidly crumbled. Counter-attacks by the central reserve failed to throw the Japanese back from their footholds into the water. By 15 February the reservoirs in the middle of the island which supplied Singapore city, the population of which had been swollen by the influx of refugees to over a million, had fallen into Japanese hands. Percival faced the prospect of an urban disaster. Late that evening he marched into Japanese lines to offer surrender. He was photographed carrying the Union Jack beside a white flag borne by a staff officer. According to the historian Basil Collier, it was for 'the British the greatest military disaster in their history', entailing the capitulation of more than 130,000 British, Indian, Australian and local volunteer troops to a Japanese force half their number. Most of the captured Indians, seduced by the appeals of the mesmeric Hindu nationalist, Subhas Chandra Bose, would shortly go over to the Japanese to form an Indian National Army which would fight on Japan's side against the British in Burma in the cause of Indian independence. The Indian defection and the white flag incident were two of many reasons why Percival was never forgiven by Churchill's government or its successor for his catastrophic mismanagement of the Malaya campaign. After liberation in 1945 he became a 'non-person', shunned by all in official life and excluded from every commemoration of Britain's belated Asian victory.

Admiral H. E. Kimmel, the Pearl Harbor commander, was to suffer much the same official oblivion, though with less justification. As the coming turn of events in the Dutch East Indies was to demonstrate, no Western commander who stood in the path of Japan's surprise attack of December 1941 could preserve his professional honour, in a theatre hopelessly unprepared for the conduct of modern war, except by death in the face of the enemy. Admiral Karel Doorman, the senior Dutch naval officer in the East Indies, has gone down in history a hero – but only because he died on the bridge of his sinking cruiser in battle against fearful odds with the Japanese fleet. The Dutch East Indies were even less ready to resist attack than Hawaii or Pearl Harbor; Doorman may have regarded death as a merciful release from catastrophe, for which he bore no more responsibility in his sector of the 'Southern Area' than Percival and Kimmel in theirs.

Unlocking the East Indies treasure-house
Japanese attacks against the East Indies had opened on the British enclave in Borneo on 16 December. It was clear that they would shortly be extended to the whole of the island chain which stretches eastward from Malaya through New Guinea to the northern coast of Australia. In 1941 Australia was almost without defences, since the bulk of its army had been shipped overseas to fight with the British in the Middle East and south-east Asia. A frantic effort ensued to concentrate such Australian, Dutch, British and American forces as existed in the region into a coherent command. It was dubbed ABDA (American-British-Dutch-Australian) and placed under the authority of General Wavell. The strength at his disposal consisted of the small United States Asiatic Fleet, the Royal Australian Navy and the home defence elements of the Australian army, the remnants of the British Eastern Fleet, the units of the Dutch navy in East

Left: A German armoured column on the southern front, autumn 1941, with a Panzer Mark III on the right.

Right: Women citizens of Moscow digging an anti-tank ditch in the path of Army Group Centre's advance in October 1941.

Left: A Soviet propaganda photograph of the defenders of Stalingrad during the autumn fighting of 1942. But the battlescape is authentic – central Stalingrad was reduced to rubble in the most bitter city fighting of the war.

A German mortar crew prepares to advance at Stalingrad. The soldier in the centre carries the base-plate and the soldier on the left a rack of bombs.

Survivors of the German Sixth Army march into captivity, January 1943.

Surrender in Singapore, 15 February 1942, the single most catastrophic defeat in British military history. Over 130,000 troops were taken prisoner.

Hitler's 'New Order': a mass grave dug at the Bergen–Belsen concentration camp
after its liberation by British troops on 15 April 1945.

US Marines race for cover at Tarawa, November 1943. The almost suicidal enemy resistance, and the heavy casualties incurred in taking the atoll, prompted a major rethink of the Marine Corps' amphibious equipment and tactics.

A B-17 bomber of the US Eighth Air Force over Germany.
Note the vapour trails of the escorting P-47 Thunderbolt fighters.

Indian waters and the Dutch East Indies Army. The latter numbered some 140,000, the vast majority locals, unequipped and untrained for modern war; unlike the best of Britain's highly professional Indian army, it had never even fought a war. ABDA's naval force included eleven cruisers, twenty-seven destroyers and forty submarines. The United States hastily rushed a hundred modern aircraft to Java; the Dutch had only obsolescent models, and the British air component was wholly engaged in – and did not survive – the fighting in Malaya.

The Japanese strategy for the conquest of the East Indies – for them a treasure-house of oil, rubber and non-ferrous metal production, as well as rice and timber – was excellently conceived. They planned to use their plentiful naval and amphibious forces to attack in close succession at widely separated points across the 2000-mile length of the archipelago: Borneo and the Celebes in January, Timor and Sumatra in February, Java in March. An important subordinate aim of the attack on Timor, which lies only 300 miles from Australia's northernmost port of Darwin, was to cut the air link between Australia and Java. All forces were eventually to combine for the capture of Batavia (today Jakarta) on Java, the capital of the East Indies.

The Japanese landing troops found little difficulty in overcoming the Dutch local forces (which the population showed little inclination to support) wherever they were met. The Australians – whose will to fight was stiffened further by an air raid on Darwin, mounted from four of the carriers which had attacked Pearl Harbor, on 19 February – proved a tougher case. However, they were too few to check the trend of events. The only substantial counter in ABDA's hands was its fleet, a formidable force as long as the Japanese did not employ airpower against it. It enjoyed some early successes. On 24 January American destroyers and a Dutch submarine sank transports off Borneo and on 19 February Dutch and American destroyers engaged others off Bali. Admiral Doorman's test came on 27 February, when the ABDA command launched a Combined Striking Force against the Japanese invasion fleet approaching Java. Doorman's ships included two heavy and three light cruisers and nine destroyers, drawn from the Dutch, British, Australian and American navies. Admiral Takeo Takagi, his Japanese opponent, commanded two heavy and two light cruisers and fourteen destroyers. Numerically the encounter looked an even match; and in resolution, as Doorman was to display, the Japanese had no edge at all. However, they possessed a superior item of equipment, their 24-inch 'long-lance' torpedo which was a far more advanced weapon than its Allied equivalents.

The Battle of the Java Sea opened late in the afternoon of 27 February, with little daylight left. The initial stage of the largest naval engagement since Jutland took the form of a gunnery duel at long range. When the Japanese closed to launch torpedoes, however, they quickly scored hits, and Doorman was forced to turn away to protect his casualties. As darkness fell he lost contact with the Japanese and shortly afterwards had to detach most of his destroyers to refuel. He nevertheless remained determined to prevent the Japanese fleet putting its troops ashore and so turned back in darkness to where he judged it to be. His force was now reduced to one heavy and three light cruisers and one destroyer, and the moon was bright. At 10.30 pm he found the Japanese again; more accurately, the Japanese found him. While he engaged one part of their fleet, another approached unseen and launched the deadly torpedoes. Both surviving Dutch cruisers went down almost at once, *De Ruyter* taking Doorman with her. The USS *Houston* and HMAS *Perth* escaped, only to be sunk the following night after a heroic fight; misaimed Japanese torpedoes sank four of the transports they had been trying to intercept. All the major units of the force on which ABDA ultimately counted to repel the Japanese from the southern Pacific and the approaches to Australia had ceased to exist.

Beaten at sea, the Dutch were also quickly forced to surrender on land. On 12 March a formal Allied capitulation was signed at Bandung, on Java; the Imperial Guards Division, which had taken Singapore, landed the same day in Sumatra, the last of the large Dutch islands remaining outside Japanese control. The Japanese were by no means unwelcome in the East Indies: the Dutch, unlike the French, had never found the knack of tempering colonial rule by offering cultural and intellectual equality to a subject people's educated class. Educated young Indonesians – as they were shortly to call themselves – responded readily to the message that the Japanese brought 'co-prosperity', as they certainly brought liberation from Dutch subjection, and were to prove among the most enthusiastic of collaborators in Japan's New Order.

Another people who had always resented colonial subjection were the Burmese, whose intractability was at odds with the much more complex mixture of love and hate their Indian neighbours felt for the British Empire. Britain had always had difficulty in ruling Burma, which they had finally conquered only in 1886 (the young Rudyard Kipling's Tommies, drawn from life, had marched on the road to Mandalay). Few Burmese had ever accepted the outcome of the war and conquest and in early 1941 a group of young dissidents, later to become famous as 'the Thirty', had gone to Japan, under the leadership of Aung San, to be trained in fomenting resistance to British rule. Their opportunity was to come sooner than they had expected. During December the Japanese Fifteenth Army, which had entered Thailand at the beginning of the month, crossed the Burmese border to seize the airfields at Tenasserim. It was clear that a major offensive would follow shortly.

Burma was defended by a single locally enlisted division; part of the 17th Indian Division joined it in January. The only other Allied forces to hand were Chiang Kai-shek's Sixty-Sixth Army, based on the Burma Road and (like most Chinese formations) of doubtful value, and two Chinese divisions commanded by the redoubtable American 'Vinegar Joe' Stilwell on the Burma-China border. The commander of the Fifteenth Army, General Shojira Iida, had only two divisions, the 33rd and 55th, but they were well trained and supported by 300 aircraft; the British troops were not well trained and had almost no air support at all.

The campaign went wrong for the British from the start. Required to defend a wide front with few troops, the 17th Indian Division soon lost its forward defensive line on the Salween river on 14 February, pulled back to the Sittang river, guarding the capital, Rangoon, held there briefly and then, through a misunderstanding, blew the only bridge across it while most of the fighting troops were on the wrong side.

Things quickly went from bad to worse. General Alexander, who had arrived from Britain to stop the rot on 5 March, decided that the remnants of 'Burcorps', as his force was called, would have to retreat to the Irrawaddy valley in the centre of the country if a stand were to be made. The Japanese Fifteenth Army, now reinforced by the 18th and 56th Divisions and 100 aircraft, followed on his heels. Alexander hoped to hold south of Mandalay, Burma's second city, on a line between Prome and Toungoo, where a Chinese division had arrived; but he was pushed out of it on 21 March and forced into further retreat. His British and Indian troops were now short of supplies and exhausted, his Burmese troops had started to desert *en masse*. He was threatened with being outflanked both to the west and to the east, where the Japanese were driving the Chinese back towards the mountains of the China border. Faced with the dilemma of following the Chinese Sixty-Sixth Army (in reality about a division strong) along the Burma Road, which led from north-east Burma into China, where he had no assurance of supply, or of embarking on a trek across the roadless mountains of north-west Burma into India, he opted for the latter course. On 21 April he agreed with Chiang Kai-shek's liaison officer in Burma that their two beaten armies should go their separate ways and set off to lead his troops, accompanied by thousands of civilian refugees, on 'the longest retreat in British military history'. On 19 May, having traversed 600 miles of Burma in nine weeks, the survivors of 'Burcorps' crossed the Indian frontier at Tamu, in the Chin Hills, just as the arrival of the monsoon made further retreat impossible – but also, fortunately, denied the Japanese the possibility of pushing their pursuit into India itself.

About 4000 of the 30,000 British troops who had begun the campaign had perished; some 9000 were missing, most of them Burmese who had left the ranks. Only one Burmese battalion, largely recruited from one of the country's ethnic minorities, arrived in India. Many of the fugitives accepted Aung San's call to arms and joined his Burma National Army, which under Japanese colours briefly fought on Japan's side in 1944 and after the war provided the nucleus of his successful independence movement. There were other survivors of the rout. 'Vinegar Joe' Stilwell trekked back into China, whence he sallied into Burma again in 1944. General Bill Slim, Alexander's subordinate, reached India; he too returned to Burma in 1944, at the head of the victorious

Fourteenth Army, which he rebuilt from the debris of the rout. Among its units were the 4th Burma Rifles, the sole surviving element of the original 1st Burma Division.

The victory in Burma almost completed the first stage of Japan's offensive into the 'Southern Area'. It had profited brilliantly from its occupation of a central strategic position – in Indo-China, Formosa, the Marianas, Marshalls and Carolines – to strike east, south and west against the scattered colonial possessions of its chosen enemies and their divided forces and to overwhelm them one by one. On 22 April, when Alexander accepted defeat and set out across the mountains into India, only one Allied stronghold still resisted the Japanese inside the 'Southern Area'. It was the American foothold in the Philippines.

The fall of the Philippines

America's presence in the Philippines, which were never an American colony and in 1941 not quite yet a sovereign state, had come about through victory over Spain in the war of 1898 (the Philippines had been Spanish since the sixteenth century). America had extended a protectorate over the islands, introduced a democratic form of government, raised a Filipino army – in 1941 commanded by the old Filipino hand, General Douglas MacArthur – and put the archipelago under the shelter of the Pacific Fleet. In December 1941 American forces in the island numbered 16,000 combat troops, but only two formed regiments, about 150 operational aircraft, sixteen surface ships and twenty-nine submarines. On 26 July 1941 the Filipino army had been taken into the service of the United States, under the terms of the 1934 Act of Congress granting provisional independence; but its ten embryo divisions were as yet unfit for operations. The only combat-ready Filipino force was the Philippine Scouts Division, American-trained but only 12,000 strong.

Against these troops, which MacArthur had concentrated near the capital, Manila, in the northern island of Luzon, the Japanese intended to deploy the Fourteenth Army from Formosa (Taiwan). It consisted of two very strong divisions, the 16th and 48th, which had fought in China, and was supported by the Third Fleet, which included five cruisers and fourteen destroyers, the Second Fleet of two battleships, three cruisers and four destroyers, and a force of two carriers, five cruisers and thirteen destroyers. The air groups of the carriers were to be supplemented by the land-based Eleventh Air Fleet and the 5th Air Division.

The first disaster suffered by the Americans came from the air. As at Hawaii, they were provided with radar but failed to act on the warning it gave; as at Hawaii, their aircraft were packed wing-to-wing as a protection against sabotage and were destroyed almost to the last machine in the first Japanese air strike, which fell at noon on 8 December. On 12 December Admiral Thomas Hart, commanding the Asiatic Fleet in Filipino waters, felt compelled by lack of air cover to dispatch his surface ships for safety to the Dutch East Indies, where, under ABDA's command, they were to be destroyed in the Battle of the Java Sea.

By that date the Fourteenth Army's landings had already begun. Scorning an indirect approach through any other of the 7000 Filipino islands, General Masaharu Homma put his troops ashore on Luzon on 10 December and began an advance directly on the capital. He had hoped, by landing at separate points, to draw MacArthur's units away from Manila; when the defenders declined to respond, he put in another large-scale landing close to the capital on 22 December and forced MacArthur to fall back into a strong position on the Bataan peninsula covering Manila Bay and its offshore island of Corregidor.

Bataan, some thirty miles long and fifteen wide, is dominated by two high jungle-covered mountains. Properly defended, it should have resisted attack indefinitely, even though the garrison was short of supplies. In forming their line on the first mountain position, however, MacArthur's troops made the same mistake as the British were simultaneously making in Malaya. They failed to extend their flanks into the jungle on the mountain's slopes; in consequence their flanks were quickly turned by Japanese infiltrators. Retiring to the second mountain position, they avoided that error; but they had surrendered half their territory and were now crowded into an area ten miles square. In addition to the 83,000 soldiers within the lines, moreover, there were 26,000 civilian refugees, many of whom had fled from Manila, which the Japanese had heavily bombed, even though it had been declared an open city. All were placed on half-rations, but these rapidly dwindled, despite occasional blockade running by American submarines. By 12 March, when MacArthur left for Australia on Roosevelt's orders (with the famous promise, 'I shall return'), the garrison was on one-third rations. On 3 April, when Homma opened a final offensive, most of the Americans and Filipinos within the Bataan pocket were suffering from beriberi or other deficiency diseases and rations had been reduced to one-quarter. Five days later General Jonathan Wainwright, MacArthur's successor, offered his surrender. About 9300 Americans and 45,000 Filipinos arrived in prison camp after a notorious 'death march'. Some 25,000 had died of wounds, disease or mistreatment. The last survivors of the Philippines garrison, who occupied the island of Corregidor, were shelled into surrender between 14 April and 6 May; on 4 May alone more than 16,000 Japanese shells fell on the tiny outpost, making further resistance impossible. With the island's capitulation the whole of the Philippines fell into Japanese hands. The population, however, unlike those of the Dutch East Indies and Burma, were not disposed to regard the Japanese victory as cause for satisfaction. They had trusted, rightly, in America's promise to bring them to full independence and rightly also feared that Japanese occupation presaged oppression and exploitation. The Philippines Commonwealth was to be the only component of the Greater East Asia Co-Prosperity Sphere in which Japan would encounter popular resistance to its rule.

The prospect of Filipino resistance was, however, at best an irrelevance to the Japanese at the moment Corregidor fell on 6 May 1942. Their strategic horizon now ran around the whole western Pacific and deep into China and south-east Asia too. The historic European empires of the East – Burma, Malaya, the East Indies, the Philippines and effectively French Indo-China also – had been drawn into their sphere. To the Chinese dependencies in which they had established rights of occupation between 1895 and 1931 – Formosa, Korea and Manchuria – they had added since 1937 vast swathes of conquered land in China proper. All the oceanic archipelagos north of the equator were theirs, and they had made inroads into those to the south. Between the west coast of the United States and the British dominions of Australia and New Zealand lay largely empty ocean, dotted by a few islands too remote or too tiny to provide their enemies with bases for a strategic riposte. From the perimeter of the 'Southern Area' the Japanese fleet and naval air forces were poised to strike deep into the Indian Ocean, towards the British Andaman and Nicobar islands (captured in March 1942), towards Ceylon (raided in April, at the cost of a British aircraft carrier), perhaps even as far away as the coast of East Africa (the appearance of a Japanese submarine off Madagascar in May would, in fact, prompt the British to occupy the island later in the year). Above all, their great amphibious – better, triphibious – fleet remained intact. Not one of their eleven battleships, ten carriers or eighteen heavy and twenty light cruisers had been even seriously damaged in the war thus far; while the United States Pacific and Asiatic Fleets had lost – or lost the use of – all its battleships and large numbers of its cruisers and destroyers, the British and Dutch Far Eastern fleets had been destroyed and the Royal Australian Navy had been driven back to port.

All that remained to the Allies to set in the strategic balance against Japan's astonishing triumph and overpowering strategic position was the surviving naval base of Hawaii, with its remote dependency of Midway Island, and the US Pacific Fleet's handful of carriers, three, perhaps four at most. Little wonder that hubris gripped even such doubters as Yamamoto; at the beginning of May 1942, the consummation of victory, a prospect he had long warned hovered at the very margin of possibility, seemed to lie only one battle away.

FOURTEEN

Carrier Battle: Midway

In the context of the Pacific war in May 1942, one more battle meant a battle between aircraft carriers. There had never been such a battle before; but the Japanese navy's victory at Pearl Harbor ensured that such a battle was inevitable, if the United States were not altogether to abdicate control of the Pacific to Japan. The destruction at Battleship Row had left the American Pacific Fleet with only its aircraft carriers among its capital ships afloat, and it must find a way of using those carriers to fight the might of eleven Japanese battleships, ten carriers and thirty-eight cruisers, wherever they might next appear. Battleships, even in the numbers in which the Japanese deployed them, could not challenge a well-handled carrier force. 'Command of the sea', therefore, now rested on winning command of the air, as both navies had long recognised. Somewhere in the depths of the Pacific, the largest space on the surface of the globe, the Japanese and American carrier fleets must meet and battle it out for a decision. If the decision went in favour of the Japanese, as probabilities implied, their New Order in Asia would be safe for years to come.

The Japanese carrier fleet outnumbered the American by ten to three; if its light carriers were excluded, its navy still enjoyed a superiority of six to three. Moreover, the Japanese carriers and – even more important – their air groups were of the first quality. Before December 1941 the Americans had dismissed the Japanese carrier force as an inferior imitation of its own. Pearl Harbor had revealed that Japanese admirals handled their ships with superb competence and that Japanese naval pilots flew advanced aircraft, dropping lethal ordnance, with deadly skill. The Zero had established itself as the finest embarked fighter in any navy; the Kate and Val torpedo- and dive-bombers, though slower than their American counterparts, carried heavy loads over long ranges.

The Imperial Japanese Navy had not built and trained a carrier fleet as a second best to its battleship force. On the contrary, its carrier fleet was a national elite. For that the Americans – and the British – had only themselves to blame. At the Washington Naval Conference of 1921 they had forced the Japanese to accept a severe restriction on the number of capital ships they were allowed to possess. The ratio fixed was three Japanese ships to five British or American. The object was to limit the number of Imperial Japanese Navy battleships in the Pacific, which was a secondary theatre for the two Western navies, who at that time were locked in unspoken conflict over which was to enjoy primacy in the Atlantic. Aircraft carriers were subject to the restriction, but the purpose of including them was to guard against the danger that any power could launch ships in the guise of carriers which might subsequently be converted to battleships. Japan went the other way about. Already persuaded that the carrier was likely to be a dominant naval weapon of the future, it not only converted a number of battleships and battlecruisers into carriers, as it was allowed to do under the 1921 treaty (and Britain and America were doing likewise, to preserve seaworthy hulls they would otherwise have had to scrap). It also launched a number of seaplane carriers, a category the Washington Treaty did not recognise, with the object of converting them into aircraft carriers at a later date.

By conversion and new building they had succeeded by 1941 in creating the largest carrier fleet in the world, which not only embarked the largest naval air force, of 500 aircraft, but was also grouped – the analogy might be with the German Panzer divisions – in a single striking force, the First Air Fleet. The four light carriers could be detached for peripheral operations. The six large carriers – *Akagi* (Red Castle), *Kaga* (Increased Joy), *Hiryu* (Flying Dragon), *Soryu* (Green Dragon), *Shokaku* (Soaring Crane) and *Zuikaku* (Happy Crane) – were kept together for strategic offensives. They formed the group which had devastated Pearl Harbor. In May 1942 they stood ready to engage the American carrier group in battle and consummate Japan's victory in the Pacific.

The American carriers, though few in number, equally did not represent their navy's second best. *Lexington* and *Saratoga*, completed on battlecruiser hulls in 1927, were in their time the largest warships in the world and were still formidable ships in 1942; *Enterprise* was a later but purpose-built carrier; *Yorktown* and *Hornet*, which were to join her in the Pacific from the Atlantic Fleet, were sister ships. The aircraft they embarked were not the equal of the Japanese. In particular, in 1942 the Americans lacked a good shipborne fighter. But their aircrew, even by comparison with the First Air Fleet's elite, were outstanding. America, after all, was the birthplace of the aeroplane, her youth had conceived a passion for flying from the start, and the US Navy's carrier pilots were leaders of the breed.

Carrier flying excluded all but the best. The technique of launching and 'landing on' was extremely rigorous: at take-off, without catapult, aircraft dipped beneath the bows of the ship and frequently crashed into the sea; at landing pilots were obliged to drive at full power into the arrester wires lest the hook missed contact and they were forced into involuntary take-off, the alternatives being a crash on the flight deck or a probably fatal ditching. Flight away from the ship was quite as perilous as launching and landing. In 1942 there was no airborne radar. The gunner of a 'multi-seat' torpedo- or dive-bomber could keep a rough check of bearings headed and distance flown, and so guide his pilot back to the sea area in which they might hope to find the mother ship by eyesight – from high altitude in clear weather. A fighter pilot alone in his aircraft, once out of sight of the mother ship, was lost in infinity and found his way home by guess or good luck. Extreme visual range in the Pacific, from 10,000 feet on a cloudless day, was a hundred miles; but strike missions might carry aircraft 200 miles from the carrier, to the limit of their endurance – and perhaps beyond. If the carrier reversed course, or a pilot was tempted by a target to press on beyond his point of no return, a homing aircraft could exhaust its fuel on the homeward leg and have to ditch into the sea, where its crew in their dinghy would become a dot in an ocean 25 million miles square. Only the bravest were embarked on carriers as aircrew.

To bravery the air groups of the Japanese carriers added experience. By May 1942 they had not only devastated Pearl Harbor but bombed Darwin in northern Australia and operated against shore targets in the East Indies. In April they had crossed the Indian Ocean to find the British Eastern Fleet in Ceylon, attacked the naval bases at Colombo and Trincomalee and chased its old battleships to refuge in the ports of East Africa. The American carrier crews, by contrast, were lacking in battle experience. They had attempted, but failed, to relieve Wake Island. They had made raids on the Marshall Islands (Kwajalein), the Gilberts, the Solomons and New Guinea. A few of the aircrew had met Japanese fighters or anti-aircraft fire; but, apart from sinking a minesweeper, they had not yet fulfilled the purpose for which they had been trained and embarked – to carry bombs or torpedoes against an enemy battle fleet.

The Doolittle raid

Suddenly, during May, the opportunity came their way not once but twice. The circumstances which provoked this outcome were unusual in the extreme. During March the American Chiefs of Staff had discussed with President Roosevelt a means of striking back at Japan for the outrage of Pearl Harbor and 'bringing the war home to the Japanese'. To do so they would have to attack the Japanese home islands, an apparently impossible mission, since the islands lay far beyond aircraft range of the United States' Pacific bases, while to send carriers close enough to launch their embarked aircraft would be to put them at fatal risk. The only solution was to embark long-range land bombers on a carrier, in the hope that they could be got into the air and deliver bombs on a target precious to Japanese national pride: Tokyo. The mission was theoretically possible but barely practicable. Nevertheless Washington resolved to try it. On 2 April USS *Hornet* left San Francisco, with sixteen B-25 medium-range bombers on its flight deck, under the command of the notable airman Colonel James Doolittle.

The plan was that *Hornet* should approach to within 500 miles of Japan, launch its aircraft and then retire, while the aircraft bombed Tokyo and then flew on to land in China in areas still controlled by Chiang Kai-shek – who was told to expect the B-25s but not informed of the mission they were flying. On

18 April *Hornet* and its escorts were 650 miles from Tokyo, having approached on a route which ran between Midway and the northern Aleutian islands, when a Japanese naval picket was sighted. Admiral William Halsey, commanding the task force, decided to launch the B-25s instantly, even though they would be at the extreme limit of their endurance, and run for the security of the deep Pacific. All Doolittle's bombers lurched safely off the flight deck, and thirteen bombed Tokyo and three other Japanese targets; four landed in China, one landed in the Soviet Union, and the remainder were abandoned by their crews, who took to parachutes over China. Of the eighty fliers who departed on this reckless venture, seventy-one survived to return to the United States.

The Doolittle raid might nevertheless have been judged a fiasco if it had not registered with the Japanese high command. The citizens of Tokyo, to whom no public acknowledgement of the raid was made by the government, did not associate the scattering of explosions with an American attack; but the generals and admirals, as servants of the god-emperor, were horrified by the threat to his person the bombing represented. At that very moment a debate was raging between the Naval General Staff and the Combined Fleet over the future development of the Pacific war. The shore-based staff officers had committed themselves to an advance into the 'Southern Area', with the object of capturing more of New Guinea and additional footholds in the Solomons and New Caledonia, from which to attack Australia and menace her long seaward flank of communication with the west coast of the United States. The Combined Fleet, represented by Yamamoto, wanted not the acquisition of additional territory, however strategically valuable, but a strategic victory. They believed they could provoke a decisive battle with the United States Navy's remaining carriers by mounting an invasion of Hawaii's outlier, Midway Island, for which they were sure the Americans would be bound to fight.

Doolittle's raid ended the argument. *Hornet* had reached its launch point through the Midway 'keyhole' in Japan's defensive perimeter. As no senior officer in Japan could publicly countenance allowing the keyhole to remain open, since to do so implied unconcern for the emperor's well-being, the Naval General Staff at once withdrew their objection to the Combined Fleet's plan and accepted Yamamoto's Midway proposal. An operation already scheduled for early May, to extend the footholds in New Guinea that they had established between 8 and 10 March, was to go forward; by putting troops ashore at Port Moresby, on the great island's southern shore, Japan would further menace Darwin and Australia's Northern Territory. However, as soon as the carriers needed to cover the landings had withdrawn, they would be concentrated in the central Pacific for an offensive against Midway.

'Magic', for the first significant time in the Pacific war, here came to the Americans' aid. Interception and decryption of careless Japanese signals – a symptom of their infection with the 'victory disease' with which they would later reproach themselves – alerted the Pacific Fleet to the impending Japanese strike on Port Moresby. It accordingly dispatched the carriers *Lexington* and *Yorktown* to intercept the Japanese invasion fleet. The Japanese fleet was protected by three carriers, the new but small *Shoho* and the large *Shokaku* and *Zuikaku*. *Shoho* was sunk by bombing in a lucky encounter on 7 May. Next day, after complex manoeuvres, the aircraft of the two remaining Japanese and the two American carriers found each other's mother ships, separated by 175 miles of sea, and delivered fierce attacks. *Shokaku* suffered heavy damage; *Yorktown* was only lightly damaged, but *Lexington* was set on fire by a leak from her aviation fuel lines and had to be abandoned.

This Battle of the Coral Sea had two salutary effects for the Americans. It checked the Japanese army's advance to positions offshore of Australia and confined it thereafter to northern New Guinea. It also reassured the American carrier and air group crews that they were at least the equals of their opponents, besides bringing *Yorktown* valuable combat experience. *Yorktown*'s damage was repaired at Pearl Harbor in forty-five hours, after she had arrived on 27 May for what her captain estimated was a necessary 'ninety-day refit'. On 30 May she sailed to join *Enterprise* and *Hornet*, as yet uninitiated in carrier-to-carrier combat, for battle off Midway.

The fatal five minutes

All Americans concerned with the battle recognised that it must be a desperate affair. Of the five American carriers in commission, *Wasp* was returning from the Mediterranean where she had helped deliver aircraft to Malta and *Saratoga* was working up after completing repairs. The three remaining constituted the nucleus of two task forces, 17 (*Yorktown*) and 16 (*Enterprise* and *Hornet*), and put to sea with an impressive escort of cruisers and destroyers. However, the admirals commanding, Frank John Fletcher (Task Force 17) and Raymond Spruance (Task Force 16), knew that they would fight at a severe disadvantage. The Japanese First Air Fleet had six large carriers (in the event four – *Akagi*, *Kaga*, *Hiryu* and *Soryu* - were to voyage to Midway), was supported by battleships and would have a clear superiority in aircraft: for Midway Japanese carriers embarked seventy each, American only sixty. On the day, 272 Japanese bombers and fighters would confront 180 American. These were extreme odds.

The odds were to be shortened, however, as before the Coral Sea, by the operation of Magic. Japanese radio security for the Midway operation was rigorous. The operation itself was designated MI, which might have meant anything, despite the apparent reference to Midway Island; and its target was designated AF. Much traffic containing these designations was intercepted by the American Magic cryptanalysts but the decrypts gave no clue where the First Air Fleet was bound. One of the cryptanalysts on Hawaii had nevertheless convinced himself that Midway was the target and he contrived to set a trap for the Japanese. On a secure telegraphic link between Hawaii and Midway he instructed the garrison at Midway to radio in clear that it was running short of fresh water, an innocuous administrative message which he believed would not arouse Japanese suspicions. His confidence was justified. An Australian antenna of the Magic network shortly intercepted a Japanese cipher transmission which signalled that AF had reported a shortage of fresh water. The trick had therefore revealed the Japanese target; subsequent decrypts established that the operation designated MI would take place on 4 June. Task Forces 16 and 17 therefore sailed to position themselves north-east of Midway in time for the Japanese arrival.

Naval officers had originally conceived of the aeroplane as an adjunct to a battleship fleet with the ability to scout for the enemy and spot for the fall of shot once action was joined. The Royal Navy, even in 1942, clung to that view of the naval aircraft's role. In both the American and Japanese fleets, however, naval aviators had achieved an authority which relegated the traditionalists' obsession with the battleship to second place. They rightly judged that the carrier and its air groups had become queen of the oceans. No battle, not even the Coral Sea, fought in confined waters in support of landing forces, had yet borne out their judgement. Now they were to put their judgement to the test in the landless expanses of the central Pacific, 2000 miles from a continent in any direction.

The First Air Fleet, accompanied by the battleships and cruisers of the Midway Occupation Force, drew within range of the tiny island on 4 June. Many other Japanese naval forces, some dispatched as far northward as the Aleutians, were in motion at the same time, their mission being to confuse the commanders of the American Pacific Fleet and compel them to disperse their strength. This was over-subtle. The United States Navy was so weak in the Pacific in mid-1942 that it had to keep all its capital units together in order to concentrate a strategic force. It did, however, dispose of a force of which there was no equivalent in Yamamoto's armada: land-based aircraft, the Catalina amphibious flying-boats and Flying Fortresses stationed at Midway atoll itself. They could operate against the Japanese carriers and escape homeward without fear of finding their landing platform moved out of range or, worse, sunk. The land-based aircraft were to exercise an important influence on the unfolding of the Battle of Midway.

Indeed, the first move against the Japanese carriers was made by a Catalina flying-boat from Midway flying reconnaissance on 3 June. It spotted the invasion fleet heading towards the island, thus confirming that the Magic intelligence was correct. Next morning Flying Fortresses bombed but missed the fleet, while four Catalinas actually sank one of the invasion ships; it was only a humble oiler, but the blow was sufficient to convince the carrier admiral, here as at Pearl Harbor the redoubtable Nagumo, that he must overcome the island's defences before the landing began. At 4.30 am all four of his carriers launched nine squadrons of bombers, armed with fragmentation bombs for a ground attack and escorted by four squadrons of Zero fighters. Radar gave the

Americans warning, but that could not compensate for the inferiority of the old fighters based on the atoll. Two-thirds of the American fighters were destroyed, heavy damage was done to Midway's installations, and the Japanese bombers arrived back at the carriers without loss.

The leader of the raiding force nevertheless reported to Nagumo on his return that Midway ought to be bombed again – which was not in the admiral's plan. While he cogitated, Midway struck back. One of its dawn patrols had identified the position of Nagumo's carriers, reported it and prompted Admiral Chester Nimitz, the Pacific Fleet commander, to order Midway to launch its aircraft again; Fletcher, in overall command of the task forces, simultaneously ordered *Enterprise* and *Hornet* to manoeuvre into an attacking positon and did likewise with *Yorktown*. The arrival of Midway's second wave of land-based bombers made Nagumo reach a decision. Though his carriers' decks were cluttered with torpedoes brought up from below to arm his returned aircraft for a strike against American surface ships – should they be identified in the area – he now cancelled that mission and ordered them to be armed again with fragmentation bombs for a second strike against Midway.

That would take time. While time was running out, *Hornet* and *Enterprise* reached positions from which they could launch their own torpedo- and dive-bombers, and did so at 7 am. *Yorktown* launched its bombers an hour later. By 9 am the sky to the north-east of Midway was filled with 150 American aircraft winging their way across the 175 miles of ocean that separated the two fleets.

Nagumo already knew that danger threatened. At 7.28 one of his reconnaissance aircraft had reported a sighting of enemy ships. Infuriatingly it did not identify what type. It was not until 8.20 that it tentatively signalled the presence of a carrier, and not until 8.55 that it warned that torpedo aircraft were in the air and heading Nagumo's way. The Japanese admiral now recognised that he had made a serious mistake and quickly countermanded his order to rearm with bombs; but, given his superiority of numbers, his error was not necessarily grievous. In any case he was fully occupied with other matters. First the Midway bombers Nimitz had ordered into the air were massacred by the guns of his Zeros. Then, between 8.40 and 9 am, the bombers sent on his own second strike against Midway returned and had to be landed on. As soon as they had been recovered, crews swarmed about them with refuelling hoses and ordnance trolleys loaded with torpedoes for their next mission, this time against ships.

It was with their decks crowded with refuelling and rearming bombers that the first arrivals from *Enterprise* and *Hornet* found the four Japanese carriers, sailing in tight box formation and protected overhead by their combat air patrol of Zeros, just before 9.30 am. Nagumo had recently changed course, so the encounter initially contained an element of luck for the Americans. It lasted but briefly. By 9.36 all *Hornet*'s and ten of *Enterprise*'s torpedo-bombers had been shot down; their dive-bombers had been foxed by Nagumo's change of course and, failing to find a target, either turned home if they had enough fuel, landed in Midway or simply ditched; all the fighters escorts ran out of fuel and fell into the sea.

Nagumo had had a lucky escape. He had counted on his superiority of numbers to see him through even if attacked when at a disadvantage, and that, with his well-calculated change of course, had extricated him from danger. Two-thirds of his enemy's strike aircraft had been repelled or destroyed. There was not even a probability that the remainder would now be able to find him.

Yorktown's torpedo-bombers, navigating by hunch and then drawn towards the smoke of combat between Nagumo's carriers and Task Force 16's aircraft, found him none the less. However, forced to fly straight, low and level to deliver their torpedoes, seven out of twelve were shot down by his combat air patrol and none of the torpedoes launched found a mark. By ten o'clock Nagumo had repelled what appeared to be the final American attack and was preparing to launch his own aircraft to find and destroy his disarmed antagonist somewhere on the far side of Midway. His formation was somewhat scattered, but none of his ships was damaged and his fighter force was intact.

Unfortunately the fighters were temporarily at the wrong altitude. Drawn down to sea level to fight off *Yorktown*'s torpedo-bombers, they had left the sky open to any dive-bomber force that might appear. One of *Enterprise*'s dive-bomber groups had overflown 175 miles of sea, lost contact with the rest of its mother ship's aircraft, taken a wrong course and been guided to its target by luck and shrewd guesswork. At 10.25 on the morning of 4 June 1942 it was exactly placed to deliver the most stunning and decisive blow in the history of naval warfare. Its leader, Lieutenant-Commander Wade McClusky, turned to attack and from 14,500 feet led his thirty-seven Dauntless dive-bombers seaward in a plunge at the Japanese carriers' flight decks.

All were cluttered with aircraft and the paraphernalia of refuelling and rearming. High-octane fuel hoses ran between piles of discarded bombs, which stood beside aircraft running their engines for take-off; they were ingredients of catastrophe. *Akagi*, Nagumo's flagship, was the first to go. A bomb started a fire in a torpedo store and within twenty minutes raged so fiercely that the admiral had to shift his flag to a destroyer. *Kaga*, hit by four bombs, was set ablaze by its own aviation fuel and had to be abandoned even more rapidly. *Soryu* suffered three hits; one started a fire among aircraft parked on deck, stopped her engines and left her victim to a shadowing American submarine which sank her at noon.

Within exactly five minutes, between 10.25 and 10.30, the whole course of the war in the Pacific had been reversed. The First Air Fleet, its magnificent ships, modern aircraft and superb pilots, had been devastated. And the disaster was not at an end. *Hiryu* had evaded attack and got away – but only temporarily. At five in the afternoon she was found racing furiously away from Midway by dive-bombers from *Enterprise*, was hit with four bombs, set on fire and left to be scuttled by her crew when the flames took hold from stem to stern.

Thus the whole of Nagumo's fleet and, with it, the dream of empire perished. Yamamoto's prophecy of 'running wild' for six months had been fulfilled almost to the day. Not only did the balance in the Pacific between fleet carriers now stand equal (*Yorktown* was sunk by a submarine on 6 June); the advantage the Japanese had lost could never be made good – as Yamamoto knew, having seen American industry at first hand. Six fleet carriers would join the Japanese navy in 1942-4; America would launch fourteen, as well as nine light carriers and sixty-six escort carriers, creating a fleet against which Japan could not stand. It was now to be condemned to the defensive, though in waging a defensive war it would test the courage and resources of the United States and its allies to the utmost.

FIFTEEN

Occupation and Repression

Midway, though a catastrophic defeat, lost Japan not a foot of territory and in no way altered the boundaries of its new empire. The true consequences of Midway lay far in the future, when the Americans penetrated the Japanese defence perimeter in the southern and central Pacific at the end of 1943 and forced the Japanese once again to wage mobile warfare – by which time Japan had lost its superiority in naval airpower. In the meantime they retained their enormous area of conquest – eastern China and Manchuria, the Philippines, French Indo-China, British Burma and Malaya, the Dutch East Indies, together with their ally Thailand – and administered it undisturbed by Western Allied interference.

Conquest brings problems as difficult, if not as acute, as those of organising victory itself. Order must be maintained, governments replaced, currencies supported, markets revived, economies sustained and exploited for the conqueror's profit. The Japanese did not come to empire unprepared, however. They had had ten years' experience of administering conquered territory in Manchuria. More important, they had a theory of empire which was by no means hostile to or unpopular with all the peoples whose sovereignty they had wrested from the former colonial powers. The idea of a 'Greater East Asia Co-Prosperity Sphere' which had taken root in the army, navy and nationalist circles in Japan before the war was at one level a cloak for imperial amibition; at another it clothed a genuine belief in the mission of Japan, as the first Asian great power, to lead other Asians to independence from foreign rule. Many in Asia were enthused and inspired by the Japanese triumph of 1942 and were ready, even eager, to co-operate with it.

The establishment of a 'New Order' in Asia had been adopted as an aim by the second Konoye cabinet in July 1940. In February 1942 the Tojo cabinet set up a Greater Asia Council and in November a Ministry for Greater Asia. The high point of Japan's pan-Asian policy came a year later when the first – and only – Greater East Asia Conference was convened in Tokyo in November 1943. Its composition revealed the varied character of administration that Japan had imposed within the Co-Prosperity Sphere.

The conference was attended by the Prime Ministers of Manchuria (Manchukuo), Chang Chung-hui, and Japanese-controlled China, Wang Ching-wei, a representative of the Thai government, Prince Wan Waithayakon, the President of the occupied Philippines, José Laurel, the head of state of occupied Burma, Ba Maw, and Subhas Chandra Bose, the leader of the 'Free Indian Government'. The Prime Ministers of Manchuria and 'China' were Japanese puppets, entirely without power and effectively instruments of Japanese exploitation of their annexed territories. Subhas Chandra Bose, a messianic Indian nationalist, had deliberately exiled himself from British India in order to raise the standard of revolt and arrived in Japan in mid-1943 in a German U-boat. He had no standing in official nationalist circles in India, where he had broken with Gandhi and the Indian National Congress, but had a considerable popular following and in the Indian National Army, raised from prisoners taken in Malaya, a sizeable military force under his influence – though not under his command, which remained in Japanese hands. Ba Maw was a genuine enthusiast for the Co-Prosperity Sphere; as the head of a state whose declaration of independence had been sponsored by Japan on 1 August 1942, he had on the same day declared war against Britain and the United States. The Burma Independence (later National) Army was led by young nationalists who were prepared to fight on Japan's side. The Thai prince represented an independent state which had allied itself with Japan (in circumstances which would have made any other policy difficult) and had been rewarded by grants of territory from neighbouring Burma and Laos. José Laurel was a political associate of the legitimate but exiled President of the Philippines, Manuel Quezon, who had charged him to pretend co-operation with the Japanese. However, Laurel had subsequently been converted to pan-Asianism, and under his presidency the Philippines had declared its independence – already conditionally conceded by the United States – on 14 October 1942.

The territories excluded from the conference were Indo-China, Malaya and the Dutch East Indies. Indo-China – Vietnam, Laos and Cambodia – though under Japanese occupation, remained under Vichy French colonial administration until March 1945, when the Japanese rightly suspected it had turned to the Free French and overthrew it. Malaya's large Chinese population, among which a small communist guerrilla movement flourished, made it an unsuitable country in which to experiment with self-rule – though a sizeable Muslim Malay element was not hostile to the Japanese, who had promised them independence at a later stage. The Muslims of the Dutch East Indies had been given a similar promise; while fighting with the Americans and Australians continued in New Guinea it was not judged opportune to grant them independence, but numbers of nationalist leaders including the future president, Sukarno, agreed to join a Central Consultative Committee, or quasi-government, in September 1943. Certain other territories, including Hong Kong, Singapore and Dutch, British or Australian Timor, Borneo and occupied New Guinea, had such strategic importance that they were simply annexed to the Japanese empire and placed under military government.

Although at the outset Japan's devolution of power worked successfully, the

Filipinos, who had no dislike for the Americans and were proud of their heavily Westernised culture, accepted occupation grudgingly and sheltered the only large-scale popular anti-Japanese guerrilla movement to flourish within the Co-Prosperity Sphere. In parts of Burma and Thailand local labour was conscripted to work in harsh conditions with Allied prisoners of war; of these, 12,000 out of 61,000 prisoners and 90,000 out of 270,000 Asian labourers died on the construction of the Burma railway. In other areas, however, the occupied population found little to resent in the change from colonial administration. The educated classes initially welcomed it and were only gradually alienated by the discovery that the Japanese were as racially arrogant as their former European masters.

Such was not the case in China. An independent state, however ineffective its government before 1937, it had been invaded without provocation and then systematically exploited for profit wherever the Japanese armies were strong enough to impose their control. The educated classes, deeply conscious of the antiquity and grandeur of their vanished imperial system and consistently resentful of the West's commercial penetration and diplomatic aggression in the previous century, were as disdainful of the Japanese as of all other foreign cultures. However, the area still controlled by the Kuomintang was ravaged by inflation, which had reached 125,000 per cent by 1945, and by constant conscription drives; the communist north-east was gripped by an austerity deeply inimical to the national character; and the rural areas within the Japanese sphere of occupation, which contained about 40 per cent of China's agricultural land, were scourged by constant 'rice offensives'. In the prevailing chaos many Chinese made an accommodation with the enemy. In 1942 the Chinese communists publicised the fact that twenty-seven Kuomintang generals had gone over to the Japanese; more significantly a vast system of local trade had sprung up between occupiers and occupied. Since the Japanese armies generally controlled only the towns, and were driven to raise necessities from the surrounding countryside by force – a wearisome and inefficient practice – landowners and merchants rapidly entered into trading arrangements with the local Japanese commanders, and this market relationship proved more satisfactory to both sides. This accommodation prevailed throughout most of western China until the institution of the Ichi-Go offensive, at Tokyo's decree, in the spring of 1944.

Hitler's 'New Order'
The ethos of Japan's 'New Order' was co-operative; yet it was self-deluding and insincere in conception, and frequently harsh and extortionate in practice. Japanese excesses and atrocities, however, were arbitrary and sporadic. The pattern of its German ally's occupation policies was exactly contrary. Hitler's 'New Order' was designed to serve Greater Germany's interests exclusively; but the system of coercion used to make it work was calculated and methodic,

while the recourse to punishment, reprisal and terror which underlay it was governed by rules and procedures implemented from the centre.

By the end of 1942 German troops occupied the territory of fourteen European sovereign states: France, Belgium, Holland, Luxembourg, Denmark, Norway, Austria, Czechoslovakia, Poland, Greece, Yugoslavia and the three Baltic states of Estonia, Lithuania and Latvia. Austria had since 1938 been amalgamated with the Reich as sovereign territory, together with the Sudetenland province of Czechoslovakia and, since 1939, the former German provinces of Poland. Luxembourg, the French provinces of Alsace and Lorraine and the Yugoslav provinces of southern Styria and Carinthia, occupied in 1940-1, were deemed territories whose 'union with the Reich' impended and were under 'special civil administration' by the Ministry of the Interior. Denmark retained its elected government and monarchy under German Foreign Office supervision. Norway and Holland were under the supervision of Reich commissioners who answered directly to Hitler; their monarchs and governments had gone into exile but their civil administrations remained in place. Belgium was placed under military government by the occupying Wehrmacht, as was north-eastern France; but the French Vichy government remained the civil administrative authority throughout the country, even after the extension of Wehrmacht occupation into the 'Free Zone' in November 1942.

There were no devolved arrangements in eastern or southern Europe except in Slovakia, which had been detached as a puppet state. The troubled Serbian province of Yugoslavia was under military government, as was Greece (though parts of Greece and Yugoslavia were Italian-occupied until September 1943, while certain Yugoslav border provinces were annexed to their neighbours). Eastern Poland, White Russia and the Baltic states were designated 'Ostland' and, with the Ukraine, run by Reich commissioners effectively as colonies. The Czech parts of Czechoslovakia had been denominated a Reich protectorate, as Bohemia-Moravia, and were directly ruled by Germany, as was the rump of Poland, known as the 'General Government'. Areas of Russia immediately behind the battlefront were held under military government.

Special economic arrangements prevailed, as they had done during the First World War, in Belgium and the French northern departments, whose coal and iron industries were run as a single unit, its output co-ordinated with that of the Ruhr and occupied Lorraine (the success of this military 'iron, steel and coal community' was to plant the seed of the European Economic Community of the post-war period). However, in a wider sense, the whole of Hitler's European empire was administered for economic return. In the industrialised West, delegates of the Economic and Armaments Ministries established working arrangements with the existing managements of individual factories and larger enterprises to agree output quotas and purchasing arrangements. Similar agreements were made with agricultural marketing agencies and ministries.

Western Europe's intensive agriculture was a magnet to economic planners in Germany, where 26 per cent of the population in 1939 had been employed on the land without managing to meet the country's foodstuff needs. Underpopulated France, a major pre-war exporting country, was expected to provide an important part of the shortfall, particularly after mobilisation had removed one in three of Germany's male population. Danish farming, famously efficient, was also considered a prime source of agricultural imports, particularly of pork and dairy products. Denmark, partly because it was ruled on a light rein, responded well to German demands; the 4 million Danes provided rations for 8.2 million Germans for most of the war. France, though it fed the sixty occupying German divisions and found surpluses for export, did so only at the expense of reducing its own intake; French agricultural productivity actually declined during the war, largely through a shortage of artificial fertiliser which affected farming throughout the Nazi empire.

Purchases of all commodities from France were financed throughout the war largely through credits levied by the so-called 'occupation costs', an arbitrary annual levy on the French revenue whose amount was dictated by Germany at an artificially low exchange rate between the franc and the mark which favoured Germany by as much as 63 per cent. Similar arrangements were imposed upon other occupied countries; but it was in France, the largest and most industrialised of the occupied countries, that they bit hardest and produced the largest returns – from 1940 to 1944 no less than 16 per cent of the Reich treasury's income. To a certain extent 'occupation costs' were offset by direct German investment in French war industry, often by private enterprises such as Krupp and IG Farben; but such investment was entirely self-serving, a mere priming of the pump to enable French industrialists to maintain or increase their supply of goods to a single market for sale at a price ultimately fixed by the German buyers.

German purchase of French, Dutch and Belgian industrial products – which included items for military use such as aero-engines and radio equipment, as well as finished steel and unprocessed raw materials – were acquired in a rigged market; it was a market none the less, and the German purchasing authorities, such as the officials of the Franco-German Armistice Commission, were careful to preserve the autonomy of their opposite numbers. Such was not the case with German intrusions into the western European labour market. During the war German industry and agriculture developed an insatiable appetite for foreign labour. Since military requirements reduced the size of the domestic labour force by one-third, and Nazi policy precluded the large-scale employment of German women, the shortfall had to be made good from outside the borders of the Reich. Prisoners of war supplied some of the numbers, over a million Frenchmen being employed on German farms and in German mines and factories between 1940 and 1945; but even military captivity failed as a source of labour supply. As early as mid-1940 economic inducement was offered to tempt skilled workers from home and by December 220,000 Western workers were employed in Germany. However, as local economies recovered after the catastrophe of 1940, others resisted the lure; by October 1941 the Western foreign labour force in Germany had not risen above 300,000, of whom 272,000 were from allied Italy.

The Germans therefore resorted to conscription. Fritz Sauckel, Reich Plenipotentiary-General for Labour, required the administrations of the occupied countries of the West to produce stated numbers of workers and thereby raised the number of foreign workers in Germany between January and October 1942 by 2.6 million. (In France the Obligatory Labour Service eventually proved a prime impetus for the defection of the young to the Maquis.) The rate of increase was sustained into 1943, largely as a result of Italy's defection from the Axis in September which allowed the imposition of labour conscription there, yielding another million and a half young men.

Nevertheless recruitment in the West, whether by incentive or compulsion, could still not ultimately satisfy German requirements. Western workers had to be paid, fed and housed at western European standards, and the consequent charge on the German war economy grew progressively burdensome. The solution Sauckel introduced was conscription in the East. An immediate source of Eastern labour had been found in 1942 in the millions of Red Army men made prisoner in the encirclement battles at Minsk, Smolensk and Kiev. Out of the eventual total of 5,160,000 Soviet soldiers captured during the war, 3,300,000 died by neglect or murder at German hands; in May 1944 only 875,000 were recorded as 'working'. Most worked in slave conditions; so too did the 2.8 million Russian civilians, mostly Ukrainians, whom the Germans brought within the Reich between March 1942 and the Wehrmacht's expulsion from Russia in the summer of 1944. Originally invited to 'volunteer', the first recruits found themselves treated so badly that news of their virtual enslavement deterred others from following, and Sauckel had to resort to labour conscription to make up the numbers. A similar policy was imposed in the Polish 'General-Government'. The SS became the instrument of enslavement. Its leader, Heinrich Himmler, outlined the principles by which it worked in his infamous Posen address of October 1943: 'It is a matter of total indifference to me how the Russians, how the Czechs fare. . . . Whether the other peoples live in plenty, whether they croak from hunger, interests me only to the extent that we need them as slaves for our culture; otherwise it does not interest me.'

The exploitation of the East
The Reich commissioners for the Ukraine and the Ostland and the Government-General of Poland adopted a similar attitude to the exploitation of the economies under their control. In Poland private enterprises were either taken over by German managers or subjected to German managerial control; in the Soviet Union, where all production had been state-owned before the

invasion, the first priority was to restore war damage, contingent or more usually deliberate, which had, for example, resulted in the wrecking of three out of four of all electricity-generating stations in the conquered area. Once damage had been repaired, the operation of the whole industrial structure, including mines, oil-wells, mills and factories, was consigned to state corporations – in particular the Berg-und Hüttenwerk Company for mining, the Kontinentale Oel company for oil and the Ostfaser Company for wool fibre – which operated as extensions of Germany's nationalised industry. Later, when state corporations proved unequal to the task of managing all the plant that had been captured, private companies, including Krupp, Flick and Mannesmann, were allocated enterprises to oversee as part of their existing empires. The one Soviet economic system with which the Germans did not tamper was the collective farm. Inefficient though it was, and despite long-term plans favoured particularly by Himmler to settle the 'black earth' region with German soldier-peasants, the supply of both native and ethnic German settlers for the occupied area was too small to permit a wholesale transformation to private agriculture. In western Poland and other areas on the fringes of the Greater Reich, native cultivators were expropriated and replaced with Teutons; throughout the Ostland and the Ukraine there was an effort at reprivatisation, but the collective system was generally judged too well established to unravel.

Such changes as the Germans imposed were cosmetic. The *Agrarerlass* (Agricultural Edict) of February 1942 reconstituted collectives as agricultural communes, allegedly equivalent to the village societies which had existed before the Revolution, in which cultivators were granted rights of property over private lots and the German occupiers assumed the role of landlords to whom a proportion of the crop was owed as rent. In practice, as the cultivators quickly discovered, the Germans were as exigent as the commissars in exacting tribute, and failure to deliver it entailed loss of the private holding, expropriation and exposure to recruitment for forced labour.

In short, German agricultural policy in the East rested upon the principle of coercion, as did its whole *Ostpolitik*. Nazi Germany was not interested in winning the goodwill or even the co-operation of peoples it deemed by ideological edict to be inferior – *Untermenschen*. What was true in the East, moreover, was true throughout Hitler's empire. Coercion, repression, punishment, reprisal, terror, extermination – the chain of measures by which Nazi Germany exercised its power over occupied Europe – were inflicted with more circumspection west of the Rhine than east of the Oder. They were, nevertheless, the common instruments of control wherever the swastika flag flew, unrestrained by the writ of civil law, and pitiless in effect whenever the will of the Führer gave their agents licence.

That had been true first of all in Germany itself. Immediately after his appointment to the chancellorship of Germany in January 1933, Hitler had broadened the existing legal provision of *Schutzhaft* – protective custody of the person concerned, to protect him or her, for example, from mob violence – to embrace 'police detention' for political activity. To hold 'police detainees' detention centres were established at Dachau near Munich and Oranienburg in March 1933 and soon other such 'concentration camps', a term borrowed from the Spanish pacification of Cuba in the 1890s and later adopted by the British during the Boer War, had been established in other parts of Germany. Their first inmates were communists, held for terms determined by the Führer's pleasure; later other political and conscientious opponents of the regime, active or merely suspect, were detained, and by 1937 'anti-socials', including homosexuals, beggars and gypsies, were sent there. At the beginning of the war the number of concentration camp detainees was about 25,000.

No concentration camp was yet an extermination camp; all were merely places of arbitrary imprisonment. However, they were administered by a special 'Death's Head' branch of the SS, whose chief, Heinrich Himmler, was since 1936 also chief of the German police. This particular stroke of *Gleichschaltung* brought under the unified control of a Nazi official the political (Gestapo) and criminal police forces of the Reich, together with the ordinary civil police, but also the security organs (*Sicherheitsdienst* or SD) of the Nazi Party. Thereafter a German citizen was liable to arrest by the Gestapo, consignment to 'police detention' by an official of the SD and imprisonment by SS 'Death's Head' guards, without any intervention by the judicial authorities whatsoever.

The great conquests of 1939-41 brought the extension of SS/Gestapo power, now allied with that of the military police (the *Feldgendarmerie*), into the occupied territories. The effect was first felt in Poland, where acts of aggression against the leaders of society began immediately after occupation: professionals such as doctors, lawyers, professors, teachers and priests were arrested under the 'police detention' provisions and confined in concentration camps. Few were ever to emerge. Forced labour was a founding principle of the concentration camp system: the Nazi slogan *Arbeit macht frei*, 'Labour wins freedom', was the precept by which they operated. As concentration camps multiplied in occupied territory and their populations increased, rations dwindled, the pace of work accelerated, disease proliferated, and forced labour thus became a death sentence. The Poles were the first to die in large numbers, and those who did not survive *Schutzhaft* represented a significant proportion of the nation's loss of a quarter of its population during the war; thereafter few peoples were spared. The penalty for resistance, even dissidence, for Czechs, Yugoslavs, Danes, Norwegians, Belgians, Netherlanders and French was not arrest and imprisonment but deportation without trial, often ending in death. The most poignant of all the memorials on the great medieval battlefield of Agincourt is not the monument over the mass graves of the French knights who fell in 1415 but the modest *calvaire* at the gates of Agincourt château, which commemorates the squire and his two sons, 'morts en transportation à Natzweiler en 1944'.

Natzweiler, to which the three Frenchmen were transported to death, was one of eighteen main concentration camps run by the SS in and outside Germany. Tens of thousands died in those west of the Oder, worked or starved to death, killed by diseases of privation or, in individual cases, executed by decree. The western concentration camps were not, however, extermination camps; the appalling spectacle of death on which the British army stumbled at Belsen in April 1945 was the result of a sudden epidemic among the chronically underfed inmates, not of massacre. Massacre, however, was the ultimate horror which underlay the concentration camp system, and those camps which lay east of the Oder including particularly Chelmno, Belzec, Treblinka, Sobibor and Majdanek, had been built and run exclusively for that purpose.

Massacre is endemic to campaigns of conquest; it had been the hallmark of the Mongols and had been practised in their time by the Romans in Gaul and the Spaniards in South America. It was an index, however, of the degree to which Western civilisation had advanced that massacre had effectively been outlawed from warfare in Europe since the seventeenth century; it was a consonant index of Nazi Germany's return to barbarism that it made massacre a principle of its imperialism in its conquered lands. The chief victims of its revival of massacre as an instrument of oppression, however, were not those who opposed German power by offering resistance – resistance was what had chiefly invoked the cruel excesses of conquerors in the past – but a people, the Jews, whose very existence Nazi ideology deemed to be a challenge, threat and obstacle to its triumph.

The fate of the Jews

Jews had been legally disadvantaged in Germany immediately after the Nazi seizure of power; after 15 September 1935, under the so-called Nuremberg Laws, Jews were deprived of full German citizenship. By November 1938 some 150,000 of Germany's half-million Jews had managed to emigrate; but many did not reach countries which lay outside the Wehrmacht's impending reach, while the great concentration of Europe's Jews, as yet unmotivated to flight, lived within it. That included the Jews of the historic area of settlement in eastern Poland and western Russia, some 9 million in number, as well as the great Jewish populations of Warsaw, Budapest, Prague, Salonica and Lithuanian Vilna, the centre of Jewish religious scholarship. The diplomatic and military victories of 1938-9 put many of these East European Jews under Nazi control; Barbarossa engulfed the rest of them. Himmler, though he persisted in trying to establish his legal right to do so, began to massacre them at once. Four 'task groups' (*Einsatzgruppen*), divided into 'special commands' (*Sonderkommandos*) composed of German SS and securitymen and locally enlisted militias, had already killed one million Jews in the new area of conquest between June and November 1941. Most, however, had been killed by mass shooting, a method Himmler regarded as inefficient. In January 1942, at a meeting held at the headquarters of Interpol, of which Himmler was president, in the Berlin suburb of Wannsee, his deputy Heydrich proposed and received authority to institutionalise the massacre of the Jews, a measure to be known as the 'Final Solution' (*Endlösung*). Jews had been obliged to live in defined ghettos in Poland since the moment of occupation, and the order had subsequently been extended to other occupied areas. It was therefore not difficult to round up and 'transport' Jews for 'resettlement' in the east. Those sent to camps associated with an industrial plant run by the SS economic branch were usually worked to a state of enfeeblement before being sent to the gas chambers, though the old, the weak and the young might be gassed immediately; Auschwitz, the large camp in southern Poland, served both purposes. Those sent to the extermination camps, like Treblinka and Sobibor, were gassed on arrival. In this way, by the end of 1943, about 40 per cent of the world's Jewish population, some 6 million people, had been put to death; of the last large European Jewish community to survive, the 800,000 in Hungary, 450,000 were delivered to the SS between March and June 1944 and gassed at Auschwitz.

By that time, the head of the SS economic branch reported to Himmler on 5 April, there were twenty concentration camps and 165 subsidiary labour camps; in August 1944 the population was 524,286, of whom 145,119 were women. In January 1945 the total had risen to 714,211, of whom 202,674 were women. There were few Jews among them, for the simple and ghastly reason that the Final Solution was effectively complete. It seems possible, however, that Jews never formed a majority of the camp populations, since it was normally their fate to die on or soon after arrival; non-Jewish forced labourers, who were kept alive as long as they could work, may always have outnumbered them. In that irony lay a chilling dimension of Nazi racial policy. For the removal and transportation of Europe's Jews was a fact known to every inhabitant of the continent between 1942 and 1945. Their disappearance defined the barbaric ruthlessness of Nazi rule, offered an unspoken menace to every individual who defied or transgressed Nazi authority and warned that what had been done to one people might be done to another. In a profound sense, the machinery of the Final Solution and of the Nazi empire were one and the same: because systematic massacre underlay the exercise of Nazi authority at every turn, Hitler needed to rule his conquered subjects scarcely at all. The knowledge of the concentration camp system was in itself enough to hold all but a handful of heroic resisters abject during five years of terror.

SIXTEEN

The War for the Islands

The victory of Midway transformed the climate of war in the Pacific not only objectively but subjectively. From now on, the reheartened American chiefs of staff recognised, they could go over to the offensive. The question was: along which axis? The ultimate objective was the home islands of Japan, unless Tojo and his government could be brought to concede defeat before invasion became necessary. However, the home islands lay 2000 miles from America's remaining Pacific bases in Hawaii and Australia, between each of which a formidable chain of Japanese island fortresses interposed to block an American amphibious advance. The ground which had been lost so quickly by unprepared garrisons – or through the absence of any garrison at all – between December 1941 and May 1942 would now have to be recovered step by step at painful cost. Was it better to proceed along the pathway of the great islands of the East Indies or to leap across the stepping-stones of the tiny, isolated atolls of the north Pacific?

Choice of route implied choice of commander and of service. On 30 March 1942 the Joint Chiefs of Staff, General Marshall and Admiral King, had agreed on a division of strategic responsibilities in the Pacific. The new arrangement abolished ABDA and put Nimitz, commander of the Pacific Fleet with headquarters at Hawaii, in charge of the Pacific Ocean Area and MacArthur, commander of army forces in the region with headquarters in Australia, in charge of the South-West Pacific Area. To choose the northern route would be to make Nimitz and the navy paramount – a logical step, since the Pacific had always been the navy's interest. However, the small Marine Corps was its only military arm, and as yet it lacked the shipping, warships and men to stride across the atolls towards Japan. The army, by contrast, had the men, who were being shipped from the training camps to Australia in growing numbers; while the South-West Pacific Area route, which began close to Australia and proceeded along large islands that yielded at least some of the resources an offensive force required, demanded proportionately smaller shipping resources. To choose it, however, was to make paramount not only the army but its commander too. Although MacArthur had become a hero to the American people for his defence of Bataan, he was not popular with the nation's admirals. A prima donna among subordinates and a man who brooked no equals, he would, they feared, usurp the direction of strategy by subordinating naval to army operations if the South-West Pacific Area was made the primary zone of the counter-offensive.

Through stormy inter-service negotiations a compromise was reached. The services would take the southern route; but the area would be subdivided to allot part of the theatre to Nimitz and the navy, part to MacArthur and the army, which would have strictly limited call on the navy's transports, carriers and bombardment fleet. The compromise, agreed on 2 July 1942, consigned Task One, the capture of the island of Guadalcanal, east of New Guinea, to the navy. Task Two, an advance into New Guinea and its offshore island of New Britain, where Japan had a major base at Rabaul, would go to MacArthur; so eventually would Task Three, a final assault on Rabaul.

Guadalcanal, in the Solomons, committed both the United States Navy and the United States Marine Corps to a desperate struggle. Though safely approachable from New Zealand, the departure point of the operation, it was surrounded on three sides by other islands in the Solomons group, which together formed a confined channel that was to become known to the American sailors as 'the Slot'. Once troops were ashore, the navy was committed to resupplying them through these confined waters and so to risking battle with the Japanese in circumstances where manoeuvre was difficult and surprise all too easy for the enemy to achieve.

The 1st Marine Division, a regular formation of high quality, was landed without difficulty on 7 August and also took the offshore islands of Tulagi, Gavutu and Tanambogo. The Japanese garrison numbered only 2200 and was swiftly overcome. However, the appearance of the Marines on Guadalcanal provoked the Japanese high command to frenzy; 'success or failure in recapturing Guadalcanal', a document later captured read, 'is the fork in the road which leads to victory for them or us.' Since the Japanese recognised that a breach in their defensive perimeter at Guadalcanal would put the whole of their Southern Area at risk, they resolved on extreme efforts to retake it. On the night of 8/9 August off Savo Island they surprised the American fleet supporting the Guadalcanal landings, sank four cruisers and damaged one cruiser and two destroyers. From 18 August they poured reinforcements into the island, supported by naval guns and aircraft which continuously attacked its airfield (renamed Henderson Field in honour of a Marine pilot killed at Midway). On 24 August a fleet carrying the largest reinforcement yet dispatched was intercepted by the American navy east of Guadalcanal and the second of five battles fought in its waters ensued. This Battle of the Eastern Solomons was an American victory; though *Enterprise* was damaged, the Japanese lost a carrier, a cruiser and a destroyer and about sixty aircraft to the Americans' twenty.

Though repelled at sea, the Japanese were fighting furiously on land. The Marines, elite troops though they were, learned on Guadalcanal both the professional respect and ethnic hatred they were to feel for the Japanese throughout the Pacific war. A feature near Henderson Field became a focus of particularly fierce fighting; the Marines called it 'Bloody Ridge'. The navy meanwhile christened the nightly convoys of Japanese destroyers which ran reinforcements to the island the 'Tokyo Express'. It made regular efforts to intercept and on the night of 11/12 October caught and surprised a Japanese cruiser force in darkness. In this Battle of Cape Esperance the Americans came off best. However, on 26 October two much larger fleets met again in the Battle of Santa Cruz, south-east of Guadalcanal, and the decision went the other way. The Japanese had four carriers present, and 100 of their aircraft were shot down. Yet though the Americans had only two carriers at risk, and suffered half the total of Japanese aircraft losses, *Enterprise* was damaged and *Hornet*, the heroine of the Doolittle raid on Tokyo, went down.

The epic struggle for Guadalcanal

Before the Battle of Santa Cruz the Japanese had launched a violent offensive against the American defenders of Guadalcanal between 23 and 26 October, days of torrential rain which grounded the American aircraft operating from Henderson Field but allowed Japanese aircraft based elsewhere to deliver a succession of attacks. The Marines held out, counter-attacked and even received reinforcements, though in the teeth of Japanese efforts to close Guadalcanal's waters to American transports. Between 12 and 15 November, in three days of heavy fighting in 'the Slot' now known as the Battle of Guadalcanal, battleships clashed with battleships in the first classic duel of capital ships since Jutland – but on this occasion action was joined at night and radar proved the decisive factor. On the night of 12 November the Japanese flagship *Hiei* was so badly damaged that next morning she fell victim to aircraft from *Enterprise* and was sunk. On the night of 14/15 November the battleship *Kirishima* inflicted forty-two hits on the *South Dakota*; but *South Dakota* was brand-new and *Kirishima* old. *South Dakota* survived, while the *Washington* sent *Kirishima* to the bottom with nine 16-inch shell strikes delivered in seven minutes. A fortnight later, in the Battle of Tassafaronga on 30 November, an American cruiser force came off less well, but there, as in the fighting in 'the Slot' (also known as 'Ironbottomed Sound' from the number of ships sunk there), the Japanese covering force failed to run its troop transports to land. Thousands of Japanese soldiers had drowned in the course of the battles to win command of Guadalcanal's waters.

Starved of reinforcements and supplies, the Japanese garrison of Guadalcanal now began to falter. The island was plagued by leeches, tropical wasps and malarial mosquitoes, and as rations dwindled the Japanese troops fell prey to disease. The Americans too became ill – pilots at Henderson Field lasted only thirty days before losing the quickness of hand and eye necessary to do battle – but the tide of battle was now running their way. In January 1943 the Japanese commander on Guadalcanal withdrew his headquarters to the neighbouring island of Bougainville. In early February the 'Tokyo Express' began to operate in reverse, evacuating the sickly and exhausted defenders to

New Guinea. By 9 February Japanese resistance on Guadalcanal had formally ceased.

For the Marines Guadalcanal was remembered as an epic struggle. Men who had fought there bore an aura of endurance which veterans of almost no other Pacific campaign acquired. In terms of casualties it had nevertheless been a comparatively cheap victory. The Japanese had lost 22,000 killed or missing, the 1st and 2nd Marine Divisions, which bore the brunt of the fighting, only a little over a thousand dead. On Guadalcanal the American forces had established the tactical method they would employ across the width of the Pacific to beat the Japanese into subjection. It entailed the commitment of elite landing troops, heavily supported by ground-attack aircraft and naval gunfire, to take and hold key islands at the perimeter of Japan's area of conquest, as stepping-stones towards the home islands. As conceived and executed, it brought about a contest between morale and material. Both sides were to display supreme bravery; but, while the emperor's soldiers were ultimately dependent upon their concept of honour in sustaining their resistance, the Americans could call up overwhelming firepower to kill them in thousands. It was an unequal contest which in the long run the Americans were bound to win.

They were about to win another Pacific victory far from the steamy shores of Guadalcanal. In June 1942, in the only successful subsidiary of the Midway offensive, a Japanese force had landed on the two westernmost islands of the Aleutian chain, the American archipelago which runs from Alaska towards Japan. The Americans, preoccupied elsewhere, had let them bide; but in May 1943 Nimitz gathered a force, landed it on Attu and confronted the occupiers. He also sent three battleships in support, since the Japanese had fought a spirited heavy-cruiser action off the islands in March. The occupiers were few in number (2500) but inflicted 1000 dead on their American attackers before running out of ammunition and launching a suicidal bayonet charge. In August an even larger force recaptured Kiska, from which the Japanese prudently withdrew before they were attacked.

On New Guinea, in the equatorial belt, the Japanese had by contrast dug in to stay, in terrain which strongly favoured the defence. They had landed in the Papuan 'tail' of the New Guinea 'bird' on 22 July 1942, after the American victory in the Battle of the Coral Sea had deflected their effort to pass round it by sea to Port Moresby from north to south. Their attempts to take Port Moresby by an overland advance through the passes of the Owen Stanley range were checked by Australian troops and they were forced to fall back on their landing places at Buna and Gona. When the Australians, with American support, went over to the offensive, however, the Owen Stanley became a barrier to the Allies' advance, since the only route through the mountains was the tortuous Kokoda Trail. It was with the greatest difficulty that an attacking force was established in position outside Gona and Buna, where the Japanese were deeply entrenched. Fierce and painful fighting ensued throughout

November and December 1942. Though the Japanese were starving, the Australians and Americans were disheartened by the appalling conditions in which they had to fight. On 2 December, however, a new general, the American Robert Eichelberger, arrived and revitalised the offensive. By 2 January 1943 Buna had been taken; the Australian 7th and US 32nd Divisions had meanwhile captured Gona, which fell on 9 December. The casualties were again grossly to the disadvantage of the Japanese: they lost 12,000 dead in the campaign, the Allies 2850, mostly Australians.

Operation Cartwheel

Victory in Papua, though it left the Japanese with footholds in the rest of New Guinea, ended the threat to Australia and cleared the way for MacArthur to concentrate his efforts on breaking back along the southern route towards the Philippines through the Solomon and Bismarck archipelagos. His strategic concept, however, even though it entailed much 'island-hopping' – an essential by-product of which was to seize airstrips from which to establish overlapping zones of air control – which would leave bypassed Japanese garrisons to 'wither on the vine', included so many landing operations that its demands for men, ships and particularly aircraft threatened to exhaust the resources available. His ultimate objective was Rabaul, Japan's strong *place d'armes* in New Britain, the largest island of the Bismarck group; but his programme of advance would require five extra divisions and forty-five additional air groups, or about 1800 aircraft. As was pointed out at the Casablanca conference of January 1943, at that time there were already 460,000 American troops in the Pacific but only 380,000 in the European theatre, where preparations for the Second Front had already begun with the invasion of North Africa. MacArthur's demands provoked a severe inter-service dispute in Washington which lasted until March 1943. While the Japanese were locked in combat with the American and Australian soldiers and Marines in Guadalcanal and New Guinea, army and air force generals and admirals pitted the interests of their rival services over a decision about the development of the Pacific war. At the end of April 1943 a plan finally emerged. It was codenamed Cartwheel and, though it preserved the spirit of the agreement of 2 July 1942, it included a significant modification. Nimitz was now made theatre commander for the whole Pacific, and, while MacArthur was left in charge of the South-West Pacific Area, Admiral William Halsey was entrusted with operations in the South Pacific which would include an advance on MacArthur's flank. In short, MacArthur was to envelop Rabaul from the south, Halsey from the north. New Guinea and the southern Bismarcks were to be the former's responsibility, the Solomons the latter's. Once MacArthur had taken the north shore of New Guinea and the hinterland of New Britain, on which Rabaul stood, and Halsey had advanced along the Solomons chain to Bougainville, they would descend on Rabaul by pincer movement.

While the Joint Chiefs of Staff and their service subordinates were conducting this Pacific Military Conference in Washington, the Japanese, acutely aware that they had been forced on to the strategic defensive in the southern Pacific, were busy reinforcing and reorganising their garrisons there to withstand the expected assault. The overall commander was General Hitoshi Imamura, whose headquarters were at Rabaul; under his command was the Seventeenth Army in the Solomons. Imperial headquarters now decided to add to this a new army, the Eighteenth, to defend northern New Guinea. General Hatazo Adachi, commanding the Eighteenth Army, brought two new divisions from Korea and north China. Landing first at Rabaul, one of the divisions, the 51st, with Adachi's headquarters, then took ship for its new station at Lae in New Guinea. En route it was intercepted by American aircraft, which inflicted on it the first of two spectacular aerial successes achieved that spring.

In August 1942, MacArthur had been given a new air commander, General George Kenney, who had wrought a revolution in the USAAF's anti-ship tactics. Previously, though army pilots had reported numerous successes against the Japanese navy, after-action analysis had revealed that they had hit very few targets at all. Kenney transformed their methods. Recognising that the USAAF's chosen method of precision bombing from high altitude lay at the root of the failure, he trained his medium-range bomber pilots to attack at low level with guns and fragmentation bombs. When the 51st Division left Rabaul for Lae on 2 March 1943 it was first intercepted by Flying Fortresses employing the old high-level technique, which sank only one ship. Next day, however, a hundred medium-range B-25s, A-20s and Australian Beaufighters found it again, skimmed in at sea level, escaped the attention of the Zeros patrolling at high altitude to deal with the expected Fortresses and sank all the transports and four of the eight destroyer escorts.

The Battle of the Bismarck Sea was a significant material victory. Next month MacArthur's air force achieved a psychological victory of perhaps even greater importance. A recent addition to its strength was the long-range, twin-engined Lightning fighter, the P-38. Since it was no match for the Zero in dogfighting, the Lightning was chiefly reserved for strategic strikes against major formations of Japanese aircraft, diving against them from a high altitude. The P-38 became an object of terror and loathing, and the sound of its engines soon became familiar to Japanese airmen in the South Pacific. In an effort to reverse the success that the Lightnings and B-25s were achieving, Yamamoto assembled the largest available force of his own aircraft and committed them against Guadalcanal and its offshore island of Tulagi in early April 1943. This 'I-Go' operation, flown in early April, failed in its object, which was to sink as much shipping as possible, but the pilots reported differently. Like the American Flying Fortress pilots previously, they believed they had sunk ships which had in fact not been touched except by the waterspouts of their bombs. Yamamoto was nevertheless convinced and decided to visit his men to encourage them to further efforts.

Imprudently, notice of his intended arrival was circulated to the Eighth Area Army from Rabaul by cipher, which the American cryptographers at Pearl Harbor quickly broke. Nimitz decided to 'try to get him'. A squadron of Lightnings was hastily equipped with drop-tanks, to give them the extra range a successful ambush flight required, and on the morning of 18 April, as Yamamoto approached the airfield of Kahili in Bougainville, his aircraft was destroyed with a burst of 20 mm cannon fire and fell burning into the jungle.

Yamamoto's ashes were buried in Tokyo on 5 June. Later in the month began the great dual drive up the Solomons and New Guinea towards Rabaul which it had been one of the purposes of the 'I-Go' operation to check. At the end of June, Woodlark and the other Trobriand islands – the latter the focus of a famous ethnographic inquiry among its primitive inhabitants – were captured, thus securing the seaward approaches to the 'tail' of the New Guinea 'bird'. In June also an amphibious hook was made towards Lae on the northern New Guinea coast; it fell on 16 September and the Americans then moved on via Finschhafen to seize Saidor, opposite Cape Gloucester on New Britain, which they assaulted on 26 December 1943; a subsidiary landing was made on New Britain at Arawe, closer to Rabaul, on 15 December.

Meanwhile Halsey had been keeping pace with MacArthur in his own advance along the Solomons chain. The Russell islands next to Guadalcanal had been taken in February, the New Georgia group in June and July and Vella Lavella in August. The Japanese attempted both land and sea counter-offensives at New Georgia and Vella Lavella but were unsuccessful. By October 1943 Halsey was ready to assault Bougainville, the westernmost and largest of the islands in the Solomons, and only 200 miles from Rabaul at the narrowest sea crossing. The landings were preceded by a fierce but unsuccessful air battle to check the American advance; the plan, codenamed 'RO' by the Japanese, was devised by Admiral Mineichi Koga, who had succeeded Yamamoto as commander of the Combined Fleet. As soon as the battle was over, Halsey launched an amphibious assault on the small Treasury Islands off Bougainville's southern coast on 27 October and then a main assault at Empress Augusta Bay on 1 November. Koga sent a strong force of two heavy and two light cruisers to oppose the landings – hoping, as off Savo Island in the Battle of Guadalcanal, to inflict damage on the American fleet – and twice forced Halsey to risk unsupported carriers against them. However, the gamble paid off; the Japanese lost heavily in aircraft (fifty-five to the Americans' twelve) and suffered damage to three of their cruisers. By 21 November, the 3rd Marine and 37th Divisions were firmly established on Bougainville; from there, in conjunction with MacArthur's advance up New Guinea, the pincers now threatened to close about Rabaul.

The threat to Rabaul opened up the prospect of a seaborne advance along the northern shore of New Guinea from which MacArthur's and Halsey's forces might leap towards the East Indian islands of the Moluccas and so towards the

Philippines. Even as the trap began to close about Rabaul, however, the character of the war in the Pacific was taking another turn. At the Casablanca conference in January 1943 Roosevelt, Churchill and the Combined Chiefs of Staff had given assent to the American navy's cherished plan, proposed by the Chief of Naval Operations, Admiral King, for an advance towards the Philippines through the central Pacific to assault the Caroline and Marshall Islands. The decision of 2 July 1942, which had allotted naval support to MacArthur for his drive towards Rabaul, had been taken when the United States Navy was still painfully rebuilding its resources after the losses suffered at Pearl Harbor and in the victories of the Coral Sea and Midway. Then there had been few carriers and no battleships. By the beginning of 1943 American shipyards had begun to make good the gaps; by mid-1943 new battleships – essential for ship-to-shore bombardment in preparation and support of amphibious landings and to provide heavy anti-aircraft support for the carriers – and new carriers, of both the fleet and light classes, had arrived or were promised in plenty. The Anglo-American Washington conference in May 1943 had agreed that, as long as the forthcoming invasion of Europe was the first charge on the now rapidly expanding output of Allied war material, the offensive against Japan could be extended. On 20 July, therefore, enlarging on the spirit of that decision, the Joint Chiefs of Staff authorised Nimitz to prepare a landing operation against Japanese conquests in the Gilberts and to plan for subsequent landings in the Marshalls.

These were dramatic prospects. MacArthur's and Halsey's campaign in the south Pacific, though amphibious in character, was in essence a traditional land-sea advance. Navy supported army, and vice versa, in a series of leaps comparatively short in span. The longest leap that MacArthur had so far taken was 150 miles between Buna and Salamua, the longest by Halsey 100 miles from Guadalcanal to New Georgia. Distances in the central Pacific, by contrast, were of a different order. Between Tarawa in the Gilberts and Luzon, the main island of the Philippines, stretched 2000 miles of sea. It was not entirely empty: the atolls of the central Pacific number over a thousand. However, while MacArthur's and Halsey's islands were great platforms of dry land, New Guinea being almost as large as Alaska and twice the size of France, the Pacific atolls were mere spits of sand and shelves of coral surrounding a lagoon and bearing a few palm groves which barely found roots above high-water mark. There had been many campaigns like MacArthur's in previous centuries, notably in the Mediterranean and in Japan's own inland sea. There had never been a campaign such as Nimitz now contemplated – a giant's leap between stepping-stones so separated that they would stretch the United States Navy to breaking-point.

What made the central Pacific offensive a feasible undertaking was the transformation the Pacific Fleet had undergone in the two years since the catastrophe at Pearl Harbor. It was no longer a 'battle-wagon' navy, a Jutland-style train of slow, old, heavy-gun platforms dedicated to finding and fighting the enemy in battering duels at 20,000 yards. Even its battleships were new, faster and stronger by far than those which still lay on the bottom of Pearl Harbor or those which had been raised and refurbished from it. Its carriers, which now formed its cutting edge, were a new breed of ship: the light carriers of the Independence class, converted from fast cruisers, embarked fifty aircraft and could manoeuvre at over 30 knots; the new Essex-class fleet carriers were equally fast, embarked a hundred aircraft and were heavily armed with 5-inch and 40- and 20-mm anti-aircraft guns. By October 1943 there were six Essex-class carriers at Pearl Harbor, ready to lead Nimitz's Pacific Fleet into battle; they were to form 'fast carrier task forces', which would protect the newly built fast 'attack transports' and their destroyer, cruiser and battleship escorts in nine atoll landings on the approach to the Philippines.

First of the atolls to be taken under attack were Makin and Tarawa in the Gilberts, British islands lying at the extreme edge of Japan's defensive perimeter. Makin, lightly garrisoned by the Japanese, fell quickly when Admiral Charles Pownall's Task Force 30 landed Marines and army units on 21 November 1943. Tarawa was a different matter. More heavily garrisoned (by 5000 Japanese), it was also surrounded by a high reef over which the new Marine amphibious armoured vehicles (amphtracs) passed easily but on which the landing craft in which most of the assault force were embarked stuck. The Marines suffered very heavy casualties getting ashore on 21 November and then found themselves pinned beneath beach obstacles which offered the only cover. Some 5000 men landed; by nightfall 500 were dead and 1000 wounded. Even direct hits from battleship guns failed to destroy the Japanese strongpoints, whose defenders ceased resistance only when killed. It was not until the following day, when a second force landed with tanks on an undefended beach and attacked from the rear, that headway was made – but in barbaric circumstances. Tarawa was the battle which taught the Marine Corps how ferocious the struggle even for the smallest Japanese-held island could be. Robert Sherrod, a war correspondent, recorded:

A Marine jumped over the seawall and began throwing blocks of TNT into a coconut-log pillbox. Two more Marines scaled the seawall [with a flamethrower]. As another charge of TNT boomed inside the pillbox, causing smoke and dust to billow out, a khaki-clad figure ran out from the side entrance. The flame thrower, waiting for him, caught him in its withering flame of intense fire. As soon as it touched him the Jap flared up like a piece of celluloid. He was dead instantly but the bullets in his cartridge belt exploded for a full sixty seconds after he had been charred almost to nothingness.

Despite such evidence of the Marines' material superiority – or perhaps, in desperation, because of it – during the night the Japanese made a 'death charge', as they had done on the Aleutians, and ran on to the American guns; next morning the bodies of 325 were found in an area a few hundred yards square. At noon the battle was over: 1000 Marines were dead and 2000 wounded; almost all the Japanese had perished. To spare their men such horrors in the next fight, commanders initiated a crash building programme of amphtracs, earmarked naval vessels to act as specialised command ships to control air and sea bombardment and co-ordinate it with the landings, and had exact copies of the Tarawa defences built so that instructors could practise against them and train Marines in the best methods of overcoming them.

Tarawa had another immediate and positive effect on the development of the central Pacific campaign. Because the Japanese fleet had not intervened or even shown its face in the area, and because Japanese land-based aircraft from other islands had also not interfered, Nimitz concluded that it would be safe to leave the garrisons of the other Marshalls to 'wither on the vine' and press forward to the westernmost in the group, Kwajalein and Eniwetok. Kwajalein was so heavily pounded by ships and aircraft before the Marines landed on 1 February 1944 that they secured its northern islets in two days, and the army's 7th Division took the southern atoll in four days, neither incurring heavy loss. As a preliminary to the invasion of Eniwetok and to complete the neutralisation of Japanese airpower in the region, Nimitz decided to launch Task Force 58 against the more remote atoll of Truk, a forward anchorage of the Japanese Combined Fleet, with room to accommodate up to 400 aircraft. Task Force 58 was really four separate task forces, each with three carriers which between them embarked 650 aircraft. In a high-speed assault on Truk on 17–18 February, its commander, Vice-Admiral Marc Mitscher, mounted thirty raids, each more powerful than either of the Japanese strikes on Pearl Harbor, destroyed 275 aircraft and left 39 merchantmen and warships sinking. The raid established Mitscher's reputation as the master of fast carrier operations. It also ensured that Eniwetok fell by 21 February, though it took five days of fighting to overcome the suicidal Japanese defence.

The fall of the Marshalls opened the way to the Marianas, among which the large islands of Saipan and Guam were obvious landing places. Nimitz was in a hurry. Far to the south, in New Guinea, MacArthur was accelerating the pace of his advance. At the Anglo-American Quebec conference in August 1943 it was agreed that the projected pace of progress towards the Philippines was too slow, that Rabaul was not to be attacked but to be neutralised by air attack, and that MacArthur should advance along the northern coast of New Guinea by a series of amphibious hooks. The Cairo conference in November, which specifically approved Nimitz's offensive into the Marshalls, appeared to MacArthur to downgrade his campaign. When his staff reported in February that they believed Rabaul could be left far to the rear by a descent north of New Guinea on the Admiralty Islands, which appeared largely undefended, he leapt at the chance. Between 29 February and 18 March 1944 the islands were secured and MacArthur at once decided to make his longest leap yet – 580 miles – to Hollandia, halfway along New Guinea's north coast. There the Japanese, when surprised on 22 April, uncharacteristically fled in panic. Thence MacArthur drove forward throughout May, to Wakde and Biak off the north-west coast of New Guinea. The Japanese fought so hard for Biak that the battle was still in progress at the end of June and it was not until the following month that MacArthur could complete his strategic programme and, on 30 July, seize the Vogelkop peninsula, in the 'head' of the New Guinea 'bird', as a departure-point for his return to the Philippines.

The intensification of MacArthur's offensive in the south had an unintended, indirect but crucial effect on the conduct of the central Pacific campaign. So alarmed were the Japanese by the landing at Biak that they determined to call a halt to it by concentrating the Combined Fleet in East Indies waters to recapture the island; at the end of May its ships, including the new giant battleships *Yamato* and *Musashi*, were already at sea. Then clear evidence that Nimitz was preparing to spring forward from the Marshalls to the Marianas and approach the Philippines obliged the Japanese to cancel the operation, and the Combined Fleet prepared to move to the central Pacific to fight a decisive battle in great waters.

Before it could arrive, Nimitz's Marines and the army's 27th Division had debarked at Saipan in the Marianas. Saipan was a large island with a garrison of 32,000 men; the American operation against it was proportionately large also. Seven battleships fired 2400 16-inch shells into the landing zone before the troops touched down on 15 June, and eight older battleships kept up the bombardment during the landing, strongly supported by aircraft. Over 20,000 American troops were put ashore on the first day, by far the largest force yet delivered in a Pacific amphibious operation, and equivalent in size to those debarked in 1943 in the Mediterranean. However, the Japanese defenders resisted fiercely and meanwhile the First Mobile Fleet – the carrier element of the Combined Fleet – was approaching to deliver its strike against Task Force 58. Fortunately the American submarine *Flying Fish*, on patrol off the Philippines, saw it clearing the San Bernardino strait and gave Mitscher warning. He at once turned to the attack, with fifteen carriers against nine, and prepared to mount an aerial offensive. In the event the Japanese established Mitscher's position before he did theirs; but because of the superiority of his radar, fighter control and now aircraft – the new Hellcat was faster and better armed than the Zero – all four of Admiral Jisaburo Ozawa's attacks failed, either in dogfighting above the carriers or against the guns of the ships. When this 'Great Marianas Turkey Shoot' was over on the evening of 19 June, 243 out of 373 Japanese aircraft had been shot down, for the loss of 29 American; and in the course of the action American submarines torpedoed and sank the veteran *Shokaku* and the new *Taiho*, Ozawa's flagship and the largest carrier in the Japanese navy.

This was not the end of the affair. Next day Task Force 58 found the First Mobile Fleet refuelling, sank the carrier *Hiyo* with bombs and damaged two others and two heavy cruisers. The Battle of the Philippine Sea, as the two days of action were called by the Americans (the Japanese named it the 'A-Go' offensive), therefore halved the operational strength of the Japanese carrier force, reduced its aircraft strength by two-thirds – perhaps an even more damaging blow, since pilots emerged very slowly from the Japanese training system – and left Task Force 58 almost intact.

Disaster at sea for the Japanese was followed by disaster on land. After a bitter fight on Saipan, the defenders began to run out of ammunition and chose suicide rather than surrender; among the Japanese on the island were 22,000 civilians, of whom a large number are alleged to have joined the survivors of the 30,000 combatants in killing themselves rather than capitulate. Saipan was declared secured on 9 July. The neighbouring island of Tinian, where resistance was much lighter, fell on 1 August and Guam, whose garrison was battered into defeat despite its desperate resistance by an overwhelming American bombardment, on 11 August. All the territory that the Americans then coveted in the Marianas was theirs. From it their new bomber, the B-29 Superfortress, would be able to reach out to attack the home islands directly. Even more important, from the Marianas the Pacific Fleet could begin preparing the assault on the northern islands of the Philippines, whose southern islands were also threatened by MacArthur's advance on the East Indies.

PART IV
THE WAR IN
THE WEST
1940–1943

THE WESTERN FRONT, 1944

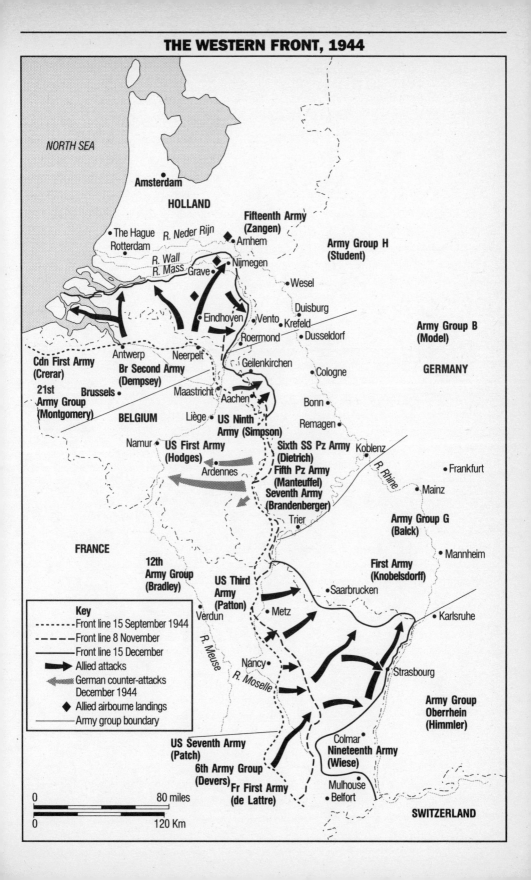

NORTH SEA

HOLLAND

- Amsterdam
- The Hague
- Rotterdam
- Arnhem
- Grave
- Nijmegen
- Eindhoven
- Vento
- Roermond
- Antwerp
- Neerpelt
- Geilenkirchen
- Maastricht
- Aachen
- Brussels
- Liège
- Namur
- Ardennes
- Trier
- Verdun
- Metz
- Nancy
- Colmar
- Mulhouse
- Belfort

R. Neder Rijn
R. Waal
R. Maas

Wesel
Duisburg
Krefeld
Dusseldorf
Cologne
Bonn
Remagen
Koblenz
R. Rhine
Frankfurt
Mainz
Mannheim
Saarbrucken
Karlsruhe
Strasbourg

Fifteenth Army (Zangen)

Army Group H (Student)

Army Group B (Model)

GERMANY

Cdn First Army (Crerar)

Br Second Army (Dempsey)

21st Army Group (Montgomery)

BELGIUM

US Ninth Army (Simpson)

US First Army (Hodges)

Sixth SS Pz Army (Dietrich)
Fifth Pz Army (Manteuffel)
Seventh Army (Brandenberger)

Army Group G (Balck)

First Army (Knobelsdorff)

FRANCE

12th Army Group (Bradley)

US Third Army (Patton)

R. Meuse
R. Moselle

US Seventh Army (Patch)

6th Army Group (Devers)

Fr First Army (de Lattre)

Army Group Oberrhein (Himmler)

Nineteenth Army (Wiese)

SWITZERLAND

Key
- - - - - - Front line 15 September 1944
– – – – – Front line 8 November
———— Front line 15 December
➤ Allied attacks
➤ German counter-attacks December 1944
◆ Allied airborne landings
——— Army group boundary

0 ————— 80 miles
0 ————— 120 Km

SEVENTEEN

Churchill's Strategic Dilemma

The coming of the Pacific war had changed the dimensions of Winston Churchill's strategy. Intimations of defeat had been replaced by the certainty of victory. 'So we had won after all!' he recalled reflecting at the news of Pearl Harbor. 'Yes, after Dunkirk; after the fall of France; after the horrible episode of Oran; after the threat of invasion, when, apart from the Air and the Navy, we were an almost unarmed people; after the deadly struggle of the U-boat war – the first Battle of the Atlantic, gained by a hand's breadth; after seventeen months of lonely fighting and nineteen months of my responsibility in dire stress. We had won the war.'

The news of the Japanese attack on Pearl Harbor, like that of the victory of Alamein, the withdrawal of Dönitz's U-boats from the Atlantic in May 1943 and the safe landing of the liberation armies on D-Day, was one of the high points of Churchill's war. Many low points awaited, including the loss of the *Prince of Wales* and *Repulse* off the coast of Malaya – 'In all the war I never received a more direct shock' – the surrender of Singapore and the fall of Tobruk. After Pearl Harbor, however, Churchill never doubted that the Western Alliance would defeat Hitler and subsequently Japan. Perhaps the sentences of his magnificent victory broadcast of 8 May 1945 were already framing themselves on the evening of 7 December 1941.

The conduct of no war is ever simple, however, and the conduct of any coalition war is always unusually difficult. The anti-Axis coalition of the Second World War, as Hitler constantly consoled himself and his entourage by emphasising, was almost unmanageably disparate. Two capitalist democracies, united by language but divided by profoundly different philosophies of international relations, had been driven by the force of events into an unexpected and unsought co-belligerency with a Marxist state which not only preached the inevitable, necessary and desirable downfall of the capitalist system but until June 1941 had been freely bound by a pact of non-aggression and economic co-operation to the common enemy. The co-ordination of a common strategy involving not merely the means but also the aims of making war was therefore destined to be difficult. How difficult, in December 1941, Winston Churchill could not foresee.

At the outset the gravity of the crisis which gripped the Soviet Union itself simplified Anglo-American strategic choice. With the German army at the gates of Moscow, there was no direct military help that either of the Western powers could lend to Russia. Britain was still scarcely armed; the United States had only just begun to emerge from two decades of disarmament. At the instant of the German attack in June 1941, acting on the principle that 'my enemy's enemy is my friend', Churchill assured Stalin that every weapon and item of essential equipment that Britain could spare would be sent to Russia, and the north Russian convoys began at once. During the meeting in August at Placentia Bay, Newfoundland, which produced the Atlantic Charter on democratic freedoms, Churchill and Roosevelt reinforced the offer, and as a result United States Lend-Lease was extended to Russia on generous terms in September. Stalin, however, wanted nothing less than the opening of a Second Front, a demand first made to Churchill on 19 July, and he was to repeat and heighten that demand throughout the next three years. In 1941 there was no chance of a Second Front. Britain and the United States could only hope for Russia's survival while they calculated how best they could together distract Hitler from his campaign of conquest in the east and weaken the Wehrmacht at the periphery of the German empire.

Calculating the location and intensity of thrusts at the periphery of Hitler's empire was to preoccupy Churchill during the next two years. He was already running one such campaign, in the Western Desert, had triumphed in another – the destruction of Mussolini's empire in East Africa – and, though he had failed in a third, the intervention in Greece, he retained the power to strike again. Norway was a sector he kept constantly in mind; after America's entry it could be only a matter of time before they did indeed jointly open a Second Front. Had Germany been America's only enemy, there might have been less delay in opening a Second Front directly against the Atlantic Wall that Hitler was building on the north coast of France. However, for most Americans, Japan was the enemy which deserved the more rapid retribution. The United States Navy, which had been granted primacy of command in the conduct of the Pacific war, was deeply committed to making its main effort in those waters. In the Japanese navy, moreover, it recognised an opponent of equal calibre and thirsted for victory over it in a great fleet action; many American soldiers, including the celebrated MacArthur, shared the navy's desire to settle with the Japanese, to take revenge for the defeats at Wake, Guam and in the Philippines, and to drive on Tokyo.

Throughout the first year of the Pacific war, therefore, Churchill found himself in an unfamiliar situation. Though no longer oppressed by the fear of defeat, he was equally no longer overlord of his country's strategy. Because Britain could win only in concert with the United States, he had to bend his will to the wishes of strategy-makers in the White House and the Joint Chiefs of Staff Committee. Roosevelt was still inclined to follow Churchill's lead. General Marshall and Admiral King were not. King was interested in the Pacific to the exclusion of all other theatres. Marshall remained committed to

Europe but believed the Second Front should be mounted on the shortest route into Germany and at the earliest possible date, and therefore was deeply suspicious of all attempts to postpone or divert effort away from this.

Churchill shrank from such a commitment. 'Remember that on my breast there are the medals of the Dardanelles, Antwerp, Dakar and Greece,' he exclaimed to Anthony Eden on 5 July 1941, in a reference to four disastrous amphibious operations of the First and Second World Wars which he had directed. It was all very well, as 1942 drew on, for the Americans to commit the Marine Corps and a handful of army divisions to island fighting in the Pacific and to contemplate wider amphibious leaps in 1943. Their campaign was fought against tiny garrisons separated by thousands of miles of ocean from their home base. A Second Front would commit the whole of the British and American expeditionary forces, not easily to be replaced if lost, to an assault on the fortified frontier of a continent within which stood an army of 300 divisions and a war-making machine without equivalent in the world. Throughout the course of 1942, therefore, Churchill found himself treading an increasingly narrow and slippery path. On the one hand, he dared not play down Britain's commitment to the Second Front, lest the Americans conclude that their strength be better deployed in the Pacific (as were a majority of American troops sent overseas in 1942); on the other, he dared not play up Britain's commitment, lest he found himself swept up in an American rush to invade the continent before the chance of success had ripened. He had agreed with Roosevelt at their meeting in Placentia Bay, four months before Pearl Harbor, that if the United States entered the war the democracies' joint strategy would be 'Germany First'; in the eighteen months after Pearl Harbor he dedicated his efforts to persuading Roosevelt, but particularly Marshall and his fellow American generals, that Allied strategy should be 'Germany First – but not quite yet.'

Temporising with military men was for Churchill a new experience. Thitherto he had dealt with generals and admirals – indeed with all in government – as an autocrat, sacking commanders with a readiness which even Hitler thought extreme and brought to the attention of his senior officers as an example of how much more reasonable he was as Führer than Churchill was as Prime Minister. 'Between 1939 and 1943,' the official historian of the Royal Navy observed, 'there was not one admiral in an important sea command . . . whom Churchill . . . did not attempt to have relieved – and in several cases he succeeded.' His dismissals of generals are notorious. In June 1941 he dismissed Wavell from command in the Middle East; fourteen months later he dismissed his replacement, Auchinleck, both in peremptory fashion; he also endorsed the dismissals of three commanders of the Eighth Army, Cunningham, Leese and Ritchie. He was difficult and demanding with those he left in office, particularly Alan Brooke, his chief of staff, with whom he was in daily contact throughout the war, but also with Montgomery, though rebuking that prima donna was to risk repayment in kind. Only General Sir Harold Alexander could do no wrong in his eyes: his famous courage and chivalrous manner excused him from reproach even for the dilatory conduct of the campaign in Italy in 1944, for which the blame attached to no one else.

Churchill could not treat the Americans thus, least of all King or Marshall. King was as tough as leather; Marshall seemed as impassive as a marble statue and intimidated even Roosevelt (as he intended – Marshall had made a resolution never to laugh at any of the President's jokes). As a guest at the table of American Lend-Lease largesse, moreover, it was not merely for diplomatic reasons that Churchill had to dissimulate, reason and prevaricate where, on his own ground, he would have demanded and dictated. The product of American war industry would have gone elsewhere – as landing ships and craft, which belonged to Admiral King's empire, did during 1942-3 – if he had not succeeded in falsely persuading Marshall during 1942 that the British War Cabinet was as eager as the American Joint Chiefs of Staff to launch a Second Front at the earliest possible moment. Churchill's exercise in inter-allied diplomacy had to be based on an entirely different approach from that which he used in managing and manipulating his cabinet and Chiefs of Staff Committee in England. By a brilliant stroke of perception, he found it in the methods his own staff officers used against him when they wished to delay one of his favoured schemes or dissuade him from a plan they judged impracticable – to agree in principle at the outset and then to drown the idea in a sea of reasoned objection.

Churchill feared the Second Front because it would succeed only if it was launched in such overwhelming force, under such devastating bombardment from the sea and air, that the Atlantic Wall and its defenders would be crushed by the impact; and he knew that neither the force nor the support would be available in 1942. In December 1941 he visited Washington for the Arcadia conference, where for the first time the British and Americans met as joint combatants to agree strategic aims. From the tone of the meeting Churchill judged that Marshall was hostile to his own inclination to sustain pressure on Germany in the Mediterranean (the one sector in which the British had found success) but favourably disposed towards maintaining a strong Allied military presence in the Pacific (the ill-fated ABDA) for which he actually proposed a British general, Wavell, as commander. The best outcome of the Arcadia conference was that it led to the establishment of a Combined Chiefs of Staff, composed of the British Chiefs of Staff and the American Joint Chiefs of Staff; the worst, in a paradoxical sense, was that it ensured American military endorsement of Churchill and Roosevelt's private 'Germany First' agreement and so brought Marshall to London in April 1942 ardent to agree a timetable. Churchill and Alan Brooke, depressed by German successes in the desert and pessimistic about Russia's ability to survive the fall of the Crimea and the Donetz basin, temporised to the best of their ability. By reasoned argument they talked down Marshall's support for Operation Sledgehammer, an invasion of France in 1942; by more devious means they won support for Bolero, the continuing American build-up of forces in Britain. Despite 'winning charm, cold persuasion, rude insistence, eloquent flow of language, flashes of anger and sentiment close to tears', Churchill failed to engage Marshall's enthusiasm for the operation later to be known as Torch, an invasion of North Africa. Marshall's 'reiteration, pressure and determination' commited both sides to Roundup, a Second Front in 1943.

Churchill and the Americans

Churchill had conceded much ground to Marshall in April, but he won some of it back when he visited Washington in June. Because of the prevailing disparity between German and Allied strengths, a cross-Channel invasion in 1942 would certainly have ended in catastrophe, and he rightly remained opposed to any such undertaking. By reasoned argument he made Sledgehammer look naively reckless and so engaged Roosevelt's interest in Torch (at that time codenamed Gymnast). Churchill argued that, if Bolero brought large numbers of American troops to Britain, Roosevelt's electors would expect them to be employed. Since they could not take part in a Second Front in 1942, why not use them in an interim operation in North Africa before the moment for Roundup came about in 1943? Roosevelt was half persuaded and in July sent Marshall to London again to thrash the matter out. Marshall was now in a headstrong mood. British resistance to an early Second Front had so incensed him that he had considered throwing his weight behind King's and MacArthur's commitment to the Pacific. Although that was only a bargaining manoeuvre on Marshall's part, King meant business and as he accompanied Marshall to London on 16 July Churchill found the next four days were devoted to perhaps the hardest-fought strategic debate in the war.

It produced deadlock, with the American Joint Chiefs of Staff demanding the Second Front that year and the British Chiefs of Staff and War Cabinet refusing to relent. The two sides agreed to lay their cases before Roosevelt, thus confronting the President with a requirement for a decision of a sort he did not usually take; in straightforward military matters he normally allowed himself to be guided by Marshall. Marshall ought therefore to have carried the day. Churchill, however, had got round his flank. Not only had he planted much doubt in the President's mind during his visit to Washington in June. He had subsequently reinforced it through the unofficial channel of communication provided by the comings-and-goings of Roosevelt's private emissary, Harry Hopkins. Hopkins had originally had reservations about British wholeheartedness that were almost as severe as Marshall's; he had been won round, however, by a concerted diplomatic offensive waged by Churchill, cabinet and Chiefs of Staff together. Lobbied by Churchill and Hopkins, Roosevelt now decided to present his Joint Chiefs of Staff with a range of choices which excluded a Second Front and among which Torch was the most attractive. When Marshall settled for this North African landing, Roosevelt enthusiastically endorsed his choice and then and there set the target date for 30 October (in the event it was launched on 8 November).

The Casablanca Conference

Churchill had therefore got his way. As he realised all too well, however, his victory was only an intermediate one. He was still committed to a Second Front in 1943 and, unless German strength declined or Allied strength increased by an improbable degree, he also knew that he would have to find a way of extricating Britain from that commitment in the coming year. For the moment the heat was off; but he knew the temperature of debate would shortly rise again, all the more so because for the first time since 1940 operations had begun to run the Allies' way. Though the Wehrmacht was driving deep into southern Russia, the Japanese had nevertheless been checked and, at Midway, defeated, the U-boats' 'Happy Time' off the American east coast had been brought to an end, the desert army had held Rommel on the border of Egypt and the bombing campaign was gathering weight against Germany. This run of success was to continue. In October General Bernard Montgomery – not Churchill's first choice for command of the Eighth Army – won the Battle of Alamein, in November the Anglo-American army made its landing in North Africa and in the same month Paulus's Sixth Army was surrounded at Stalingrad. By the time that Churchill, Roosevelt and their chiefs of staff met again at Casablanca, on ground just won by Eisenhower, the weakening of Germany which Churchill had conceded would be grounds for launching the Second Front in 1943 was a fact.

Moreover, since last meeting Roosevelt and Marshall, he had been to Moscow during August and there had also given hostages to Stalin, not exactly a promise to invade France in 1943 but a strong indication that an Anglo-American army would. Casablanca, the conference codenamed Symbol, therefore proved almost as difficult a meeting for Churchill as that in London the previous July. He realised that if King and the 'Pacific school' were to be defeated – and despite 'Germany First' the number of troops under MacArthur's command now equalled the number under Eisenhower's command in Europe, at about 350,000 in each theatre – he would have to enthuse Marshall for a 'follow-on' operation to Torch, preferably the invasion of Sicily; yet he could do so only if he succeeded in convincing him that Sicily would not obstruct a Second Front and that Britain held good to its promise of the previous year. It was an almost insoluble diplomatic problem, since Churchill could not frankly reveal to his allies his fears that a cross-Channel invasion might still fail, even in 1943. The fact that the problem was solved, after five days of disagreement, was due almost exclusively to superior British diplomatic technique. The British party had come prepared. They had brought their own floating communications centre, a fully equipped signals ship, so that they operated as an extension of the government machine in London. Long experience in the administration of empire had taught them the pitfalls which await politicians, officials and service chiefs who have not agreed a common position in anticipation of events; unlike the Americans, they did not have to thrash out their internal disagreements as they went along. Finally, they were masters of words. Air Chief Marshall Sir Charles Portal, Chief of the Air Staff and probably the cleverest service leader on either side, eventually devised a verbal formula which seemed to concede what everyone wanted. Since he knew what was in Churchill's mind and the Americans were still muddled in theirs, the Americans grasped too eagerly at his formula and went home satisfied – to repent at leisure. The statement permitted the Torch armies to proceed to Sicily as soon as the North African campaign was terminated. That was almost the only provision about which Churchill cared, since he understood that an involvement in Sicily would preclude the launching of the Second Front in 1943. The Americans regarded the Sicily commitment as only one among many and persisted in the delusion that a Mediterranean strategy need not detract from the attack on the Atlantic Wall. It would take them nearly a year to discover that even their enormous and expanding war machine could not yield enough resources to sustain both commitments.

Casablanca yielded other decisions of importance, including the proclamation, at Roosevelt's insistence, that the only terms the Allies would accept from Germany, Japan and Italy were those of 'unconditional surrender'. The Sicily decision, however, was the crucial provision and one, moreover, which the Americans would find it increasingly difficult to modify as 1943 unfolded. The course of events, rather than British diplomatic ingenuity, was to be the cause of that. At the Trident conference in Washington in May 1943, the Americans arrived 'armed to anticipate and counter every imaginable argument of the British and backed by ranks of experts whose briefcases bulged with studies and statistics', according to General Albert Wedemeyer of the US Army War Plans Division. Wedemeyer, who had been at Casablanca, summarised the American experience there: 'We lost our shirts . . . we came, we listened and we were conquered.' They were determined not to lose again and would in future outdo the British at the game of preordination. Their detailed preparations ought to have won them the match at Trident, but during the course of the conference Alexander signalled from Tunis that the Anglo-American army was victorious and that its soldiers were 'masters of North Africa's shores'. This euphoric signal, and Churchill's skilful over-bargaining for an extension of the Mediterranean campaign into the Balkans, persuaded the Americans to endorse the Sicilian expedition as a safer alternative. The Sicily campaign began in July, and events there determined that they should then give their agreement to the invasion of mainland Italy. Marshall and his colleagues approached the Quadrant conference, held at Quebec in August, in what he had laid down should be 'a spirit of winning': no further diversion from the Second Front whatsoever. However, during the course of Quadrant news arrived from Sicily of Italy's impending offer of surrender. This first outright defeat of one of the Axis partners, and the prospect it offered of being able to establish a front on the Italian mainland close to one of Germany's frontiers, undermined the Americans' commitment to the purity of a Second Front strategy yet again. Eisenhower was authorised to launch the operation, sketched in at Trident in Washington, to put an Anglo-American army ashore in Italy; but it was to be limited to the south and its purpose was to divert German strength from the sector chosen for the Second Front, now codenamed Overlord.

The Quadrant decision was not quite the end of Churchill's protracted effort to put back the landing on the north coast of France until such time as he felt sure it would succeed without grievous loss. Eisenhower's advance up the Italian peninsula went further than Marshall intended, before troops were finally withdrawn to take part in the invasion of France via the southern route. Quadrant was, however, the last occasion on which Churchill could propose any diversion of force at all from the Second Front. The Americans had absolutely rightly set their face against Balkan adventures, since not only geography but the Wehrmacht's own difficulties in campaigning against Tito should have dissuaded him from such notions. Nevertheless the Americans ought to have set even stricter limits on the Italian campaign, which ultimately came to serve Germany's purpose better than that of the Allies. After Quadrant, they did quash all Churchill's efforts to diversify the Mediterranean strategy. Thereafter it was to be Overlord and only Overlord, and Churchill could wriggle away from it no further. At Trident he had agreed to the appointment of a chief of staff to the Supreme Allied Commander, charged to prepare the Overlord plan. At Quebec he had conceded that the Supreme Allied Commander should be American. The irony was, however, that as events and American insistence drove him ever closer to biting the bullet Britain's teeth grew blunter. 'The problem is', Churchill minuted to his Chiefs of Staff on 1 November 1943, 'no longer one of closing a gap between supply and requirements. Our manpower is now fully mobilised for the war effort. We cannot add to the total; on the contrary it is already dwindling.' Oppressed by this sense of decline, Churchill could still not bring himself to name the date for an event he accepted could no longer be postponed. Neither Roosevelt nor even the stony-faced Marshall as yet pressed him to face the inevitable. That would be left to the implacable Stalin, whom all three were to meet at Tehran in November.

EIGHTEEN

Three Wars in Africa

The First World War came to Africa three days after the outbreak in Europe when the German west coast colony of Togoland was invaded and swiftly occupied by British and French forces from the Gold Coast and Senegal; the Kaiser's three other colonies, with the exception of German East Africa in which the redoubtable von Lettow-Vorbeck sustained a guerrilla resistance to the end, were brought under Allied control soon afterwards. The Second World War, by contrast, came to Africa piecemeal and with delay. For that there was good reason: one result of Versailles had been to transfer sovereignty over Germany's former African colonies to Britain, France and South Africa by League of Nations mandate; and while Italy, which had extensive African possessions on the Mediterranean and Red Sea coasts, was allied to Germany, nevertheless it did not enter the war against Britain and France, both also major colonial powers within the continent, until June 1940. Although Hitler retained a colonial governor-in-waiting on his ministerial staff, he had made no move in the meantime to extend his war-making southward across the Mediterranean. Indeed, until Italy declared for him, he had no means with which to mount offensive operations into Africa, and, unless Mussolini tried but failed there, he had no cause.

Germany's defeat of France, in which Italy played an ignominious and Johnny-come-lately part, provided Mussolini with the stimulus to reach for laurels in Africa. Pétain's armistice with Hitler left Vichy in control of the French empire as well as the French navy and armed forces, and therefore neutralised the French forces on the fringes of Mussolini's empire – the *Troupes spéciales du Levant* in Syria and Lebanon and the great *Armée d'Afrique* in Tunisia, Algeria and Morocco. Further, when the armistice provoked the British into attacking and crippling the French main fleet at its moorings at Mers-el-Kebir on 3 July 1940, killing 1300 French sailors in the process, after its admirals had refused to sail it out of Pétain's hands, the resulting bitterness ensured that the French forces would lend no support at all to their former allies. In July, therefore, Mussolini struck at the British where the Italian forces in Africa were strongest and theirs weakest. On 4 July units from the Italian garrison in Ethiopia occupied frontier towns in the Anglo-Egyptian condominium of the Sudan, on 15 July they penetrated the British colony of Kenya, and between 5 and 19 August they occupied the whole of British Somaliland on the Gulf of Aden.

Italy's ability to move so audaciously against the East African territories of what was still the world's greatest imperial power was determined by the otherwise uncharacteristic disparity of strength prevailing between the two in that corner of the continent. After the recent conquest of Ethiopia, still only superficially pacified, Italy maintained there and in its older colonies of Eritrea and Somaliland an army of 92,000 Italians and 250,000 native troops supported by 323 aircraft. The British, by contrast, deployed only 40,000 troops, most of them local, and 100 aircraft. Britain's local forces included the soldierly and loyal units of the Somaliland Camel Corps, the Sudan Defence Force and the Kenya battalions of the King's African Rifles, but they were wholly outnumbered by the enemy and outclassed in equipment. The 10,000 troops in the French enclave of Djibouti were loyal to Vichy (and would remain so until they were persuaded to come over after the North African landings in November 1942).

Britain was limited in its ability to reinforce its East African garrison from Egypt, where it had maintained an army since its annexation of that semi-autonomous fief of the Ottoman Empire in 1882, because of the need to defend Egypt's western frontier against the army of 200,000 men, mostly Italians, that Italy maintained in Libya (which it had ruled since also annexing it from Turkey in 1912). Britain's strategic difficulty cast a long shadow. Douglas Newbold, Civil Secretary in the Sudan, writing home on 19 May 1940, gloomily anticipated the outcome of the approaching war: 'Kassala is Italy's for the asking. Port Sudan probably, Khartoum perhaps. Bang goes 40 years' patient work in the Sudan and we abandon the trusting Sudanese to a totalitarian conqueror.'

Newbold's fears for the security of Britain's hold on all East African territories were fortunately to prove over-pessimistic. Although strong on the ground, Italy's Ethiopian army suffered from disabling weaknesses. It was timidly led – though the Duke of Aosta, the Italian viceroy, was a man of personal courage and distinction – it was isolated from resupply and it could not be reinforced. The British, by contrast, were at liberty to build up their forces in the region by the transfer of troops from India and South Africa through the chain of ports they controlled along the littoral of the Indian Ocean. In April 1940, General Sir Archibald Wavell, commander-in-chief in the Middle East, had visited Jan Smuts, the Prime Minister of South Africa – whose parliament had narrowly voted to enter the war the previous September – and brought back the guarantee that the dominion would raise a brigade and three squadrons of aircraft for service in Kenya. The force was to be commanded by Dan Pienaar, like Smuts a veteran of the Boer War against the British but, like him, now also a devoted supporter of the imperial cause. In September Wavell risked transferring the 5th Indian Division from Egypt to the Sudan, to join a British brigade there. During the autumn two extra South African brigades arrived in Kenya to form the 1st South African Division. In December, following the success of Wavell's counter-offensive in Egypt against the Italian Libyan army, Wavell sent to the Sudan the additional reinforcement of the 4th Indian Division. By the beginning of January 1941, therefore, the new British commander in East Africa, General Alan Cunningham, brother of the admiral commanding the Mediterranean fleet, disposed of sufficient force to contemplate expelling the Italians from their footholds on British territory and carrying the war into their Ethiopian empire.

The Ethiopian campaign

The British had been to Ethiopia before, on a punitive campaign against the Emperor Theodore in 1867-8; the difficulties of campaigning among its towering mountains had wisely persuaded them not to stay. The Italians, by the deployment of aircraft, tanks and overwhelming numbers, in 1936-7 had broken the primitive army of the Emperor Haile Selassie and thereby also avenged themselves for their defeat at the hands of the Emperor Menelek at Adowa during their first attempt to establish an Ethiopian empire in 1896. The coming Ethiopian campaign, though fought between the European powers, was to partake of the spirit of those preceding it. It was to be essentially colonial in character; many of the troops engaged were non-European; and the mountainous terrain and the absence of roads, railways and all the rest of the infrastructure upon which European armies depended for movement and supply imposed a colonial rhythm on its course.

The British plan for their counter-offensive against the Duke of Aosta's command had been fixed at Khartoum at the end of October 1940. Anthony Eden, the British war minister, had arrived there on 28 October to join Haile Selassie, returned from exile in England in expectation of reinheriting his kingdom, Wavell, Cunningham, who was to take command on 1 November, and Smuts, who had flown from South Africa. Smuts and Eden had strong political motives for urging an offensive. Smuts needed a victory to overcome opposition by his anti-British nationalists to South Africa's participation in the war; that opposition, though not as strong as in 1914 when unreconciled Boers had actually taken up arms in revolt, was still a challenge to his leadership. Eden, for his part, was anxious for the British success at this point of juncture between the African and Arabian corners of the Islamic world, because he needed to offset growing German influence over such Muslim leaders as the Mufti of Jerusalem and Rashid Ali in Iraq, who saw in Britain's time of adversity an opportunity to repay her for such grievances as the maintenance of an imperial garrison at Baghdad and the sponsorship of Zionist settlement in Palestine.

Haile Selassie, a diplomatist of subtlety, persuaded Eden at Khartoum that despite Foreign Office representations to the contrary his return to Ethiopia, where resistance to the Italian occupation was beginning to revive, offered the best prospect of undermining their common enemy's grip on the country.

Ethiopian 'patriot' units, armed by the British, were already in existence on the Sudanese border. On 6 November a British officer, Orde Wingate, representative of a tradition of irregular soldiering which reached back to the early days of Indian conquest and had been most recently embodied by T. E. Lawrence, arrived in Khartoum with a million pounds to spend and a fervent belief that he could restore Haile Selassie, the Lion of Judah, to his throne. He immediately took the 'patriot' units under command, flew into Ethiopia to make contact with the internal resistance and, on his return, began preparations to escort the emperor across the frontier.

On 20 January 1941, in the words of an official imperial propagandist, 'His Majesty the Emperor Haile Selassie I accompanied by the Crown Prince . . . and two powerful Ethiopian and English armies crossed the frontier of the Sudan and Ethiopia and entered into his own.' The exigencies of long exile excused the exaggeration; Wingate's column was almost comically weak, camel-mounted and bereft of modern equipment. However, it was at least in motion towards the capital of Addis Ababa; and so too, after some inconclusive border skirmishes, were the main British forces which constituted the real threat to Italy's Abyssinian empire. On 19 January the 4th and 5th Indian Divisions crossed the frontier north of the Blue Nile, heading for the fabled city of Gondar; they met little resistance, though at one point a force of local horsemen, the Amharic Cavalry Band, led by an Italian officer on a white horse, attempted a death-or-glory charge against their machine-guns. On 20 January the Sudan Defence Force, whose officers included the famous anthropologist Edward Evans-Pritchard (the equally famous Arabist, Wilfred Thesiger, was on the staff of Wingate's 'Gideon Force' accompanying the emperor), crossed into Ethiopia south of the Blue Nile. Finally, on 11 February, Cunningham's army of South Africans, the King's African Rifles and the Royal West African Frontier Force marched out of Kenya into southern Ethiopia and Italian Somaliland.

The Duke of Aosta correctly estimated that the most dangerous of these incursions was that of the 4th and 5th Indian Divisions in the north and accordingly concentrated the best of his troops around Keren, a small town in Eritrea defended by high peaks and approachable only along a deep and narrow gorge. The Indian divisions attacked it on 10 February and were driven off, attacked again on 15 March and were counter-attacked; but, when their engineers undertook a systematic dismantling of the obstacles with which the approaches to Keren had been surrounded, the Italians decided that they were beaten and retreated into the hinterland. The whole of Eritrea was occupied by 2 April. By then the Italian position in the south had also collapsed. General Cunningham's army, advancing from Kenya into Italian Somaliland, found it difficult to keep up with the enemy, so keen were the local troops to desert their Italian officers and make for home with their rifles and ammunition, rich prizes in that territory of endemic banditry. In late March, having swung north-west from Somaliland towards central Ethiopia, he was forced to fight a battle to open the road to the ancient walled city of Harar, which was won by the black Nigerians of the Royal West African Frontier Force – soldiers in whom Cunningham had previously but wrongly reposed little trust. Thereafter the Italians' hold over their local units began to collapse irretrievably; by early April only a thin screen of Savoy Grenadiers stood between Cunningham and Addis Ababa. They were brushed aside, and on 5 April the capital fell to the British. Haile Selassie, escorted by Wingate's 'Gideon Force', made a triumphal entry on 5 May. Meanwhile the Duke of Aosta had retreated to the mountain fastness of Amba Alagi, where he surrendered in late May. He was to die of tuberculosis in British captivity the following year.

The war in Ethiopia was now effectively over. British Somaliland had been recaptured by an amphibious landing launched from Aden on 16 March; the Italian commander of Berbera, the capital, burst into tears on surrendering his revolver to a British officer, who comforted him with the thought that 'war can be very embarrassing'. A handful of Italian diehards escaped westward to surrender to a Belgian force advancing from the Congo on 3 July. In the course of the campaign Italy lost some 289,000 troops, mostly locals, and the majority being taken prisoner. The victors were at once dispersed to other fronts where they were more urgently needed – the Indians and South Africans to the Western Desert, the West and East Africans to their home stations, whence they would be shipped in 1944 to fight the war against the Japanese in Burma, in which Wingate would win a legendary reputation. A Free French force which had come from the Middle East to fight returned there. General Sir William Platt, the commander of the Sudan Defence Force, would go on to capture Madagascar from its Vichy garrison – which Churchill feared it could or would not hold against the Japanese – in November 1942. Cunningham, the conqueror of Ethiopia, departed for Egypt, where he would lose his reputation as a successful soldier in the struggle against Rommel.

The Ethiopian campaign was an oddity among those of the Second World War, strategically a footnote to the nineteenth-century 'scramble for Africa', tactically a Beau Geste episode of long camel treks and short bitter conflicts for mountain strongpoints and desert forts. It was appropriate that among the colourful variety of colonial units which had taken part – Mahratta Light Infantry, Rajputana Rifles, Gold Coast Regiment, *Gruppo Banda Frontiere* - the Foreign Legion should have been one. Committed at the personal insistence of General de Gaulle, who at that time was urgently seeking means to turn his declared revolt against Pétain and Vichy into a reality, the Legion had fought vigorously and effectively in the Battle of Keren before returning to the Middle East to take part in the Battle of Bir Hacheim, with its great reputation yet further enhanced.

Ethiopia was not the only front south of the Mediterranean on which de Gaulle sought, in the aftermath of the fall of France, to establish an alternative to the Vichy regime. During September 1940 he had led a Free French force, embarked together with British units of the Royal Navy, against Dakar in Senegal, the cornerstone of the French presence in West Africa. His aim, which was to rally the garrison to the Free French cause, failed; so too did the Royal Navy's, which was to immobilise units of the French fleet which had arrived to defend the harbour. However, though on 25 September de Gaulle was forced to withdraw discomfited, this Free French effort at penetrating West Africa was not without results. On 27 August the resolute follower of de Gaulle, Philippe Leclerc, had succeeded in rallying the colony of Cameroon; on hearing that news the black governor of Chad also came over and the French Congo rallied shortly afterwards. With Cameroon, Chad, Congolese and some rallied Senegalese troops, Leclerc invaded Gabon on 12 October and with his confr're, Pierre Koenig, led columns against the capital Libreville, which surrendered on 12 November. It was evidence of how bitterly ideological this fratricidal war between Frenchmen had become that the governor, Masson, hanged himself rather than surrender; his successor capitulated the same day.

The Syrian war

De Gaulle now controlled a solid wedge of territory in the great West African bight and also disposed of four independent military forces on the continent; a brigade in Egypt and a 'division' in East Africa (the two soon to be united as part of the British Western Desert Force); a garrison in West Africa and, in Chad, Leclerc's *Groupe Nomade de Tibesti*. Leclerc, by far the most dynamic of de Gaulle's followers, led his tiny command northward into Italian Libya in the spring of 1941, made contact with the British Long Range Desert Group and then independently captured the oasis of Kufra on 1 March. It was the first single-handed Free French success against the Axis. Conscious of the significance of his victory, Leclerc at once prompted his little band of white and black French soldiers to take a solemn oath ('Le serment de Kufra') not to lay down arms until the French flag should once more fly over the German-annexed cities of Metz and Strasbourg; Leclerc, a former cadet at Saint-Cyr, belonged to the graduating class of 'Metz et Strasbourg'. In the spring of 1941 it must have seemed a bold gesture to cast down such a challenge. Not even the indomitable Leclerc might have dared to foresee that three years later he would be leading French soldiers down the Champs-Elysées to a solemn Te Deum of gratitude in Notre-Dame de Paris for the liberation of the city, or that by November 1944 his 2nd Armoured Division would indeed be present to watch the tricolour rise over Metz and Strasbourg.

In the spring of 1941, it was the spectre of further fratricidal wars rather than any vision of liberation which exercised those Frenchmen who had taken sides over the issue of the armistice. The largest concentration of Vichy French troops, General Maxime Weygand's great *Armée d'Afrique* in Morocco, Algeria and Tunisia, lay as yet outside the strategic ambit; but General Henri Dentz's Army of the Levant in Syria and Lebanon was a natural target for subversion by Axis agents. Its bases outflanked from the east those of the British in Egypt, where their desert war with the Italians had broken out in earnest in December; it also provided a bridgehead through which Britain's Arab enemies, Rashid Ali in Iraq and the Mufti of Jerusalem in Palestine, could be supported. Dentz, like Weygand, was bound to neutrality by the terms of the armistice; but because of the relative weakness of his force (38,000 to Weygand's 100,000), its isolation from France and its proximity to the Axis power-base in Italy and the Balkans he could be put under pressure to which Weygand was impervious. Early in April British intelligence decrypts revealed that the Germans and Italians were jointly planning to use Syria as a staging and basing area from which to supply Rashid Ali in Iraq, where that general had overthrown the pro-British regent on 3 April. By 13 May new decrypts revealed that German aircraft with Iraqi markings had arrived in Syria, and next day they began bombing the British forces which were entering Iraq to put down Rashid Ali's coup. Rashid Ali's action had been intemperate and premature. His army was not strong or resolute enough either to overcome the British garrison, which by treaty occupied the large air base of Habbaniya outside Baghdad, or to prevent British troops also exercising their treaty right to enter and transit Iraq through the port of Basra. His siege of Habbaniya, begun on 30 April, was actually broken by the besieged, who chased the investing force away from the aerodrome on 5 May. Reinforced by the hastily organised 'Habforce' of units from Palestine, which made a trans-desert march, and by the 10th Indian Division landed at Basra, British forces in Iraq entered the city and restored the regent on 31 May.

Evidence of Dentz's complicity, however unwilling, in the Iraq episode clinched the British decision (for which de Gaulle had been pressing) to turn against the Army of the Levant; the danger it offered to the rear of the Western Desert Force operating in Libya was too great to be tolerated. On 23 June, therefore, four British columns moved against it – the 10th Indian Division and Habforce from Iraq against Palmyra and Aleppo, the British 6th Division from northern Palestine against Damascus and the 7th Australian Division from Haifa against Beirut. The short war which ensued was not pleasant; on the border of northern Palestine the involvement of the Free French division resulted in Frenchmen fighting Frenchmen, in the bitterest yet of the internecine struggles between the followers of Pétain and de Gaulle. On all fronts the fighting was imbued with resentment: the British believed they were spilling blood better saved for the Germans; the Vichy French felt the war had been unfairly forced upon them. The French Army of the Levant put up so good a fight that only the 7th Australian Division succeeded in breaking the defences it encountered, and then because it benefited from heavy naval gunfire support south of Beirut. Once it broke through, however, as it did on 9 July, Dentz accepted that his position was untenable and sued for terms. They were granted on 11 July, and allowed all Vichy troops who rejected de Gaulle's offer of a place in the Free French forces to return home; only 5700 of Dentz's defeated 38,000 rallied to de Gaulle. The majority, including Foreign Legionnaires who had fought Foreign Legionnaires in an almost sacrilegious outturn of events, were transhipped to North Africa, where Allied troops would meet them again in the Torch landings of November 1942.

Sour, costly and regrettable though the little Syrian war had been – 3500 Allied soldiers were killed or wounded in its course – the effect of its outcome on British strategy in Africa was wholly beneficent. Following on the heels of Italy's defeat in Ethiopia and the crushing of the pro-Axis party in Iraq, it ensured the security of Britain's *place d'armes* in Egypt from the landward side and liberated the commander of the Western Desert Force from all other preoccupations but that of beating the Axis in Libya.

The Libyan-Egyptian war had begun in earnest in September 1940. It was the second of the three wars fought on African territory between 1939 and 1945, since its outbreak slightly postdated the Ethiopian campaign and antedated the Tunisian war by over two years. At the time it bulked very large in British eyes, being the only focus of engagement on land between a British army and the enemy anywhere in the theatre of hostilities. Tactically, however, it was a very small war indeed, and, though its strategic implications were considerable, that dimension could not be developed while local British weakness was offset by Italian military incompetence, and those conditions determined its character during its six opening months.

Victory in Libya

The Italian army in Libya, commanded by Marshal Rodolfo Graziani, numbered some 200,000, organised in twelve divisions and based on Tripoli, at the end of the short sea route from Sicily. General Archibald Wavell, with 63,000 troops, had his main base at Alexandria, which was also that of the Mediterranean fleet, since Malta had been effectively relegated to the status of an air base in June, immediately after the collapse of France and Italy's declaration of war. Thitherto Italy's Libyan army had been held in check by the French Army of Africa beyond the Tunisian border; the combination of the French Toulon fleet with the British Malta fleet had also sufficed to nullify Italy's considerable maritime strength. After 24 June, however, when Pétain signed terms with Mussolini, Italy's six battleships suddenly became the largest capital force in the Mediterranean, held at risk by the Royal Navy's five only because it also deployed two aircraft carriers, while Graziani's army four times outnumbered Wavell's.

Apparently parity at sea and incontestable numerical superiority on land prompted Mussolini unwisely to order an offensive into Egypt on 13 September 1940. Three days later and sixty miles into Egypt, Graziani halted his forces to construct a firm base. They were to remain there, building camps and forts, for the next three months. However, Mussolini had certainly misread the signs, and his assumption of the offensive had abashed the Royal Navy not at all. On 8-9 July its Force H (based on Gibraltar) and the Mediterranean fleet (based on Alexandria) had engaged the Italian battle fleet in its entirety between Sardinia and Calabria, inflicted damage on it and forced it to retire. Four months later, on 11 November, the air group of HM Carrier *Illustrious*, operating with Admiral Sir Andrew Cunningham's Alexandria fleet, caught the Italian battleships in the harbour of Taranto in the heel of Italy and seriously damaged four of them at their moorings. The Royal Navy's superiority over the Italian surface fleet was established by these engagements and was to be reinforced by its destruction of three heavy cruisers in the night battle of Cape Matapan (Tainaron) on 28 March 1941 at the outset of the campaign in Greece. Thereafter, though the Italian navy intermittently succeeded in running convoys across the narrows between Sicily and Tripoli, and its light forces of motor torpedo-boats and midget submarines achieved some daring successes against the Mediterranean fleet, Mussolini's battleships kept to port. The British Admiralty's fear in June 1940 that it might have to abandon the Mediterranean, as it did at the nadir of its fortunes in 1796, thereafter receded. Axis airpower, punishingly deployed against the emergency convoys run to Malta and Alexandria during 1941, denied it free use but could not break its command of the inner sea.

The Italian army, which ought to have operated as an amphibious extension of the Italian fleet in Libya, was thereby reduced, like the British army in Egypt, to the status of an expeditionary force capable of mounting offensive operations only in so far as it was supplied and reinforced from Sicily, through its main base of Tripoli. Its advance into Egypt in September 1940 had overextended its line of communications and when on 9 December the Western Desert Force, under General Richard O'Connor, launched a surprise counter-offensive against it in its ill-constructed outposts at Sidi Barrani its defences crumbled and it was sent tumbling backward along the coast towards Tripoli in a retreat that did not stop until it reached Beda Fomm, 400 miles to the west, in early February.

'Wavell's offensive', as the counter-thrust was called, set the pattern for the fighting that was to typify the war in the Western Desert for the next two years. It was unusual in the haul of prisoners it yielded – over 130,000, a total which went far to equalise the odds between Graziani's army (200,000) and Wavell's (63,000). It was characteristic in that it took the form of a pell-mell retreat along the single coast road by the defeated party, hotly pursued by the main body of the victor, who meanwhile mounted a series of 'hooks' inland through the desert, designed to unseat the enemy from his defended positions at one port after another (from east to west, Sollum, Bardia, Tobruk, Gazala, Derna, Benghazi, El Agheila, Tripoli) and, if possible, to pincer him between the desert 'hook' and the coastal thrust.

At Beda Fomm on 7 February the Western Desert Force achieved that result. Its 7th Armoured Division had got ahead of the Italians by a breakneck

march across the desert neck of the bulge of Cyrenaica, to block the retreat of
the Italian Tenth Army, whose rearguards were being pressed by the 6th
Australian Division on the coast road. When it recognised that it was caught
between two fires it surrendered – an outcome that crowned the daring of
'Wavell's offensive' with a crushing success.

It was, however, to be short-lived, for two reasons. One was that Churchill's
decision to intervene in Greece robbed Wavell of the strength necessary to
sustain his advance as far as Tripoli; the second was that Hitler sent a German
general and a small armoured force to rescue Graziani's army from its
misfortunes. While British, New Zealand and Australian divisions were leaving
for Athens, Rommel and the Afrikakorps, consisting initially of the 5th Light
and 15th Panzer Divisions, were arriving in Tripoli. Though wholly new to
desert warfare, Rommel and his troops were prepared to embark on an offensive
by 24 March, only forty days after the advance guard arrived at Tripoli. Its
opening stages tossed the British out of their weakly defended positions at Beda
Fomm, by 3 April Rommel had captured Benghazi and by 11 April he was near
the line from which O'Connor (who had been captured during the course of the
fighting) had launched 'Wavell's offensive' four months earlier; Tobruk, held
as a fortress by the 9th Australian Division, was surrounded inside the German-
Italian rear.

On this sudden and brilliant reversal of advantages, however, Rommel could
not improve. For all his dynamism, he was a prisoner of the geographical and
territorial determinants of the desert campaign: the desert yielded nothing, and
over long stretches the landward edge of the coastal plain was bounded by high
ground or a steep depression, effectively confining the movement of the armies
to a strip forty or fewer miles wide. In that strip, which extended for 1200 miles
between Tripoli in the west and Alexandria in the east, the chain of small ports
were the only, but essential, points of military value. Campaigning necessarily
took the form, therefore, of a dash from one point of maritime resupply to the
next, in the hope that its impetus would topple the enemy off balance and allow
his destruction when he was bereft of water, fuel, ammunition, food and
reinforcements – the essentials, in that order, of desert warfare.

Rommel's advance had dangerously attenuated his line of supply from
Tripoli; that port's connections with Sicily were themselves harried by British
surface, submarine and air attack. During April he tried but failed to capture
Tobruk, to shorten his resupply route; meanwhile the Royal Navy had
successfully run a convoy (codenamed Tiger) past Malta, its vital mid-
Mediterranean stronghold, from Gibraltar to Alexandria, bringing a strong
reinforcement of tanks to the Western Desert Force. With this accretion of
strength, Wavell went over to the counter-offensive and in an operation
codenamed Battleaxe tried to unseat Rommel from his advanced position.

Battleaxe was a costly failure, largely because the British threw tanks against
carefully positioned screens of German anti-tank guns – the superlative 88-mm

gun came into its own over the long clear fields of fire which desert terrain offered – until their armoured formations were sufficiently weakened for the German Panzer units to counter-attack. The failure of Battleaxe undermined Wavell's position; he was dispatched to India and replaced by the Indian army's leading soldier, Claude Auchinleck, on 5 July.

Auchinleck launches Crusader

A period of stalemate now descended on the desert war. Britain, not yet the full beneficiary of American Lend-Lease, could not find the means to reinforce its desert army to a decisively battle-winning level; Germany, committed since June to the conquest of Russia, could spare nothing to the Afrikakorps. The only clear-cut shift of advantage in the African war during the summer of 1941 occurred far away from the focus of the fighting, in Iran, where Germany's attempts to repeat the success it had nearly achieved in Iraq in April were checked by an Anglo-Russian ultimatum to the Shah's government, issued on 17 August; it demanded the granting of rights to move men but particularly supplies, including vital Lend-Lease shipments, into and through Iran's Gulf ports to southern Russia and the Middle East. When the Shah's army showed resistance to the British troops that arrived on 25 August to lend force to the ultimatum, it was overcome and he was exiled to South Africa; Soviet troops, who had entered northern Iran, met the British in Tehran on 17 September, after which the country was effectively divided and administered by the two governments until 1946.

While Iran was being firmly incorporated within the anti-Axis sphere of influence, Auchinleck had been preparing his own offensive riposte to Rommel on the borders of Egypt. Tobruk, garrisoned by the 9th Australian Division, still held out; so too, despite unrelenting Axis air attacks, did Malta, which was resupplied by offensive convoy action three times during 1941 – Excess in January, Substance in July and Halberd in September. Auchinleck's aim was now to relieve Tobruk and recapture the bulge of Cyrenaica, as a preliminary to driving Rommel and his Italian satellites – who supplied the bulk of his troops if not fighting power – out of Libya. Crusader, as his winter offensive was codenamed, began on 18 November with nearly 700 tanks against 400 German-Italian. A first attempt to raise the siege of Tobruk failed, but on 10 December, after Auchinleck had relieved General Alan Cunningham of his command of the Eighth Army (as Western Desert Force had been retitled on 18 September), the Eighth Army linked arms with the British-Polish force which had replaced the Australian garrison; among the Australians' triumphs during the eight months of siege was their repayment in kind to the Germans' technique of drawing attacking tanks down into a destructive anti-tank screen.

Their defeat at Tobruk forced the Germans to retire as far as El Agheila, from which Rommel had commenced his offensive the previous March; but the factors of 'overstretch' which had left him so exposed in November now worked against the British – as did their need to transfer troops to the Far East – and when he counter-attacked on 21 January 1942 they in their turn were obliged to surrender much of the coastal strip so recently won and retire halfway back along the Cyrenaica bulge to the Gazala-Bir Hacheim position, which they reached on 28 January 1942 and then fortified.

Both sides were now tired and paused to recuperate; during Crusader the British had lost some 18,000 men killed and wounded and 440 tanks, the German-Italian army 38,000 men and 340 tanks; aircraft losses were about equal, some 300 on each side. During the spring these losses were gradually made good and by May Auchinleck came under pressure from Churchill to resume the offensive; while he prepared to do so, Rommel anticipated him and attacked on 27 May. The battle which followed, known as Gazala, was among the most reckless and costly fought during the desert war. At one stage Rommel personally led a strong tank raid into the British lines, trusting to the enemy's own minefields to secure his flanks and rear. While he sat defiantly inside the British position, repelling all assaults made upon him at heavy tank loss to the British, his 90th Light and the Italian Ariete Divisions were overcoming the gallant resistance of the force to which Auchinleck had entrusted the security of his desert flank, Koenig's Free French Brigade at Bir Hacheim. On 10 June Koenig's survivors were forced to surrender, their attackers turned north to assist Rommel in his 'cauldron' battle, and on 14 June Auchinleck decided to withdraw from Gazala to a stronger position further east, at Alam Halfa, near Alamein, where the impassable Qattara Depression most closely approaches the sea. Tobruk was left garrisoned in his rear as a fortress, and he expected it to hold out as a thorn in the enemy's side.

On 21 June, however, after only a week of siege, the 2nd South African Division surrendered Tobruk to the enemy; the capitulation came as a grievous blow above all to Churchill, then in Washington to confer with Roosevelt on plans for a Second Front. 'I did not attempt to hide from the President the shock I received,' he wrote. 'It was a bitter moment. Defeat is one thing; disgrace is another.' Although the surrender reawoke Churchill's doubt of the fighting spirit of his soldiers which was first aroused by the collapse at Singapore four months earlier, it instantly drew from the Americans the generous offer to divert supplies of their new Sherman tank (the first produced by the Allies in the war that matched the Panzer Mark IV in gunpower) from their own armoured divisions, then in process of formation, to the Western Desert. Accordingly, 300 Shermans and 100 self-propelled guns were shipped by sea around the Cape and arrived in Egypt in September. The strength of the Axis air forces in Sicily still precluded the use of the Mediterranean as a supply route to the desert army – as was demonstrated by the devastation of the Pedestal convoy running supplies to Malta in August. In order to bring bare necessities of fuel and food to the island's garrison and population (who had been collectively awarded the George Cross for their stoicism under relentless air attack), the Royal Navy lost one aircraft carrier and two cruisers sunk, and eleven out of sixteen convoyed merchant ships. By way of reaction, however, the British Desert Air Force was currently interrupting three out of four convoys sailing from Italy to Tripoli, and inflicting losses which threatened almost totally to deprive Rommel of tank and aviation fuel.

The desert war would not, however, be decided by balance of logistic advantage. After the humiliation of Tobruk, Churchill was determined on a victory in the field, which was now urgently required to boost Britain's standing as an ally of the United States – flushed with the triumph of Midway – and the Soviet Union – still tenaciously contesting the Wehrmacht's advance into southern Russia. Between 4 and 10 August Churchill visited the British Middle East headquarters in Cairo to confer with Smuts, the South African premier, Wavell, the commander-in-chief India, Alan Brooke, Chief of the Imperial General Staff, and Auchinleck. The Prime Minister had decided it was time for a purge. On 15 August he replaced Auchinleck with General Harold Alexander as commander-in-chief Middle East; General Bernard Montgomery was simultaneously appointed to command the Eighth Army.

Dismissing Auchinleck, Churchill reflected, was like 'shooting a noble stag'. With a magnificent physical presence, Auchinleck had every soldierly quality except the killer instinct. Churchill, however, was currently as close to desperation as he reached during the war; on 1 July he had to defend himself against a motion of censure in the House of Commons and he feared that any protraction of stalemate in the desert war would undermine his domestic and international leadership still further. Montgomery, though he lacked Auchinleck's stature and reputation, had a name for ruthless efficiency, and Churchill counted on him to pit his undoubted killer instinct against Rommel's in a decisive contest for victory in the desert.

Numbers – of men, tanks and aircraft – were for the first time turning conclusively in Britain's favour. In August Rommel still had an advantage in numbers of divisions, ten to seven, and with these he launched a local offensive on 31 August against the position Montgomery had inherited from Auchinleck at Alam Halfa. In his first weeks of command, however, Montgomery had done much to strengthen it and had also fiercely impressed upon his subordinate commanders that he would tolerate no retreat. Nor was there any in this bitter but brief battle of Alam Halfa. By 2 September Rommel accepted that he could not break through and, having lost fifty tanks, many in the dense British minefields, withdrew to his original position. A lull now descended during which Montgomery retrained his veteran divisions for offensive action and integrated his new divisions, including the 51st Highland, into the Eighth Army's structure. By October he deployed eleven altogether, with four armoured divisions, the 1st, 7th, 8th and 10th, which between them operated 1030 tanks (including 250 Shermans) supported by 900 guns and 530 aircraft. Panzer Army Africa was supported by 500 guns and 350 aircraft but of its ten divisions only four (two armoured) were German. The Italian divisions, of which two also were armoured, did not command Rommel's confidence. They were dispirited by heavy losses and earlier defeats, shaken in their commitment to the Axis cause by America's entry into the war, badly equipped, intermittently supplied and conscious that their lack of mechanised transport condemned them to the role of Rommel's cannon fodder. Their readiness to stand in the fore of what Rommel now recognised would be a major British offensive was so questionable that he decided to 'corset' them with German units, so that no long section of his line was held by Italians alone.

Rommel was troubled by much else – militarily by his over-extension at the extreme end of his line of communications, 1200 miles from Tripoli, personally by his health. For all his force of character, Rommel was not robust. He suffered from a recurrent stomach ailment, perhaps psychosomatic, and on 22 September was invalided to Germany. He was replaced by a Panzer general from Russia, Georg Stumme, and was told that when fit he would be given an army group in the Ukraine; but on 24 October he was telephoned in hospital by Hitler with the words: 'There is bad news from Africa. The situation looks very black. . . . Do you feel well enough to go back?' He was not, but left the following day and arrived at the headquarters of the Panzer Army Africa that evening, 25 October, to find a battle furiously raging at Alamein and the German-Italian front already creaking under the strain of the Eighth Army's assault.

The 'dogfight'

Montgomery had conceived his offensive in a style altogether different from that of his predecessors, who had been consistently tempted by the freedom of manoeuvre the desert terrain offered into using their tanks as the principal tactical instrument, in the hope of achieving a Panzer-style *Blitzkrieg*. Montgomery rightly judged that the British armoured divisions lacked the flair to out-German the Germans, and in any case he was not prepared to settle for a mere advantage of manoeuvre. Rather than chase the Panzer Army out of its position back towards Tripoli, as had happened three times before, he wished to inflict on it a crushing defeat in a set-piece battle so as to destroy its offensive power for good.

Accordingly he had laid his plan for the Battle of Alamein as a deliberate infantry-artillery assault, supported by some heavy tanks, which would destroy the enemy's fixed defences and their garrisons. Only after what he grimly forecast would be a 'dogfight' did he intend to launch the main body of his armour into and through the position. The battle began at midnight on 23 October with a bombardment by 456 guns, concentrated to support an infantry drive down the coast road but supported by a diversionary thrust in the desert further south. The diversionary thrust failed to draw enemy forces away from the crucial sector and on 26 October, Rommel's first day back in command, Montgomery reinforced the main assault with armour. In a week of bitter fighting, which reduced German tank strength to thirty-five, he succeeded in carving two 'corridors' through the Panzer Army's coastal position and on 2 November stood poised to break through. Rommel was now prepared to retreat but was refused permission to do so by Hitler and committed the last of his strength to hold the northernmost of the two corridors on the coast. Montgomery, who was being kept informed by Enigma of the fluctuations in German intentions, accordingly decided on 4 November to commit the bulk of his armour into the southern corridor. By mid-afternoon the 7th and 10th Armoured Divisions had destroyed the unfortunate Italian Ariete Division, whose obsolete tanks were completely outclassed, and were streaming into the Panzer Army's rear. Rommel, unable to implement Hitler's 'stand fast' order even had he so wished, knew that the battle was lost and directed all the units that could still move to retreat post-haste along the coast road to the west. It was the start of a harrowing 2000-mile retreat.

Montgomery has been reproved by post-war critics for an alleged failure to harry Panzer Army Africa to its destruction in the days and weeks after Alamein. It is true that his immediate pursuit was cautious; but he attempted a pursuit none the less, and at Fuka, late on 5 November, the 2nd New Zealand Division nearly succeeded in outflanking the retreating enemy and establishing a roadblock in his rear. Thereafter, however, heavy rain made off-road movement difficult and Rommel's beaten army succeeded in keeping ahead of its pursuers. In any case, it is doubtful whether an attempt at annihilation would have been possible or even wise. Certainly none of Montgomery's predecessors, with the exception of O'Connor in February 1941, had ever succeeded in getting ahead of an enemy retreating along the single coast road. O'Connor's success, moreover, had been won against only a portion of the thoroughly demoralised army – and Rommel's Afrikakorps at least was not demoralised. More important, the rationale of the battle Montgomery had fought precluded a sudden transformation of effort from dogged assault to headlong chase. 'This battle', he had warned in his orders before its inception, 'will involve hard and prolonged fighting. Our troops must not think that, because we have a good tank and very powerful artillery support, the enemy will all surrender. The enemy will NOT surrender and there will be bitter fighting. The infantry must be prepared to fight and kill, and to continue doing so over a prolonged period.' There had been bitter fighting and much killing: the number of British soldiers killed or wounded was 13,500 (a figure almost exactly predicted by Montgomery), by far the highest toll suffered by a British army in the war thus far; it amounted to 5 per cent of the Eighth Army but about a quarter of its infantry. Such losses could be justified only by a clear-cut victory. If Montgomery had mounted a confused and costly battle of pursuit, Rommel and the Afrikakorps might have profited by their cunning in mobile operations to muddy the outcome of Alamein, and Montgomery would have incurred criticism far more severe than he has suffered retrospectively at the pens of literary strategists.

His strategy after Alamein – correctly, it may be judged – was the eighteenth-century one of leaving his beaten enemy 'a golden bridge', the coast road to Tripoli. Along it Rommel beat a passage, under constant attack by the Desert Air Force, to reach Benghazi on 20 November and Tripoli on 23 January 1943, having made a stand at Wadi Zem Zem from 26 December to 16 January. He received no reinforcements and few supplies en route, had left 40,000 of his 100,000 men (mostly Italians) as prisoners in British hands and had only eighty tanks still running. The Panzer Army Africa, by every token of military failure or success, had been beaten at Alamein. Montgomery's début on the battlefield had been one of the most brilliant in the history of generalship.

What now saved the Panzer Army Africa from immediate extinction was a development which should have ensured its destruction. The appearance in its rear of the Anglo-American army committed to the Torch landings was to initiate the Allies' third war in Africa. Torch had been agreed upon by the Americans and British in London in July as a second best to the cross-Channel invasion they were then persuaded could not be risked in 1942. Until a Second Front was launched in 1943, as the Americans hoped, Torch provided employment for the American army which had begun to gather in the United Kingdom that spring. It also provided employment for part of Britain's home reserve, surplus to strategic need now that the danger of a German invasion had receded, and for the first of the ninety divisions which were being mobilised in the United States. When complete, the Torch army consisted of three task forces, Western, Central and Eastern, destined to land respectively at Casablanca on the Atlantic coast of Morocco and at Oran and Algiers inside the Mediterranean. Western Task Force, commanded by General George Patton, consisted of the 2nd Armoured, 3rd and 9th Divisions, transported direct from the United States; Central Task Force comprised the American 1st Armoured Division and part of the future 82nd Airborne Division from Britain; and Eastern Task Force was composed of the British 78th and American 34th Divisions. The whole was embarked in an inter-Allied armada of American and British ships. Sailing at high speed under strong air cover, the convoys reached their pre-assault positions without interception by U-boats. Until Central and Eastern Task Forces passed the Straits of Gibraltar during 5-6 November, German naval intelligence assured Hitler that the fleet was another Pedestal-style convoy assembling to rush to Malta; then it switched to the view that the fleet would land troops at Tripoli. On 7 November fresh indications suggested that it would land in North Africa, until then the least likely destination because Hitler clung to the belief that the Americans would do nothing to drive Vichy more deeply into his arms. Here was a double misapprehension. It was certainly true that the Americans accepted the reality of Vichy's hostility to Britain, but they had persuaded themselves nevertheless that they were regarded by many of Pétain's supporters in a different light. It was equally the case that many in Vichy France clung to the terms of the armistice only as long as Hitler remained clearly the master of Europe; at the merest appearance of any diminution of his power they held themselves ready to defend the long-term interests of France by a change of allegiance.

The North African landings forced such a change of allegiance. The Americans had arranged contact with local anti-Pétainists through General Mark Clark, who landed from a British submarine at Cherchell, ninety miles from Algiers, on 21 October. However, American over-caution in preserving the security of their plans prompted their supporters to premature action, which resulted in Vichy adherents resecuring control of Algiers and Casablanca, where the task forces began to land on 8 November (at Oran a British naval assault was botched). A fortuitous event then worked to reverse the Allied setback. Admiral Darlan, Pétain's commander-in-chief, happened to be in Algiers on a private visit; when it became clear that the Frenchman chosen by the Americans to assume local control, General Henri Giraud, lacked the authority to establish it, the Americans opened direct negotiation with Darlan, who was persuaded by the evidence of Allied strength to change sides, and declared an armistice on the evening of 8 November. This enabled the British and Americans swiftly to take possession of coastal Morocco and Algeria. Pétain immediately disowned Darlan. The Vichy Prime Minister, Pierre Laval, visited Hitler at his headquarters on 10 November and assured him that Darlan was acting illegitimately, but his protestations availed the Vichy regime not at all. Hitler demanded rights of free access to Tunisia for his forces, proceeded to take it of his own accord and simultaneously ordered his troops to enter the French metropolitan 'unoccupied' zone the next morning (Operation Attila). By the evening of 11 November, the whole of France was under German military occupation and Pétain's government at Vichy had been reduced to a cipher. The marshal would linger on in the office of head of state until driven into exile in Germany in September 1944; but after November 1942 his two-year pretence of sustaining French autonomy stood revealed as a sham.

The German counter-stroke

The balance of military advantage between the Axis and the Allies in North Africa ought now to have swung decisively in the latter's favour. Two large Allied armies dominated most of the coastline, Montgomery's Eighth Army in Libya, Eisenhower's First Army in Algeria and Morocco; the *Armée d'Afrique* was meanwhile veering to the Allied side. As late as a week after the landing, the only Axis force still operational in Africa was Rommel's battered Panzer Army, hastening northward from Alamein and as yet a thousand miles from the Tunisian border. Hitler now acted with dispatch to deprive the Allies of their advantage. On 12 November Pétain formally denounced the North African armistice, thus obliging the French commanders in Tunisia, the only sector of French North Africa not yet occupied by the Western Allies, to open its ports and airfields to Vichy's Axis allies. The first German forces began to arrive on

16 November from France; they consisted of the 10th Panzer, Hermann Goering Panzer Parachute and 334th Divisions, together constituting the Fifth Panzer Army, and were at once deployed westward to hold the line of the eastern Atlas mountains against Eisenhower's advancing troops.

The Atlas mountains in Tunisia form a doubly strong military position, since, a little way south of Tunis, the chain divides into the Western and Eastern Dorsals; seen on the map the Dorsals resemble an inverted Y with the tail at Tunis. The Fifth Panzer Army (commanded by Walther Nehring until 9 December, Jürgen von Arnim thereafter) at first lacked the force to hold the Western Dorsal, and British and American troops had advanced there in patrol strength by 17 November. It also had to fight hard to hold off a determined push by the British First Army, with French support, on Tunis and Bizerta, their ports of entry. The arrival of the American II Corps, with armour, allowed the Allies to fix their line on the Eastern Dorsal at the end of January 1943. They were also drawing larger reinforcements from the *Armée d'Afrique*, now under the command of Giraud; at the Casablanca conference in January he had made an uneasy accommodation with de Gaulle which was to last until April 1944.

However, the Germans had meanwhile been improving their position in Tunisia: more troops and aircraft had been transferred from Sicily, and Rommel was approaching the Mareth Line via Tripoli. The Mareth Line was a fortification system on the Libya-Tunisia border built by the French against the Italian army in Libya before 1939; its occupation by Rommel's troops in early February secured the Germans' back against Montgomery, while their holding of the Eastern Dorsal protected them from frontal attack by Eisenhower. Indeed, in the short term at least, the strategic situation in North Africa had been reversed. Rommel, instead of finding himself caught between the pincers of the First and Eighth Armies, had retired to join an army which could now strike at either or even both its enemies from a strong central position. It was about to do so.

The Fifth Panzer Army had used its mobility and armoured strength to keep the Allied forces off balance along the Eastern Dorsal, striking at the weak French XIX and inexperienced American II Corps in turn – at Fondouk on 2 January, at Bou Arada on 18 January and at Faid on 30 January. These attacks disorganised the French, essentially a colonial force quite unequipped to contest the issue with modern tanks, and forced the dispersion of the American armour. Arnim, in colloquy with Rommel, decided in early February that the enemy's situation in southern Tunisia was ripe for a counter-stroke. A dispute between them over how it was to be launched was settled by their superior, Kesselring, Supreme Commander South, and in February one each of their Panzer divisions, 10th and 21st (refitted since Alamein), drove into the American II Corps at the Faid pass through the Eastern Dorsal and further south, panicked the defenders, and by 19 February were pressing at the Kasserine pass through the Western Dorsal. The Allied position in Tunisia was threatened by a 'roll-up' operation from south to north and the threat was only averted by the intervention of the British 6th Armoured Division, supported by the artillery of the American 9th Division. The terrain also favoured the defence, confining the German tanks to narrow valleys as they tried to force their way forward; on 22 January, when Rommel met Kesselring, he confessed that he had misjudged the situation, could not widen the attack swiftly enough to exploit his initial advantage and must now return to Mareth to meet Montgomery's offensive which was being prepared in his rear.

Arnim and Rommel (appointed commander of Army Group Africa on 23 February) now both mounted spoiling attacks against the First and Eighth Armies respectively, but with limited success. The Americans had learned battle wisdom at Kasserine and been brought, moreover, under the command of Patton, who did not tolerate amateurism; the two British armies were battle-hardened and commanded by experienced generals. On 20 March, while Patton was probing at Army Group Africa's rear, Montgomery launched a breaching assault on the Mareth Line, found a way round it when his direct attack was held and drove the remnants of the old Panzer Army Africa back to the tail of the Eastern Dorsal by 31 March.

After this setback the Germans and Italians still fielded a considerable force in Tunisia, amounting to over eleven divisions when reinforcements were included with the survivors of the old Panzer Army Africa. However, their supply situation was critical: twenty-two out of fifty-one ships had been sunk during January, and the airlift mounted to supplement the sea convoys had delivered only 25,000 of the necessary 80,000 tons during February, despite the employment of the MC323 Gigant motorised gliders; on 22 April Allied fighters intercepted and shot down sixteen out of twenty-one Gigants flying petrol to Tunisian airfields. Not even secret weapons sufficed to offset the German disadvantage. Many of the first formidable Tiger tanks, rushed to Tunisia to oppose the Allied preponderance in armour, were lost in swampy ground and some were even penetrated by Allied anti-tank weapons. Moreover, Hitler did not have his heart in this battle, coming so soon after Stalingrad, a fortress position he had also vainly hoped to sustain by airlift. Tunisia seemed to him doomed as early as 4 March: 'This is the end,' he forecast then; 'Army Group Africa might just as well be brought back.' Characteristically, though he ordered Rommel home on 6 March, he could not bring himself to liquidate the front while something might yet be saved but charged Arnim with fighting it out to the last.

By the end of April Arnim had only seventy-six tanks still running and was trying to distil fuel for their engines from locally produced wines and spirits. On 8 May the Luftwaffe, confronted by an Allied air force of 4500 combat aircraft, abandoned its Tunisian bases altogether. Army Group Africa, which had been hustled from the Eastern Dorsal into the northern tail of the Dorsals by the Eighth Army between 7 and 13 April, was then confined to a small pocket covering Tunis and Bizerta. Its front had been broken in a set-piece assault by the First Army opposite Tunis on 6 May. Both Tunis and Bizerta fell next day. Rearguards kept up resistance during the next week as the remnants of Army Group Africa, short of ammunition and bereft of fuel, tried to withdraw into the final sanctuary of Cape Bon. However, on 13 May no territory remained for it to defend, and its last elements surrendered; 275,000 Axis soldiers including both the German and Italian commanders, Arnim and Messe, passed into Allied captivity. It was the largest capitulation yet imposed by an Allied force upon the Axis, a grave humiliation for Hitler and a disaster for Mussolini, who had committed his destiny to the creation and maintenance of a great Italian empire in Africa. Each of his three wars on the continent had now ended in catastrophe. Hitler, who had participated in two of them, could survive the aftermath; he had risked only enough force to demonstrate loyalty to his fellow dictator and profit by the strategic diversion which his intervention achieved. Mussolini could contemplate the aftermath in no such sanguine spirit. In Africa he had lost both the greater part of the Italian army and his reputation. Whether he and his regime could survive at all now depended upon Hitler.

NINETEEN

Italy and the Balkans

'Happy Austria,' the seventeenth-century tag went. 'Others wage war, you wage weddings.' The Habsburgs did indeed have a habit of marrying into property, and this eventually brought them the greatest landholdings of any monarchy in Europe. Italy, parts of which remained in Habsburg possession until 1918, was Austria's antithesis – unlucky in both love and war. Its north and south, unified only in 1866 under the House of Savoy, never achieved a proper marriage; its wars for independence from the Habsburgs in the mid–nineteenth century, and to win itself colonies in Africa later, turned out at best unvictorious, at most inglorious. The Italian expeditionary force which met the Ethiopians at Adowa in 1896 was one of the few European armies to suffer defeat at the hands of indigenous forces throughout the course of the imperial conquest of the continent; while its avenging of Adowa in the successful campaign against the Emperor Haile Selassie in 1936 brought it international odium.

No war cost Italy more than the First World War, its experience of which explains almost everything about its domestic and international conduct in the years that followed. Although their efforts were disparaged, the Italians fought with tenacity and courage against the Austrians on the most difficult of all fronts contested by the Allies between 1914 and 1918. Beginning in May 1915, when Italy threw in its lot with Britain, France and Russia, the Italians mounted eleven successive offensives into the Julian Alps, winning little ground but suffering heavy casualties. Surprised in a twelfth battle in November 1917 by a German intervention force, in which the young Rommel was one of the most enterprising junior officers, the Italian army was thrown back into the plain of Venice but recovered enough by late 1918 to go over to the attack and end the war with its self-esteem restored.

There was the rub. Italy had won its place among the victors; but, although 600,000 young Italians had given their lives to the Allied cause, neither Britain nor France would allow Italy the spoils it felt it had won. France and Britain divided between themselves Germany's colonies and Turkey's Arabian dominions, Syria, Lebanon, Palestine, Iraq and Transjordan. All Italy got was a small slice of former Austrian territory and a foothold in the Near East which proved untenable. Moreover, when the United States and Britain decided in 1921 to fix treaty limits on the size of fleet which the Allied powers were to be allowed to operate, Italy was obliged to accept constraints which effectively set its naval strength at the same level as the Royal Navy's in the Mediterranean – a sea in which it reasonably felt it had claims to be predominant.

The disparity between Italy's entitlement, as Italians perceived it, and her post-war inheritance lay at the root of the fascist revolution which overwhelmed established order in the kingdom in 1922. Mussolini's appeal to the Italian working and lower-middle class was only partly economic; it was equally that of a veteran to veterans. At a time of recession, unemployment and financial turmoil, he not only offered work and security of savings but also promised honour to ex-servicemen and the territorial recompense to the nation that it had not received at the peace conference. The transformation of Libya, annexed from Turkey during the Balkan wars of 1912-13, into an overseas 'empire' was followed by the conquest of Abyssinia in 1936 and the annexation of Albania in 1939. Italy's intervention in the Spanish Civil War was part and parcel of Mussolini's assurance to Italians that their country would cut a figure on the world stage; and that was ultimately the motivation also for his decision to enter the Second World War on the German side in June 1940. His efforts to build an alliance centred on Austria, as an alternative to the Italo-German Axis, had collapsed when Austria was incorporated into the Reich by the *Anschluss* of 1938, which automatically devalued his bilateral treaties with Hungary and Yugoslavia. The *Anschluss* determined that Mussolini should become Hitler's partner in the Second World War.

Circumstances dictated, however, that Italy should never be an equal partner, hard though Mussolini strove to make himself one. It was not only that Italy's economy could support only one-tenth of the military expenditure met by Germany (Italy $746 million, Germany $7415 million in 1938); it was also that Italy's military strength had declined absolutely during the inter-war period, so that it was less a match for Britain and France in 1940 (as long as the war with France lasted) than it had been for Austria in 1915. Italian divisions were weaker in infantry and artillery than twenty-five years earlier, partly because numbers were diverted, entirely for Mussolini's political conceit, into the Fascist Party's dubiously valuable Blackshirt formations. Italian manpower had continued to decline through the surge of emigration to the United States. Italian equipment, though elegant and brilliantly engineered, was produced by artisan methods which could not match the output of British – and eventually American – factories working to volume demands. The Italian services also suffered from the disadvantage of having been driven by Mussolini's urge to national aggrandisement into rearming too early. Italian tanks and aircraft were a whole generation outdated by their British equivalents; when confronted by American equipment, which reached the British in 1942, they appeared antediluvian.

There was a final and ultimately disabling impediment to Italy's effective commitment to war on Germany's side: the Italians harboured little or no hostility towards the enemies Hitler had chosen for them. Mild Francophobia may be an Italian sentiment; but the Italian upper class is notably Anglophile, while Italy's peasants and artisans have high regard for the United States, whose known hostility to Nazism influenced the national outlook from the start – and decisively so after the American entry into the war. Consequently it was a half-hearted Italian army which crossed swords with the British in East Africa and the Western Desert in 1940-1. Its confidence had not been improved by its poor showing against the Greeks in October-November 1940. It was severely shaken by Wavell's counter-offensive in December and, despite the arrival of the Afrikakorps to its assistance in February 1941, it never really recovered. Brilliant though Rommel was as a general, and notably *simpatico* though the ordinary Italian soldiers found him, their commanders could not but remember that the origins of his reputation lay in his exploits at Caporetto in November 1917, when he had captured several thousand Italians at the head of 200 Württemberg mountaineers.

By the end of the campaign in Africa in May 1943, the total number of Italians who had become prisoners of the Allies – in East Africa in 1941, in Libya in 1941-2 and in Tunisia in 1943 – exceeded 350,000, more than the number of those who had garrisoned Mussolini's African empire at the start of the war. Even before the Tunisian débâcle, the Italian army, which Mussolini had been planning the year before to raise to a strength of ninety divisions, had equipment for only thirty-one. The loss of so many of the best divisions in Africa, so soon after the catastrophe suffered by the Italian Eighth Army (220,000 strong) at Stalingrad, reduced it to a shadow; and these twin crises drove the Italian high command to examine the wisdom of continuing to lend Mussolini and the fascist regime its support. Italy's generals were disproportionately drawn from the northern society of Savoy-Piedmont, seat of the royal house where their loyalty ultimately lay. They had acquiesced in fascism as long as it favoured the monarchy's and the army's interests. Once it became clear that it was failing to do so, they began to reconsider their position. During the summer of 1943, and particularly as Italian cities began to feel the weight of Allied air attack, they were driven into plotting Mussolini's removal. The trigger to action was the appearance of Allied landing forces on the southern coast of Sicily on 9-10 July 1943.

The decision to invade Sicily after the expulsion of the Axis from Tunisia had not been taken without disagreement between the British and Americans. To the Americans, Husky, as the operation was to be known, risked diverting forces from and even setting back the Second Front. To the British it seemed to promise highly desirable if intangible benefits: the domination of the central Mediterranean, from which threats could be levelled at the 'soft underbelly' of the Axis in southern France and the Balkans; the humiliation of Mussolini, perhaps leading to his downfall; the acquisition of a stepping-stone towards the location of the invasion of Italy itself, if that subsequently proved easy, desirable or necessary. The British eventually had their way, at the Trident conference in

Washington in May 1943, but then only because the changing circumstances persuaded the Americans that a Second Front could not be opened that year. In the event, the invasion took Hitler even more by surprise than Mussolini – or his Italian enemies. Hitler harboured no illusions about the sympathies of the Italian ruling class. On 14 May he had told his generals:

> In Italy we can rely only on the Duce. There are strong fears that he may be got rid of or neutralised in some way. The royal family, all leading members of the officer corps, the clergy, the Jews [still at liberty; for all Mussolini's faults he was not anti-Semitic] and broad sectors of the civil service are hostile or negative towards us. . . . The broad masses are apathetic and lacking in leadership. The Duce is now marshalling his fascist guard about him. But the real power is in the hands of others. Moreover he is uncertain of himself in military affairs and has to rely on his hostile or incompetent generals as is evident from the incomprehensible reply – at least coming from the Duce – turning down or evading [my] offer of troops.

Hitler had just offered Mussolini five German divisions, to join the four reformed in Sicily and southern Italy from the rear parties of those lost in Tunisia, but his offer had been refused. As a precaution, plans had been prepared for the occupation of Italy (Operation Alarich, so named after the fifth-century Teutonic conqueror of Rome). However, although Mussolini warned that he expected the Allied army released by its victory in Tunisia to attack Sicily, Hitler insisted that the island was too heavily defended to be taken easily and that the Anglo-American descent would fall on Sardinia, Corsica or the Greek Peloponnese. The spectre of a landing in Greece aroused Hitler's worst forebodings; it threatened not only the opening of a 'third front' in the rear of the *Ostheer* but also the interruption of supply of Germany's most vital raw materials, bauxite, copper and chrome from the Balkans and, most precious of all, oil from Romania's wells at Ploesti.

Operation Husky
A remarkable Allied deception plan involving the planting of a corpse bearing fabricated top-secret papers had further helped to convince Hitler that any enemy invasion fleet detected in the Mediterranean would be heading for Greece, Corsica or Sardinia, not Italy. Even when an earthquake bombardment of Sicily's offshore island, Pantelleria, forced its commander to capitulate to the Allies on 11 June, he still refused to consider the possibility of an invasion of Italy. Hitler, moreover, was distracted by events elsewhere – by the intensification of the combined bomber offensive against the Reich, by the worsening of the German situation in the Battle of the Atlantic and by last-minute decisions over the launching of the Kursk offensive (Operation Citadel) in Russia. He had also just changed headquarters again. Since March, after a

prolonged sojourn at his Werwolf headquarters in the Ukraine, he had been at his holiday house, the Berghof in Berchtesgaden. He left there only at the end of June for his gloomy forest retreat, Wolfschanze at Rastenburg in East Prussia, and was re-established there a bare four days before Citadel began on 5 July. Since it was on the outcome of Citadel, designed to destroy the Red Army's offensive potential, that the course of the war on the Eastern Front in 1943 depended, it was understandable that his attention should have been divided at the moment when Patton's and Montgomery's divisions began their descent west and east of Cape Passero on 9 July.

The Allies had brought eight seaborne and two airborne divisions to the assault – an armada which greatly exceeded not only OKW's forecast of their amphibious capability but also the Axis force deployed on the island. Alfredo Guzzoni, the Italian general in overall command, disposed of twelve divisions, but of these six were static Italian divisions of negligible worth; four other Italian divisions, though capable of manoeuvre, were no match for the Allies; only the 15th Panzergrenadier and the newly raised Hermann Goering Panzer Division (the elite of the Luftwaffe's ground troops) were first class. Despite the disparity in strength and the surprise the invaders achieved, however, Operation Husky, as the Sicily landing was codenamed, went less smoothly than planned. The Allied airborne forces, drawn from the US 82nd and British 1st Airborne Divisions, suffered enormous casualties when inexperienced pilots dropped them into the sea and nervous anti-aircraft gunners shot down their aircraft. A key operation by British paratroopers to seize the Primosole bridge south of Mount Etna on the fourth day of the invasion proved particularly costly when the German 1st Parachute Division counter-attacked.

However, the seaborne landings mounted against Italian 'coast' units were uniformly successful, and some of the 'defenders' even helped unload the invaders' landing craft. On 15 July Major-General Sir Harold Alexander, Patton's and Montgomery's superior, was able to issue a directive for the final elimination of Axis forces on the island. While Patton occupied the western half, Montgomery was to advance each side of Mount Etna and secure Messina at the north-eastern tip, thus cutting off the Axis garrison's line of retreat into the toe of Italy. In the event, Patton made rapid progress against light resistance, but Montgomery, opposed by the Hermann Goering Division, found it impossible to pass east of Mount Etna on the short route to Messina and was forced to redeploy his divisions to pass to the west. On 20 July Alexander accordingly ordered Patton to delay his assault on Palermo and Trapani and instead turn eastward to drive along the coast road to Messina. Hitler, who had sent a German liaison officer, Frido von Senger und Etterlin, to oversee Guzzoni's conduct of the battle, and five German divisions as reinforcements to the Italian army, now ordered two of them, the 1st Parachute and the 29th Panzergrenadier, into Sicily to stiffen the defence.

Confronted by these forces, the Allied advance slowed. It was not until 2

August that Patton and Montgomery had formed a line running south-east and north-west between Mount Etna and the north coast of the island. Even then they moved forward only by using seaborne forces in a series of amphibious hooks (8, 11, 15 and 16 August) to unseat the enemy from his strong defensive positions. Nevertheless, Guzzoni had accepted as early as 3 August that his situation was ultimately indefensible and had begun to evacuate his Italian units across the Straits of Messina. The Germans began to evacuate on 11 August; sailing at night, they largely evaded Allied air attack and were even able to save a large portion of their equipment (9800 vehicles). The Allies made a triumphal entry into Messina on 17 August; but the enemy had escaped.

Although Operation Husky failed to inflict much damage on the enemy's troops, it had indeed secured the Allied line of communications through the Mediterranean to the Middle East; but, since the wars there and in North Africa were now over, that was a hollow achievement. It had not by any visible sign brought Turkey nearer to joining the Allies; it had not diverted German divisions from Russia, since all those sent (after 24 July) to Italy, the 16th and 26th Panzer, 3rd and 29th Panzergrenadier and 1st Parachute Divisions, had come from the west. It remained to be seen whether it would exert sufficient pressure on the anti-fascist forces in Italy to bring about a reversal of alliances.

The Americans, as represented by General George Marshall, the chief of staff, in any case doubted the value of a reversal of alliances. As always they held to the view that direct assault into north-west Europe was the only quick and certain means of toppling Hitler. They had been deflected from this position by practicalities in 1942 but had never been converted to it by argument. They suspected (in retrospect, rightly so) the logic of Churchill's commitment to a 'peripheral' strategy against what he called the 'soft underbelly' of Hitler's Europe, better seen as its dewlap. Hitler valued Italy because its loss would be a blow to his prestige and because it offered flank protection to the Balkans, where he had genuinely vital economic and strategic interests. However, if he had been able to eavesdrop on General Marshall's assessment of Italy as a secondary front where operations would 'create a vacuum into which it is essential to pour more and more means', he would have wholeheartedly agreed.

A reversal of alliances was nevertheless at hand. The arrival of the Allies in Sicily and the incontrovertible evidence of how limply the Italian forces in the island had opposed them now persuaded Italy's ruling class that it must change sides. Churchill, in conference with Roosevelt at Quebec (Quadrant, 14–23 August), remarked when he heard the first news of approaches from Mussolini's enemies: 'Badoglio [the senior Italian general] admits he is going to double-cross someone . . . it is . . . likely that Hitler will be the one to be tricked.' Hitler himself had formed the same impression on 19 July. While the battles of Sicily and Kursk were both in progress, he had made the long flight to Italy to see his fellow dictator and assure him of his support, in a form of words intended to disguise his intention of neutralising the Italian army and seizing the defensible portion of the peninsula with his own troops at the first sign of treachery. On 25 July a meeting of the Fascist Grand Council requested Mussolini's resignation as Prime Minister. When he meekly obeyed a summons to the royal palace by the king, he was arrested and imprisoned. King Victor Emmanuel assumed direct command of the armed forces and Marshal Pietro Badoglio became Prime Minister.

The new government publicly announced that it would remain in the war on Hitler's side but secretly entered immediately into direct negotiations with the Allies. The first meeting took place in Sicily on 5 August, the day before Raffaele Guariglia, the new Italian Foreign Minister, gave the German ambassador his word of honour that Italy was not negotiating with the Allies. Eisenhower was soon afterwards empowered by Roosevelt and Churchill to conclude an armistice, but on terms much harsher than Badoglio expected. While the Italians quibbled, preparation for a landing on the mainland went forward. The Italians hoped that the Allies would land north of Rome and seize the capital by parachute landing, thus forestalling the moves they guessed Hitler had in train to occupy the peninsula himself. Eventually, on 31 August, they were presented with an ultimatum: either to accept the terms, which were in effect unconditional – as Churchill on 28 July had told the House of Commons they would be – or to suffer the consequences, which meant German occupation. On 3 September the Italians signed, believing that they were being given time to prepare themselves against the German intervention they knew must follow as soon as news of the armistice became public. Only five days later, however, on 8 September, Eisenhower made the announcement, just a few hours before his troops began landing south of Naples at Salerno.

Hitler's countermeasures

The Salerno landing (Operation Avalanche) was not the first by the Allies on the Italian mainland. On 3 September Montgomery's Eighth Army had crossed the Straits of Messina to take Reggio Calabria as a preliminary to the occupation of the toe of Italy. Hitler had nevertheless decided to discount this move as unimportant, a view shared by Montgomery, who was disgruntled at being shunted into a secondary role. The Salerno landing, by contrast, stirred Hitler to order Operation Alarich to begin. Although he failed to prevent the sailing of the Italian fleet to Malta as required by the armistice terms, the Luftwaffe did succeed in sinking the battleship *Roma* en route by release of one of its new weapons, a guided glider bomb. In almost every other respect, Operation Alarich (now codenamed Achse) worked with smoothness.

Washington was reluctant to commit forces to Italy because it was determined the Alliance should launch an invasion of north-west Europe without avoidable delay. Accordingly, it was very much to Hitler's advantage that Eisenhower's lack of landing craft and divisions had obliged him to go ashore so far to the south. In consequence Hitler was able to use the divisions which had escaped from Sicily to concentrate against the Avalanche forces, while he deployed those brought from France and elsewhere (the 1st SS Panzer Division was temporarily transferred from Russia for the mission) to occupy Rome and subdue the Italian army in the centre and north of the peninsula. Before the invasion he had received contradictory advice: Rommel, one of his favourites, had warned against trying to hold the south; Kesselring, the general on the spot and an acute strategic analyst, had assured him that a line could safely be established below Rome. He now employed both men's talents. While Rommel took charge of the divisions which had been rushed through the Alps to put down military and civilian resistance in Milan and Turin (and to recapture the tens of thousands of Allied prisoners liberated from captivity on Italy's defection), Kesselring organised the Tenth Army in the south to check and contain the Salerno landing.

German troops elsewhere moved rapidly to disarm and imprison Italian troops or extinguish their resistance when it was offered. The areas of Yugoslavia under Italian occupation were brought under German control or that of their Croat (Ustashi) puppets. Italian-occupied France was taken over by German troops (with tragic consequences for the Jews who had found refuge there). Sardinia and Corsica, regarded as indefensible, were both skilfully evacuated, the former on 9 September, the latter by 1 October after a Free French invading force had come to the rescue of the local insurgents, who had risen in revolt on news of the armistice. In the Italian-occupied sectors of Greece the Germans actually scored a remarkable success against the run of strategic events. Encouraged by an outbreak of fighting between the Germans and the Italian garrison of the Ionian islands on 9 September (brutally put down by the Germans, who shot all the Italian officers they captured), the British, in the teeth of strong and wise American discouragement, invaded the Italian-occupied Dodecanese islands on 12 September and, with Italian acquiescence, took Kos, Samos and Leros. Sensing an easy success, offered by their local command of the air – as the Americans had perceived but the British refused to acknowledge – the Germans assembled a superior triphibious force, retook Kos on 4 October, forced the evacuation of Samos and seized back Leros by 16 November. The Dodecanese operation, painfully humiliating to the British, was then extended into the Cyclades. By the end of November the Germans directly controlled the whole of the Aegean, had taken over 40,000 Italian and several thousand British prisoners, and had actually set back the likelihood of Turkey's entering the war on the Allied side – Churchill's justification for mounting his second Greek adventure in the first place.

These were not the only chestnuts plucked by Hitler from the fire raised by Italy's defection. On 16 September an airborne task force, led by an SS officer, Otto Skorzeny, rescued Mussolini from the mountain resort in the Gran Sasso where he was currently confined. Mussolini at once proclaimed the existence of an 'Italian Social Republic' in the north of the country; after 9 October it was to have its own army, formed from soldiers still loyal to him and led by Marshal Rudolfo Graziani, once governor of Libya and Wavell's opponent in Egypt. The creation of Mussolini's successor state to fascist Italy ensured that the growing resistance to German occupation of the north would swell into a civil war, with brutal and tragic consequences. To those consequences Hitler was entirely indifferent. Italy's change of alliance relieved him of the obligation to supply a large part of the country with the coal on which it depended for energy; it added a captive labour force to the body of Italian volunteer workers in German industry; and it brought him nearly a million military prisoners who could also be set to work for the Reich.

Meanwhile the strategic effort to minimise the effect of the Allied invasion of the mainland was developing to his satisfaction. Rommel's deprecation of the chances of defending Italy south of Rome had proved ill founded. Kesselring, by affiliation an officer of the Luftwaffe but by training and background a product of the general staff elite, had correctly argued that the Italian terrain lent itself admirably to defence. The peninsula's central mountain spine, rising in places to nearly 10,000 feet, throws numerous spurs east and west towards the Adriatic and Mediterranean. Between the spurs, rivers flow rapidly in deep valleys to the sea. Rivers, spurs and mountain spine together offer a succession of defensible lines at close intervals, made all the more difficult to breach because the spine pushes the north-south highways into the coastal strip, where the bridges that carry them are dominated by natural strongpoints on the spurs above.

Salerno, the spot chosen by the Central Mediterranean Force staff for the main landing in Italy, falls exactly within this topographical pattern. Although the coastal strip is unusually wide and level (the factor recommending the beaches to the planners), it is dominated on all sides by high ground and the exit northward is blocked by the massif of Mount Vesuvius. Had Kesselring had sufficient force available at the outset, he might have formed the line across the peninsula that he had assured Hitler was militarily feasible, perhaps as far south as Naples. Commanding as he did, however, only seven divisions in his Tenth Army, of which the 16th Panzer alone was at full strength, he was obliged to commit what strength he had against the northern edge of the bridgehead with the aim of denying the invaders a swift exit towards Naples, and thus win time to construct a front above the city (eventually to be known as the Winter Position).

Despite the Tenth Army's immediate weakness, it nevertheless gave the Avalanche force a bad time in the first week of the invasion. Mark Clark, the American general commanding the Fifth Army, had two corps under his command, the British X and US VI. Supported by overwhelming naval and air bombardments, both got easily ashore on 9 September. They were slow to exploit their initial superiority, however, and next day came under heavy counter-attack from German reserves, including those from the toe of Italy who had escaped Montgomery's army. Counter-attacks by the 16th Panzer Division were particularly effective. On 12 September it retook from the British the key village of Battipaglia, close to their boundary with the Americans; the next day, together with 29th Panzergrenadier Division, it redoubled the pressure, threatening to break the bridgehead in half and cut the British off from the Americans who lost Altavilla and Persano and were preparing to re-embark their assault divisions. The Allies managed to stabilise the bridgehead by unleashing a tremendous weight of firepower on the advancing Germans. While the infantry of the US 45th Division took to their heels, the division's artillerymen stuck to their guns and, with naval and air support, eventually halted the German Panzergrenadiers in their tracks.

By 15 September, thanks to the landing of British armour and American airborne infantry in the bridgehead, the crisis had passed. General Heinrich von Vietinghoff, in direct command of the Tenth Army, recognised that the balance of force had now shifted against him, and Kesselring accordingly sanctioned a fighting withdrawal towards the first of his chosen mountain lines further north. Montgomery's Eighth Army had been reinforced by the British 1st Airborne Division, which on 9 September had landed at Taranto. On 16 September spearheads of the Eighth Army, advancing from Calabria, made contact with the Americans in the bridgehead south of Salerno. Two days later the Germans began their withdrawal, covering it by blowing the bridges in their rear as the Fifth Army pursued them. On 1 October British troops entered Naples. Meanwhile the Eighth Army had pushed two divisions, including the 1st Canadian, up the Adriatic coast to capture the complex of airfields at Foggia, from which it was intended to mount strategic bombing raids into southern Germany. In early October the Fifth and Eighth Armies established a continuous line across the peninsula, 120 miles long, running along the Volturno river north of Naples and the Biferno river which flows into the Adriatic at Termoli.

Kesselring's Winter Position

Now began the bitter and costly winter campaign to breach the line by which the Germans defended the approaches to Rome. Advance along the central mountain spine being impossible, the Eighth and Fifth Armies' offensive efforts were confined to short stretches on either coast, on fronts at most twenty miles long. This – and the failure of the British and Americans to co-ordinate their offensives – greatly simplified Kesselring's strategy, since it allowed him to leave his central sector almost undefended while concentrating his best divisions on the Mediterranean and Adriatic flanks. German troops in Italy, because they had been drawn from OKW's central mobile reserve, were of high quality and would remain so throughout the Italian war. In October Kesselring deployed the 3rd and 15th Panzergrenadier Divisions against the Fifth Army, with the Hermann Goering in reserve, and the 16th and 26th Panzer, 29th Panzergrenadier and 1st Parachute, together with two infantry divisions, on the Adriatic flank. Against these nine divisions the Allies could deploy only nine of their own, of which one alone was armoured; and, although Clark and Montgomery had additional tank resources in independent units, they did not enjoy material superiority, nor could they count on their total command of the air to unseat the Germans from their fortified positions. Airpower has its limitations, which the topography of Italy made all too evident. The Allied air forces posed no threat to the defenders: established on and behind steep, rocky hillsides, they had no need to manoeuvre and required only the barest of essentials to sustain their resistance. Historians might have recalled that Italy had only twice in modern times been overrun in a rapid offensive, first by Charles VIII of France in 1494 and second by Napoleon after Marengo in 1800. In the first case the French had brought a revolutionary weapon, mobile cannon, to the campaign, and in the second they had been confronted by inept and divided opponents. Neither condition obtained in the winter of 1943. The Allies enjoyed at best material parity in a battle with a resolute and skilful enemy who had nothing to lose and much to gain by standing his ground. The effort to make him loosen his grip on the crags and outcrops of the Apennines was to involve the British and Americans in the bitterest and bloodiest of their struggles with the Wehrmacht on any front of the Second World War.

The bloodiness of the Italian fighting was felt all the harder by the Allied Mediterranean Force because, by a chance of assignment, so many of its divisions were drawn from narrowly localised recruiting areas. The US 36th and 45th Divisions were respectively Texas and Oklahoma formations of the National Guard, while the British 56th and 46th Divisions came from London and the North Midlands. The two Indian divisions, 4th and 8th, were raised from the 'martial race' minority of the Raj, while the 1st Canadian was formed of volunteers from a dominion which, after the tragedy of a failed raid on Dieppe in August 1942, harboured ill-concealed suspicions about the freedom with which British generals shed its soldiers' blood. Three other groups of soldiers under Alexander's command, the 2nd New Zealand Division and the French Moroccan and the Polish II Corps, were renowned for their hardihood; the Poles in particular demonstrated the fiercest determination to pay back the enemy for the sufferings inflicted on their country since 1939. However, in the prevailing circumstances, all three lacked any easy means to make good the losses they suffered at the front. Recognition of the human fragility of the instrument under their command afflicted all the Allied generals throughout the battle for Italy and deeply affected their conduct of it.

Some of the most harassing fighting was to follow immediately on the Salerno success, as the Allies drove forward to attack the Winter Position which Kesselring was busily fortifying between Gaeta and Pescara. Its western end, hinged on the great fortress abbey of Monte Cassino, where Benedict had established the roots of European monasticism in the sixth century, was known as the Gustav Line and was the strongest section of the whole position. Its approaches were strong also and were to cost the Allies heavily in the five offensives they launched between 12 October and 17 January to reach it. From 12 to 15 October the Fifth Army established bridgeheads across the Volturno, just north of Naples. Meanwhile on the Adriatic coast the Eighth Army crossed the river Trigno beyond Termoli, which had been captured on 6 October, and then breasted up to the line of the river Sangro. The Sangro battle (20 November to 2 December) proved particularly difficult. Winter rains turned the river to spate and forced both sides into inactivity during the first week. When Montgomery got his army across he was prevented from exploiting his success by the tenacious German defence of the coastal town of Ortona, where the 1st Canadian Division suffered heavy casualties in house-to-house fighting. Sangro was Montgomery's last Mediterranean theatre battle before he left to assume command of the Overlord forces.

While the Sangro campaign was in progress, the Fifth Army had been inching forward, through a maze of broken country and enemy demolitions, to the river Garigliano, from which the valley of the Liri led past the Monte Cassino massif towards Rome. The approaches to the Liri were, however, dominated by the peaks of Monte Camino, Rotondo and Sammucro, each of which had to be scaled and conquered in a succession of bitter actions between 29 November and 21 December. Winter snowstorms then imposed a pause until 5 January, when the American and French divisions of the Fifth Army attacked again to reach the Rapido river, which flows into the Liri below the Cassino heights. As a final move in his drive to enter the Liri valley, Clark ordered the 36th (Texas) Division to make an assault crossing of the Rapido, on the seaward side between Cassino and its junction with the Liri, on 20 January 1944.

The American engineer commander responsible for clearing the mines with which the Germans had strewn the battlefield, and in charge of bridging the watercourse once the infantry had crossed in assault boats, warned beforehand that 'an attack through a muddy valley that was without suitable approach routes and exit roads and that was blocked by organised defences behind an unfordable river [would] create an impossible situation and result in a great loss of life.' His prediction was gruesomely borne out in practice. The Texans tried for three days to cross the river; some did, but all help failed to reach them, and most of them swam back to the near side. When the operation was abandoned, 1000 were dead, out of an infantry strength of less than 6000. The after-action report of the 15th Panzergrenadier Division which opposed them conveyed no sense of the disaster it had inflicted, merely stating that it had 'prevented enemy troops crossing'. The repulse of the Texan attack ended all Mark Clark's hopes for an early breakthrough to Rome up Highway 6, the main north-south route on the Mediterranean coast. He did not despair of capturing Rome quickly, however, for since 3 November a plan, sponsored by Eisenhower, had been afoot to unhinge the Winter Position by an amphibious landing in the Fifth Army's rear at Anzio, close to Rome. The genesis of the plan was not entirely military; it partook of the politics of the Second Front, in particular the controversial plan to match Overlord in Normandy with another landing (Anvil, later Dragoon) in the south of France. General Walter Bedell Smith, Eisenhower's chief of staff, personally regarded Anvil as a wasteful diversion. However, it was his duty to facilitate it, and he recognised that for its support it required the retention by the Central Mediterranean Force of a considerable portion of its landing fleet scheduled to leave Italy for England at the end of 1944, since Anvil could only be launched from northern Italy. Possession of a line running from Pisa to Rimini was regarded as essential for a successful launching of Anvil; to reach it by mid-1944 the Fifth Army would have to get north of Rome quickly; and to advance beyond Rome it would require landing craft to make a descent behind the Winter Position at once – hence Anzio and Operation Shingle.

The logistic calculation was flawless, the operational practice was lamentable. According to Bedell Smith's plan, sixty Landing Ships Tank (the key amphibious vessel) were detained in the Mediterranean until 15 January, a terminal date later extended to 6 February. On 22 January the US VI Corps, which included a large complement of British troops as well as the American 1st Armoured and 3rd Divisions, commanded by General John P. Lucas, debarked at Anzio thirty miles south of Rome. The landing achieved complete surprise; neither the Abwehr nor Kesselring's staff had detected any sign of its preparation. Had Lucas risked rushing at Rome the first day, his spearheads would probably have arrived, though they would have soon been crushed; nevertheless he might have 'staked out claims well inland', as Montgomery was to try to do in Normandy. In the event he did neither but confined himself to landing large numbers of men and vehicles and securing the perimeter of a tiny bridgehead. He thus achieved the worst of both worlds, exposing his force to risk without imposing any on the enemy. The Germans, rescued from crisis by his inactivity, hastily assembled 'emergency units' (*Alarmeinheiten*) from soldiers returning from leave, and these were rushed to Anzio while formed units were transferred from the north and quiet sectors of the Winter Position. When Lucas tried to move inland on 30 January he found the way barred; and on 15 February the newly formed Fourteenth Army counter-attacked him. This offensive, codenamed *Fischgang*, was undertaken in great strength on Hitler's orders as a warning to the Allies that an Anglo-American landing could be thrown into the sea, and as a reassurance to the German people of the fate that awaited the invaders of northern Europe. *Fischgang* failed; but it left Lucas's men besieged in squalor and danger. He was relieved on 23 February and his successor, General Lucius Truscott, was left to sustain the defence for the next three months.

A crisis in Allied strategy

Having failed to take Rome both via the Liri valley and via Anzio, General Mark Clark now found himself confronted by the necessity to smash his way forward past the great fortress monastery of Cassino which dominated Highway 6. It had been chosen by St Benedict 1400 years earlier as a place of impregnable refuge for his contemplative monks; the monks remained, assailed on three sides by the clamour of war; the monastery was as impregnable as ever. Its immediate environs were garrisoned by the 1st Parachute Division, one of the best in the Wehrmacht. Frido von Senger und Etterlin, the local corps commander and a lay member of St Benedict's Order, would not allow them to use the monastery buildings for defence; but the crags and re-entrants of the mountain provided all the defences they needed to hold the Allies at bay.

Four times in the next three months, between 12 February and 17 May, Allied troops came forward to the assault and three times they were repulsed. In the First Battle of Cassino the US 34th Division merely learned the painful lesson of how naturally strong and how strongly defended the Cassino position was. In the Second Battle the 2nd New Zealand and 4th Indian Division, commanded by Bernard Freyberg, the veteran of Crete, assaulted the monastery and the town at its foot between 15 and 18 February; their attack was preceded by the bombing of the monastery by 135 Flying Fortresses which reduced it to ruins, but both bombers and troops failed to dislodge the German parachutists from their positions. In the Third Battle, 15-23 March, Freyberg's divisions tried again, with even heavier air support. Again the attack failed, leaving the Cassino position still more impregnable than it had been at the outset: constant bombing and shelling had tumbled the monastery and the town below into a heap of ruins, into which the German parachutists burrowed to form tunnels and bunkers.

By April the conduct of Allied strategy in Italy was almost in crisis. Churchill had become openly scathing at the lack of progress. Hitler exulted in the success of the Tenth and Fourteenth Armies; although vast sectors of the Eastern Front were falling to Russian attack and German cities rocked nightly under Bomber Command's assault, in Italy his Anglo–American enemies had advanced only seventy miles in eight months. Mark Clark, perhaps the most egocentric Allied general of the Second World War, feared for his career, his temperamental antipathy to the British having been fed by their double failure at Cassino. Alexander, the theatre commander since Eisenhower's assumption of the Supreme Allied Command in Britain in January, could see no way forward, and even Churchill, who revered him as the model of the military aristocrat, had begun to doubt his will and capacity to unlock the stalemate. What was needed was a plan and a new impetus to relaunch Allied Armies Italy on to the path of victory.

Behind the locked front the Allied air forces were playing their part. They were commanded by Ira C. Eaker, who had been transferred from Britain, where he had directed the first (unsuccessful) stage of the American strategic bombing attack on Germany. From March onwards, they had been prosecuting Operation Strangle, designed to destroy the logistic network which supplied the Tenth and Fourteenth Armies at Anzio and in the Winter Position. Although the terrain precluded successful ground-attack missions against units in the front line, the Italian roads and railways presented profitable strategic targets for aircraft. Eaker's interdiction plan was a model of military logic; then, in April, Alexander's chief of staff, John Harding, began to construct an equally logical plan to exploit the Allied capacity for manoeuvre on the ground.

Since the end of the previous year Allied Armies Italy had been significantly reinforced. The Polish II Corps was now present in its full strength. The Eighth Army (commanded by Oliver Leese after Montgomery's departure to England for Overlord in late December) had been joined by an additional Indian division, a South African armoured division and another hard-fighting Canadian formation, the 5th Armoured Division. Truscott's corps in the Anzio bridgehead had doubled in size. Further, a French Expeditionary Force, formed largely of Moroccan hill tribesmen to whom mountain warfare was second nature, had taken over the sector between Cassino and the coastal plain. These reinforcements largely compensated for the withdrawal to Britain, in preparation for Overlord, of the six experienced British and American divisions which had fought the Italian campaign thus far. Out of their disparate but complementary qualities, Harding began to construct an operational plan (Diadem) designed to turn the Cassino position, open up the Liri valley and draw in the Anzio force, with the object of encircling the Germans south of Rome and delivering the city into Allied hands.

Harding's plan was that, covered by an elaborate deception (Dunton) designed to persuade the Germans of the danger of another amphibious descent in their rear, nearer the Pisa-Rimini position which marked Kesselring's ultimate line of retreat in the peninsula, the Poles would attack and seize Cassino in a Fourth Battle from the north, while the French infiltrated the mountains from the south. This move would open the Liri valley to the Canadian and South African armour, while the Americans on the west coast drove across the Garigliano to link up with the Anzio corps, which would break out from its bridgehead to block the Germans' line of retreat to Rome. A major encirclement victory promised to stand in the offing.

Much of the initiative for this plan came from General Alphonse Juin, commanding the French Expeditionary Force, who promised Harding and Alexander that his North Africans had the experience to find a way through the mountains to which Anglo-Saxons were blind. When Diadem opened on 11 May they were indeed able to do so. The Poles, opposed by the German 1st Parachute Division, at first failed to match the North Africans' progress; but after Juin's mountaineers, led by his Moroccan irregulars, had wriggled their way through into the entrance to the Liri valley by 17 May, the Poles carried Monte Cassino in a final and self-sacrificial assault. The mouth of the Liri valley and the coastal zone thus being opened, the American infantry and British armoured divisions started forward on 23 May, the same day on which

Truscott's VI Corps broke out of the Anzio bridgehead.

Both the German Tenth Army and Rome now stood within the Allies' grasp, the encirclement of the first inevitably determining the occupation of the second. Rome, declared an 'open city', and thronging with escaped Allied prisoners of war who circulated openly under the noses of the few Germans who remained, awaited liberation. The prospect of a triumphal entry overcame Clark's strategic sense. Always impatient of Alexander, whose style of command was advisory rather than emphatic, and increasingly suspicious of what he conceived to be his British allies' intention to rob him of the laurels of victory, he issued orders of his own on 26 May for his American troops to abandon their northward drive across the rear of the retreating Germans, thus surrendering the chance to encircle them, and drive directly into the capital. The realignment played directly into Kesselring's hands. While his rearguards fought effective delaying actions at Valmontone and Velletri in the Alban hills south of Rome, he hurried the intact formations of the Tenth Army across the Tiber and made post-haste for the first of a series of defensive positions on the Gothic Line, which his engineers were fortifying between Rome and Rimini.

Clark's entry into Rome on 4 June proved, therefore, a hollow triumph. Even the crowds were absent; fearing a last-ditch stand by the departing Germans, the Romans kept behind locked doors, thus depriving the supremely publicity-conscious (and photogenic) 'American Eagle', as Churchill called him, of his lap of honour.

Kesselring's Tenth and Fourteenth Armies were nevertheless in retreat, and would conduct a fighting withdrawal as they made their way to the Pisa-Rimini Line, 150 miles to the north, which he had identified as the next most defensible position across the Italian peninsula. Allied Armies Italy followed as best they could; but the withdrawal of seven divisions – four out of the seven French divisions that had come from North Africa and the American 3rd, 36th and 45th – to mount the Operation Anvil/Dragoon landing in the south of France, scheduled for mid-August, prevented Clark from pressing the retreat. Kesselring succeeded in fighting two delaying actions, first on the so-called Viterbo Line and then on the Trasimene Line, before safely reaching sanctuary on the Gothic Line in early August.

The focus of action in the Mediterranean now shifted to the coast of southern France, defended by the German Nineteenth Army of Army Group G. It was already depleted by withdrawals of troops to Army Group B which was locked in struggle in Normandy, and, although initially it contained four good divisions, the eight divisions which remained were dispersed so widely between Nice and Marseille that they could not adequately deny the Allies landing places. Churchill had long opposed the operation as militarily valueless, but Marshall's staff in Washington had insisted that Marseille was vital to the logistic support of the Anglo-American invasion of the north of France, while Roosevelt, sensitive to Alliance politics, had argued that it could not be cancelled without giving offence to Stalin. On 15 August, therefore, the newly constituted American Seventh Army under General Alexander Patch debarked between Cannes and Toulon, preceded by a brilliantly successful airborne landing, and supported by air and sea bombardment. The army, which had been collected from ports as far afield as Taranto, Naples, Corsica and Oran, got ashore with little loss and, though it had to fight hard for Toulon and Marseille, meanwhile launched a thrust up the valley of the Rhône which drove the mobile elements of the Nineteenth Army, including the 11th Panzer Division, pell-mell past Avignon, Orange and Montélimar towards Lyon and Dijon. As Army Group B was itself in full retreat by late August, the Nineteenth Army did not tarry. The spearheads of Patch's Seventh and Patton's Third Armies, the latter advancing from Normandy, met north of Dijon on 11 September, but by 14 September about half of the Nineteenth Army had found refuge in southern Alsace, where it stood ready to defend the approaches to Germany's West Wall.

The loss of the south of France was not in itself significant in Hitler's view; the course of the campaign in Italy, even though it had entailed the surrender of a broad band of territory, might actually be counted as strategically advantageous to the Germans, for it left the bulk of the Allied forces in Italy lodged against the strong defences of the Gothic Line, at a safe distance from the Italian industrial area and the Alpine approaches to the borders of the Greater Reich, while Anvil had actually diverted the Allies' amphibious fleet and the bulk of their disposable reserve into an operationally vacant zone and away from the Balkans, which still bulked so large in importance for his conduct of the war.

The Balkans

British support for the resistance in Yugoslavia had thus far troubled the Wehrmacht little. Although up to thirty Axis divisions had been engaged in internal security operations in the Yugoslav mountains, including Italian, Bulgarian, Hungarian and Croat (Ustashi) formations, only twelve were German, most of a military value too low to permit their employment on the major battlefronts. Even after the British had definitively transferred their sponsorship of Yugoslav resistance in December 1943 from the royalist Chetniks to Tito's communist guerrillas, who then numbered over 100,000, the Germans were able to keep the resistance forces constantly on the move, forcing them to migrate from Bosnia to Montenegro and then back again during the campaigning season of 1943 and in the process inflicting 20,000 casualties on their troops, as well as untold suffering on the rural population. The capitulation of Italy in September 1943 had eased Tito's situation. It brought him large quantities of surrendered arms and equipment and even allowed him to take control of much of the area relinquished by the Italians, including the Dalmatian coast and the Adriatic islands. However, as long as the Germans continued to isolate the Partisans from direct contact with external regular forces, the rules of guerrilla warfare applied: Tito had a strong nuisance value but an insignificant strategic effect on Germany's lines of communication with Greece and the areas from which it drew essential supplies of minerals.

In the autumn of 1944, however, Germany's position in the Balkans began to weaken, so threatening to elevate Tito from the role of nuisance to menace. Hitler's Balkan satellites, Bulgaria, Romania and Hungary, had been brought into the war on his side by a combination of threat and inducement. Hitler could no longer offer inducement, while the principal threat to these states' welfare and sovereignty was now prescribed by the Red Army, which between March and August had reconquered the western Ukraine and advanced to the foothills of the Carpathians, southern Europe's natural frontier with the Russian lands. Much earlier in the year the satellites had begun to think better of their alliance with Hitler. Antonescu, the ruler of Romania, had been in touch with the Western Allies since March; his Foreign Minister had even attempted to draw Mussolini into a scheme for making a separate peace as early as May 1943. Bulgaria – whose staunchly pro-German King Boris died by poisoning on 24 August 1943 – had made approaches to London and Washington in January 1944 and then placed its hopes in coming to an understanding with Stalin. Hungary, which had benefited so greatly at Romania's expense by the Vienna Award of August 1940, was meanwhile playing its own game: Kallay, the Prime Minister, had made contact with the West in September 1943 with the aim of arranging through them a surrender to the Russians, while the chief of staff suggested to Keitel, head of OKW, that the Carpathians be defended by Hungarian troops only – a device intended to keep not so much German as Romanian troops off the national territory.

Even while the German troops were in full retreat in Italy and the Russians were advancing irresistibly to the Carpathians, Hitler could deal with Hungary. He had easily put down a revolt in the puppet state of Slovakia, raised by dissident soldiers in July when they imminently but over-optimistically expected the arrival of the Red Army on their doorstep. In March he had quelled the Hungarians' initial display of independence by requiring Admiral Horthy, the Hungarian dictator, to dismiss Kallay and grant Germany full control of the Hungarian economy and communications system and rights of free movement into and through the country by the Wehrmacht. Horthy's dismissal of his pro-German cabinet on 29 August alerted Hitler to the revived danger of Hungary's defection. When on 15 October, therefore, Horthy revealed to the German embassy in Budapest that he had signed an armistice with Russia, German sympathisers in Horthy's Arrow Cross party and in the army were ready to take control of the government. Horthy was isolated in his residence, where he was persuaded to deliver himself into German hands after Skorzeny, the rescuer of Mussolini, had kidnapped his son as a hostage.

The occupation of Hungary, though smoothly achieved, could not at that stage halt the unravelling of the Balkan skein. Hungary had ultimately been driven into opening negotiations with the Russians because it feared, quite correctly, that Romania might otherwise make its own deal with Stalin and secure the return of Transylvania, which it had been forced to cede to Horthy under the Vienna Award. However, it was Hungary that had been forestalled; as soon as the Red Army crossed the Dniester from the Ukraine on 20 August, King Michael had had Antonescu arrested, thus provoking Hitler to order the bombing of Bucharest on 23 August and so allowing Romania to declare war on Germany next day. This change of sides forced the German Sixth Army (reconstituted since Stalingrad) into precipitate retreat towards the passes of the Carpathians. Few of its 200,000 men escaped. Bulgaria, into which they might have fled southward, was now closed to them because on 5 September the government had opened negotiations with the Russians (with whom it had never been at war) and promptly turned its army against Hitler. In Romania, reported Friesner, the commander of the Sixth Army, 'there's no longer any general staff and nothing but chaos, everyone, from general to clerk, has got a rifle and is fighting to the last bullet.'

The defection of Romania immediately entailed the loss of access to the Ploesti oilfields, fear of which had so deeply influenced Hitler's strategic decision-making throughout the war. It was that fear which, in large measure, had driven him to take control of the Balkans in the first place, to contemplate the attack on Russia, and to hold the Crimea long after it was militarily sound to do so. Now that the synthetic oil plants which had subsequently come on stream within Germany had been brought under disabling attack by the US Eighth Air Force, the loss of Ploesti was doubly disastrous. However, Hitler could not hope to recover them by counter-attack, for not only did the Russian Ukrainian Fronts which entered Romania on its defection enormously outnumber his own local forces; the simultaneous defection of Bulgaria put the German forces in Greece at risk also and on 18 October they evacuated the country and began a difficult withdrawal through the Macedonian mountains into southern Yugoslavia. Tolbukhin, commanding the Third Ukrainian Front, entered Belgrade, the Yugoslav capital, on 4 October, having made his way there through Romania and Bulgaria. The 350,000 Germans under the command of General Löhr's Army Group E thus had to make their escape from Greece past the flank of a menacing Soviet concentration, through mountain valleys infested with Tito's Partisans and overflown by the Allied air forces operating across the Adriatic from their bases in Italy.

The security of the other German forces – Army Group F – in what remained to Hitler of his Balkan occupation area now closely depended upon Kesselring's ability to defend northern Italy. Should it fall, Allied Armies Italy would be free both to strike eastward through the 'gaps', notably the Ljubljana gap which led into northern Yugoslavia and so towards Hungary, and also to launch major amphibious operations from the northern Italian ports across the Adriatic, as the commanders of Land Forces Adriatic, supported by the Balkan

Air Force (established at Bari in June 1944), had already begun to do on a small scale. At a meeting with Stalin in Moscow in October 1944, Churchill concluded a remarkable, if largely unenforceable, agreement advocating 'proportions of influence' between Russia and Britain in the Balkans. Unlike the Americans, Churchill continued to be fascinated by the opportunities that a Balkan venture offered. In the event it was not Allied scheming but German force allocation that decided the issue. By the time the Fifth and Eighth Armies reached the Gothic Line, their strength stood at only twenty-one divisions, while that of the German Tenth and Fourteenth, thanks to the transfer of five fresh formations and the manpower for three others, had increased to twenty-six. Although the Gothic Line was eighty miles longer than the Winter Position, it was backed by an excellent lateral road, the old Roman Emilian Way from Bologna to Rimini, which allowed reinforcements to be sped from one point of danger to the other, and on the Adriatic coast was backed by no fewer than thirteen rivers flowing to the sea, each of which formed a major military obstacle.

This terrain and the onset of Italy's autumn rains now ensured that Kesselring's hold on northern Italy, if not the whole of the Gothic Line itself, could not be broken. Alexander, correctly assessing that the route towards the great open plain of the river Po was more easily negotiable on the right than on the left, secretly had transferred the bulk of the Eighth Army to the Adriatic coast during August. On 25 August it attacked, broke the Gothic Line and advanced to within ten miles of Rimini before being halted on the Couca river. While it paused to regroup, Vietinghoff, commanding the Fourteenth Army, rushed reinforcements along the Emilian Way to check its advance. The British renewed the offensive on 12 September but were fiercely opposed; the 1st Armoured Division lost so many of its tanks that it had to be withdrawn from offensive operations. In order to divert enemy strength from the British front, Alexander ordered Clark to open his own offensive on the opposite coast on 17 September, through the much less promising territory north of Pisa. So narrow is the coastal plain there, dominated by heights reminiscent of Cassino, that it made very slow progress. During October and into November, as rains turned the whole battlefield into a slough and raised rivers in unbridgeable spate, the campaign dragged on, while ground was won in miles and lives lost in thousands. The Eighth Army lost 14,000 killed and wounded in the autumn fighting on the Adriatic coast, the Canadians bearing the heaviest share, for they were in the forefront. The Canadian II Corps took Ravenna on 5 December and pushed onwards to reach the Senio river by 4 January 1945. The Fifth Army, attacking through the mountains of the centre, reached to within nine miles of Bologna by 23 October; but it had also lost very heavily – over 15,000 killed and wounded – and was confronted by terrain even more difficult than that on the Eighth Army's front. So weakened was it that a surprise German offensive in December won back some of the ground it had captured in September north of Pisa.

Losses, terrain and winter weather determined that at Christmas 1944 the campaign in Italy came to a halt. It had been a gruelling passage of fighting, almost from the first optimistic weeks of landing and the easy advances south of Rome sixteen months earlier. The spectacular beauty of Italy, natural and man-made, its scenery of crags and mountain-top villages, ruined castles and fast-flowing rivers, threatened danger at every turn to soldiers bent on conquest. The painters whose landscapes had delighted European collectors had left warnings to any general with a sharp eye of how difficult an advance across the topography they depicted must be to an army, particularly a modern army encumbered with artillery and wheeled and tracked vehicles. Salvator Rosa's savage mountain landscapes and battle scenes spoke for themselves. Claude Lorrain's deceptively serene vistas of gentle plains and blue distances were equally imbued with menace; painted from points of dominance that an artillery officer would automatically choose as his observation post, they demonstrate at a glance how easily and regularly ground can be commanded by the defender in Italy and what a wealth of obstacles – streams, lakes, free-standing hills, mountain spurs and abrupt defiles – the countryside offers. The engineers were the consistent heroes of the campaign in Italy in 1943-4; it was they who rebuilt under fire the blown bridges the Allied armies encountered at five- or ten-mile intervals in the course of their advance up the peninsula, who dismantled the demolition charges and booby traps the Germans strewed in their wake, who bulldozed a way through the ruined towns which straddled the north-south roads, who cleared the harbours choked by the destruction of battle. The infantry too proved heroic: no campaign in the west cost the infantry more than Italy, in lives lost and wounds suffered in bitter small-scale fighting around strongpoints at the Winter Position, the Anzio perimeter and the Gothic Line. Such losses were shared equally by the Allies and the Germans, as were the natural hardships of the campaign, above all the bleakness of the Italian winter. As S. Bidwell and D. Graham put it in their history of the campaign: 'A post on some craggy knife-edge would be held by four or five men . . . if one of them were wounded he would have to remain with the squad or find his own way down the mountain to an aid post . . . if he stayed he was a burden to his friends and would freeze to death or die from loss of blood. If he tried to find his way down the mountain it was all too easy . . . to rest in a sheltered spot . . . or lose his way . . . and die of exposure.' Many of the Germans of the 1st Parachute Division who held Cassino so tenaciously must have come to such an end; many, too, of the Americans, British, Indians, South Africans, Canadians, New Zealanders, Poles, Frenchmen and (later) Brazilians who opposed them there and at the Gothic Line.

Losses and hardships were made the more difficult to bear, particularly by the Allies, because of the campaign's marginality. The Germans knew that they were holding the enemy at arm's length from the southern borders of the Reich. The Allies, after D-Day, were denied any sense of fighting a decisive campaign.

At best they were sustaining the threat to the 'soft underbelly' (Churchill's phrase) of Hitler's Europe, at worst merely tying down enemy divisions. Mark Clark, commander of the Fifth Army and, under Alexander, of Allied Armies Italy, sustained his sense of personal mission throughout. Convinced of his greatness as a general, he drove his subordinates hard, and his frustration at the deliberation of British methods poisoned relations between the staffs of the Fifth and Eighth Armies – a deplorable but undeniable ingredient of the campaign. More junior commanders and the common soldiers were sustained, once the spirit of resistance to German occupation had taken root among the Italians, by the emotions of fighting a war of liberation. No great vision of victory drew them onward, however, as it did their comrades who landed in France. Their war was not a crusade but, in almost every respect, an old-fashioned one of strategic diversion on the maritime flank of a continental enemy, the 'Peninsular War' of 1939–45. That they were continuing to fight it so hard when winter brought the campaigning season to an end at Christmas 1944 was a tribute to their sense of purpose and stoutness of heart.

TWENTY

Overlord

Until November 1943 Hitler refused to concede to his generals or associates that the Greater Reich was threatened by the opening of a Second Front in the west. Although from the first weeks of Barbarossa Stalin had pinned his hopes on Britain's rescuing the Soviet Union from defeat and, after December 1941, on an Anglo-American counter-invasion of western Europe, Hitler would have none of it. In June 1942 he told the staff of the army's western headquarters that, having thrown the British out of the continent once, he no longer feared them, while he relished the opportunity, should it arise, of teaching the Americans a lesson. Moreover, on 19 August a major Allied reconnaissance-in-force raided the port of Dieppe in northern France, and only 2500 of the 6000 largely Canadian troops committed managed to return to Britain. This defeat reinforced Hitler's confidence. Although the raid had been planned as an experiment to test how difficult it would be to seize a harbour for the opening of a Second Front, Hitler understandably chose to believe he had inflicted a severe blow that would deter the British and Americans from staging a full-scale invasion. In September, during the course of a three-hour speech to Goering, Albert Speer, his Armaments Minister, and Rundstedt, the Commander-in-Chief West (*Oberbefehlshaber West* – OB West), he told them that, if an invasion could be delayed beyond the spring of 1943, when the Atlantic Wall would be complete, 'nothing can happen to us any longer'. He went on: 'We have got over the worst of our foodstuffs shortage. By increased production of anti-aircraft guns and ammunition the home base will be protected against air raids. In the spring we shall march with our finest divisions down into Mesopotamia [Iraq] and then one day we shall force our enemies to make peace where and as we want.'

By November 1943 the bloom had gone off the apple. The dismissiveness expressed in 1942 had been rooted in reality. Then the British army was indeed still reeling from the shock of the defeat of 1940; the Americans were not yet hardened to the rigours of warfare against the Wehrmacht. His skilful penetration of the weak spots in an adversary's position rightly convinced him, even in the absence of objective evidence, that there would be no Second Front in 1942 and probably not in 1943 either. However, by the autumn of that year, his blithe minimisation of Germany's difficulties no longer held good. The Anglo-American air offensive against the homeland was growing in weight. The German armies had been driven not only far from the approaches to Iraq but also out of the richest food-producing areas of western Russia (Kiev, capital of the 'black earth' region, fell to the Red Army on 6 November 1943). The British had regained and the Americans won their self-confidence as combat soldiers. Worst of all, the Atlantic Wall had not been completed, in many sectors not even built.

On 3 November 1943, therefore, Hitler issued Führer Directive No. 51, one of the half-dozen most important of his instructions to the Wehrmacht of the whole war.

> The hard and costly struggle against Bolshevism has demanded extreme exertions. . . . The danger in the east remains, but a greater danger now appears in the west: an Anglo-Saxon landing. The vast extent of territory in the east makes it possible for us to lose ground, even on a large scale, without a fatal blow being struck to the nervous system of Germany. It is very different in the west. Should the enemy succeed in breaching our defences on a wide front here the immediate consequences would be unpredictable. Everything indicates that the enemy will launch an offensive against the Western Front of Europe, at the latest in the spring, perhaps even earlier. I can therefore no longer take responsibility for further weakening the west, in favour of other theatres of war. I have therefore decided to reinforce its defences, particularly those places from which long-range bombardment of England [with pilotless missiles] will begin.

Führer Directive No. 51 went on to specify the particular measures for strengthening OB West's forces. They included the reinforcing of Panzer and Panzergrenadier divisions in his zone of operations and a guarantee that no formation would be withdrawn from it except with Hitler's personal approval. In November 1943 OB West (Rundstedt) commanded all German ground forces in Belgium and France, organised into the Fifteenth and Seventh Armies (Army Group B) and the First and Nineteenth Armies (Army Group G), from his headquarters at Saint-Germain near Paris. The boundary between the army groups ran west-east along the Loire, with the First Army defending the Biscay and the Nineteenth the Mediterranean coast, the Fifteenth Army in Belgium and northern France and the Seventh in Normandy. Unbeknown to all in Germany, it was in Normandy that the Allied stroke was destined to fall.

Rundstedt's divisional strength stood at forty-six, soon to be raised to sixty, including ten Panzer and Panzergrenadier divisions. Six of the armoured divisions were north of the Loire, four south. That was entirely appropriate. Jodl, Hitler's operations officer, warned an assembly of the Nazi Party's Gauleiters at the time Directive No. 51 was issued that 'along a front of 2600 kilometres it is impossible to reinforce the coastal front with a system of fortification in depth at all points. . . . Hence it is essential to have strong mobile and specially well-equipped reserves in the west for the purpose of forming

Schwerpunkte [centres of military effort].' Strategic analysis revealed that the Allies' own *Schwerpunkte* against the *Westheer*, even if reinforced by another in the Mediterranean, must be formed by forces assembling in Britain and lie on the Channel coast. Hence the Panzer concentration north of the Loire.

The Panzer concentration was critical because the rest of OB West's divisions were barely mobile. The two parachute divisions stationed in Brittany and the army divisions with numbers in the 271-278 and 349-367 series were of high and adequate quality respectively, though lacking mechanised transport. The rest were not only of average to low quality but were wholly dependent on the French railway system if they were to leave their permanent bases for the invasion front. Their artillery and supply units were horse-drawn; their infantry units, except for bicycle reconnaissance companies, manoeuvred by marching at a speed no faster than Napoleon's or, indeed, Charlemagne's. Moreover, they would have to move under the threat of Allied airpower, which, he had already conceded on 29 September 1942, would be absolutely supreme. Railway and even road movement would be severely inhibited. It was therefore vital that the Panzer divisions, which alone had the capability for rapid, off-road movement, should be positioned close to the invasion zone, to hold a line until the infantry reinforcements arrived. The coast itself would be garrisoned by 'ground-holding' (*bodenständige*) divisions, unable to manoeuvre but protected from Allied air and naval bombardment by concrete fortifications. The beaches that their positions overlooked were to be mined, wired and entangled with obstacles; much of this defensive material was to be stripped from the Belgian fortified zone and the Maginot Line which had survived the onslaught of the Wehrmacht in 1940.

The Atlantic Wall scheme was excellent in theory. When complete it would go far to offset the feebleness in the west of the Luftwaffe, which at the end of 1943 deployed only 300 fighters in France (to hold in check Allied air forces whose strength would total 12,000 aircraft of all types on the day of the invasion); but on the day that Führer Directive No. 51 was issued the Atlantic Wall had still far to go before completion. During the two years when Hitler had discountenanced the invasion danger, the *Westheer* had led a bucolic life. Its commander, Gerd von Rundstedt, was not a firebrand. After his removal from the Eastern Front in December 1941 he had settled into a comfortable routine at Saint-Germain reading detective stories and allowing his staff officers to practise English conversation, a mark of the 'aristocratic' style that Wehrmacht traditionalists cultivated to differentiate themselves from the 'Nazi' generals Hitler favoured in the *Ostheer*. The lower ranks behaved accordingly. Life in France was agreeable. Live and let live characterised relations with the population, which, if not actively collaborationist, lent little support to the embryo resistance movement. Forced labour (*service du travail obligatoire*), introduced in 1942, was unpopular because it conscripted young Frenchmen to factories in Germany, to join the million French prisoners of war still held there in 1943; so, too, was *la Milice*, Vichy's paramilitary police force, which punished the contempt of fellow countrymen by exceeding its powers. The cost of occupation rankled; the German levy on the French treasury, exacted at a 50 per cent overvaluation of the mark, not only forced France to pay for the indignity of having a German army in its territory but allowed the Reichsbank to make a profit on the transaction. However, these were aspects of defeat which did not affect the French people at large. Most accepted the presence of the ('very correct') German soldiers with resignation; the Germans, more than content to be posted to the only easy billet in the Wehrmacht's zone of operations, gathered roses while they might, ate butter and cream, and worked no harder than their officers drove them.

The cosy life ended with the arrival of Rommel in December 1943, first to inspect the defences, then to take command of Army Group B. Since his invaliding from Tunisia in March he had held an undemanding post in northern Italy, but on the promulgation of Directive No. 51 he was selected by Hitler to put fire and steel into the western defences. According to his biographer, Desmond Young, 'to the snug staffs of the coastal sectors [he] blew in like an icy and unwelcome wind off the North Sea.' Rommel found that since 1941 only 1.7 million mines had been laid – he reminded his staff that the British had laid a million in two months during his campaign against them in North Africa – though explosive held in France was sufficient to manufacture 11 million. Within weeks of his arrival, mine-laying had increased from a rate of 40,000 to over a million a month and by 20 May over 4 million were in place. Between November and 11 May half a million obstacles were laid on the beaches and likely airborne landing-grounds, and he had ordered the delivery of an additional 2 million mines a month from Germany. On 5 May he dictated to his secretary: 'I am more confident than ever before. If the British give us just two more weeks, I won't have any more doubt about it.'

The defence of the French coast could not, however, be assured by the Atlantic Wall alone. Rommel, a master of mobile warfare but also a respectful veteran of campaigns fought under conditions of Western Allied air superiority, knew that he would have to get tanks to the water's edge at the moment the Allies disembarked if they were to be defeated. To do so he must solve two problems: the first was to identify where they would land; the second was to establish the shortest possible chain of command between himself and his armoured units. The problems were interconnected. To justify taking personal command of the Panzer divisions under OB West he must be able to show that he knew where they could be best used; but he could not credibly lay claim to the divisions as long as the Allies wreathed their intentions in a mist of misinformation and deception.

The war of mirrors

The Allied deception plan for Operation Overlord, as the invasion of north-

Rommel in the desert. Having made his name commanding 7th Panzer Division in France in 1940, Rommel arrived at Tripoli on 12 February 1941, and soon demonstrated his mastery of mobile operations against his British opponents.

A British soldier brings in two German prisoners under Castle Hill at Cassino, May 1942. Cassino was taken by Polish II Corps after a savage battle.

Facing page: Members of the French resistance take cover from a German sniper during the liberation of Paris.

Above: A German view of the Allied assault on the Normandy beaches.

Right: Commandos wade ashore on D–Day, 6 June 1944.

Left: An uncharacteristically genial exchange between Patton (left) and Montgomery. Relations between them were frequently strained. Patton and Montgomery were instinctive showmen, as evidenced by the former's famous pearl-handled Colt revolvers and the latter's cap badges. Standing between them, General Bradley assumes a more sober style of command.

Above: Advanced units of 1st SS Panzer Division at a critical road crossing, 17 December 1944. Their vehicle is the amphibious *Schwimmwagen*.

Right: Mud, armour and endless spaces, three constant factors in the spring and autumn campaigns fought on the Eastern Front. A German Stug III self-propelled gun negotiates the mire.

Facing page: A stricken B-17 bomber of the 94th Bombardment Group over Berlin, its tailplane mangled by bombs falling from aircraft higher in the formation. The 94th flew on the ill-fated Regensburg raid of 17 August 1943.

Left: Deep lines of exhaustion are etched on the face of a German infantryman during the fighting in the Kiev salient in December 1943.

Left: The bodies of Soviet civilians are cut down from the gallows by German security forces.

Above: Hitler celebrates his 56th birthday, tweaking the cheeks of the boy defenders of the Third Reich in the rubble-choked garden of the Chancellery.

Below: A fearsome salvo fired by the Red Army's *Katyusha* rocket launchers, dubbed 'Stalin Organs' by German troops on the Eastern Front.

Left: The Red Flag flies
over the Reichstag.
The Red Army lost
over 300,000 men in
the Battle for Berlin.

Below: The Big Three
at Yalta, February 1945,
where the shape of
postwar eastern and
central Europe was
decided.

west Europe was codenamed at the Washington Trident Conference in May 1943, was deliberately conceived to persuade the enemy that the landing would fall in the Pas de Calais, where the Channel is narrowest, rather than in Normandy or Brittany (though Hitler's fears of a descent on Norway, to which he was acutely sensitive, were also kept alive, with the profitable result of fixing eleven German divisions there throughout 1944-5). A Pas de Calais landing made military sense: it entailed a quick crossing to level and sandy beaches, which were not closed off from the hinterland by high cliffs, whence the exploitation route into the Low Countries and Germany was short. Operation Fortitude, as the deception plan was codenamed, centred on the implantation in the consciousness of German intelligence – the Wehrmacht's Abwehr and the army's Foreign Armies West section – of the existence, wholly fictitious, of a First US Army Group (FUSAG), located opposite the Pas de Calais in Kent and Sussex. False radio transmissions from FUSAG were sent over the air; false references were made to it in bona-fide messages. General Patton, whose reputation as a hard-driving army leader was known to the Germans, was mentioned as its commander. Moreover, to reinforce the notion that FUSAG would debark on the short route to the Reich, the Allied air forces in their programme of bombardment preparatory to Overlord dropped three times the tonnage east of the Seine as they did to the west. By 9 January 1944 the deception had borne fruit: an Ultra intercept referred to FUSAG on that day and others followed. It was the proof the Fortitude operators needed that their plan was working. They could not, of course, expect to distract the attention of the Germans from Normandy, the chosen landing site, for good; but they hoped to minimise German anticipation of a Normandy landing until it was actually mounted, and thereafter keep alive the anxiety that the 'real' invasion would follow in the Pas de Calais at a later stage.

Hitler was only partially deluded. On 4 and 20 March and 6 April he alluded to the likelihood of a Normandy landing. 'I am for bringing all our strength in here,' he said on 6 April, and on 6 May he had Jodl telephone Günther Blumentritt, Rundstedt's chief of staff, to warn that he 'attached particular importance to Normandy'. However, apart from allocating Panzer Lehr and 116th Panzer Divisions to Normandy in the early spring, he made no decisive alteration of OB West's dispositions; indeed until he allowed divisions to cross the Seine into Normandy from the Pas de Calais at the very end of July, he himself remained prisoner to the delusion of a 'second' invasion throughout the crucial weeks of the Overlord battle.

His concern to back both horses nevertheless compromised Rommel's urge to disperse the mist of deception by direct assault. Rommel's argument was that it was better to have some armour on the right beach, even if the rest was wrongly disposed, than to keep armour in central reserve and then fail to move it when Allied airpower descended. At the end of January 1944 he was translated from the post of inspector of the Atlantic Wall to commander of

Army Group B (Seventh and Fifteenth Armies), as Rundstedt's direct subordinate for defence of the invasion zone. Almost at once he fell into dispute with his chief. Rundstedt had never experienced a battle in which the Luftwaffe was not dominant. He therefore believed that there would be time, even after the enemy landing craft had arrived, to make a deliberate assessment of the military situation and then commit reserves to a counter-attack. Rommel knew that an unhurried counter-attack would be destroyed by enemy aircraft. From personal experience in Egypt and Tunisia he knew how great was the power of the Allied air forces and was convinced that only by holding armour 'forward' and committing it immediately could the invasion be met and defeated.

The Rommel–Rundstedt dispute, in which personal experience favoured one general, conventional military wisdom the other, eventually reached the ears of Hitler. He resolved it on his own terms, to neither subordinate's liking, when the two visited him at Berchtesgaden on 19 March 1944. Panzer Group West, which oversaw the six armoured divisions of Army Group B, was split; three of its divisions were allocated to Rommel, three to Rundstedt – but with the proviso that Rundstedt's divisions (21st, 116th and 2nd) were not to be committed without the direct approval of Hitler's operations staff at OKW, with the attendant risk of even greater delay than Rommel had feared in the first place.

As the 21st Panzer Division was the only armoured division close to the beaches chosen by the Overlord planners, Rommel's intention to launch a quick counter-attack was thus compromised from the start. Montgomery, his old desert opponent, had warned on 15 May in his pre-invasion assessment:

[Rommel] will do his level best to 'Dunkirk' us – not to fight the armoured battle on ground of his choosing but to avoid it altogether and prevent our tanks landing by using his own tanks well forward. On D-Day he will try (a) to force us from the beaches; (b) to secure Caen, Bayeux, Carentan. . . . We must blast our way onshore and get a good lodgement before he can bring up sufficient reserves to turn us out. . . . While we are engaged in doing this, the air must hold the ring and must make very difficult the movement of enemy reserves by train or road towards the lodgement areas.

Had Montgomery known, at the time he wrote this assessment, how grievously the Rommel–Rundstedt–Hitler dispute on armoured deployment had harmed the *Westheer*'s prospect of defeating the landing force, his fears for the successful outcome of D-Day would have been greatly relieved.

Montgomery was appointed to the command of the landing force only on 2 January 1944. Until the Stalin–Roosevelt–Churchill conference at Tehran in November 1943 no commander for Overlord had been nominated at all. Both the American and British chiefs of staff, General George Marshall and General Sir Alan Brooke, had been promised the appointment by their heads of government, though since August Brooke had known that for reasons of international politics it must go to an American. However, it was only at Tehran that the issue of nomination had been brought to a head. Stalin had there made it the test of Anglo-American dedication to the alliance's Second Front. 'Do the British really believe in "Overlord",' he had asked, 'or are they only saying so to reassure the Soviet Union?' In the face of Churchill's protestations of commitment, he demanded that a commander be nominated not later than one week after the conference ended. Churchill acquiesced and Roosevelt agreed to make the choice. On 5 December, however, at the end of the time limit, Roosevelt recognised that he could not spare his helpmate, Marshall, from Washington, and told him so; the Supreme Command of the Allied Expeditionary Force would therefore go to Eisenhower. Because Eisenhower's talents were strategic rather than tactical, however, operational authority would be vested in a ground commander, Montgomery, until the 'foothold' on the soil of France had been consolidated into a 'lodgement' from which the Wehrmacht could not displace the Allied liberation army.

Montgomery, arriving in England direct from Italy where he had been commanding the Eighth Army, threw himself into the rationalisation of the Overlord plan with an energy, familiar to his staff in the Mediterranean, that left the COSSAC headquarters breathless. General Sir Frederick Morgan, Chief of Staff to the Supreme Allied Commander (Designate), had been putting together a scheme for a landing in north-west Europe since the Churchill-Roosevelt meeting at Casablanca in January 1943. COSSAC's proceedings had not been dilatory; but they had been deliberate. Morgan had set himself the task of presenting the Supreme Commander, when nominated, with a flawless military appreciation. His Anglo-American staff, proceeding from first principles, had first of all identified where landings would be possible. The operational radius of a Spitfire, the most numerous Allied fighter, was used to delimit the zone in which the Allies would enjoy unchallenged air superiority. It reached from the Pas de Calais to the Cotentin peninsula in Normandy; the coast east and west of those places could be eliminated. Within the zone, however, long stretches of coastline were topographically unsuitable: the chalk cliffs of the Pays de Caux were too steep, the mouth of the Seine estuary was too indented, the Cotentin itself was too easily sealed off at its base. By reduction, therefore, only two coastal stretches recommended themselves: the Pas de Calais, with its gently shelving, sandy beaches, and the Normandy coast between the Seine and the Cotentin. The Pas de Calais had the attraction of proximity both to the English coast and to the 'short route' into Germany; but for those reasons it could be judged the sector where the Germans would expect to be attacked and would defend most heavily. COSSAC therefore plumped for Normandy.

Because the chosen stretch of Normandy had no ports, but also because the Germans could be counted on to fight to deny nearby Cherbourg and Le Havre to the enemy, it was decided to construct two artificial floating harbours ('Mulberries') and tow them to the beaches once they had been seized. The initial landing would be made by three divisions, disembarked from landing craft under heavy air and naval bombardment; airborne troops would be dropped at either end of the chosen bridgehead to secure 'blocking positions' on the flanks. As soon as the bridgehead was consolidated, seaborne reinforcements would be poured in to transform it into a 'lodgement area' from which a break-out into Brittany and then the west of France would be mounted. Eventually a hundred divisions would pass through Normandy; the main strength of the American army, which would supply the majority of divisions, would be shipped directly from the United States.

Success depended, however, on minimising the strength the Germans could oppose to the landing. Although an intelligence blackout over the invasion fleet itself could be guaranteed, and German air and naval interference be discounted, COSSAC agreed it was vital that near Caen, the *Schwerpünkte* of the invasion zone, there should be 'no more than three [German divisions] on D-Day, five by D plus 2 and nine by D plus 8'. The first week of the landings, in short, would be a race between the Allied and German armies' capacity to build up forces in and against the bridgehead. The Germans could not prevent the Allied build-up; the Allies could, by contrast, prevent the German. A crucial element of the invasion effort, therefore, would be the bringing to bear of Allied airpower against the roads, railways and bridges by which Rundstedt's sixty divisions would march to the battlefield. The greater the devastation Allied airpower could inflict on the infrastructure of the French transport system – at whatever subsequent cost to the Allies' own capacity to supply its armies in mainland France – the more certainly would the seaborne divisions survive the landing and the shock of initial combat in the lodgement area.

Montgomery, on his arrival in London in January 1944, dissented from none of COSSAC's broad criteria. However, he and Eisenhower, who was eventually to succeed him in command on the ground, had both briefly seen the operational plan when en route to England via Marrakesh (where Churchill was recovering from pneumonia), and they jointly judged that the attack would have to be launched 'in greater weight and on a broader front'. In brief, they wanted the American landing to be separated from the British, both to be made in heavier weight, and the airborne contribution to be much increased. Montgomery warned that, as things stood, '[German] reserve formations might succeed in containing us within a shallow covering position with our beaches under continual covering fire.' He remembered Salerno, where a well-planned assault had almost come to naught because of the rapidity of the German reaction.

By 21 January, therefore, he had proposed a major amplification of the landing. It was to be mounted by five seaborne divisions abreast, two American to the west, two British and a Canadian to the east; the original 'two airborne brigades' were to be increased to two American airborne divisions, dropped astride the river Vire at the base of Cotentin peninsula, and the British 6th Airborne Division, dropped astride the river Orne between Caen and the sea. The creation of airheads on the Vire and the Orne would prevent the Germans from 'rolling up' the amphibious bridgehead in between; within it the five seaborne divisions, reinforced by two others pre-loaded in landing craft, would win ground for the post-invasion reinforcements to be landed and deployed. Specialist armour, including 'swimming' Sherman tanks, would accompany the assault infantry to their debarkation; the 79th (British) Armoured Division, composed of obstacle-clearing tanks, would open the way out of the beaches for the assault battalions to move inland.

Eisenhower, as Supreme Commander, at once endorsed these proposals. The only difficulty that remained was how to accumulate the craft necessary for the enlarged landing. Admiral King, Chief of (US) Naval Operations, both an Anglophobe and a devotee of the amphibious war in the Pacific, directly controlled the lion's share of Allied landing-craft production, since the vast majority were launched from American yards (82,000 were built in the USA throughout the war). A near-doubling of the D-Day assault divisions required a proportionate accretion of vessels in which to deliver and support them. These included the Landing Ship Tank (LST), Landing Craft Tank (LCT), Landing Craft Infantry (LCI), Landing Craft Mechanised (LCM) and Landing Craft Vehicle and Personnel (LCVP) as well as the versatile amphibious truck (DUKW or 'Duck'). King had a surplus of such vessels, particularly the crucial Landing Ship Tank, in the Pacific, but proved unwilling either to transfer any from one ocean to the other or to make available craft no longer needed in the Mediterranean. As a result SHAEF, the Supreme Headquarters Allied Expeditionary Force (as COSSAC was renamed after Eisenhower's nomination), was obliged to accept a postponement of Overlord from May to June, while its staff scrambled to find landing craft where they could. In addition, Operation Anvil, the landing in the south of France originally scheduled to coincide with Overlord, was set back a month further.

Subsequent investigation has revealed that the shortage of landing craft was illusory rather than objective. By 1943 the output of LSTs from British yards alone already sufficed to land the D-Day divisions; American LSTs were a bonus. The COSSAC staff had convinced itself that the US Navy's anti-Japanese imperative was depriving it of its just allocation; but the truth seems to be that the shortage was the result of faulty allocation in Europe, not of deliberate starvation by SHAEF's Pacific rivals. The postponement of Anvil, moreover, though undoubtedly caused by the lack of landing craft, may actually have helped rather than hindered the success of Overlord. Although it was initially conceived as the answering blow to the northern operation (hence 'Anvil') which would crush the *Westheer* by concentric action, the force dedicated to Anvil – four French and three American divisions – was not strong enough to mount a major attack on the rear of the *Westheer* and, because of the conflicting demands of the Italian campaign, could not have been increased, however many landing craft might have been assembled in the Mediterranean. Anvil's real value proved to be diversionary; as we shall see, the mere menace of a 'third' landing, like that of a 'second' in the Pas de Calais, succeeded in retaining German divisions in Provence throughout the weeks when they were desperately needed for the north to fight the real landing in Normandy.

Allied strength, German weakness

To the invasion army assembling in southern England during the spring of 1944, the notion that it might lack for anything would have defied all appearances. The great natural anchorages in which the Channel coast abounds – Chichester, Portsmouth, Southampton, Poole, Portland, Plymouth, Falmouth – were filling with warships and transports. So vast was the gathering armada – which could only have been assembled off Normandy where the Channel is widest – that two of the seven seaborne forces into which it was divided had to be harboured as far away as South Wales and East Anglia. These, Forces B and L, were to sail the day before invasion and join the other five under cover of darkness on the night of D-Day in the mid-Channel 'Area Z' from which, through channels cleared by a vanguard of minesweepers, they were to proceed in parallel columns to the five beaches on which the assaulting infantry and swimming tanks would debark; the beaches were codenamed from west to east Utah, Omaha, Gold, Juno and Sword. Operation Neptune, the Naval plan, provided for 6483 vessels to make the voyage, including 4000 landing craft, hundreds of 'attack transports', and a bombardment force of 7 battleships, 2 monitors, 23 cruisers and 104 destroyers. Their role was to engage and destroy the coastal batteries of the Atlantic Wall. Close fire support was to be provided by squadrons of rocket-firing landing craft, while others embarked self-propelled artillery which would 'shoot itself in' against the German shore positions before rolling up the beach to follow the seaborne infantry. Behind the bombardment and amphibious squadrons sailed the craft bringing the infrastructure required by the assault waves – the 'beach parties' which would set up traffic control and signal stations, organise obstacle clearance and evacuate casualties. Assault engineers, manning amphibious bulldozers, demolition tanks and fabric road layers, were also to follow the assault waves at close interval. And in the very forefront would land forward air controllers, to call in rocket, bomb and machine-gun strikes from the fighters and ground-attack aircraft among the 12,000-strong British and American air forces that were to support the landings.

Of these 12,000, over 5000 were fighters; to oppose them General Hugo Sperrle's Third Air Fleet had only 169 available on the Channel coast on 6 June 1944. A thousand Dakotas were to fly the parachute battalions of the three airborne divisions to their destinations, and hundreds of other transport aircraft were to tow gliders filled with airlanding infantry, artillery and engineers. The mightiest element of the air forces, however, was provided by RAF Bomber Command and the US Eighth Air Force, temporarily diverted from the strategic campaign against Germany to prepare and support the invasion. In the weeks beforehand the 'heavies' – Lancasters and Fortresses – with the medium bombers of the British Second and US Ninth Air Forces had largely destroyed the French northern railway system. On the night and morning of D-Day, Bomber Command and Eighth Air Force, each dropping the unprecedented weight of 5000 tons of bombs – the short haul allowed them to substitute bombs for fuel – were targeted against German defences in the immediate vicinity of the beaches.

The Allies' overwhelming air superiority guaranteed not only fire support at the moment of assault but security from surveillance beforehand. In the first six months of 1944 only thirty-two Luftwaffe daytime flights over England were recorded; there was only one in the first week of June – on 7 June, a day too late – and this at a time when Allied intrusions into French air space were as common as the flight of swallows. Ultra was meanwhile monitoring the movement of units to and within France on an hourly basis, while the Abwehr had no access whatsoever to the meaning of Allied signals; the volume of such signals, however, was carefully controlled to disguise the presence of the invasion army in the west of England and enhance belief in the fictitious existence of FUSAG in Kent. The Abwehr could, in compensation, draw on the reports of its network of agents in Britain, and these were eagerly assessed for indications of the strength, timing and above all objectives of the invasion. However, since every single one of the agents apparently at liberty had in fact been 'turned' by British counter-espionage (the 'Double-Cross System'), their reports were not only valueless but actively misleading. The British entertained fears that agents outside their control in Lisbon and Ankara might succeed in hitting on the truth by speculation, but none did so; the only serious leak of secrets, sold to the Abwehr out of the ambassador's safe in Ankara by his Turkish valet, contained references to an 'Operation Overlord' but was bereft of details (this was the much misunderstood and over-inflated 'Cicero' affair).

The *Westheer*, OKW and Hitler were thus denied any useful foreknowledge of Overlord in the weeks before its launching; last-minute intelligence was distorted by the jamming of selected German coastal radar stations and the simulation of a bogus invasion fleet and air armada in the Channel narrows opposite the Pas de Calais during the night of 5/6 June itself. In the weeks before D-Day, however, Hitler and the *Westheer* did succeed in materially reinforcing the anti-invasion forces, including those in Normandy. Between April and June the excellent Panzer Lehr Division was returned from Hungary to Chartres, only a day's drive from the beaches, and the 21st Panzer Division was brought from Brittany to Caen; while the 352nd and 91st Infantry Divisions, both of good quality, were put in coastal positions, the 352nd above

Omaha beach, where it would inflict heavy casualties on the American 1st Division on D-Day. When these redispositions were complete, the chosen beaches were defended by three instead of two infantry divisions, with another in close support; four instead of three Panzer divisions stood at close hand, one almost directly behind the British beaches themselves. It was prudence, not prescience, that dictated these new deployments, but the effect was to strengthen the *Westheer*'s capacity to resist at the key point, precisely as if correct intelligence had guided their new location.

By the first week of June, however, there was no more SHAEF could do to soften enemy resistance until the commitment of the invading troops. Throughout that week they were confined to camp, isolated from civilian contact and entertained by cinema shows and record concerts. The belief was that D-Day casualties would be high – the troops' commanders believed very high indeed. Most of the Americans and some of the British had no battle experience and contemplated the coming ordeal with sang-froid; those British divisions which had been brought home from three years of fighting in the desert and Italy were altogether less insouciant. They knew the ferocity with which the Wehrmacht fought and did not relish meeting it in the defence of the approaches to the Reich. Lieutenant Edwin Bramall, a new subaltern with the veteran 2nd King's Royal Rifle Corps (and a future British chief of staff), thought the battalion 'worn out': 'They had shot their bolt. Everybody who was any good had been promoted or become a casualty.' By contrast, Eisenhower's naval aide found the young American officers who had not seen action 'as green as growing corn', and asked himself, 'How will they act in battle and how will they look in three months' time?' Commander Butcher's and Lieutenant Bramall's anxieties were to prove equally unfounded. Most British troops, however battle-weary, rose to the challenge of Normandy; the Americans grew into it almost overnight, once again demonstrating that three minutes of combat exceeds in value three years of training in making a soldier. No military formation, moreover, was to win a more ferocious reputation in Normandy than the 12th SS Panzer Division 'Hitler Jugend', whose soldiers had been recruited direct from the Nazi youth movement at the age of sixteen in 1943.

Sea and sky turned stormy in the Channel at the end of the first week of June. The good weather on which Eisenhower and Montgomery had counted to coincide with favourable mid-month tides failed them; 4 June, the day chosen to launch the invasion, produced winds and waves which made landing by sea or air impossible. The airborne divisions stood down, the seaborne divisions which had sailed from the further ports turned back, the main armada kept to harbour. It was not until the evening of 5 June that the weather was judged to have abated enough for D-Day to be set for the following morning.

When it dawned, the spectacle that confronted those embarked – and those ashore – was perhaps more dramatic than any soldiers, sailors or airmen had ever seen at the beginning of any battle. On the Normandy coast the sea from east to west and as far north as the seaward horizon was filled with ships, literally by the thousand; the sky thundered with the passage of aircraft; and the coastline had begun to disappear in gouts of smoke and dust as the bombardment bit into it. 'The villages of La Breche and Lion-sur-Mer', reported Captain Hendrie Bruce of the Royal Artillery, 'are smothered with bursts, and enormous dirty clouds of smoke and brick dust rise from the target area and drift out to sea, completely obscuring our target for a time.' Under these angry clouds the British, Canadian and American infantry were debarking from their landing craft, picking their way between the shore obstacles, diving to cover from enemy fire and struggling to reach the shelter of the cliffs and dunes at the head of the beaches.

The time (H-Hour), depending on the set of the tide from beach to beach, was between 6.00 and 7.30, and the early minutes of the landing, for all except the Americans doomed to the agony of Omaha beach, were the worst. However, the wet and frightened infantrymen struggling through the surf along sixty miles of Normandy coastline were not the first Allied soldiers to have landed in France that day. In the darkness of the early morning the parachute units of the three airborne divisions, spearheads of the glider battalions that were to follow, had already dropped across the lower reaches of the two rivers, Vire and Orne, that demarcated the bridgehead's outer flanks. The British 6th Airborne Division, compactly released by experienced pilots on to open pasture, had made a good drop, rallied quickly and moved rapidly to their objectives. These were the bridges of the Orne and its eastward neighbour, the Dives, which were to be respectively held and blown, in the latter case to prevent German armour 'rolling up' the British seaborne bridgehead by a drive along the coast. The American 82nd and 101st Airborne Divisions had been less lucky. Their pilots were inexperienced, the narrow neck of the Cotentin peninsula was easy to overshoot and the valley of the Vire was heavily flooded by deliberate defensive inundation. Some American parachutists fell into the sea, many drowned in the floods, many others, scattered by bad navigation and fear of flak, dropped miles from their objectives; the 101st Division's 'spread' was 'twenty-five miles by fifteen, with stray "sticks" even further afield'. Twenty-four hours later only 3000 men of the 'Screaming Eagles' had rallied, and some were to roam for days behind enemy lines, refusing to surrender while rations and ammunition lasted.

Confusion in the German camp

The scattering of the American parachutists was thought a calamity at the time, most of all by their tidy-minded commanders. In retrospect it can be seen materially to have added to the confusion and disorientation the invasion was inflicting on their German opposite numbers. The general commanding 91st Division, for example, was ambushed and killed by wandering American parachutists while returning from an anti-invasion conference in the early hours of 6 June, before he had even grasped that the event had begun.

Elsewhere it sometimes took hours for German commanders to comprehend that the reports they were receiving from units actually under attack by Overlord forces were different from the bombardments and commando raids that had disturbed their occupation of France during the previous three years. On the day before, Luftwaffe meteorologists had discounted the possibility of an imminent invasion because of bad weather forecasts. By ill luck, Rommel was temporarily absent in Germany on leave, Rundstedt was sleeping the sleep of the old campaigner at Saint-Germain (he had been chief of staff of one of the divisions sent to invade France in 1914 and had a thick skin for alarms and excursions), while Hitler was preparing for bed at his holiday house at Berchtesgaden on the Obersalzberg and would not be presented with the firm evidence that the invasion had begun until his noon conference six hours after the assault waves had touched down.

Local commanders nevertheless made such reactions as their authority allowed when they got firm indication that a landing had begun. Such indications were too soon to arrive. Because only eighteen out of ninety-two radar stations were working – those the Allied electronic-warfare teams had left unjammed in the Pas de Calais region – the pitifully small number of German night-fighters available (most were permanently defending the Reich) were scrambled to deal with the bogus air armada approaching from the Channel narrows. The real parachute fly-in was not attacked at all, since it was out of range of any working radar stations. And the seaborne armada was eventually detected by sound at two in the morning, twelve miles off the Cotentin. At 4 am Blumentritt telephoned Jodl at Berchtesgaden for permission to move Panzer Lehr Division towards the beaches but was told to wait until daylight reconnaissance clarified the situation. As late at 6 am, when the naval bombardment was already devastating the beaches, LXXXIV Corps, which commanded the threatened sector, reported to the Seventh Army that it 'appears to be a covering action in conjunction with attacks to be made at other points later'.

Three German divisions, the 709th, 352nd and 716th, were thus to undergo attack by eight Allied divisions without any immediate support from their higher headquarters. The 709th and 716th Divisions found themselves in particularly desperate straits. Neither was of good quality and both lacked any means of manoeuvre. The first was defending the area on which the 82nd and 101st Airborne Divisions were dropping as well as Utah beach, where the US 4th Division was assaulting from the sea. It was an almost impossible mission. The US 4th Division was an excellent formation which put nine good infantry battalions on to the beach in the first wave. The 82nd and 101st were the cream of the American army, trained to a knife-edge and prepared for battle; their eighteen battalions, though scattered, were the equal of an ordinary force twice their size. The 709th Division was very ordinary indeed; its six battalions, finding themselves surrounded and outnumbered, put up scarcely any resistance. The three battalions on the beach surrendered after firing a few shots. Allied casualties at Utah numbered 197, the lowest of D-Day, and insignificant when set against the total of 23,000 men landed on that beach.

The 716th Division, confronting the British 50th, Canadian 3rd and British 3rd Divisions on Gold, Juno and Sword beaches at the eastern end of the bridgehead, was of no better quality than 709th and was also disorientated by the descent of the 6th Airborne Division in its rear area. The British had additionally brought two commando brigades to the landing and three brigades of assault armour; their swimming Sherman tanks were briefed to leave their landing craft as close to the beaches as possible, so that the infantry would have covering fire from the moment of touchdown. The effect of launching such large numbers of well-supported infantry against the scattered German defenders was notable. At Sword and Juno the British and Canadians got ashore with little loss and quickly pressed inland; the British 3rd Division joined up with the 6th Airborne later that morning. On Gold the 50th Division had mixed fortunes: one of its two landing brigades debarked in front of dunes and crossed with little difficulty; the other was confronted by a fortified beach village which the naval bombardment had spared. By bad luck its swimming Shermans were late arriving, and in the meantime the two leading battalions, the 1st Hampshires and 1st Dorsets (which 185 years earlier had been the first British regiment to set foot in India), suffered heavy casualties.

One of the brigade's supporting artillerymen, Gunner Charles Wilson, supplies a picture of the extraordinary confusion of the last moments of 'run-in' and first moments of 'touchdown' which, in its mixture of dreadful fatality and hair's breadth survival at a few yards' distance, holds good for the incidents of D-Day from one end of the bridgehead to the other. His Landing Craft Tank (LCT) was carrying four self-propelled 25-pounder guns which were firing at targets ashore throughout the approach:

We hit two mines going in – bottle mines on stakes. They didn't stop us, although our ramp was damaged and an officer standing on it was killed. We grounded on a sandbank. The first man off was a commando sergeant in full kit. He disappeared like a stone in six feet of water. We grasped the ropes of the net over which the guns were to drive ashore and plunged down the ramp into icy water. The net was quite unmanageable in the rough water and dragged us away towards some mines. We let go the ropes and scrambled ashore. I lost my shoes and vest in the struggle and had only my shorts. Somebody offered cigarettes but they were soaking wet. George in the bren carrier was first vehicle off the LCT. It floated for a moment, drifted on to a mine and sank. George dived overboard and swam ashore. The battery command-post half-track got off with one running behind. The beach was strewn with wreckage, a blazing tank, bundles of blankets and kit, bodies and bits of bodies. One bloke near me was blown in half by a shell and his lower part collapsed in a bloody heap in the sand. The half-track stopped and I managed to struggle into my clothes.

The 50th Division overcame its initial difficulties by mid-morning and by nightfall had advanced almost to the outskirts of Bayeux, closer to its prescribed objectives than any other Allied formation on D-Day. Closer by far than the US 1st Infantry Division, which on Omaha beach had undergone the worst of the invasion ordeals, one nearly as costly in lives as the planners had feared would be the lot of all divisions landing on the morning of the invasion. The 1st Division had been opposed by the 352nd, the best German formation in coastal positions on 6 June. Moreover, it defended beaches backed in places by steep shingle banks and overlooked at either end by steep cliffs. Exit from the beaches was difficult, while the cliffs provided commanding positions from which fire was directed on to the seaborne infantry below as the landing craft neared the shore and even as they touched ground. Their swimming Shermans, launched too far from shore in rough seas, had foundered. They had no direct fire support. The results were lamentable. The ordeal of the 1st Battalion, 116th Infantry Regiment, conveys the experience:

Within ten minutes of the ramps being lowered, [the leading] company had become inert, leaderless and almost incapable of action. Every officer and sergeant had been killed or wounded. . . . It had become a struggle for survival and rescue. The men in the water pushed wounded men ashore ahead of them, and those who had reached the sands crawled back into the water pulling others to land to save them from drowning. Within 20 minutes of striking the beach A Company had ceased to be an assault company and had become a forlorn little rescue party bent upon survival and the saving of lives.

Had all the German defenders of Normandy been as well trained and resolute as those of the 352nd Division and had accident overtaken more of the swimming Shermans, the débâcle at Omaha might have been repeated up and down all five beaches, with catastrophic results. Luckily, the fate of the 1st/116th Infantry was extreme. The Omaha landing as a whole was costly. Most of the 4649 casualties suffered by the American army on D-Day occurred there. Yet some of the Omaha battalions got ashore unscathed and even those worst afflicted eventually gathered their survivors and got away from the water's edge. By the end of D-Day all chosen landing places were in Allied hands, even if the bridgehead was in places less than a mile deep. The question which loomed as evening drew in was whether the separate footholds could be united and in what strength the Germans would counter-attack.

The battle of the build-up
Because of the immobility of all the German infantry divisions, it was the

handling of the Panzer divisions by Army Group B and OKW which alone threatened the invaders with riposte. Of the four within or close to the invasion area, only the 21st Panzer Division, positioned near Caen on the eastern flank of the British Sword beach and 'airborne bridgehead', was close enough to the scene of action to exert a decisive effect. Its commander, like Rommel, was absent on the morning of D-Day (Rommel, by furious driving from Ulm, would arrive at Army Group B headquarters at 10.30 in the evening). Rommel's chief of staff, Hans Speidel, succeeded in extracting permission from OKW for the 21st Panzer Division to intervene at 6.45 am, but it was two hours before General Erich Marcks, next in the chain of command, passed on an operational order. That order required the tanks to probe into the gap between Sword and Juno beaches, to halt the British advance on Caen, which was only eight miles from the sea, and 'roll up' the bridgehead.

Advancing on Caen from Sword beach was a brigade of British infantry led by the 2nd Battalion of the King's Shropshire Light Infantry. It should have been accompanied by the tanks of the Staffordshire Yeomanry but they were trapped in a giant traffic jam on the beach. The Shropshires had therefore to take each German defensive position as they came to it by orthodox fire and movement. Progress was slow. In the afternoon the tanks caught them up, but at six o'clock the column ran into the vanguards of the 21st Panzer Division, which had been delayed by one time-wasting mission after another on its way to the front. Its guns at once forced the Shropshires to go to ground, three miles short of Caen, and its 22nd Panzer Regiment moved forward to attack the bridgehead. 'If you don't succeed in throwing the British into the sea', Marcks warned the regimental commander with all too exact foresight, 'we shall have lost the war.' However, its Mark IV tanks ran into the Staffordshire Yeomanry's anti-tank Fireflies (Shermans armed with the long 17-pounder) and suffered heavy losses. A few made contact with the infantry of the 716th Division still holding out at Lion-sur-Mer; but, when they were overflown by the 250 gliders of 6th Airborne Division bringing reinforcements to the parachutists across the Orne, they concluded that they risked being cut off and withdrew. By nightfall, though Caen remained in German hands, the Sword bridgehead perimeter was intact and the crisis of D-Day had passed.

Though the Germans could not know it – Marcks's gloomy prognosis was an inspired guess – their opportunity to extinguish the invasion at its outset had now passed. On 7 and 8 June the next nearest Panzer division, the 12th SS (Hitler Youth), came forward to assault the Canadians in their bridgehead west of Caen and inflicted heavy losses on them, but the Hitler Youth failed to break through to the sea; a Germany army officer reported seeing some 'crying with frustration'. Meanwhile the invaders were linking hands across the gaps that separated Sword from Juno and Gold and the British from the American beaches (the British were joined to Omaha on 10 June, Omaha to Utah by 13 June), as their navies simultaneously outstripped the enemy in the race to bring reinforcements to the battlefront. The explanation of the Allies' success in 'the battle of the build-up' is simple. The Channel was a broad highway, wholly under Allied control; only a few ships were lost to mines and E-boat attack; and, although some of the new 'schnorkel' submarines succeeded in reaching the Channel from Brittany, they suffered heavy losses, so the general effect was trifling. By contrast, not only was the carrying capacity of the French roads and railways grossly inferior to that of the Allied transport fleet, but the whole interior of northern France lay under the eye of the Allied air forces, which, from 6 June onwards, redoubled their pre-invasion efforts to destroy the transport infrastructure and shot at anything that moved in daylight. Rommel himself was to be severely wounded in a ground attack by a British fighter while driving in his staff car on 17 July.

Even if Hitler had allowed reinforcements for the Seventh Army to be drawn wholesale from the Fifteenth, First and Nineteenth, they would have found great difficulty in reaching the battlefield at any rapid pace. As it was, he forbade the transfer of units from the Fifteenth, the nearest army, until the end of July, lest the 'second invasion' materialise in the Pas de Calais, and he only grudgingly released others from the First and Nineteenth. The Panzer divisions moved first; the 9th and 10th SS Panzer, returning from a counter-offensive mission in Poland, took four days to cross Germany but a further eleven days to reach Normandy from the French frontier, entirely as a result of air attack. The march of the unmechanised divisions was even more laborious. The 275th Division, for example, took three days to cover thirty miles from Brittany to Normandy (6-8 June) and then another three days to reach its battle positions. Allied reinforcement divisions were meanwhile moving from southern England to Normandy in less than twenty-four hours. The first month of the Normandy battle therefore resolved itself into a struggle between arriving Allied formations that were attempting to seize ground deemed essential to the development of a successful offensive and break-out, and German mobile divisions seeking to nail and wear them down. The essential ground for which the Americans struggled, the Cotentin and the port of Cherbourg, lay within the bridgehead (they reached the Atlantic coast of the peninsula on 18 June); for the British the essential ground was Caen and its environs, from which they could plunge into the open plains that led directly to Paris a hundred miles away.

Montgomery had hoped to take Caen on 6 June; when the effort failed, he launched three separate attacks to take the city. A local offensive by the Canadians was contained by the 12th SS Panzer Division on 7-8 June; an armoured attack west of Caen on 13 June was largely defeated by one of the few Tiger tank battalions in Normandy; finally a large offensive by the 15th Scottish Division (26 June to 1 July), codenamed Epsom, was blunted by the recently arrived 9th and 10th SS Panzer Divisions. At its last gasp, Operation Epsom secured ground across the river Odon, a tributary of the Orne which joined it at

Caen. The most advanced position, the village of Gavrus, was taken and held by the 2nd Argyll and Sutherland Highlanders (the 'thin red line' of Balaclava), which had been forced to surrender at Saint-Valery four years earlier and been reconstituted in Britain since. However, an attempt by the supporting tanks of the 11th Armoured Division to break out into the open country south of the Odon failed and after five days of costly fighting – 4020 men had been killed or wounded – Epsom was called off.

The fight in the bocage

The Americans meanwhile were overcoming the German defenders of the Cotentin, the 243rd, 709th, 91st and 77th Divisions. Two of these, the 77th and 91st, were good-quality formations; the attacking American formations, the 4th, 9th, 29th and 90th Divisions, were inexperienced and unprepared for the difficulty of the terrain. The hedgerows, backbone of the soon to be infamous *bocage* country, were field boundaries planted by the Celtic farmers 2000 years earlier. Over two millennia their entangled roots had collected earth to form banks as much as ten feet thick. 'Although there had been some talk in the UK before D-Day,' wrote General James Gavin of the 82nd Airborne Division, 'none of us had really anticipated how difficult they would be.' Later in the campaign, the Americans fitted their Shermans with 'hedgedozers', but in June 1944 each hedgerow was impenetrable to tanks, as well as to fire and view. To the Germans they offered almost impregnable defensive lines at intervals of 100 or 200 yards. To the attacking American infantry they were death traps. Before them the green American infantry lost heart, forcing Bradley, the First Army commander, to call too often on the overtired parachutists to lead the assault. The 'All American' and 'Screaming Eagles' never flinched from the task; but the cumulative effect of losses in their ranks threatened these superb formations with dissolution. Lieutenant Sidney Eichen, of the incoming 30th Division, encountering a group of paratroopers in the Cotentin asked, '"Where are your officers?", and they answered: "All dead." He asked, "Who's in charge, then?", and some sergeant said, "I am". I looked at the unshaven, red-eyed GIs, the dirty clothes and the droop in their walk, and I wondered: is this how we are going to look after a few days of combat?'

Step by step, however, the Germans were forced back into the perimeter of Cherbourg. Hitler planned to hold the French ports as fortresses – as he had done in the Crimea and was to do in the Baltic states – so as to deny them to the enemy, whatever ground was lost in the hinterland. On 21 June he signalled to General Karl Wilhelm von Schlieben, the port commander, 'I expect you to fight this battle as Gneisenau once fought in the defence of Colberg' (one of the epics of Prussia's resistance against Napoleon in 1807). Five days later Cherbourg fell; the commander of the citadel requested the Americans to fire artillery at the main gate, to give him a pretext for surrender. Immediately afterwards he and all his men marched out under the white flag. On 26 June

Hitler demanded that Rundstedt inaugurate court-martial investigations against all who could be held responsible. General Friedrich Dollmann, commander of the Seventh Army, whose headquarters directed the Normandy battle, took poison the same night. There had been many suicides in the Red Army in 1941 but few so far in the Wehrmacht; as the shades drew in around the Reich the number would grow.

Mid-June 1944 was a time of desperate crisis for Hitler, the worst he had faced since the surrender of Stalingrad seventeen months earlier. Although on 12 June he had at last opened his secret-weapons campaign against Britain, the launch rate of the V-1s was much lower than hoped, about eighty a day, of which only half reached London, and there were many misfires. One rogue flying-bomb crashed directly on to Hitler's command bunker at Margival on 17 June during the course of the only visit he made to France throughout the Normandy battle. Moreover, although the danger in the west was great, a crisis on the Eastern Front now suddenly compounded his strategic difficulties. On 22 June, the third anniversary of Barbarossa, the Red Army had opened Operation Bagration, which, in six weeks of relentless armoured attack, destroyed Army Group Centre and carried the Russian line 300 miles westward from White Russia to the banks of the Vistula outside Warsaw; thirty divisions, 350,000 German soldiers, were killed, wounded or captured in the catastrophe.

During those terrible weeks of Bagration, the *Westheer* in Normandy continued to lose men in thousands but eventually succeeded in holding a line of defence. This illusory stability on the Normandy front after the fall of Cherbourg therefore brought a welcome sense of relief to Hitler's twice-daily situation conferences at Rastenburg. In early July, despite a continuing erosion of the Seventh Army's infantry strength, which was being ground away by incessant attrition in the hedgerow fighting, the perimeter of the bridgehead seemed to have been 'nailed down'. Montgomery had commited himself to the capture of Caen; having failed to capture it on 6 June, he now conceived the scheme of using it as a focal point for successive blows which would destroy the German mobile forces while the Allies accumulated reserves for the break-out. On 19-21 June reinforcement of the bridgehead was interrupted by a great Channel gale, which wrecked the American and damaged the British Mulberry harbours. Improvisation, however, soon made good the capacity, so that by 26 June there were already twenty-five Allied divisions ashore, with another fifteen in Britain on their way, to oppose fourteen German. That represented not only a quarter of the *Westheer* but two-thirds, eight out of twelve, of its Panzer divisions. Hitler may have been able to convince himself that the invasion had been brought under control. Rundstedt could not; on 5 July he advised Hitler to 'make peace' and was at once relieved as OB West by Kluge. Montgomery, daily informed by Ultra intelligence of the rising losses suffered by the *Westheer*, stuck resolutely to his scheme of making Caen 'the crucible' of the Normandy battle.

On 7 July, after the RAF dropped 2500 tons of bombs on Caen, virtually completing the destruction of William the Conqueror's ancient capital, the British 3rd and 59th and Canadian 3rd Divisions advanced on the city. They failed to take the centre but occupied all its outskirts. This operation, codenamed Charnwood, almost isolated Caen from the rest of the German positions in Normandy. There was evidence too that continuing American pressure was also drawing enemy armour away towards the base of the Cotentin, where it was planned that the ultimate break-out should erupt. Montgomery therefore decided that one more blow would bring on the climactic struggle with the Panzers that he sought and clear the way into the open country that led towards Paris. This new offensive was to be called Goodwood and would be mounted from the 'airborne bridgehead' east of the Orne into the corridor between that river and the Dives. Only one stretch of high ground, the Bourguébus ridge, closed the exit from that corridor to the high road towards Paris.

Goodwood, involving all three British armoured divisions in Normandy, the Guards, 7th and 11th, began on 18 July. It was preceded by the heaviest aerial 'carpet' bombardment yet staged in the campaign, took the defenders completely by surprise, and left the survivors trembling with shock. German tanks were overturned by the concussions and prisoners collected in the early stages of the advance stumbled to the rear as if drunk. By mid-morning the British tanks were halfway to their objectives and success seemed certain. Then the German army's extraordinary qualities of resilience and improvisation were asserted: Hans von Luck, a regimental commander of the 21st Panzer Division, arrived on the battlefield straight from leave in Paris to find pockets of artillery and armour which had escaped the bombardment and hastily co-ordinated a defence. While the gunners, including those of a Luftwaffe anti-aircraft battery, began to engage and slow the advancing British tanks, the engineer battalions of the 1st SS Panzer Division – German engineers (*Pioniere*) were used to acting as infantry in an emergency – hastily dug in on the crest of Bourguébus ridge, while the tanks of both the 1st SS and the 12th SS Panzer Divisions were hurried forward to form an anti-tank screen. By the time the British 11th Armoured Division had forced its way through to the foot of the ridge it was mid-afternoon; and, as the British tanks began to deploy to climb it, they were caught by salvoes of 75-mm and 88-mm fire from the high ground above. The leading squadron of the Fife and Forfar Yeomanry went up in flames on the spot. The 23rd Hussars, coming to their rescue, were hit as hard. 'Everywhere wounded and burning figures ran and struggled painfully for cover,' the regimental history recorded, 'while a remorseless rain of armour-piercing shot riddled the already helpless Shermans. All too clearly we were not going to "break through" that day. . . . Out of the great array of armour that had moved forward to battle that morning, one hundred and six tanks now lay crippled or out of action in the cornfields.'

The correct figure was 126 from 11th Armoured alone, more than half its strength; the Guards Armoured Division had lost another sixty in its first battle. Goodwood was close to being a disaster. Montgomery's post-battle protestations that it had not really been expected to produce a break-out were treated with impatience by both Churchill and Eisenhower. Churchill's patience in any case had been wearing thin at the slow pace of the advance inland. It was D + 43 on 20 July, the day the Goodwood fighting finally spluttered out, and the 'phase lines' drawn on the planners' maps before D-Day had forecast that the Allies should be halfway to the Loire by that date. As it was they had not yet even reached the projected line for D + 17. Montgomery had to argue at length to Churchill to persuade him that his grand design retained its logic and that a result would not now be long delayed.

Compulsively self-justifying though he was, Montgomery was right both to put the disappointment of Goodwood behind him and to argue that it had served a purpose. For it had indeed pulled Army Group B's armoured reserves back towards the British front at the moment when they had been concentrating to meet what growing evidence indicated was a great American offensive in the making. During July the Americans had been fighting a horrible and costly battle in the *bocage* south of the Cotentin. Between 18 and 20 July the 29th and 35th Divisions had lost respectively 2000 and 3000 men in the battle for Saint-Lô – five times the number of casualties suffered by the British armoured divisions in the same period east of Caen. German losses were even worse: the 352nd Division, the Americans' principal opponent, still in action after its stubborn defence of Omaha beach, almost ceased to exist after Saint-Lô. Its casualties went to swell the total of 116,000 suffered by the Seventh Army since 6 June, for which only 10,000 replacements had come from the *Ersatzheer* (Replacement Army) in Germany. Material losses had been equally severe: against 2313 tanks produced in German factories in May–July, 1730 had been destroyed, one-third of them in France, but by the end of June only seventeen replacements had arrived. The strength of the perimeter drawn around the Allied bridgehead was stretched close to breaking-point; and it was about to be subjected to a powerful blow at its weakest point.

On the morning of 25 July – after a false start when American aircraft bombed their own infantry – four American infantry and two armoured divisions moved to the assault west of Saint-Lô behind a heavy carpet bombardment. They belonged to General 'Lightning Joe' Collins's VII Corps. He had a reputation for hard driving of subordinates which the day's events justified. General Fritz Bayerlein, commanding Panzer Lehr in VII Corps's path, testified to the weight of the attack: 'After an hour I had no communication with anybody, even by radio. By noon nothing was visible but dust and smoke. My front lines looked like the face of the moon and at least 70 per cent of my troops were knocked out – dead, wounded, crazed or numbed.' The next day opened with another carpet bombardment. Progress, less than a mile the day before, increased to three and the American 2nd Armoured Division reached positions from which it stood poised to break out. Kluge, OB West and also the new commander of Army Group B, 'sent word', Bayerlein recalled, 'that the line along the Saint-Lô-Périers road must be held at all costs, but it was already broken.' He promised reinforcement by an SS tank battalion with sixty Tigers; it arrived with five. 'That night', Bayerlein went on, 'I assembled the remnants of my division south-west of Canisy. I had fourteen tanks in all. We could do nothing but retreat.' Panzer Lehr had once been perhaps the best and certainly the strongest armoured division in the German army. Its condition was an index of the state to which the *Westheer* had been reduced by six weeks of fighting in Normandy. Hitler was nevertheless adamant that the crumbling front must be restored and the situation reversed.

The July bomb plot

Five days before Cobra, as the American breakthrough operation was codenamed, a group of army officers had made an attempt to assassinate Hitler in his headquarters. On 20 July, Colonel Claus von Stauffenberg, a disabled veteran who held a staff appointment with the *Ersatzheer*, placed a bomb under the conference table at Rastenburg and then escaped to fly to Berlin and direct a conspiracy designed to replace the Nazi leadership throughout Germany with military appointees. By a succession of mischances the conspiracy miscarried. The bomb wounded but did not kill Hitler. An early misapprehension that the explosion was an act of sabotage was corrected. The signals officer who belonged to the conspiracy was accordingly prevented from interrupting outward communication from Rastenburg. Goebbels was thus able to mobilise soldiers loyal to Hitler in a military reaction against the conspiracy in Berlin. The conspirators were quickly arrested, and several of them, including Stauffenberg, were shot the same evening. By nightfall the danger of a coup had been averted and Hitler, even though isolated in his Rastenburg fortress, was once again secured in power. However, the 20 July Plot understandably reinforced every one of his deep-laid prejudices against the higher ranks of the army of which Stauffenberg was the epitome. An aristocrat, a devout Christian, a cavalryman – Hitler hated not only the church and the nobility but also horses, riding, equestrian apparel and everything they represented – Stauffenberg had been drawn into the anti-Hitler conspiracy because he recognised the mortal danger of defeat into which the Führer had led the fatherland and anticipated the disgrace and punishment that the iniquity of Nazism would bring to his countrymen in its wake. Stauffenberg's motives, in short, were patriotic rather than moralistic, though his moral sense was deeply engaged by the conspiracy. For both his patriotism and his morality Hitler had only hatred and contempt, feelings which he automatically transferred to all he identified as belonging to Stauffenberg's social class and professional caste. Far too many of them, he believed, officered the *Westheer*. General Heinrich Graf von Stülpnagel, the military governor of France, was certainly in the plot; so too, Hitler believed, was Rommel, even though he came from outside the 'old' officer class and since 17 July had been lying seriously injured in hospital. He had also a suspicion, though not proof, of the complicity of Kluge, since 4 July the linchpin of the battle against the 'Anglo-Saxons' as both OB West and direct commander of Army Group B. Only a resolute – and successful – riposte to the American breakthrough at Saint-Lô would convince him that his suspicions were misfounded and restore his belief in the dedication of the *Westheer* to the National Socialist revolution.

The test of the *Westheer*'s loyalty – also designed to produce a strategic reversal of the military situation in the west – was to be a counter-attack with all available armour into the flank of the American spearhead which was driving south from Saint-Lô between the *bocage* country and the sea towards the interior of Brittany. On 2 August an emissary from Rastenburg, Walter Warlimont, deputy chief of OKW's operations staff, reached Kluge's headquarters at La Roche-Guyon, believing he was to discuss with the field marshal the question of withdrawing to a defensive position deeper within France. On his arrival, however, he discovered that Hitler had meanwhile sent in orders to begin a counter-offensive as soon as possible, to start from Mortain and drive to the sea, and that he expected its results – as Warlimont discovered when he returned to Rastenburg on 8 August – to lead to a 'rolling up of the entire Allied position in Normandy'.

The Mortain counter-attack began on 7 August. It involved, immediately, four Panzer divisions, the 116th, 2nd, 1st SS and 2nd SS, and was intended to draw in four more, the 11th and 9th from the south of France, which Hitler had already promised to Kluge on 27 July, and the 9th and 10th SS from the Caen sector. Together these eight divisions, deploying 1400 tanks, would lead the *Westheer*, in an operation codenamed Lüttich (Liège), towards a great counter-encirclement of the invaders, the consequences of which would match Ludendorff's breakthrough into the rear of the French armies at Liège exactly thirty years earlier to the day. As Hitler had told Warlimont on the eve of his mission to Kluge, 'The object remains to keep the enemy confined to his bridgehead and there to inflict serious losses upon him in order to wear him down and finally destroy him.'

TWENTY-ONE

Tank Battle: Falaise

The battle of which Operation Lüttich marked the opening stage was to develop into the largest clash of armour in any of the campaigns fought on the Western Front, if not the largest of the war. Only the Battle of Kursk, fought the previous July, had assembled a larger number of German Panzer divisions – twelve, against ten in Normandy; but at Kursk the German offensive had been defeated by minefields and anti-tank guns rather than by mobile riposte. The Battle of the Falaise Gap, by contrast, took the form of a gigantic manoeuvre of twenty armoured divisions (ten German, ten Allied), tank against tank, over 800 square miles of countryside and extending through the two weeks of frenzied movement and violent combat.

By the summer of 1944 the mystique of *Blitzkrieg* had been long overlaid. In the summer of 1940 Kleist and Guderian had been able to count on the mere appearance – or even rumour – of tanks to panic infantry into flight or surrender. Now commanders could no longer have such expectations. Green or shaken infantry would, of course, still run at the approach of tanks, and had done so on numerous occasions in Normandy; but experienced infantrymen had learned for themselves, as they had been taught, that flight was even more dangerous than keeping to their positions in the face of armoured attack. For by 1944 tanks did not operate, as they had in their palmy days, as independent spearheads; they advanced only in close concert with specialist infantry of their own, the Panzergrenadiers, and under the protective bombardment of supporting artillery. Defending infantry who left their trenches to make for safety thus exposed themselves to several sorts of fire: that of the tanks themselves, that of the tanks' foot soldiers and that of their associated gunners. In the face of this most terrifying of all assaults, therefore, the defenders struggled to hold their ground, counting on their own anti-tank weapons to hold the attackers at bay while calling for their own artillery support and air strike – if that were available – and hoping that friendly tanks would come forward to do battle with those of the enemy. In short, by 1944 the tank had ceased to be an autonomous instrument of strategy but had taken its place in an elaborate machinery of tactical attrition, which achieved its effects by a cumulative wearing-down of resistance rather than by a rapier-like penetration of the enemy's front.

The dethronement of the tank from the status of revolutionary 'war-winning' weapon to that of workaday tool of tactics followed a pattern long established in the history of armaments. The ironclad, the torpedo and the machine-gun had each at first appearance been hailed as making defence, even war itself, 'impossible'; to each, in turn, an antidote had been found and the 'revolutionary' weapon subsumed within a slightly altered and more complex system of war-making than had prevailed before. However, although the tank had undergone a similar displacement, its autonomy had been doubted from the outset and disputed energetically between the two great theorists whose names will always be associated with its development. Major-General J.F.C. Fuller, who had masterminded the first great tank offensive at Cambrai in 1917, saw no future place for any but the tank arm on the battlefield; Basil Liddell Hart, his friendly rival in their paper debates of the 1920s and 1930s, argued that the tank would not win battles single-handed and that all arms, including infantry and artillery, would in future be mechanised, to produce armies which would resemble fleets of larger and smaller armoured and mobile 'land ships'.

Liddell Hart looked too far into the future; not until forty years after the end of the Second World War would even the most advanced states command the wealth and industrial resources to mechanise their field armies completely. It was nevertheless he, not Fuller, who saw the future true. Already by 1944 'land fleets' existed in embryo. It was with a land fleet of Panzer and Panzergrenadier divisions that OB West had striven to defeat the Allied invasion; and it was with a land fleet of armoured and mechanised divisions that Montgomery and Bradley would achieve the encirclement and destruction of Army Group B. Allied tanks were to play the leading part in blunting the German attack which initiated the Battle of the Falaise Gap and then in making the advances which drew the line of encirclement around the enemy; but the ground they won was consolidated and held by their accompanying infantry and the work of destruction within the Falaise pocket was completed by their supporting artillery and air squadrons. Falaise was an all-arms battle, and its nature exactly depicts the extent to which armoured tactics had been rationalised since the early days of the war, when the Panzer generals had acted as if invincible.

The diminution of the tank can in fact be traced to an early date in the war's development. Gamelin, for all his ineptitude of decision, correctly perceived the proper riposte to the Panzer thrust immediately after the crossing of the Meuse on 13 May 1940; it was to aim armoured counter-thrusts at the neck joining tank spearhead to infantry shaft. Two such counter-thrusts were mounted by de Gaulle's 4th Armoured Division at Laon on 18 May and by Frankforce at Arras on 21 May. Unco-ordinated in time, however, and unsupported by large resources of infantry and artillery, both counter-strokes failed. It was the Germans rather than the Allies who profited by the experience of these engagements. At Arras, Rommel, commanding the 7th Panzer Division, rescued himself from over-exposure by calling into service the heavy (88-mm) anti-aircraft guns of his flak battalion to halt and turn back the charge of the Royal Tank Regiment into his divisional centre after it had evaded his armoured screen. The eighty-eights stopped the heavy British tanks – which his lightly gunned Panzers had failed to do – and saved him from a defeat which might have extinguished his career then and there.

Arras emphasised to the Germans that the most effective means of waging armoured warfare against an equal enemy was to use a combination of tanks and anti-tank weapons; and they were to learn in their desert war against the British that these tactics worked in offence as well as defence. At the First Battle of Tobruk in April 1941 the Afrikakorps broke the perimeter of the fortified port with its tanks, but many of these were quickly lost to Australian tank-hunting parties because the defenders closed the gaps behind the intruders and prevented the German infantry from following in support. These tactics of divide-and-win were shortly to be used by Rommel against the British desert army itself. In the open desert battles of Sidi Rezegh and Gazala, from November 1941 to June 1942, Rommel perfected a method whereby he engaged British tanks with his own, retreated to draw the enemy on to a screen of anti-tank guns and then advanced when the losses inflicted had robbed the British of the means to conduct a mobile defence. Motorised infantry and self-propelled artillery accompanied his advancing tanks, thus ensuring that British positions overrun could be held and consolidated.

It was a positive advantage to Rommel in his adoption of these tactics that, for extraneous reasons, the number of tanks in German Panzer divisions had been halved since 1940. Hitler's purpose in reducing divisional tank strengths in late 1940 was to accumulate a surplus on which new Panzer divisions could be built; the number of divisions was in fact doubled between the fall of France and the opening of Barbarossa. The indirect effect of this bisection was to force German commanders to make better use of the non-tank elements of their Panzer divisions, particularly the mechanised infantry (Panzergrenadiers) and self-propelled artillery. Out of this necessity was born a true doctrine of tank-infantry-artillery co-operation which the Panzer divisions brought to a high level of practicality in 1943-4, as they found themselves progressively outnumbered by the enemy, particularly on the Russian front. Even when divisional tank strengths fell below 200 (from the 400 that had been standard in 1940), German Panzer divisions proved themselves equal or superior to much stronger Allied formations – as, for example, the 10th Panzer Division demonstrated when it routed the US 1st Armoured Division at Kasserine in Tunisia in February 1943. British and American armoured divisions followed the German pattern of organisation from mid-war onwards, shedding tank battalions and acquiring larger complements of motorised infantry and self-propelled anti-tank artillery to achieve a better balance of arms. The Americans, out of their large automotive capacity, were actually able to put their 'armoured infantry' into tracked carriers, with a notable improvement in mobility. Even so, the best German Panzer formations – those of the privileged SS and such

favoured divisions as the army's Lehr and the Luftwaffe's Hermann Goering –
remained superior to their Allied counterparts until, after the battles of
Normandy and White Russia, the relentless effects of attrition, imposed at the
front by combat and at the rear by aerial bombardment, began to depress their
strengths below the level at which losses of men and equipment could be made
good from the Replacement Army and the tank factories.

The technology of armoured warfare

Superior organisation and experience alone, however, did not explain
Germany's ability to wage armoured warfare on equal terms against a coalition
of industrially superior powers from early 1942 until late 1944. The quality of
German armour also counted significantly in the balance. German armoured
vehicles were, with only one or two exceptions, better than their equivalents in
the opposing armies. British armour in particular was lamentably inferior to the
German products. Though the British had invented the tank, first deployed it
in action in September 1916, and largely conceived the theoretical basis of
armoured warfare, they did not succeed in building an effective tank in the
Second World War. That crucial balance between firepower, protection and
mobility which underlies successful tank design eluded them. Their Infantry
Tank Mark I, which at Arras Rommel found he could penetrate only with his
eighty-eights, was strong but almost immobile. The Churchill was equally
tough but scarcely faster. Only the Cromwell, which appeared in 1944 to equip
the reconnaissance battalions of British armoured divisions, had speed and
protection; its gun remained inadequate. As a result the British divisions of
1944 were dependent on the American Sherman for their main tank strength,
but the Sherman too had defects: though fast, reliable and easy to maintain, it
burnt readily and lacked gunpower. Britain's most successful contribution to
Anglo-American armoured capability was to fit its fearsome 17-pounder anti-
tank gun to specially adapted Shermans, called Fireflies, which provided
British armoured divisions with their principal if not only antidote to heavy
German armour in 1944-5. The great merit of the Sherman, and a tribute to
America's industrial power, was that it could be manufactured in quantity. In
1943-4 the USA produced 47,000 tanks, almost all Shermans, while Germany
produced 29,600 tanks and assault guns. Britain, in 1944, produced only 5000
tanks.

It was Russia, alone among Germany's enemies, which matched its output of
tanks in quality and quantity. In 1944 Soviet tank production totalled 29,000,
most of which comprised the remarkable T-34. This tank owed much of its
technology to the independent American designer, Walter Christie, from whom
the Soviet Union bought prototypes at a time when fiscal stringency kept the
US Army on a shoestring budget. To Christie's chassis and suspension the
Russians added an all-weather engine and sloped armour, as well as an effective
gun, thus producing such a well-balanced design that in 1942, when Albert

Speer succeeded Dr Fritz Todt as head of the German armaments industry, the German army actually considered copying it wholesale as a successor to the ageing Panzer Mark IV. That humiliating concession of technical inferiority was ultimately avoided by the production of the Mark V Panther. Although the new tank failed at Kursk – as late as January 1944 Hitler was calling it 'the crawling Heinkel 177' in an allusion to a disappointing bomber – it eventually justified the effort put into its development in Normandy, where it equipped many of the Panzer battalions of the SS divisions. Even in 1944, however, the Mark IV remained the Panzer arm's mainstay. It had originated before 1939 as the final series in a ladder of designs, each larger than the one before. The Mark IV in particular had proved remarkably adaptable and had been progressively improved; finally, when equipped with a 'long' 75-mm gun as its main armament, it became almost a match for the T-34 itself.

Its predecessors, notably the Panzer Mark III, had also been readily adaptable as self-propelled anti-tank and 'assault' guns, and from February 1943 onwards, when the tank pioneer Heinz Guderian became Inspector-General of Panzer Troops, it was incorporated with the tanks into a new 'Panzer arm'. Such weapons suffered from the disadvantage that their guns fired only in the direction that the vehicle itself was pointing; but, because they lacked complex turret machinery, they were cheap to produce, and their low profile made them difficult to detect when deployed in well-chosen defensive positions on the battlefield. Their design was rational enough to be widely imitated by the Russians as well as the Americans and British, and they largely provided the mobile firepower of the Panzergrenadier divisions like the 17th SS, which the Allies found such formidable opponents in France. The German army itself distinguished little between tanks and assault guns. Indeed, Guderian's main difficulty in reorganising the Panzer arm in 1943 lay in overcoming the reluctance of the artillery to relinquish control of its assault guns, which, according to its senior officers, provided gunners with their sole opportunity to win the Knight's Cross, the Wehrmacht's ultimate decoration for bravery.

The only instrument of armoured warfare which German commanders regarded as qualitatively different from the rest was the Mark VI Tiger, which was not allotted to divisions but organised in independent battalions, kept under central control and committed to crucial offensive and counter-offensive missions. The Tiger had defects – its enormous weight was symptomatic of creeping 'gigantism' in German tank design which robbed it of speed while its turret traversed with ponderous deliberation: but, with its 88-mm gun and 100-mm-thick armour, it proved consistently superior, in static if not mobile operations, to every other tank of the war. The cough of the Tiger's engine starting up in the distance was something all Allied soldiers remembered with respect.

Tigers, Panthers, Mark IVs and assault guns were all to play their part in the great armoured battle in Normandy which, culminating in the holocaust of the

Falaise Gap, was about to begin at Mortain on the night of 6-7 August. Hitler, inspired as so often at the map table by a pictorial glimpse of opportunity, had decided that the outpouring of the American armies from Normandy into the narrow corridor between Mortain and the sea laid them open to a decisive counter-stroke. 'We must strike like lightning,' he had announced to his OKW operations staff on 2 August. 'When we reach the sea the American spearheads will be cut off. . . . We might even be able to cut off their entire beachhead. We must not get bogged down in cutting off the Americans who have broken through. Their turn will come later. We must wheel north like lightning and turn the entire enemy front from the rear.'

It was this decision which had brought the 116th, 2nd, 1st SS and 2nd SS Panzer Divisions to stand shoulder to shoulder at Mortain, only twenty miles from the Atlantic, on the flank of General Omar Bradley's US First Army which was streaming southwards into Brittany. Disastrously for them, for the *Westheer* and for Hitler, however, the signals which had directed their deployment had been monitored by the Ultra decryption service since 5 August; their objectives, Brécey and Montigny, were passed to Montgomery's headquarters and four American divisions, the 3rd Armoured, 30th and 4th, with the 2nd Armoured in support, were directed to block their path down the valley of the river Sée which Hitler had designated as their avenue to the ocean.

The *Westheer*'s ordeal

Some 200 German tanks (in first line), attacking without artillery preparation to assist surprise, advanced in two columns either side of the Sée during the night of 6-7 August. The southern column overran the outposts of the 30th Division but was stopped when the American infantry coolly dug in on high ground, called forward the divisional tank-destroyer battalion equipped with assault-gun-type weapons, which destroyed fourteen tanks, and waited for daylight and better weather conditions to bring out the tactical aircraft which would wreak even greater damage. Thus did a quite average American infantry division deal with the vanguard of the 2nd SS Panzer Division, almost invincible sword of the Panzer arm.

On the north bank the 2nd Panzer and 1st SS Panzer (the Adolf Hitler Division, which had never failed the Führer) were stopped even more easily by the infantry of the US 9th Division; the commander of the 116th Panzer Division declined to intervene and was relieved of command. At daybreak the US 2nd Armoured Division counter-attacked; it 'appeared to materialise out of thin air', noted the official historian, writing at a time when the Ultra secret was still jealously guarded. The 2nd Armoured Division and rocket-firing Typhoons of the Second British Tactical Air Force, which flew 294 sorties on 7 August, reduced the 2nd Panzer Division's tank strength to thirty that day. From Rastenburg Hitler demanded that the attack 'be prosecuted daringly and recklessly. . . . Each and every man must believe in victory.' As dusk fell on the

Mortain battlefield, however, defeat confronted each unit which had been committed to Operation Lüttich.

Other events of 7 August increased the *Westheer*'s ordeal. On that day Montgomery had launched a new drive into the German lines at the opposite end of the bridgehead, aimed towards Falaise. It followed two recent but aborted thrusts, by the Canadians down the track of the Goodwood offensive on 25 July and by the British towards Caumont (Operation Bluecoat) on 2 August. Operation Totalise, mounted on 7 August, was not the outright success Montgomery hoped it would be, even though preceded by a carpet bombardment as heavy as that before the Americans' Cobra two weeks before. It was again mounted by the Canadians, who met heavy resistance from their sworn enemy, the 12th SS Panzer (after a massacre of Canadian prisoners by the Hitler Youth Division early in the campaign, few of its soldiers survived capture at Canadian hands). However, the Canadians were now stronger than at any stage of the campaign; they had recently been joined by their own 4th Armoured Division and also by the émigré 1st Polish Armoured Division, which had a particular quarrel to settle with the Germans, made all the more bitter by the Poles' awareness of the battle currently raging in Warsaw between Bor-Komorowski's Home Army and the security troops of the German occupation force. Operation Totalise did not reach its objectives; but it thrust these two armoured divisions forward into positions from which they menaced the rear of the whole German Panzer concentration engaged in Normandy.

That concentration (less the 12th SS, which still stood on the British front) now numbered ten divisions and was grouped at the far western end of the bridgehead – in various states of disarray. Panzer Lehr, originally the German army's 'demonstration' division and before 6 June the strongest in its Panzer arm, was a shadow; all four SS Panzer divisions, the 1st, 2nd, 9th and 10th, had been grievously damaged in close combat since late June; the 17th SS Panzergrenadier, weak in armour to begin with, was a cripple; the 2nd, 21st and 116th Panzer had all suffered heavy tank losses, the last in the Mortain battle; only the 9th Panzer, which had arrived in Normandy from the south of France in August, remained largely intact. Even the 9th Panzer did not have its full complement of 176 tanks (half Mark IV, half Panther); average tank strengths were half the figure, and Panzer Lehr had almost no tanks at all.

The divisions, moreover, were in the wrong place. The surviving German infantry divisions, terribly reduced in numbers, were bunched into three groups, one group of seven standing in the path of the British and Canadians advancing on Falaise, one group of five scattered about in the path of the American break-out into Brittany, the remaining nineteen still clinging to the collapsing perimeter of the bridgehead they had defended so stoutly since 6 June. All were in imminent peril of encirclement, as the British-Canadian 21st Army Group drove south to cut off their line of retreat to the Seine, while the American 12th Army Group swung eastward to meet it behind their backs. But the Panzer divisions of what had recently been designated the Fifth Panzer Army were at the extremity of danger. Hitler's maniac dream of decapitating the American break-out at Mortain had carried them to the furthest end of the Normandy front, from which they could battle their way to safety from between the closing jaws of the Allied encirclement only at the cost of mortal combat.

How great that cost was becoming the 12th SS Panzer had learnt in Totalise, where the three Canadian infantry battalions of the 4th Armoured Division, mounted for the first time in the campaign in armoured carriers, had suffered only seven fatal casualties during their assault; so dense was the strength of their accompanying armour that it had simultaneously succeeded in bringing an end to the career of Michael Wittmann, the Wehrmacht's most renowned tank commander. He had destroyed 117 Russian tanks before arriving in Normandy; there he had been largely responsible for blunting the British attack at Villers-Bocage on 13 June. On 7 August he was cornered in his Tiger by five Shermans which destroyed it with a concerted salvo of gunfire. Given the opposed effects of Allied reinforcement of the bridgehead and German losses within it, such disparities of strength were now standard and would determine the final outcome of the Normandy battle with mathematical inevitability.

Inevitability had been hurried forward by a telephone conference held between Bradley (with Eisenhower at his side) and Montgomery on 8 August. The Americans had suggested that, since the Seventh Army and the Fifth Panzer Army were clearly no longer able to manoeuvre as a result of the blows rained on them in the recent Goodwood, Cobra and Totalise operations, strategic sense spoke for abandoning the plan, conceived before D-Day, for a 'wide' envelopment of the Wehrmacht in Normandy reaching as far south as the Loire. They proposed instead that a 'short hook' be staged by the Americans, designed to achieve rapid formations with the British and Canadians near Falaise. Montgomery agreed that 'the prospective prize was great' and left Bradley to issue the necessary orders to his subordinate, George Patton, who commanded the formations which the plans for a 'short hook' would bring into play.

Patton, the phantom with whom the authors of the D-Day deception scheme had deluded the Abwehr during the spring of 1944, was now a figure of substance and power on the Normandy scene. It was his Third Army which had assumed responsibility for the Saint-Lô breakthrough and his dynamism which had driven it through the defended zone and out into open country. 'The passage of Third Army through the corridor at Avranches', he wrote later, 'was an impossible operation. Two roads entered Avranches; only one left it over the bridge. We passed through this corridor 2 infantry and 2 armoured divisions in less than 24 hours. There was no plan because it was impossible to make a plan.' Patton characteristically exaggerated his achievements. The logistics of the Avranches manoeuvre were chaotic, and the tactical success of Operation Cobra owed more to the personal leadership of his VII Corps commander, Collins, than to his own generalship. Nevertheless, without Patton's relentless demand for action, the Third Army's *Blitzkrieg* would not have occurred.

Blitzkrieg was what Third Army's breakthrough amounted to; it was the first – and, as it would turn out, the last – true exercise in that operational form achieved by a Western army in the Second World War. *Blitzkrieg* proper entailed not merely the sudden and brutal penetration of the enemy's front by concentrated armoured force and the rapid exploitation of that success; it also required that the enemy forces lying beyond the point of break-in should be encircled and destroyed. That was the pattern of operations that the Wehrmacht had achieved in France in 1940 and in western Russia in June to October 1941. Thereafter it had fallen outside every combatant army's grasp. The Wehrmacht's great advances into southern Russia in spring and summer 1942 had not achieved encirclements on the scale which had brought the Red Army to the verge of destruction the year before, while the great eastern battles of 1943 and early 1944 had been struggles of attrition, as at Kursk, or bludgeoning Russian frontal offensives. The lightning dashes along the coast of North Africa by Rommel and his British adversaries in 1941-3 more resembled old-fashioned cavalry raids than campaigns of decision; had the Anglo-American Torch army not arrived in Algeria in November 1942, who can say how long the game might have been protracted? In Italy, where the terrain precluded breakthrough, none of the fighting had been touched by the electricity of *Blitzkrieg*; while Montgomery's efforts in the early stages of the Normandy campaign to unleash the lightning of armoured penetration against the Germans had all foundered because of their system of fixed defences and rapid counter-attack. Bagration, by which the Red Army had brought about the destruction of Army Group Centre in June 1944, was the only operation in the preceding three years of combat which replicated in its form and effects the spectacular German triumphs of Sickle Stroke and Barbarossa.

The biter bit

There was an excellent reason why *Blitzkrieg* had lapsed after the Kiev encirclement of September 1941 and why the chance to revive its form reappeared in France in August 1944: *Blitzkrieg* depended for its effect on the co-operation or, at the very least, the acquiescence of the enemy. In France in 1940 the Allies had both acquiesced and co-operated. By failing to provide their front in the Ardennes with adequate anti-tank defences – obstacles, anti-armour weapons, tank counter-attack forces – they had invited the German armoured offensive at that point; by their simultaneous advance into Belgium which carried the best of their mobile divisions eastward past the shoulder of the German Panzer divisions at the precise moment when those were hurrying westward, they actively co-operated in their isolation and ultimate encirclement.

The penalty for acquiescence and co-operation in the opponent's *Blitzkrieg* plans were quickly learned, by Germany's enemies at least. Indeed, as we have seen, both the French and the British correctly identified during the first week of the German *Blitzkrieg* of May 1940 that the right response was to attack into the flank of the armoured column as it drove towards its objective. The Russians too eventually learned the same lesson and at Kursk, a sector in which they had been given time to prepare the ground, they not only amputated the German spearheads but then ground the attacking forces to pieces in the dense minefields and network of fire positions in which they had become engulfed. Kursk may be regarded as the first battle in which the anti-tank gun, with which infantry had been equipped as early as 1918, actually performed the role intended for it – to deflect and if possible destroy attacking enemy tanks without recourse to supporting armour.

By 1944 each British and American infantry division had 60-100 anti-tank guns, as well as several hundred hand-held anti-tank missile projectors; the latter were weapons of last resort, but the former were genuine tank-destroyers. The enhanced effectiveness of the anti-tank gun derived not only from the growth in its distribution but also from the greatly increased calibre of those on issue to the infantry by mid-war – 57-mm was standard, 75-mm common and the heavier 80-mm and 90-mm available in specialist units. It is a rule of thumb that armour is penetrable by rounds equal in diameter to its thickness, and only the thickest tank armour exceeded 100 mm. Therefore, as the German armoured divisions engaged in Operation Lüttich found in their attack on the American 30th Division at Mortain, infantry could now hold their positions and inflict losses on the enemy under the weight of concentrated armoured attack.

The precariousness of the Fifth Panzer Army's position, confronted by superior enemy tank concentrations and by genuinely self-defending infantry formations, was now extreme. Its best hope was to form protective flanks along both southern and northern edges of the salient occupied by the Seventh Army behind which the battered German infantry divisions in Normandy could begin to make their withdrawal to the Seine. If Kluge, commanding both the Fifth Panzer and the Seventh Armies, as well as Army Group B, had enjoyed the freedom to make strategic decisions, there seems little doubt that he would have ordered exactly that disposition of his force. But freedom of decision was not what Hitler would concede him. On the contrary, on 10 August he sent orders to Kluge that Operation Lüttich was to be resumed the following day: 'The [Panzer] attack failed because it was launched prematurely and was thus too weak, and under weather conditions favouring the enemy. It is to be repeated elsewhere with powerful forces.' Six Panzer divisions were to engage in a more south-westerly direction under the command of General Hans Eberbach.

To attack south-westward was to commit the Panzer divisions into the pocket which Eisenhower's 'short hook' was now drawing around the Seventh Army. Hitler, the impresario of *Blitzkrieg*, was thus orchestrating exactly the manoeuvre best calculated to deliver his armoured striking force to its destruction. For all the evidence its enemies had given the Wehrmacht of the dangers of acquiescence and co-operation in *Blitzkrieg*, Hitler was now bent on tactics which more closely co-operated in a hostile *Blitzkrieg* than any employed by their enemies. Kluge, his immediate subordinate in the west, was aware of the 'incredibility of a large military force of twenty divisions blissfully planning an attack while far behind it an enemy is busily forming a noose with which to strangle it'. However, he was inhibited, even more than most German generals in the wake of the 20 July bomb plot, by the knowledge that his own complicity was suspected by the SS and Gestapo, and that their suspicions had substance. Kluge had known that a plot was in the making, since many of the plotters had previously served on his staff in Russia, but he had neither dissociated himself from it nor, when invited to join, shown loyalty by refusing; 'Yes, if the pig was dead' were the words he had used on the evening of 20 July. He now recognised that he could rescue himself from suspicion only by accepting the right of 'a command ignorant of front-line conditions', as the Seventh Army's chief of staff put it, 'to judge the situation from East Prussia'. His two immediate subordinates, General Paul Hausser of the Seventh and General Sepp Dietrich of the Fifth Panzer Armies, both SS officers recently nominated by Hitler to replace army generals, were currently taking advantage of a loophole in his orders for the renewal of the attack to draw their divisions eastward and so away from the tightening clasp of the American 'hook'. Kluge accepted the military logic of their redeployments but felt driven, none the less, to go through the motions of reviving the offensive as Hitler directed. On 15 August he set off on a tour of the pocket in which both his armies were now confined, with the object of persuading Hitler that he was carrying out his orders. The events of the day, ironically, were to produce exactly the opposite impression. Attacked in his staff car, exactly as Rommel had been twenty-nine days earlier, he spent most of the day skulking in ditches and reached the Seventh Army's headquarters only at midnight. During the hours he had remained incommunicado Hitler – for whom 15 August was 'the worst day of my life' – had convinced himself that Army Group B's commander was planning 'to lead the whole of the Western Army into capitulation'. Late in the evening he decided to relieve Kluge of command, sent for Walther Model, 'the Führer's fireman', to replace him and ordered the disgraced field marshal to return to Germany. Kluge, who rightly divined that he was to be met on arrival by the Gestapo, took poison on the homeward flight.

Kluge's suicide could not expiate the mistakes which had led Army Group B into its present predicament. Nor could Model, for all his proven expertise in reconstructing broken fronts, rescue it. Hitler's co-operation in the American *Blitzkrieg* had carried it too far into danger for anything but a pell-mell retreat to save its remnants from annihilation. And remnants were what the divisions of Army Group B now amounted to; though some 300,000 German soldiers were entrapped in the Falaise pocket, eight of the twenty surrounded divisions had disintegrated, while the tank strength of the best Panzer divisions – 1st SS, 2nd, 9th and 116th – had fallen to thirty, twenty-five, fifteen and twelve respectively. The renewal of Operation Lüttich was out of the question. Fortunately for the survivors, Hitler, in changing commanders, had also changed his tune. Model arrived in France on 17 August with orders to re-form the line on the Seine, holding enough ground to sustain the V-weapons attack on Britain and protect the frontiers of Germany from direct assault.

His mission was overtaken by events. On 19 August Patton's spearhead reached the Seine north-west of Paris at Mantes. This extension of Eisenhower's hook, ordered by Bradley on 14 August, conceded the trapped Germans a temporary breathing space, since after that date the American concentration at Argentan, forming one shoulder of the gap through which Army Group B had to escape, did not move further northwards to meet the Anglo-Canadian concentration at Falaise. The thrust to Mantes nevertheless nullified any hope the Germans had of sustaining a defensive line on the Seine, which thereafter became merely the barrier that Army Group B must cross to make good its escape out of Normandy. Meanwhile the Germans within the pocket, from which all anti-aircraft units had been evacuated in the hope of using them later elsewhere, were being devastated by constant air attack, while the British and Canadians were bearing down from Falaise to Argentan to put the stopper in the bottleneck. In the bottleneck itself a newly arrived Allied formation, the 1st Polish Armoured Division, representative in the west of the large Polish army in exile still sustaining its war effort against the Germans, took and held the commanding heights of Chambois between Falaise and Argentan in three days of desperate battle from 18 to 21 August. Its tank crews and infantrymen launched a succession of assaults against the road below where Army Group B streamed to the Seine bridges and ferries; but a defence against them was made by the equally resolute 12th SS Panzer (Hitler Youth) Division, which there performed the last of its many crucial operational missions in Normandy.

The liberation of Paris

The Hitler Youth Division's success in holding open the neck of the Falaise pocket until 21 August allowed some 300,000 soldiers to escape and, more surprisingly, 25,000 vehicles to cross floating bridges and ferries operated by German army engineers under cover of darkness between 19 and 29 August. Behind them, however, the fugitives left 200,000 prisoners, 50,000 dead and the wreck of two armies' equipment. Constant air attack into the clogged roads and fields of the pocket had left it choked with burnt and broken tanks, trucks and artillery pieces. Over 1300 tanks were lost in Normandy; of the Panzer divisions which escaped in some semblance of order none brought more than fifteen tanks out of the holocaust. Two Panzer divisions, Lehr and 9th, existed only in name; fifteen of the fifty-six infantry divisions which had fought west of the Seine had disappeared altogether.

Hitler directed some of the fugitive divisions to enter and hold the Channel coast ports as fortressess. He had already garrisoned the Atlantic ports of Lorient, Saint-Nazaire and La Rochelle, but the point of holding these ports had been nullifed as soon as Bradley had decided on 3 August to curtail his drive southward along the Atlantic coast in favour of an encirclement of Army Group B from the west. The decision to occupy the Channel ports, by contrast, was one of the highest strategic importance. It lay in the pattern of his earlier insistence on holding Baltic and Black Sea ports even after their hinterland had fallen to the Red Army, but in this case was far more strongly justified by logistic reality; for, while the Red Army depended scarcely at all upon seaborne supply, the Anglo-American armies did so almost completely. The denial to them of the Channel ports of Le Havre, Boulogne, Calais and Dunkirk gravely impeded their ability to provision their advancing forces and was to have a critical impact on the development of the campaign of liberation throughout the coming autumn and winter.

Hitler's Channel ports decision demonstrated once again his uncanny ability, even at moments of desperate crisis, to avoid the worst consequences of his acts of operational folly. It could not compensate for his wilful and egotistic co-operation in the unfolding of Patton's *Blitzkrieg*, which had culminated in the devastation of the *Westheer* within the Falaise pocket. It could certainly not compensate for the irreplaceable loss of tanks and trained soldiers which the closing of the pocket had inflicted on the German army. However, it would mitigate the immediate consequences and help to ensure that when he came to mount his next – the last – great armoured offensive of the Second World War in the west he would do so on more equal terms than expected from the outcome of the battle of Normandy in August 1944.

While the Channel ports were filling up with their garrisons of stay-behinds from the Fifteenth Army and its remaining units were joining the fugitives of the Seventh and Fifth Panzer Armies in the flight to the West Wall, the final act in the drama of the liberation epic was being played out in Paris. As the Normandy battle swelled to its climax, Hitler had conceived a plan to transform the French capital into a great defended bridgehead through which the Seventh Army could make an orderly retreat to the line of the Somme and Marne rivers and then to use the city itself as a battleground on which crippling losses might be inflicted on the Allied pursuers, even at the cost of turning it 'into a field of ruins'.

Two developments worked to allay this outcome. The first was the arrival in France on 15 August of a second Allied invasion army, not, as long anticipated, on the 'short route' to Germany from the Pas de Calais, but in the distant south, between Nice and Marseille. The Seventh Army, the instrument of Operation Anvil, mounted by three American and four French divisions, briskly overcame the resistance of General Wiese's Nineteenth Army and by 22 August had raced up the Rhône valley to reach Grenoble. Its appearance, and its peremptory unseating of the only effective manoeuvre formation remaining to General Johannes Blaskowitz's Army Group G, the 11th Panzer Division, not only threatened an attack against the West Wall from a hitherto unexpected direction, through Alsace-Lorraine. It also made nonsense of any hope of holding Paris when a new Allied thrust menaced the rearward communications of the capital with Germany from the south.

The second development was domestic to Paris itself. Its population was not overtly resistant. In March it had welcomed Pétain with tumultuous popular demonstrations; as late as 13 August, Laval had returned to the city in the hope of reconvening the Chamber of Deputies to accord him powers as legitimate head of government who might treat on sovereign terms with the liberating armies. Nevertheless a spirit of resistance to German occupation smouldered, and as soon as it became clear that the days of the occupying force were numbered armed resistance broke out in the streets. On 18 August the Paris police force had, literally, raised the standard of revolt over the prefecture on the Ile de la Cité; as soon as it did so the covert resistance, of which the left-wing Franc-Tireurs et Partisans (FTP) was the most numerous, rallied to the flag. By 20 August the German garrison found itself under such pressure to maintain control of the streets that the Commander of Greater Paris, Dietrich von Choltitz, offered and succeeded in negotiating a truce. The scale the fighting had reached, however, now worked to alter Allied plans for the city's liberation. While Hitler was ordering that the city be turned into a western Stalingrad, Eisenhower and Montgomery had set their faces against allowing their troops to penetrate its perimeter 'until it is a sound military proposition to do so'. As soon as it became clear that the city was struggling to liberate itself, however, the Allied leaders found themselves obliged to go to the insurgents' assistance. The appropriate means of intervention lay to hand. Since 1 August the French 2nd Armoured Division, which owed allegiance to General de Gaulle, had been in Normandy. On 20 August the general himself, whose title to the leadership of France the Allies would not yet admit, had also arrived – uninvited, unannounced and by a circuitous route. On 22 August Bradley transmitted orders from Eisenhower that the French 2nd Armoured Division under General Leclerc was to direct itself on Paris. De Gaulle, who had installed himself in the French President's country seat at Rambouillet, endorsed the order and prepared to travel in its wake.

23 August was spent traversing the 120 miles which separated the 2nd Armoured Division's positions from the outskirts of the city. Detained on the approaches by stiffening German resistance, Leclerc despaired of entering the capital that day. Then, stung by American allegations that the French were 'dancing to Paris' (there had been outbreaks of *fête* between episodes of fighting), he launched an infiltration by a small tank-infantry force along back routes into the centre of the city. At 9.30 on the evening of 23 August three tanks of the French 2nd Armoured Division, Montmirail, Champaubert and

Romilly, named from Napoleonic victories of 1814, stood under the walls of the Hôtel de Ville. Next day they would be joined by the bulk of the division which would fight its way into the historic heart of the city against last-ditch German resistance, and the day after by de Gaulle himself. No one more than he, the French army's apostle of armoured warfare, would grasp how appropriate it was that the capital of the country first overwhelmed by *Blitzkrieg* should be liberated by the tanks of its own renascent army.

TWENTY-TWO

Strategic Bombing

On 12 January 1944 Air Marshal Arthur Harris, chief of RAF Bomber Command, wrote:

> It is clear that the best and indeed the only efficient support which Bomber Command can give to [Operation] Overlord is the intensification of attacks on suitable industrial targets in Germany as and when the opportunity offers. If we attempt to substitute for this process attacks on gun emplacements, beach defences, communications or [ammunition] dumps in occupied territory, we shall commit the irremediable error of diverting our best weapons from the military function, for which it has been equipped and trained, to tasks which it cannot effectively carry out. Though this might give a spurious appearance of 'supporting' the Army, in reality it would be the greatest disservice we could do them.

'Bomber' Harris's prognosis of the effect of diverting his strategic bombers from the 'area' bombing of Germany to 'precision' bombing on France was to be proved dramatically incorrect. In the first place, his crews demonstrated that they had now acquired the skill to hit small targets with great accuracy and to sustain this 'precision' campaign even in the teeth of fierce German resistance. In March the objections of Harris and General Carl Spaatz, commanding the Eighth Air Force, Bomber Command's American equivalent, were overruled and both air forces were placed under Air Chief Marshal Sir Arthur Tedder, Eisenhower's deputy. From then onwards strategic air forces embarked on a campaign against the French railway system which was to cost them 2000 aircraft and 12,000 aircrew in a little over two months. In April and May Bomber Command, which had dropped 70 per cent of its bombs on Germany in March, reversed its proportional effort: in April it dropped 14,000 tons on Germany but 20,000 on France; in May it launched three-quarters of its sorties against France. During June the weight of attack on France increased again when 52,000 tons were dropped in the invasion area and on the military infrastructure surrounding it.

Moreover, in flat contradiction of Harris's forecast, RAF bombers carried out their missions with an effectiveness which not only 'supported' the army very effectively indeed but went far towards determining the Germans' defeat in Normandy. By comparison with the British and American armies, the German army belonged to a previous generation of military development. Its Panzer and motorised divisions apart, it moved over short distances on foot by road and over long distances by rail; while all its supplies and heavy equipment, even for formations which possessed their own motor transport, moved exclusively by rail. The interruption of the French railway system and the destruction of bridges therefore severely restricted its ability not only to manoeuvre but even to fight at all; from April to June, and thereafter during the course of the Normandy battle itself, French railway working was brought almost to a standstill and most bridges over the major northern French rivers were broken or at least damaged too severely to be quickly repaired.

Much of the devastation was achieved by the medium-range and fighter bombers of the British Second Tactical and the recently formed American Ninth Air Forces; American Thunderbolt and British Typhoon ground-attack fighters flying vast daylight 'sweeps' over northern France destroyed 500 locomotives between 20 and 28 May alone. However, the far more serious structural devastation – to bridges, rail yards and locomotive repair shops – was the work of the strategic bombers. By late May, French railway traffic had declined to 55 per cent of the January figure; by 6 June the destruction of the Seine bridges had reduced it to 30 per cent, and thereafter it declined to 10 per cent. As early as 3 June a despairing officer of Rundstedt's staff sent a report (decrypted by Ultra) that the railway authorities 'are seriously considering whether it is not useless to attempt further repair work', so relentless was the pressure the Allied forces were sustaining on the network.

The rail capacity that Germany's OB West succeeded in maintaining in June and July 1944 just sufficed to provide the Seventh and Fifth Panzer Armies with the irreducible minimum of food, fuel and ammunition (though not enough to revictual Paris, which was in serious danger of starvation just before its liberation). However, such supplies could be guaranteed to the fighting troops only as long as they did not attempt to manoeuvre; so fragile and so inflexible was the network of communication improvised between the Reich and Germany that the troops at the battlefront could depend upon it only if they remained fixed to its terminals. Once they moved, they risked starvation of essentials – hence their inability to 'make a fighting withdrawal in France'. When their fortified perimeter of the bridgehead was destroyed by Patton's *Blitzkrieg*, they could only retreat at the fastest possible speed to the next fortified position with which a communication system connected; and that was the West Wall on the Franco-German border.

The Normandy campaign, in both its preliminaries and its central events, therefore proved Harris wrong. Airpower used in the direct support of armies had worked with stunning success at the immediate and at the strategic level. None the less it was inevitable and also understandable that Harris should have resisted pressure from above to direct his bomber force from the attack on

German cities. After all, Bomber Command justifiably prided itself on having for three years been the only instrument of force the Western powers had brought directly to bear against the territory of the Reich (the US Eighth Air Force had more recently come to the struggle). Moreover, Harris was the spokesman of a service whose singular and unique *raison d'être* was to bomb the enemy's homeland.

The Luftwaffe, on the other hand, never espoused such an operational doctrine. Its chiefs had considered the desirability of founding it as a strategic bomber force in 1934 when it came into being, but rejected the option because they judged the German aircraft industry too underdeveloped to provide the necessary complement of large, long-range machines. Like the Red Air Force, it therefore grew to maturity as a handmaiden of the army, a role which its leaders, for the most part ex-army officers, were content to accept. Its 'strategic' campaign against Britain in 1940-1 was thus mounted with medium-range bombers designed for ground-support missions. When Günther Korten succeeded Hans Jeschonnek as Luftwaffe chief of staff in August 1943 he instituted a 'crash' effort to create a strategic bomber arm, but the attempt foundered for lack of the appropriate aircraft as a direct result of the decisions about the Luftwaffe's future taken by his predecessor a decade earlier.

Korten's belated attempt to endow the Luftwaffe with a strategic capability was motivated by the belief, which he shared with Speer, the Armaments Minister, that the Red Army's assumption of the offensive in 1943 might be offset by a counter-offensive against its industrial rear. In short the crisis had obliged him to take up the policy which a generation of British and American airmen had adopted and refined at leisure. While he was forced into the expedient of hastily adapting medium-range bombers and retraining their crews for 'penetration' operations – operations which short-term emergencies would, in the event, deny him the chance to undertake – Harris already commanded a thousand-strong fleet of four-engined bombers developed over many years specifically for penetration missions.

The command of the air

Britain's commitment to the concept of strategic bombing can, indeed, be traced to the last years of the First World War. Even though the 'Independent Air Force' of 1918 succeeded in dropping only 534 tons of bombs on German territory its strategy was already informed by the idea that the direct attack of the enemy's rear was the correct role for an air force. That idea was to be elaborated by the Italian airman, Giulio Douhet, into a coherent philosophy of airpower, equivalent in scope to Mahan's philosophy of seapower during the 1920s. Meanwhile, without benefit of elaborate theory, the Royal Air Force was creating the first 'air navy' of strategic bombers that the world had seen. The roots of its operational function lay in a study prepared by the 'father' of the Royal Air Force, Sir Hugh Trenchard, for the Allied Supreme War Council in the last months of the First World War. 'There are two factors,' he wrote then, 'moral and material effect – the object being to obtain the maximum of each. The best means to this end is to attack the industrial centres where you (a) Do military and vital damage by striking at the centres of war material; (b) Achieve the maximum effect on the morale by striking at the most sensitive part of the German population – namely the working class.'

By advocating this simple and brutal strategy – to bomb factories and terrorise those who worked there and lived nearby – Trenchard proposed to extend to general warfare a principle so far admitted by civilised nations only in the siege of cities. In siege warfare armies had always operated by the code that citizens who chose to remain within a city's walls after siege was laid thereby exposed themselves to its hardships: starvation, bombardment and, once the walls had been breached and the offer of capitulation refused, rapine and pillage. The almost uncontested generalisation of siege-warfare morality demonstrates both how closely the First World War had come to resemble siege on a continental scale and how grossly its prosecution had blunted the sensitivities of war leaders, civilian and military alike. Indeed, Trenchard's proposals went almost uncontested: they met no principled objection among the Western Allies at the time; and once the war was over they influenced governments in Britain and France by prompting policies designed to avert 'air raids', minimise their effect or maximise the capacity of their own air forces to mount such raids against a future enemy. Thus, at Versailles, the Allies insisted on the abolition of the German air force in perpetuity; but by 1932 the British Stanley Baldwin, then a prominent member of the coalition government, was gloomily conceding that 'the bomber would always get through', while the leaders of the Royal Air Force were battling relentlessly for the expansion of the bomber fleet, even at the cost of depriving the home air defences of fighter squadrons.

The RAF's commitment to bombing was rooted in the conviction that attack was the best form of defence. Air Marshal John Slessor, the Air Staff's Chief of Plans in the late thirties, expressed his service's views in classic form when he argued that an offensive against enemy territory would have the immediate effect of forcing the enemy air force on to the defensive and the secondary, indirect, but ultimately decisive effect of crushing the enemy army's capacity to wage war. In *Airpower and Armies* (1936) he wrote: 'It is difficult to resist at least the conclusion that air bombardment on anything approaching an intensive scale, if it can be maintained even at irregular intervals for any length of time, can today restrict the output from war industry to a degree which would make it quite impossible to meet the immense requirements of an army on the 1918 model, in weapons, ammunition and warlike stores of almost every kind.'

So acute and general were the fears that the prospect of strategic bombing aroused at the outset of the Second World War – fears very greatly enhanced by the international left's brilliantly orchestrated condemnation of the bombing of

Republican towns by Franco's air force and the expeditionary squadrons of his German and Italian allies during the Spanish Civil War, of which Picasso's *Guernica* is the key document – that paradoxically even Hitler joined in an unspoken agreement between the major combatants not to be the first to breach the moral (and self-interested) embargo against it.

Hitler did not extend the embargo to exclude attacks on countries unable to retaliate – hence the bombings of Warsaw in September 1939 and Rotterdam in May 1940 – or on military targets in those that could. The bombing of military targets including airfields, naval ports and railway centres was legitimate under the most traditional conventions of war. However, until midsummer 1940 all held each other's cities inviolate. Even at the outset of the Battle of Britain, Hitler insisted that attacks be confined to airfields and to targets that might be deemed military, like London Docks. Such restrictions became increasingly difficult to observe, however, as the Battle of Britain protracted without the prospect of outcome. As the argument for 'making the RAF fight' intensified, entailing direct attack on populated targets, Hitler looked for means to justify breaching the embargo. In his victory speech to the Reichstag on 19 July he had publicised the notion that Freiburg-in-Breisgau had already been bombed by the French or the British air force (Goebbels had inculpated both); in fact it had been mistakenly attacked on 10 May by an errant flight of the Luftwaffe. When on 24 August another vagrant Luftwaffe crew bombed East London in error, provoking a retaliatory raid next night by the RAF on Berlin, he seized the opportunity to announce that the gloves were off. 'When [the British] declare that they will increase their attacks on our cities [Churchill had not done so], then we will raze their cities to the ground. We will stop the handiwork of these air pirates,' he told an ecstatic audience in the Berlin Sports Palace on 4 September; 'the hour will come when one of us will break and it will not be National Socialist Germany.'

Crisis in Bomber Command

British Bomber Command altogether lacked the power to bring Germany to breaking-point when it began its bombing campaign in earnest in the winter of 1940. When it impertinently bombed Munich on the anniversary of Hitler's Beer Hall *Putsch* of 8 November 1923, the Luftwaffe retaliated by raiding the industrial city of Coventry, destroying or damaging 60,000 buildings. In an attempted escalation of tit-for-tat the RAF attacked Mannheim on the night of 20 December, but it largely missed the city and caused only a twenty-fifth of the damage Coventry had suffered, if the score is reckoned by the tally of civilian casualties – 23 dead to 568 – which, gruesomely, was to be the measure of strategic bombing success thenceforward. Since the Mannheim raid was an exercise in 'area bombing' or direct attack on civilians in all but name, Bomber Command now found itself in the unenviable position of having descended to the same moral level as the Luftwaffe, while lacking the means to equal, let alone exceed, the Luftwaffe's area bombing capacity. Throughout the 'blitz' winter of 1940-1 London and other British cities burned by the acre; on 29 December 1940 the Luftwaffe started 1500 fires in the City of London alone, destroying much of the remaining fabric of the streets familiar to Samuel Pepys, Christopher Wren and Samuel Johnson. No German city suffered equivalent damage during 1940 or even 1941. To all intents, Bomber Command, the service Churchill had told the War Cabinet on 3 September 1940 'must claim the first place over the Navy or the Army', was and would remain for months to come 'little more than a ramshackle air freight service exporting bombs to Germany'.

The most shaming index of its incapacity was the 'exchange ratio' between aircrew and German civilians killed in the course of bombing raids during 1941; the number of the former actually exceeded that of the latter. The imbalance had several explanations. One was material: the poor quality of British bombing aircraft, which as yet lacked the speed, range, height and power to deliver large bomb-loads on to distant targets. Another was geographical: to reach Germany – as yet only western Germany – the bombers had to overfly France, Belgium or Holland, where the Germans had already begun to deploy a formidable defensive screen of fighters and anti-aircraft guns. The third, and most important, explanation was technological: committed to bombing by night, since the RAF did not have the long-range fighter escorts necessary to protect bombers on daylight raids, Bomber Command lacked the navigational equipment not merely to find its designated targets – factories, marshalling yards, power stations – within the cities against which it flew but even the cities themselves. The suspicion that Bomber Command was bombing 'wide', even wild, was confirmed with exactitude by a study prepared at the suggestion of Churchill's scientific adviser, Lord Cherwell, in August 1941. The Butt Report's main findings were: 'of those aircraft attacking their targets, only one in three got within five miles . . . over the French ports the proportion was two in three; over Germany as a whole . . . one in four; over the Ruhr [the heartland of German industry and Bomber Command's principal target area] it was only one in ten.'

During 1941, when 700 aircraft failed to return from operations, Bomber Command's crews in short were dying largely in order to crater the German countryside. Set beside the hopes reposed in it by Churchill and the British people as their only means of bringing the war directly to Hitler's doorstep, this realisation was bound to precipitate a crisis. At the end of 1941 the crisis occurred. As early as 8 July 1941 Churchill had written: 'There is one thing that will bring [Hitler] down, and that is an absolutely devastating exterminating attack by very heavy bombers from this country upon the Nazi homeland.' Goaded by Churchill, the RAF first of all committed itself to a programme of building up Bomber Command to a strength of 4000 heavy bombers (when the daily total of serviceable machines was only 700); after that target was recognised to be unattainable, it brought itself to accept that the bombers it already deployed must in future be used to kill German civilians, since the factories in which they worked could not be hit with precision. On 14 February the Air Staff issued a directive emphasising that henceforward operations 'should now be focused on the morale of the enemy civilian population and in particular of industrial workers'. Lest the point not be taken, Air Chief Marshal Sir Charles Portal wrote the following day: 'I suppose it is clear that the new aiming points are to be the built-up [residential] areas, not, for instance, the dockyards or aircraft factories. . . . This is to be made quite clear if it is not already understood.'

It was appropriate that it should have been Portal, the intellectual patrician, who revealed the central idea of area bombing, for it depended ultimately upon class bias – the judgement that the latent discontents of the proletariat were the Achilles heel of an industrial state. Liddell Hart, writing in 1925, had envisaged 'the slum districts maddened into the impulse to break loose and maraud' by bombing attack, thereby dramatising Trenchard's first statement of the theory in 1918. The preconceptions of all three were determined by the ruling classes' prevailing fear of insurrection, perhaps leading to revolution, which the success of the Bolsheviks in war-torn Russia had rekindled throughout Europe after 1917. Events would prove that it was the proletariat's endurance of suffering – particularly of 'dehousing' which Cherwell advocated in an important paper of March 1942 – that the effects of area bombing would most powerfully stimulate; but in early 1942 the proletariat's class enemies – as Marx would have identified them – had contrary expectations. The 'bomber barons' embarked on their campaign against the German working class in the firm belief that they would thereby provoke the same breach between it and its rulers that the ordeal of the First World War had brought about in tsarist Russia.

There was a strong flavour of class reaction too in the Air Staff's choice of agent to implement the new policy. Arthur 'Bomber' Harris was a commander of coarse single-mindedness. He had neither intellectual doubt nor moral scruple about the rightness of the area bombing policy and was to seek by every means – increasing bomber numbers, refining technical bombing aids, elaborating deception measures – to maximise its effectiveness. 'There are a lot of people who say that bombing cannot win the war,' he told an interviewer soon after taking command at High Wycombe, the bomber headquarters, on 22 February 1942. 'My reply is that it has never been tried yet. We shall see.'

He was fortunate to assume command at a moment when the first navigational aid to more accurate bombing, 'Gee', was about to come into service. 'Gee' resembled the 'beam' system by which the Luftwaffe had been guided to British targets in 1940-1. It transmitted two pairs of radio signals which allowed a receiving aircraft to plot its precise position on a gridded chart and so release its bombs at a preordained point. 'Gee' was followed in December by the precision-bombing device 'Oboe', which was subsequently fitted to Pathfinder Mosquitos, and in January 1943 by H2S, a radar set that gave the navigator a picture of the ground beneath the aircraft with its salient landmarks.

All three navigational aids were greatly to improve Bomber Command's target-finding capacity, though it was the formation of the specialist Pathfinder squadrons in August 1942 which achieved the decisive advance. The Pathfinders, equipped with a mixture of aircraft that included the new, fast and high-flying Mosquito light bombers, preceded the bomber waves to 'mark' and 'back up' the target with incendiaries and flares, starting fires into which the main force then dropped its loads. Harris fiercely opposed the creation of the Pathfinder units. He believed they deprived the ordinary bomber squadrons of their natural leaders (the same argument was used by British generals against the formation of commando units) and also diminished the size of the area bombing force. However, he was rapidly obliged to withdraw his objections when the Pathfinders demonstrated how much more effectively they found targets than the unspecialised crews of Bomber Command.

The arrival of the 'heavies'

Harris's commitment to area bombing was also lent credibility by the appearance, at the moment he took command, of a new and greatly improved instrument of attack. The British bombers available at the beginning of the war, the Hampdens, Whitleys and elegant Wellingtons, were inadequate bomb-carriers. Their larger successors, the Stirlings and Manchesters, were also defective because they lacked altitude and power respectively. The Halifax and particularly the Lancaster, however, which appeared in 1942, were bombers of a new generation. The Lancaster, which first flew operationally in March 1942, proved to be capable of carrying enormous bomb-loads, eventually the 10-ton 'Grand Slam', over great distances and to be robust enough to withstand heavy attack by German night-fighters without falling from the sky.

At the outset, though, Harris was concerned not with quality but with quantity. His aim was to concentrate the largest possible number of bombers over a German city with the object of overwhelming its defences and fire-fighting forces. A successful raid on the Paris Renault factory in March prompted him to undertake a raid against the historic Hanseatic town of Lübeck on the Baltic on the night of 28/29 March 1942. He was cold-bloodedly frank about his intentions: 'It seemed to me better to destroy an industrial town of moderate importance than to fail to destroy a large industrial city. . . . I wanted my crews to be well "blooded" . . . to have a taste of success for a change.' Lübeck, a gem of medieval timber architecture, burned to the ground, and the raiding forces returned to base 95 per cent intact. The 'exchange ratio' persuaded Harris that he had discovered a formula for victory.

On four nights in April Bomber Command repeated its incendiary success at Rostock, another medieval Baltic town; 'These two attacks', wrote Harris, 'brought the total acreage of devastation by bombing in Germany up to 780 acres, and in regard to bombing [on Britain] about squared our account.' The Luftwaffe retaliated by so-called 'Baedeker' (tourist-guide) attacks on the historic towns of Bath, Norwich, Exeter, York and Canterbury. However, it lacked the strength to match Harris's next escalation, which took the form of an attack by a thousand bombers, the first 'thousand-bomber raid', on Cologne in May. By stripping training units and workshops of their machines, Bomber Command concentrated the largest number of aircraft yet seen in German skies over this, the third largest city in the Reich, and burned everything in its centre except the famous cathedral.

The success of Bomber Command's new tactics depended not only upon increased numbers and improved target-finding but also on a frank adoption of fire-raising methods. Thenceforward its bomb-loads were to contain small incendiaries and large high-explosive containers in the proportion of two to one. At Cologne 600 acres were burned. Thousand-bomber raids on Essen and Bremen in June achieved similar effects; Essen, in the Ruhr, Germany's industrial centre, had been already attacked eight times between March and April. In the spring and summer of 1943 Bomber Command devoted its efforts to a 'Battle of the Ruhr' which multiplied the incendiary effect many times over.

By then the strategic bombing offensive against Germany had become a two-air-force campaign. The United States Army's Eighth Air Force had arrived in Britain in the spring of 1942 and undertaken its first raid in August, when it attacked marshalling yards at Rouen. The attack was staged in daylight, in accordance with the philosophy worked out over many years before the war by the Army Air Force's officers. Exercised by the pressing need to destroy hostile naval forces operating in American waters, they had developed both an aircraft and a bombsight designed to deliver large bomb-loads on to small targets with precision in daylight. The Norden bombsight was the most accurate optical instrument yet mounted in a strategic bomber. The bomber which carried it, the B-17, was notable for its long range and heavy defensive armament, the latter central to the American belief that, in the absence of a satisfactory long-range fighter, their bombers could fight their way to and from the target without suffering unacceptable losses. However, the requirements of range and armament placed a heavy penalty on the B-17's bombload. Under normal circumstances the bombload of a B-17 seldom exceeded 4000lb and in many operations fell as low as 2600lb. Redeployed from a maritime defensive to a continental offensive role, General Ira C. Eaker's Eighth Air Force was destined for deep-penetration missions by daylight to complement Bomber Command's night raids into Germany and its occupied territories. By January 1943 Eaker had 500 B-17s available.

The combined bomber offensive
The integration of the developing American with the continuing British bombing attacks on Germany was formalised at the Casablanca conference of

January 1943 in a 'Casablanca Directive', which laid the basis for a 'combined bomber offensive' (codenamed Pointblank in May), against key targets. These were defined, in order of priority, as German submarine construction yards, the German aircraft industry, transportation, oil plants and other targets in enemy war industry. The specification of targets disguised, however, a sharp difference of opinion between the British and Americans over operating methods. Eaker rejected British arguments for committing his B-17s to area bombing. He remained convinced that they were best employed in precision attack, against what Harris contemptuously dismissed as 'panacea targets'. Harris, for his part, refused to be diverted from his chosen method. As a result, the two air forces effectively divided the Casablanca agenda between them, the RAF continuing its night attacks on 'other targets', which meant the built-up areas of the major German cities, while the USAAF committed itself to daylight raids on 'bottlenecks' in the German economy.

The first 'bottleneck' chosen by the economic analysts who advised the USAAF was the ball-bearing plant at Schweinfurt in central Germany, bombed by the Eighth Air Force on 17 August 1943. Analysis suggested that destruction of the factory, from which essential components of the gearing in aircraft, tanks and U-boats were supplied, would cripple German armaments production. The theory was only partially correct, since Germany had alternative sources of supply from another plant at Regensburg and from neutral Sweden, which not only lay outside the Allied targeting area but was also bound to Germany by dependence on coal imports. The practice was almost wholly disastrous. Forced to traverse northern France and half of Germany in daylight without fighter escort, the 'self-defending' Flying Fortress formations were devastated by fighter attack. Of the 229 B-17s that had set out, 36 were shot down, an 'attrition rate' of 16 per cent, more than three times the rate that Bomber Command had established as 'acceptable' for a single mission. When the 24 B-17s lost suffered in the complementary raid on Regensburg were added in, and heavy damage to 100 returning bombers allowed for, it became clear that 17 August had been a day of disaster. The pre-war theory of the self-defending bomber had proved to be a misconception. The Eighth Air Force suspended its deep-penetration missions into Germany for five weeks, and they would not be fully resumed until long-range fighters had been developed to escort the daylight bombers to their targets.

The Hamburg raids

While the American campaign hung fire, the British had been spreading destruction even more widely across the cities of western Germany. The 'Battle of the Ruhr', which lasted from March to July, involved nearly 800 aircraft in 18,000 sorties (individual missions) which dropped 58,000 tons of bombs on Germany's industrial heartland. In May and August Harris was also obliged to mount two 'panacea' missions, both of which were brilliant successes. In the first a specially trained squadron, 617, destroyed the Möhne and Eder dams, which supplied the Ruhr with much of its hydroelectricity; in August a major raid laid waste the laboratories and engineering workshops at Peenemünde, on the Baltic coast, where intelligence sources had revealed that Germany's arsenal of secret pilotless missiles was being built.

More to Harris's taste, however, was the four-night raid on Hamburg in July, which provoked a 'firestorm' and burned to cinders the heart of the great North German port covering 62,000 acres. A firestorm is not an effect that a bombing force can achieve at will; it requires a particular combination of prevailing weather conditions and the overwhelming of civil defences. When such circumstances are present, however, the consequences are catastrophic. A central conflagration feeds on oxygen drawn from the periphery by winds which reach cyclone speed, suffocating shelterers in cellars and bunkers, sucking debris into the vortex and raising temperatures to a level where everything inflammable burns as if by spontaneous combustion. Such conditions prevailed in Hamburg between 24 and 30 July 1943. There had been a long period of hot, dry weather, the initial bombardment broke the water mains in 847 places, and soon the core temperature of the fire reached 1500 degrees Fahrenheit. When it eventually burned itself out, only 20 per cent of Hamburg's buildings remained intact; 40 million tons of rubble clogged the city's centre, and 30,000 of its inhabitants were dead. In some areas of the city the total of fatal casualties among the inhabitants exceeded 30 per cent; 20 per cent of the dead were children, and female deaths were higher than male by 40 per cent.

When the toll of Hamburg's bombing victims throughout the war was calculated it was found to be only 13 per cent lower than the proportion of battle deaths among soldiers recruited from the city between 1939 and 1945; and the majority had died in the great raids of July 1943. Hamburg was not the RAF's only firestorm. It was to achieve the same effect, if with lower casualties, in October at Kassel, where fires burned for seven days. Later Würzburg (4000 dead), Darmstadt (6000 dead), Heilbronn (7000 dead), Wuppertal (7000 dead), Weser (9000 dead) and Magdeburg (12,000 dead) would also burn in the same way.

Hamburg, however, encouraged Harris to set his sights beyond Germany's western periphery of industrial cities and Hanseatic ports. Berlin had been one of Bomber Command's first targets when it assumed a retaliatory role during the Luftwaffe's 'blitz' on London. In November 1943 Harris decided to make it his crews' main target during the coming season of long nights, which were their best protection against German fighter attack. It had last been attacked in January 1942 but was thereafter left off the targeting list because of its long distance from Bomber Command's bases and its strong defences, which combined to make the 'attrition rate' on Berlin raids exceptionally high. Probing attacks mounted in August and September suggested, however, that the German capital had become a softer target than hitherto to Harris's greatly strengthened bombing force, and on the night of 18-19 November 1943 it committed itself to the 'Battle of Berlin'.

Between that night and 2 March 1944 it mounted sixteen major raids on the city. No more than 200 acres of its built-up area had been damaged in all the raids mounted by the RAF since August 1940, and it continued to function normally as the capital not only of the Reich but of Hitler's Europe. It remained a major industrial, administrative and cultural centre: its great hotels, restaurants and theatres flourished; so too, did life in its elegant residential districts, like Dahlem, home of *haut-bourgeois* opposition to Hitler. 'Missie' Vassiltchikov, an Anglophile White Russian refugee in Berlin, and a close friend of Adam von Trott, one of the principal conspirators in the July Plot, found pre-war life scarcely interrupted at all by 'enemy action' (the phrase used in Britain to denote the cause of bombing deaths) until late 1943. She continued to dine, dance and absent herself from work in Goebbels's Propaganda Ministry on such pretexts as attending the last great German aristocratic wedding of the war years at Hohenzollern-Sigmaringen right up until the moment when the Battle of Berlin began.

Then the clouds of war drew in fast. Goebbels, as Gauleiter of Berlin, persuaded one million of its four and a half million inhabitants to leave before Bomber Command's main attacks began. Those who remained then began to undergo the most sustained experience of air attack undergone by any city population throughout the Second World War. Berlin did not suffer firestorm; having been built largely in the nineteenth and twentieth centuries, with wide streets and many open spaces, it resisted conflagration. Nevertheless its relentless drenching with high explosive and incendiaries, six times in January alone, resulted in devastation. Although only 6000 Berliners were killed in the Battle proper, thanks to the solid construction of shelters in eleven enormous concrete 'flak towers', 1.5 million were made homeless and 2000 acres of the city were ruined by the end of March 1944.

When the battle was then called off, however, it was not only because Harris's aircraft were needed to help in the preparation for D-Day. Even he had been brought to accept that, in the 'exchange ratio' between the attrition of Berlin's fabric and defences and that of his bomber crews, Berlin had suffered less. Though by March 1944 he disposed of a daily average of 1000 serviceable bombers, losses on raids had risen above the 'acceptable' maximum of 5 per cent and had sometimes touched 10 per cent (on the most costly of all raids, ironically not against Berlin but against Nuremberg on 30 March 1944, it exceeded 11 per cent). Since bomber crews were obliged to fly thirty missions before qualifying for rest, each faced the probability, in statistical terms, of being shot down before a tour was completed. In practice, crews who had flown more than five missions achieved a much higher survival rate than novices, who figured disproportionately among the 'acceptable' 5 per cent lost. When the attrition rate rose towards 10 per cent, however, even experienced crews were killed. The survivors sensed doom, and there was a corresponding decline in morale, indicated by bombing 'short' and premature return to base.

The rise in the attrition rate testified to the short-term success of German defensive measures. As the bombers penetrated more deeply into Germany, their exposure to German flak and fighter attack increased. In the early days of night bombing the Luftwaffe had found as much difficulty in intercepting the RAF as the RAF had had in combating the German night 'blitz' of 1940-1. During 1942, however, its success rose sharply as a result of improvements in the control of fighters, as well as in their armament and equipment. Flak, though it badly frightened the bomber crews, was a lesser means of destruction. Anti-aircraft gunfire could not touch the Mosquitoes of the Pathfinder Force flying at 30,000 feet; but fighters attacked at a range of 400 feet, once they had been guided to the target. From October 1940 onwards in Holland, Bomber Command's natural approach route to the Reich, the Germans began to deploy, on the so-called Kammhuber Line, a force of radar-equipped night-fighters which were guided to the intruders by ground radar 'Würzburg' stations. The RAF retaliated by equipping their aircraft with radar detection devices, by increasing the density of their bomber streams to present fighters with a smaller target, and eventually (July 1943) by dropping metallic chaff, 'Window' – first used in the Hamburg raids – to cause radar interference. Eventually all these expedients were overcome: the Germans became adept at using Bomber Command's electronic emissions as target indicators, at refining their radar sets to overcome Window, and at increasing the density of their own fighter formations to match that of the bomber streams. At the end of 1943, 'Tame Boar' squadrons of radar-equipped night-fighters were being supplemented by strong forces of 'Wild Boar' day-fighters flying as night-fighters; lacking radar, they were guided towards the bombers by radio and light beacons and then attacked in the illumination provided by flak and searchlights.

The battle of material

Had Bomber Command been Germany's only airborne enemy it would have been close to admitting defeat in the spring of 1944. However, the Eighth Air Force was still committed to a campaign of daylight precision bombing, had now assembled a force of 1000 B-17s and B-24 Liberators in Britain and was ready to show the Germans what 'Americans meant by a *real* battle of material'. So far, apart from its costly forays to Schweinfurt and Regensburg, it had ventured few mass attacks deep into Germany. Beginning in February 1944 (the 'Big Week' of 20-26 February) under new commanders, Spaatz and James Doolittle, the latter the hero of the Tokyo raid of April 1942, it started to penetrate to targets which the Luftwaffe was bound to defend: aircraft factories and then the twelve synthetic-oil production plants. Speer, Hitler's able Armaments Minister, had robbed the enemy air forces of much of their target system in 1943 by separating manufacturing processes and dispersing the fragments to new small sites, particularly in southern Germany. However, aircraft factories and particularly oil plants defied dispersion, and they provided the 'Mighty Eighth' with prime targets.

The Eighth Air Force had, moreover, been provided with the means to reach them. Daylight bombing required fighter escorts; consequently Bomber Command had abandoned it in 1941, since the Spitfire lacked the range to reach Germany. During 1943 the range of American fighters had also largely confined the Eighth Air Force to attacks in France and the Low Countries. After August, however, at the prompting of Robert A. Lovett, US Assistant Secretary for War, the P-47 Thunderbolt and P-38 Lightning fighters were equipped with drop tanks, external auxiliary fuel tanks which could be jettisoned in an emergency; these gave them the endurance to reach beyond the Ruhr. In March 1944 there appeared in numbers a new fighter equipped with drop tanks, the P-51 Mustang, which could fly to and even beyond Berlin, 600 miles from its British bases. The P-51 was a new phenomenon: a heavy long-range fighter with the performance of a short-range interceptor. It had been delayed in production because it was an Anglo-American hybrid without a strong sponsor. Into an underpowered American airframe the British had inserted the famous Merlin engine; once its improved performance was recognised by Spaatz and Doolittle, they demanded its production in volume and 14,000 were to be built altogether. By March it was present in the German skies in great numbers and already beginning to break the strength of the Luftwaffe.

As soon as the demands imposed by preparation for Overlord ceased, and despite a temporary diversion of effort against the German secret-weapons sites in northern France, Pointblank resumed with redoubled force. The Eighth Air Force had continued its attack on German synthetic oil plants even during the Normandy battle and by September its results were even greater than anticipated. Between March and September oil production declined from 316,000 to 17,000 tons; aviation fuel output declined to 5000 tons. The Luftwaffe thereafter lived on its reserves, which by early 1945 were all but exhausted. Meanwhile the two bomber forces co-ordinated a round-the-clock campaign against German cities, with particular concentration on transport centres. By the end of October the number of rail wagons available weekly had fallen from the normal total of 900,000 to 700,000, and by December the figure was 214,000.

Under day and night attack by the USAAF and RAF, each deploying over 1000 aircraft during the autumn, winter and spring of 1944-5, German economic life was paralysed by strategic bombing. With enemy armies on its eastern and western frontiers, the Reich was no longer protected by a *cordon sanitaire* of occupied territory. The Luftwaffe was overwhelmed as well as outclassed by the daylight bombers' escorts and eventually could not get its few surviving fighters off the ground. Although the anti-aircraft system drained two million men and women out of other services – perhaps the bombing campaign's chief justification – flak dwindled into ineffectiveness as the night-bomber streams became too dense and fast-moving to engage for more than a few minutes. As bomber numbers grew over Germany in 1945, the attrition rate conversely declined to as little as one per cent per mission.

The sudden reversal of advantage between defence and attack undoubtedly derived directly from the appearance of the Mustang as an escort to the Eighth Air Force's Fortresses and Liberators and later as a unit of aggressive fighter patrols, seeking out the enemy. In late 1943 the American campaign had been defeated by the Luftwaffe's day-fighters, in early 1944 the British campaign by its night-fighters. The Mustang restored the Eighth Air Force's ability to penetrate German airspace. In so doing it starved the Luftwaffe of its fuel supply and thereby drastically undercut its ability to sustain the high attrition rate it had inflicted on Bomber Command in 1943-4. Thus it opened the way for the British to match the Americans' level of destructiveness in the round-the-clock campaign of late 1944 and so ensured that German industrial production, whether as a result of physical damage or by the strangulation of supply, should come to a halt in early 1945.

Because the peak of the bombers' success coincided with the defeat of the Wehrmacht in the field and the progressive occupation of Reich territory by the Allied armies, the claims of the strategic-bombing advocates that they possessed the secret of victory have not and can never be proved. Such claims are better supported by the results of the USAAF's bombing campaign against Japan mounted by General Curtis LeMay's XXI Bomber Command: between May and August 1945 the dropping of 158,000 tons of bombs, two-thirds incendiaries, on to the fifty-eight largest Japanese cities, all largely wooden in construction, destroyed 60 per cent of their ground area and brought their populations to destitution and despair. Even before the dropping of the atomic bombs on Hiroshima and Nagasaki and the unleashing of the Red Army's *Blitzkrieg* into Manchuria, to which Japan's decision to surrender is variously ascribed, the home islanders' will to resist had unquestionably been brought to breaking-point by the American bombers.

German civilian morale, by contrast, was never broken by bomber attack. The populations of individual cities were severely distressed by heavy raids. Dresden, overwhelmed on the night of 14 February 1945, did not begin to function again until after the war was over; but in Berlin public transport and services were maintained throughout and were still functioning during the ground battle for the city in April 1945. In Hamburg the 50,000 deaths from bombing, largely concentrated into the period in July 1943, almost equalled those suffered in Britain throughout the war (60,000), yet industrial production returned to 80 per cent of normal within five months. Nothing better vindicated the German people's reputation for discipline and hardihood than the resilience of their urban men and women under Allied air attack in 1943-5 – perhaps the women most of all, since so many were forced by the war to act as heads of families.

The cost of strategic bombing on Germany's civilian population was tragically high: 87,000 people were killed in the towns of the Ruhr, at least 50,000 in Hamburg, 50,000 in Berlin, 20,000 in Cologne, 15,000 in the comparatively small city of Magdeburg, 4,000 in that tiny baroque gem, Würzburg. Altogether some 600,000 German civilians died under bombing attack and 800,000 were seriously injured. Children represented some 20 per cent of the dead and female deaths exceeded male by as much as 40 per cent at Hamburg and 80 per cent at Darmstadt, both cities where firestorms occurred. In the aftermath, privation added to the suffering caused by bereavement and homelessness: reductions in output of up to 30 per cent of steel, 25 per cent of motor engineering, 15 per cent of electrical power, 15 per cent of chemicals and effectively 100 per cent of oil, combined with the effect of a nearly total transport standstill in May 1945, deprived the surviving population of the means to begin reconstruction; the breakdown of transport also imposed fuel shortages which reduced consumption to barest subsistence level.

Because the whole of Germany was occupied by the time of the capitulation, however, no part of the population starved, as happened during the Allies' sustainment of the wartime blockade after November 1918. The armies, even the Red Army, collected food and made themselves responsible for its distribution. The air forces which had devoted themselves to Germany's economic devastation in 1943-5 found themselves engaged, almost as soon as the war was over, in transporting essential supplies to the cities they had recently been overflying with high explosive and incendiaries in their bomb-bays.

In the course of their campaign the Allied bombing forces had suffered grievously themselves: in 1944 alone the Eighth Air Force lost 2400 bombers; throughout the war Bomber Command suffered 55,000 dead, more than the number of British army officers killed in the First World War. The dead aircrew were not, however, accorded the memorialisation given to the 'lost generation'. Their campaign, though it gave a dour satisfaction to the majority of the British people in the depths of their war against Hitler, never commanded the support of the whole nation. Its morality was publicly questioned in the House of Commons by the Labour MP Richard Stokes, more insistently in the Lords by Bishop Bell of Chichester and in private correspondence by the Marquess of Salisbury, head of the leading Conservative family in Britain. All made the point, to quote Lord Salisbury, 'that of course the Germans began it, but we do not take the devil as our example.' This accorded with a nagging self-reproach to the national conscience which, when the war was over, denied 'Bomber' Harris the peerage given to all other major British commanders and refused his aircrew a distinctive campaign medal of their own. With their backs to the wall the British people had chosen not to acknowledge that they had descended to the enemy's level. In victory they remembered that they believed in fair play. Strategic bombing, which may not even have been sound strategy, was certainly not fair play. Over its course and outcome its most consistent practitioners drew a veil.

TWENTY-THREE

The Ardennes and the Rhine

The German army, adept as always at surmounting crisis, lost no time in putting distance between itself and the disaster of Normandy. Hitler had been forced to accept the result of the Falaise battle, yet previously he had refused to allow any construction of defences on the line of the Somme and Marne rivers, tentatively designated by Army Group B as an intermediate position between the Atlantic and West Walls. In consequence, once the *Westheer* crossed the Seine between 19 and 29 August it could and did not pause in its retreat until it reached defensible positions on the great northern European waterways – the Schelde, the Meuse, the tributaries of the Rhine – in the first week of September. The British captured Brussels on 3 September, to ecstatic civic rejoicing, and Antwerp, Europe's largest port, the next day. By 14 September the whole of Belgium and Luxembourg was in Allied hands, together with a fragment of Holland, and on 11 September patrols of the American First Army actually crossed the German border near Aachen. The vanguard of the Franco-American force which had landed in Provence on 15 August linked up with Patton's Third Army near Dijon on 11 September. Thereafter, as the 6th Army Group, it went into the line in Alsace. By the end of the second week in September there was a continuous battlefront in northern Europe running from the banks of the Schelde in Belgium to the headwaters of the Rhine at Basle on the Swiss frontier.

However, Patton and Montgomery, Eisenhower's two most thrusting subordinates, arrived at the approaches of the German frontier both believing that a more clear-cut strategy and a more calculated allocation of supplies would have resulted in the West Wall's being breached. The roots of this dispute, subsequently known as the 'Broad versus Narrow Front Strategy', lay far back in the Overlord campaign, when the air campaign against the French railway system was at its height. The Allied forces had then been so successful in destroying French railway bridges, lines and rolling stock that when the armies at last broke out of the bridgehead in August the means to supply their advance could be provided only by truck and by road. It was hoped that, as the armies advanced, the truck route would be shortened by the progressive capture of ports along the Channel coast (also desirable because Hitler's flying-bomb launch sites lay in the same area); but Hitler's insistence on Army Group B's leaving garrisons to hold Le Havre, Boulogne, Calais, Dunkirk and the mouth of the Schelde vitiated that hope. Although Le Havre was captured on 12 September, Boulogne on 22 September and Calais on 30 September, Dunkirk held out until the end of the war, while, more critically for the Allies, the defences of the Schelde estuary were still in German hands at the beginning of November.

In retrospect it can be seen that the failure to clear the Schelde estuary, and thus to open the way for the Allies' fleet of cross-Channel supply vessels to deliver directly to Antwerp in the immediate rear of the Canadian First, British Second and American First Armies, was the most calamitous flaw in the post-Normandy campaign. It was, moreover, barely excusable, since Ultra was supplying Montgomery's headquarters from 5 September onwards with intelligence of Hitler's decision (of 3 September) to deny the Allies the use of the Channel ports and waterways; and as early as 12 September Montgomery's own intelligence section at 21st Army Group reported that the Germans intended to 'hold out as long as possible astride the approaches to Antwerp, without which the installations of the port, though little damaged, can be of no service to us.'

Montgomery – despite every warning, and contrary to his own military good sense, which was acute – refused to turn his troops back in their tracks to clear the Schelde estuary. Instead he determined upon using the First Allied Airborne Army (the British 1st and the American 82nd and 101st Airborne Divisions) to leap across the Meuse and the lower Rhine, establish a foothold on the North German plain and capture the Ruhr, heartland of Germany's war economy. On 10 September – the day on which formal command of ground forces in north-west Europe passed from himself to Eisenhower, and he became a field marshal in recognition of his achievements – he secured the Supreme Allied Commander's assent to the plan and on 17 September the operation, codenamed Market Garden, began.

Market, the seizure of the bridges at Eindhoven and Nijmegen by the American airborne divisions, proved a brilliant success. Garden, the descent of the British 1st Airborne Division on the more distant Rhine bridges at Arnhem, did not. Because of the experience of the German 7th Parachute Division in Crete, where it had been massacred while dropping directly into its objective, the Allied airborne forces had established the doctrine that airborne descents should be made at a distance from the chosen target, on which the parachutists should concentrate only after having assembled and collected their equipment. The 1st Airborne Division got safely to earth; but when it advanced on the Arnhem bridges it found their vicinity held by the remnants of the 9th and 10th SS Panzer Divisions, which were refitting in the district after their ordeal in Normandy. Between them the two divisions mustered only a company of tanks, some armoured cars and half-tracks; but even the remnants of a Panzer division deployed more firepower than the 1st Airborne, whose artillery support was provided by 75-mm pack howitzers which one of its own gunner officers described as 'quite unlethal'. The British parachutists, after seeing one of Arnhem's two bridges fall into the Rhine as they approached it, succeeded in seizing and holding the other. They held it steadfastly until 20 September, hourly expecting the arrival of British tanks to their relief, but the Guards Armoured Division which was advancing to join them found itself confined to a single road between inundated fields and could not move forward at its planned speed. German reinforcements had now gathered around the Arnhem perimeter, constricting it ever more closely, and on 24 September the British received orders to withdraw. Some managed to do so by improvised ferry, many swam the Rhine back to the southern bank. Just over 2000 men succeeded in escaping; 1000 were killed in the course of the battle, and 6000 became prisoners. The 1st Airborne Division had effectively ceased to exist.

Arnhem was the German army's first overt success since decamping from Normandy. It had also, however, fought a little-noticed but successful defence of Aachen and was meanwhile busily reinforcing its position along the Schelde estuary apparently unobserved and certainly unhindered by Montgomery's 21st Army Group. During its pell-mell drive to Brussels 21st Army Group had bypassed the unmechanised elements of the German Fifteenth Army left in northern France and along the south Belgian coast. Its new commander, General Gustav von Zangen, took advantage of the distraction of Arnhem to evacuate these remnants, amounting to 65,000 men of nine divisions, across the mouth of the Schelde on to the island of Walcheren and the coastal area of South Beveland, leaving a bridgehead on the south bank at Breskens. The reconstituted Fifteenth Army was left undisturbed by Montgomery until 6 October, when, at last alerted to the precariousness of the liberation armies' logistic position as long as Antwerp remained unusable – with its outlet to the sea in German hands – he set his Canadian troops to capture and clear the Schelde's waterlogged banks in what would become the most difficult and unpleasant operation fought by any of the Allied armies in the winter of 1944. When the battle was concluded on 8 November, two river minefields still had to be cleared and it was not until 29 November, eighty-five days after its capture, that Antwerp was at last open to shipping.

Logistic improvisation, including a high-speed truck route carrying 20,000 tons of supplies daily over the 400 miles separating the Normandy beaches from the zone of operations, was meanwhile permitting the resumption of offensives up and down the front. On the American front, next to Montgomery's, Bradley's 12th Army Group was confronted by the West Wall, which had fallen into disrepair since 1939 but had been hastily rehabilitated. Eisenhower hoped that a concerted drive either side of Aachen would allow a breakthrough to Cologne before the winter brought campaigning to an end. The West Wall, however, proved a still formidable obstacle when the First and Ninth Armies attacked on 16 November, and, although it was penetrated, the terrain beyond, particularly the dense thickets of the Hürtgen forest, defied their efforts to break out. At the southern end of the front, Patton, still annoyed about Eisenhower's refusal to support his 'Narrow Front' advance from the Seine the previous August, was fighting a more mobile battle in Lorraine against Balck's Army Group G, consisting of the divisions which had escaped from the south of France and hastily raised reinforcements from the German Home Army. The Germans benefited from the defensive advantages offered by a succession of river lines, the Moselle, Meurthe and Seille, and by the old French fortification zone built in 1870-1914, and they conducted a step-by-step withdrawal, denying Patton's Third Army possession of Metz in a bitter battle that lasted from 18 November until 13 December. Not until 15 December was it fully in contact with the lower reaches of the West Wall which followed the line of the Saar river. Patton's spearheads succeeded in seizing some small bridgeheads across the Saar as the first heavy snow of the winter set in. Devers's 6th Army Group, consisting of the American Seventh and French First Armies, had been more successful in clearing the Germans out of Alsace to the south, despite having to fight through the difficult mountainous sector of the Vosges. American troops entered Strasbourg on 23 November, but a pocket of resistance around Colmar, protecting the Upper Rhine and the West Wall behind it, still resisted the French army's efforts to take it in mid-December.

Germany gains a respite

The deceleration of the Allied drive against the outer defences of Germany in the autumn and early winter of 1944 was caused largely by the logistic difficulties under which they campaigned enhanced by their far greater divisional needs than those of the Germans, 700 tons a day as opposed to 200 tons a day. There was also the improved fighting power of the German army. In early September Hitler had charged Goebbels to raise within the Home Army (now commanded by Himmler since the dismissal and execution of its commander, Fromm, after the July Plot) twenty-five new *Volksgrenadier* divisions to man the western defences. The manpower was found by 'combing through' headquarters, bases and static units inside Germany, a process which also yielded replacements for the broken divisions which had struggled back to the West Wall from Normandy. Between 1 September and 15 October an additional 150,000 men were found in this way – though losses in the west in that period exactly equalled that figure – and another 90,000 from within the resources of OB West (to which post Rundstedt had again been appointed on 2 September). Moreover, despite the full resumption of the Anglo-American Pointblank bombing offensive after Normandy, German industry had achieved higher levels of output of war material in September than in any month of the war, thanks to the success of Speer's policy of dispersal of production and assembly away from the traditional centres. As a result, tank and assault-gun production during 1944 approached that of the Soviet Union during the same period. The 11,000 medium tank and assault guns, 16,000 tank destroyers and

5200 heavy tanks produced were sufficient to keep existing Panzer divisions in the field (despite their appalling losses in Normandy and White Russia) and to provide the material for thirteen new Panzer brigades, nine of which were subsequently to be reconstituted as weak Panzer divisions.

Much self-delusion was necessary at Hitler's headquarters to represent this rebuilding and re-equipment as genuine reparation for the losses suffered in the catastrophic summer of 1944. Hitler, however, was a master of self-delusion and also of the art of clutching at straws. Although adamant in his refusal to allow any of his commanders to surrender ground for whatever reason, nevertheless he always reconciled himself to the loss of ground that inevitably occurred by asserting that the enemy had thereby overreached himself and exposed himself to a counter-strike which would repay all the damage done and recover the abandoned territory into the bargain. This self-defence mechanism had allowed him to justify denying permission to Paulus to break out of Stalingrad in November 1942, refusing Arnim leave to evacuate Army Group Africa from Tunisia while time allowed in March 1943, and driving the Fifth Panzer Army to destruction in the Mortain counter-attack of the August just past. In the aftermath of Normandy, indeed before the battle was fully over, the same pattern of deception began to surface in Hitler's strategic appreciation. On 19 August, while the Seventh and Fifth Panzer Armies were still struggling out of the neck of the Falaise pocket, he summoned Keitel, Jodl and Speer and told them to begin preparing the restoration of the *Westheer* because he planned to launch a major counter-offensive in the west in November; 'night, fog and snow', he predicted on 1 September, would ground the Allied air forces and thus inaugurate the conditions for a victory.

Hitler announced his decision to undertake the offensive to his operations staff at the Wolf's Lair on 16 September, having briefed Jodl some days previously to prepare an outline plan. It was then that he first revealed both the location and objective of *Wacht am Rhein*, as the attack was codenamed. 'I have made a momentous decision,' Hitler announced. 'I shall go over to the offensive . . . out of the Ardennes, with the objective, Antwerp.' His reasoning emerged in more detail as planning progressed: Antwerp, in mid-September still unavailable for use by the Allies, was potentially their major port of supply for an offensive into Germany. If taken by the Germans its loss would set that offensive back many months. Meanwhile his V-2 rockets, the main launching sites for which lay just beyond Antwerp, would be inflicting increasingly serious damage on London, with a demoralising effect on its population. Further, in the course of the drive on Antwerp, which lay only sixty miles from the *Westheer*'s positions in the Ardennes, he would cut off the British Second and Canadian First Armies from the Americans positioned further to the south, encircle and destroy them. The balance of force on the Western Front would thus be equalised, if not actually reversed, and the growing power of his secret-weapons campaign would allow him to regain the strategic initiative. It would then be the

Ostheer's turn to strike at the Russians on the eastern borders, so that Germany, profiting by its occupation of a central position between its enemies, could recoup its theoretically intrinsic advantage and strike for victory.

Hitler's belief in the fantasy he had constructed for himself was strengthened by the fact that the natural point of departure for his forthcoming offensive lay in the Ardennes. For it was on the German side of the Ardennes, the Eifel, that he had gathered the army which had broken the French front in 1940, and through the Ardennes that his Panzer divisions had then advanced to make their surprise attack. In 1944, as in 1940, the Eifel and the Ardennes offered his soldiers the protection of thick forest and narrow valleys almost impenetrable to air surveillance; inside that maze of broken ground and dense vegetation his new army of Panzer divisions could assemble and move forward to their attack positions with the minimum of anxiety at any premature discovery of their presence and intentions. Moreover, in a feckless repetition of the strategic errors made by the French high command four years earlier, Supreme Allied Headquarters had deemed the Ardennes a secondary front during the autumn of 1944 and, by keeping the bulk of their forces, British and American, concentrated to the north and south, had allowed it to become for the second time precisely the same sector of weakness that Kleist and Guderian had been able to exploit in May 1940.

For all that, the generals with whom Hitler had entrusted the execution of *Wacht am Rhein* did not share his confidence in the plan. Rundstedt and Model, Kluge's successor as commander of Army Group B, agreed between themselves in late October that the plan did not have 'a leg to stand on'. Together they devised an alternative, which they called the 'Small Solution' in distinction from Hitler's 'Big Solution', aimed at damaging the enemy forces opposite the Ardennes rather than trying to destroy them. Hitler would have none of it. First of all he sent Jodl to see Model on 3 November with word that the plan was 'unalterable', and on 2 December he called Model and Rundstedt to the Reich Chancellery in Berlin – now his main headquarters after he had left Rastenburg for good on 20 November – to impress the point on them in person. His only concessions to them were to set back the opening date of the offensive still further (it had already been postponed from 25 November) and to give it a new codename, Autumn Mist, originally chosen by Model for the 'Small Solution'.

'All Hitler wants me to do', complained Sepp Dietrich, commander of one of the two armies earmarked for the operation, 'is to cross a river, capture Brussels, and then go on and take Antwerp. And all this in the worst time of the year through the Ardennes when the snow is waist deep and there isn't room to deploy four tanks abreast let alone armoured divisions. When it doesn't get light until eight and it's dark again at four and with re-formed divisions made up chiefly of kids and sick old men – and at Christmas.' This analysis by one of Hitler's most loyal supporters was closely exact. On paper the German order of battle for Autumn Mist appeared impressive. It consisted of two Panzer armies, the Fifth and Sixth, commanded by Manteuffel, one of the best of the younger German tank generals, and Dietrich; between them they deployed eight Panzer, one Panzergrenadier and two parachute divisions, most of which had fought the Normandy campaign, therefore enjoyed experienced leadership and had been brought up to strength again since the retreat from Falaise. They included the 1st, 2nd, 9th and 12th SS Panzer and the 2nd, 9th, 116th and Lehr Panzer Divisions, the 3rd and 15th Panzergrenadier Divisions (the latter belonging to the supporting Seventh Army) and the 3rd and 5th Parachute Divisions. However, appearance and reality diverged. Although every effort had been made to find men and equipment for these divisions, so that the 1st and 12th SS Panzer, for example, were well up to strength, even such first-line formations as the 2nd and 116th Panzer deployed only a hundred tanks each, while the Volksgrenadier divisions which provided support for the armoured spearheads were ill equipped, under strength and filled out with 'ethnic' Germans who owed their nationality to frontier changes. The 62nd Volksgrenadier Division, for example, contained many Czech and Polish conscripts from regions annexed to the Reich who spoke no German at all and belonged in sympathy to the Allied armies they were committed to attack; the 352nd Volksgrenadier Division, rebuilt on the ruins of its predecessor which had fought so stoutly at Omaha beach, was filled with airmen and sailors; and the 79th Volksgrenadier Division had been formed out of soldiers 'combed out' of rear headquarters.

Another deficiency in the plan was lack of fuel. Only a quarter of the minimum requirement was on hand when the offensive opened, much of it held east of the Rhine, while the leading attack elements were expected to capture supplies from the Americans as they advanced. Hitler nevertheless clung to the conviction that Autumn Mist would succeed. Speaking to the generals at Rundstedt's command post on 12 December, he painted a picture of an alliance of 'heterogeneous elements with divergent aims, ultra-capitalist states on the one hand, an ultra-Marxist state on the other . . . a dying empire, Britain . . . a colony bent on inheritance, the United States. . . . If now we can deliver a few more heavy blows, then at any moment this artificially bolstered common front may suddenly collapse with a gigantic clap of thunder.'

Thanks partly to the careful security measures observed by Army Group B during the preparations for Autumn Mist, and partly to Supreme Allied Headquarters' close attention on its own operations at Aachen, in the Saar and Alsace, such warning signs of the offensive as were emitted failed to alert Allied anxieties. On the morning of 16 December, D-Day for Autumn Mist, the front of attack was held by only four American divisions, the 4th, 28th and 106th Divisions supported by the inexperienced 9th Armoured Division, disposed across a space of nearly ninety miles. Two of the three infantry divisions had between them suffered 9000 casualties in the Hürtgen forest battle and had been sent to the Ardennes to rest; the third, 106th, was entirely new to battle.

On to these ill-fitted and unprepared American defenders of the Ardennes, the Sixth and Fifth Panzer Armies fell like a whirlwind on the morning of 16 December. In the centre the American 28th Division was quickly overrun and in the north the 106th Division's forward elements were surrounded; only in the south, where the 4th Division was supported by the 9th Armoured Division, did the Germans make less progress than anticipated. During this troubled day, moreoever, Bradley's 12th Army Group headquarters, in which the Ardennes sector lay, failed to appreciate the magnitude of the attack that was developing. Bereft of air reconnaissance because winter weather 'closed in' its airfields, and denied intercept intelligence because of strict German radio security, it formed the view that the attack was local and diversionary and did not react with urgency to the developing crisis.

Eisenhower, whom Bradley was by chance visiting during the day, fortunately took a more precautionary view. He decided to bring down two armoured divisions from the neighbouring formations, the 7th from the Ninth Army and the 10th from the Third, to stand on the flanks of the German attack lest it develop into a full-blown offensive. Patton, battling into the Saar and still imbued with the belief that he was on the point of breakthrough, automatically protested; but Eisenhower's caution was to be justified by events. On the second day of the offensive, 17 December, the 1st SS Panzer Division arrived at the key road junction of Saint Vith, from which a valley route led to the Meuse and so into the plains of Belgium and the approaches to Antwerp. It was to be denied a breakthrough by the appearance of the US 7th Armoured Division's spearheads and thereafter found itself turned away from access to open country – and to the vast American fuel dumps near Stavelot on which it had counted for resupply – by one American blocking move after another.

While the Sixth Panzer Army was being diverted from the direct north-westward route to Antwerp, and forced progressively due east, the Fifth Panzer Army was making better progress in the southern sector towards Monthermé, where Kleist's Panzers had crossed the Meuse in 1940. The key to its breakthrough was the road centre of Bastogne, a junction for the sparse network of highways that runs from the Eifel into the Ardennes and onward. The capture of Bastogne was essential to the successful development of Autumn Mist. At dawn on 19 December Panzer Lehr Division was only two miles from the town; but during the night the US 101st Airborne Division had arrived by truck, having driven 100 miles from Reims at breakneck speed, and was positioned to deny the Germans possession. The parachutists were quite unequipped to combat tanks; but by their resolute defence of the small town's streets they prevented Panzer Lehr's infantry from gaining entry and so turned Bastogne into an even more effective road block than Saint Vith (which fell on 23 December) on the Sixth Panzer Army's axis of advance.

By Christmas Day Bastogne was completely surrounded by German troops and the Fifth Panzer Army had moved on; Panzer Lehr had worked around a flank to appear beyond Saint-Hubert, only twenty miles from the Meuse. On

Christmas Day itself, however, the pace of the German advance across the whole front of attack began to slow and the nose of the salient which the Panzer armies had driven into Allied lines was attenuating. Allied counter-measures had begun to tell. On 20 December, in the face of Bradley's strident objections, Eisenhower had confided command of operations against Dietrich's Sixth Panzer Army, on the northern face of the 'bulge' closest to Antwerp, to Montgomery; while the intervention of divisions from Patton's Third Army against the southern face during 17-21 December matched the effect of the counter-attacks that the British commander set in train. Montgomery, who was copiously informed from 20 December by Ultra decrypts of both Sixth and Fifth Panzer Armies' intentions, took prompt steps to guard the bridges across the Meuse, towards which Dietrich's spearheads were advancing, with British troops brought down from northern Belgium. He thereafter took the view that the attackers, now opposed by nineteen American and British divisions, including such seasoned formations as the US 82nd Airborne and 2nd Armoured Divisions, would simply wear themselves out by their effort to make progress.

Montgomery's analysis proved exactly correct. Indeed, unrecognised though it was at the time, the American divisions, particularly the 28th and 106th, which had stood in the path of the initial attack, had through the dedicated and self-sacrificing resistance of many of their rifle platoons and anti-tank teams done a great deal to wear down the impetus of the German Panzer divisions on the first day of attack. They had inflicted heavy casualties, damaged if not always destroyed equipment, and delayed the timetable on which the success of the offensive too narrowly depended.

On 26 December Eisenhower's headquarters received the first evidence that Autumn Mist had lost its vital impetus. The weather cleared, allowing the Allied air forces to intervene effectively for the first time. Patton's 4th Armoured Division broke through the southern face of the 'bulge' to bring relief to the 101st Airborne Division surrounded in Bastogne, and the 2nd Armoured Division, from Hodges's First Army, found the 2nd Panzer Division immobilised for lack of fuel near Celles, five miles from the Meuse at Dinant, and destroyed its leading tanks. Indeed, in the course of its one-sided encounter with the 2nd Armoured Division, the 2nd Panzer Division lost almost all the eighty-eight tanks and twenty-eight assault guns with which it had begun the offensive.

By 28 December Montgomery was sure that Autumn Mist had failed, though he expected the Germans to persist in the offensive and even to launch a further attack. That attack, when it came, fell outside the Ardennes 'bulge', in the Saar, where Blaskowitz's Army Group G struck against Patch's Seventh Army and managed to take and briefly hold a triangle of territory on the west bank of the Rhine. This brief success reinforced Hitler's view that his aggressive strategy had been correctly conceived, even though it had been launched at a moment when the Eastern even more than the Western Front cried out for defensive reinforcement. In fact, however, North Wind (as this second offensive was codenamed) caused mild political but little military alarm and contributed to Autumn Mist not at all. On 3 January 1945 Montgomery launched a convergent counter-attack against the northern and western faces of the 'bulge', which obliged Hitler on 8 January to order the withdrawal of the four leading Panzer divisions from their exposed situation. On 13 January the American 82nd and British 1st Airborne Divisions made contact in the centre of what had been the Ardennes salient, and by 16 January the front was restored.

Between 16 December and 16 January the Fifth and Sixth Panzer Armies had inflicted some 19,000 fatal casualties on the US 12th Army Group and taken 15,000 Americans prisoner. In the first days of their offensive they had spread panic throughout the civilian population of Belgium and caused alarm among the military as far away as Paris – where precautions were taken against sabotage raids it was feared would be mounted by the small clandestine units which Otto Skorzeny infiltrated (in practice with little success) behind the Allied lines. The German offensive had also shaken the optimism prevailing in Washington and London over the early conclusion of the war. Hitler spoke to his subordinates of 'a tremendous easing of the situation . . . the enemy had had to abandon all his plans for attack. He has been obliged to regroup his forces. He has had to throw in again units which were fatigued. He is severely criticised at home. . . . Already he has had to admit that there is no chance of the war being ended before August, perhaps not before the end of next year. This means a transformation of the situation such as nobody would have believed possible a fortnight ago.'

Hitler exaggerated; he also grossly misinterpreted the true significance of the Ardennes campaign. It had, of course, inflicted losses on the enemy, but those losses could be borne and made good. The British army had come to the end of its manpower resources, but the American army had not. Since September it had shipped twenty-one divisions to France including six armoured; between January and February it was to land another seven, including three armoured, all fully equipped and up to strength. The *Westheer*, by contrast, had lost 100,000 men killed, wounded or captured in the Ardennes, 800 tanks and 1000 aircraft – many thrown away in the Luftwaffe's last offensive, *Bodenplatte*, launched against Allied airfields in Belgium on 1 January 1944. None of these losses, human or material, could be made good. The Wehrmacht's resources were exhausted, while German war industry's output could no longer keep pace with everyday attrition, let alone the surges of destruction caused by indulgence in heavy offensive activity. Steel production alone, fundamental to weapons manufacture, had been reduced from 700,000 to 400,000 tons monthly in the Ruhr by bombing between October and December, and it continued to decline; while the disruption of the transport system meant that it was increasingly diffi-cult to move weapon components from point of production to point of assembly.

All that Autumn Mist achieved was to impose a brief delay on the Western Allied armies' preparations to break into Germany, at the expense of transferring from or denying to the Eastern Front men and equipment needed to stem the continued advance of the Red Army into southern Poland and the Baltic states. During November and December 2299 tanks and assault guns and eighteen new divisions had been committed to the Western Front but only 921 tanks and five divisions to the Eastern, where 225 Soviet infantry divisions, twenty-two tank corps and twenty-nine other armoured formations confronted 133 German divisions, thirty of which were already threatened with encirclement in the Baltic states. Hitler's 'last gamble', as the Ardennes came to be described, was extremely short-sighted. It bought a little time at great cost, failed in its object of destroying Montgomery's army and won back no ground at all.

Indeed, despite the intervention of Generals January and February, who fought in 1945 on the German side, the Western armies recovered quickly from the shock of Autumn Mist and succeeded in making advances as creditable, given the more defensible nature of the terrain west of the Rhine, as the Red Army was currently achieving in Poland, Hungary and Yugoslavia. In January the two German salients west of the West Wall, the Roermond triangle north of Aachen and the Colmar pocket south of Strasbourg, were reduced. In February and March Eisenhower's armies advanced along the whole front, to reach the Rhine between Wesel and Koblenz and to seize the north bank of the Moselle between Koblenz and Trier. By the end of the first week of March it was the Rhine alone which stood between the Allies and the German hinterland.

PART V
THE WAR IN
THE EAST
1943–1945

COLLAPSE OF THE EASTERN FRONT, 1945

TWENTY-FOUR

Stalin's Strategic Dilemma

Hitler's decision to commit Germany's last army to a winter offensive in the west in 1944, rather than to use it as a counter-attack force against the encroachments of the Red Army in the east, might seem with hindsight one of the most perverse of the Second World War. In the east, Germany was defended by neither geography nor man-made fortifications. In the west, both the Siegfried Line (West Wall) and the Rhine stood between the Anglo-American armies and the interior of Germany. Comparatively weak forces committed to hold those obstacles would have sufficed to hold Eisenhower's troops at bay for months, while the Sixth SS and Fifth Panzer Armies, under which Hitler's last tank reserve was concentrated, might have won an equal amount of time had they been deployed to fight on the line of the Vistula and the Carpathians instead of being cast away in the Ardennes adventure. Hitler's rationalisation of his decision is well known: in the west the Allied armies were exposed to a counterstroke towards Antwerp, the success of which would have freed his forces to unleash in the east a subsequent offensive designed to destabilise the Red Army. In short, he chose to strike for the chance of victory rather than settle for postponing the onset of defeat. Events were to rob him of both outcomes. The choice of the Ardennes offensive, though it may have set back a little the launching of the Rhine crossing, actually thereby ensured the unimpeded advance of the Red Army's offensive in the east whenever Stalin chose to launch it.

Yet it remains generally unperceived that Hitler's plunge into double jeopardy was determined by his confrontation with not a double but a triple threat. In the west he faced the danger of an Allied assault on the Rhine. In the east the Red Army menaced the Greater Reich on two large and widely separated fronts: from Poland through Silesia towards Berlin; and also from eastern Hungary towards Budapest, Vienna and Prague. Since Hitler had no means of knowing on which of these two axes Stalin would make his major effort, strategic sense positively argued for disposing of the danger in the west first and then transferring his striking force eastward – always supposing it had survived the shock of battle – to oppose the Red Army on whichever sector, north or south of the Carpathians, that it appeared in greater strength. The ultimate validation of this judgement, though Hitler could only guess at it, was that until November 1944 Stalin himself had been in two minds about whether to strike directly for Berlin or to distract and destroy the fighting power of the *Ostheer* by a thrust elsewhere, the Budapest-Vienna axis being the most obvious choice.

Since the moment when the Red Army had been able to go over to the offensive at Stalingrad in 1942, the sheer size of the Eastern Front, the ratio of force to space, the erratic flow of supplies and the paucity of road and rail communications had time and again forced Stalin into a similar choice between fronts. Even the German army, during the summer months of Barbarossa in 1941, had been obliged to close down Army Group Centre's front for six weeks while Army Groups North and South made up ground to come abreast of it on the roads to Leningrad and Kiev. Those were armies at the height of their powers, led by commanders flushed with victory, spearheaded by superbly efficient armoured forces and backed by still ample reserves of manpower. The Red Army which went over to the offensive for the first time at Stalingrad, by contrast, had been ravaged by eighteen months of losses on a scale never experienced before in history, was led by generals whose self-confidence had been shaken by a succession of disasters and was fed from a pool of recruits in which the very young and the over-mature were now disproportionately represented. It was an army which had yet to learn how to manoeuvre; until it did so, its operations were perforce limited to responding to German thrusts and to taking up ground by frontal advance on the sectors where the Germans had weakened themselves by over-extension.

The deficiencies of the Red Army, moreover, permeated the Soviet military structure from the bottom to the top. Stalin himself was an uncertain military leader, surrounded by civilian and military subordinates who lacked the experience of directing armed forces under the strain of war, and served by a command structure he had to improvise from scratch. Because of the nature of the Soviet system and of his own devious character, moreover, Stalin could not mobilise and focus upon himself the popular support which so strengthened Churchill's hand, for example, in rallying the nation to cope with crisis. The peoples of the Soviet Union did not form a nation, the experience of industrialisation and collectivisation had alienated millions from the rule of the Communist Party, the party was further tainted by its exclusive and repressive methods of government, while Stalin himself commanded it by the use of selective terror against his comrades which was made all the more distasteful by his maintenance of the fiction that he was no more than first among equals in a collective leadership.

The spirit of patriotism could to some extent be artificially revived. The epics of Russian history could be recalled, Russian heroes of the past – Ivan the Terrible, Alexander Nevsky, Peter the Great – could be rehabilitated, decorations and orders which commemorated victorious generals of the imperial era (Kutuzov and Suvorov) could be created, distinctions of rank and dress, abolished at the Revolution, could be revived. The Orthodox Church, an object of contempt in a professedly atheist state, could even be enlisted to preach the crusade of the Great Patriotic War; its reward, in September 1943, was to be allowed to elect its first synod since the institution was suppressed after the Revolution. These, however, were mere expedients. They were no substitute for an effective organ of strategic command, which Stalin must provide or else fail as a war leader, consigning Russia to defeat and himself to extinction.

Stalin does indeed seem to have come close to breakdown in the first weeks of Barbarossa. 'For a week', describes Professor John Erickson,

> it was all the anonymity of 'the Soviet government', the 'Central Committee' and '*Sovnarkom*', the clamour of organisation, and the rattle of Party exhortations. . . . Stalin, committed irrevocably to war in spite of himself, 'locked himself in his quarters' for three days at least after the first, catastrophic week-end. When he emerged, he was, according to an officer who saw him at first hand, 'low in spirit and nervy'. . . . [He] put in no more than rare appearances at the *Stavka* in these early days; the main military administration was, for all practical purposes, seriously disorganised and the General Staff, with its specialists dispatched to the front commands, functioned with tantalisingly persistent slowness. . . . The *Stavka* discussions ground into an operational-administrative bog; while trying to formulate strategic-operational assignments, Stalin and his officers busied themselves with minutiae which devoured valuable time – the type of rifle to be issued to infantry units (standard or cavalry models), or whether bayonets were needed, and if so should they be triple-edged?

In fairness it must be said that Hitler also took refuge in the discussion of military minutiae as an escape from the pressure of crisis and often, if the crisis protracted, refused to discuss anything else, as the surviving fragments of his Stalingrad Führer conference records reveal. Stalin, by contrast, returned quickly to realities. On 3 July 1941, the eleventh day of the war, he broadcast to the Soviet people – an almost unprecedented event – and addressed the 'comrades, citizens, brothers and sisters' as his 'friends'. Moreover, he moved at once to put the government of the Soviet Union on a war footing. The way in which he did so is almost incomprehensible to Westerners, attuned as they are to a strict separation between organs of state and political parties, civil power and military authority, bureaucrats and commanders. In peacetime the Soviet system blurred such distinctions; Stalin heightened this ambiguity in the structure he erected for war. His first move, on 30 June, was to create a State Defence Committee to oversee the political, economic and military aspects of the war; its membership, later slightly broadened, consisted of himself, Molotov, the Foreign Minister, Voroshilov, who had been Commissar for Defence from 1925 to 1940, Malenkov, his right-hand man in the party organisation, and – significantly – Beria, head of the secret police (NKVD). On 19 July he nominated himself as People's Commissar for Defence and on 8 August secretly assumed the post of Supreme Commander; as Supreme Commander (though he continued to be identified only as Commissar for Defence) he controlled the Stačka, in effect the executive organ of the State Defence Committee (GKO), which oversaw both the General Staff and the operational commands or fronts. As the acts and decisions of the GKO automatically carried the authority of the council of People's Commissars, of which Stalin was head, and as he could also detach officers of the General Staff, notably and most frequently Zhukov and Vasilevsky, to run fronts or direct specified operations, Stalin dominated the direction of the Great Patriotic War from top to bottom (the designation 'patriotic' had been used in his broadcast of 3 July). Though he cautiously disguised from the Soviet people his ultimate responsibility for command decision, and would emerge as Marshal, Generalissimo and 'the great Stalin' only when a roll of substantial victories had been secured, Stalin was effectively commander-in-chief from the beginning of July 1941 onwards. He was implacable in that role. When Army Group Centre resumed its advance on Moscow in October, his confidence was shaken almost as severely as it had been in June, but he never relaxed the grip of fear in which he held his subordinates: dismissal, disgrace, even execution were the penalties which awaited failures. General Ismay, Churchill's military assistant who visited Moscow in October, noted the effect: 'as [Stalin] entered the room every Russian froze into silence and the hunted look in the eyes of the generals showed all too plainly the constant fear in which they lived. It was nauseating to see brave men reduced to such servility.'

A few held out. Zhukov was notably robust, appearing not to be frightened by Mekhlis, the chief political commissar used by Stalin to bring others down. Zhukov had the advantage of having successfully commanded tanks against the Japanese in a brief and undeclared Russo-Japanese border war in Mongolia in 1939; more important, he was naturally tough, able to accept dismissal by Stalin from the post of Chief of the General Staff and to proceed to an operational command without diminished confidence in his own abilities, which he knew Stalin recognised. Others of Zhukov's stamp were to appear, notably Rokossovsky and Konev. By the time all three were commanding fronts in 1944, Stalin's difficulties in finding able subordinates were largely solved.

In the meantime, however, he had to do most of the work of directing the Great Patriotic War and running the Red Army himself; to a greater extent than was true of the high command of any other of the combatant powers, Stalin dominated Russia's war effort. Hitler and his generals coexisted in a constant state of tension. Churchill imposed his will by argument, which prevailed less and less as the Americans took over an ever greater share of the fighting. Roosevelt largely presided over rather than directed his chiefs of staff. Stalin, however, dictated. All information flowed to him, wherever he was to be found during the day or night, whether in the Kremlin, at his country *dacha* at Kuntsevo or, while German bombs threatened Moscow, in an improvised headquarters on a platform of the Moscow underground railway; from him all orders flowed back. He held a situation conference three times a day, in a routine curiously similar to Hitler's, hearing reports first at noon, then at four in the afternoon, and finally dictating orders directly to officers of the General Staff but in the presence of the Politburo between midnight and three or four in the morning.

Vasilevsky, in effect Stalin's operations officer, playing a role equivalent to that of Jodl in Hitler's headquarters, perceptively observed and later recorded the dictator's methods of command. He noted that Stalin established his dominance over the military in the first year of the war, that is to say far more quickly than Hitler did over the Wehrmacht, perhaps because of his previous experience of operations as commissar of the First Cavalry Army during the Civil War. In the early months he took his confidence too far: in 1941 he was almost wholly responsible for the disaster at Kiev, having refused permission for withdrawal until it was too late for the defenders to escape encirclement; in 1942 he dismissed the danger of a renewed German offensive into southern Russia and committed Timoshenko's fronts to the Kharkov counter-offensive, an altogether premature seizure of the initiative which resulted in over 200,000 Russians being taken prisoner – almost a repetition of the encirclements of the year before. Thereafter he was more cautious. It was eventually Zhukov and Vasilevsky who proposed the double envelopment at Stalingrad; they outlined the concept to Stalin in his office on 13 September 1942 and he accepted it only after they had reasoned away his cautious objections.

Zhukov's highly retrospective assessment of Stalin's worth as a commander was that he excelled above all as a military economist who knew how to collect reserves even while the front was consuming manpower in gargantuan mouthfuls. Certainly his achievement both at Stalingrad and in the two years that followed was to have such reserves at hand – he estimated to the British a consistent surplus over the Germans of some sixty divisions, probably an overestimate – whenever the *Ostheer* gave him the opportunity to profit by a strategic mistake. He deployed such a reserve in a counter-attack when the Germans had exhausted themselves in the offensive phase of the Kursk operation in July 1943. He sustained the success at Kursk by using his reserve in August to recapture Kharkov, the most fought-over city in the Soviet Union. By October his autumn offensive, fuelled by the units he held in reserve, had retaken all the most valuable territory won by the *Ostheer* during its advance into Russia in the two previous years – an enormous tract 650 miles in breadth from north to south, 150 miles in depth, beyond which only the Dnieper, the last truly substantial military obstacle on the steppe, lay to oppose the Red Army's advance. By the end of November the Red Army had secured three enormous bridgeheads on the European side of the Dnieper, had cut the Crimea off from contact with the *Ostheer* and stood poised to open its advance into Poland and Romania.

Ironically, victory brought Stalin a dilemma. Until Stalingrad he had been staving off defeat; until Kursk he still faced the risk of a disabling German initiative; until the advance to the Dnieper he fed, supplied and manned the Red Army by wartime improvisation. Thereafter he knew, like Churchill, that he 'had won after all'. Germany's armoured *masse de manoeuvre* no longer existed, while he had regained possession of his country's most productive agricultural and industrial regions. Moreover, he could now count upon shifting much of the burden of destroying the Wehrmacht from the Red Army to the Allies. At Tehran on November 1943, Brooke, Churchill's chief of staff, noted that in his quick and unerring appreciation of opportunity and situation he 'stood out compared to Churchill and Roosevelt'. By one of the most brutal contrivances of public embarrassment recorded in diplomatic history, he shamed Churchill into conceding his total commitment to Overlord and to agreeing to name both a commander and a date. Thereafter he could be certain that from mid-1944 Hitler would be caught between two fires, and he could let that in the west blaze while he chose where he could most profitably apply the heat elsewhere. As events turned out, he chose to attack on his northern front, destroying Army Group Centre and driving the Germans back to the Vistula. However, that decision did not commit his hand. He still retained the option of either (like the Western Allies) committing all his strength to a final throw designed to destroy the Wehrmacht in a final battle for eastern Germany and Berlin, or diverting a major part of the Red Army's force into southern Europe, there to build a Soviet equivalent of Hitler's Tripartite Pact and so assure the Soviet Union against invasion for decades to come.

It was a tantalising choice. Stalin had not chosen to enter the Second World War; but he had chosen, even before it began, to profit from the tensions that brought it about. In the twenty-one months during which the war had raged while he stood on the sidelines, he had greatly profited from its unfolding. From his alliance with Hitler he had gained in turn eastern Poland, then – through the freedom the non-aggression pact had allowed him to attack Finland – eastern Karelia, then the Baltic states, finally Romanian Bessarabia and northern Bukovina. Barbarossa had engulfed his country in the worst of the fighting brought on by the Second World War. By the summer of 1944, however, he could begin to consider again how best the Soviet Union might profit geopolitically from the war's concluding stage. Stalin, even more than Hitler, was committed to a view of war as a political event. Between Barbarossa and Kursk the 'correlation of forces' had worked against him. Thereafter they began to operate to his advantage. Even as Hitler was laying the groundwork for his last offensive in the west, Stalin was considering where he might best seize the opportunities presented by the collapse of Hitler's strategy in the east.

TWENTY-FIVE

Kursk and the Recapture of Western Russia

The disastrous Stalingrad campaign of 1942-3 left Hitler a debilitated and shaken man. Guderian, visiting him at his Ukrainian headquarters on 21 February 1943 on his unexpected reappointment to command as Chief of Panzer Troops, found him greatly changed since their last meeting in December 1941: 'His left hand trembled, his back was bent, his gaze was fixed, his eyes protruded but lacked their former lustre, his cheeks were flecked with red. He was more excitable, easily lost his composure and was prone to angry outbursts and ill-considered decisions.'

His will to make decisions had also been weakened. In the year between the onset of the Battle of Moscow and the Russian encirclement and destruction of the Sixth Army at Stalingrad Hitler had exercised the *Führerprinzip* at its fullest. He had peremptorily dismissed generals who failed or displeased him and held the rest of the *Generalität* strictly obedient to his orders. Apart from the failure to advance boldly southward from Voronezh in July – and Bock had been sacked for that – his generals had done his will to the letter; that was precisely the problem. The triumphs of the 1942 campaign belonged exclusively to Hitler, but so too did the disasters, both the over-extension into the Caucasus and the defeat of Stalingrad. The consequent loss of twenty of the *Ostheer*'s divisions lay on his conscience, so that even two years afterwards he would confess to one of his doctors that his sleepless nights were filled with visions of the situation map marked with the positions occupied by the German divisions at the moment they were destroyed. The unspoken reproaches of his military intimates – Jodl and Keitel – were hard enough to bear; his self-recrimination was still more painful.

During the spring of 1943, therefore, in planning the *Ostheer*'s strategy Hitler conceded a freedom of action to his subordinates they had not known since his first exercises in command in 1940 and were certainly never to know again – although in other areas of policy he continued to make demands. Believing that Rommel lacked both 'optimism' and 'staying power,' he intervened heavily in the conduct of the battle against the Anglo-American armies in North Africa in the spring of 1943, releasing precious armoured units from his central reserve and requiring Goering to transfer air squadrons from Sicily to Tunisian airfields. He meanwhile hectored Goering's subordinates about the worsening of the air war over Germany – Allied 'round the clock' bombing began on 25

February and heavy British or American raids on Berlin, Nuremberg, Essen, Bremen, Kiel and the Möhne-Eder dams followed in the next weeks. He demanded and got a measure of retaliation against Britain, commissioned Guderian to multiply German tank production and approved Goebbels's programme for the promulgation of 'total war', outlined to him at a meeting with the Nazi Party's Gauleiters at Rastenburg on 7 February. However, in the immediate direction of operations on the Eastern Front, his principal theatre of war-making, during the first half of 1943 he took a curiously tentative and indecisive part.

This was not to prove wholly to the *Ostheer*'s disadvantage. In Field Marshal Erich von Manstein, commander of Army Group South, it had a battlefield commander of the highest quality, acutely sensitive to the tactical opportunities offered by the Red Army's lumbering style of manoeuvre, yet strongly resistant to the psychological intimidation by which Hitler overcame the intellectual independence of his lesser generals. During February, however, in the aftermath of the Stalingrad surrender and his own aborted attempt to relieve Paulus's Sixth Army, Manstein was discountenanced by an unexpectedly successful Soviet attack on the key city of Kharkov, west of the Don.

The Red Army's victory at Stalingrad, and the subsequent disorder caused to the whole of the German southern front, had presented the Stavka for the first time with the prospect of seizing the initiative and throwing the *Ostheer* clean out of the Ukraine, Germany's most valuable territorial acquisition in the Soviet Union. By the end of January a plan had been conceived for the Southern and South-West Fronts to advance as far as the line of the Dnieper, the third great river line beyond the Don and Donetz, by the spring thaw. Thereafter their neighbouring fronts would advance and swing north-westward to unseat Army Group Centre from the northern Ukraine and roll it back to Smolensk. The first and crucial move in the great offensive would be played by a Front Mobile Group, commanded by General M. M. Popov and consisting of four tank corps, which was to attack in the vanguard of Vatutin's South-West Front and drive on Kharkov.

The Stavka plan was superficially well judged, for the Russian victory at Stalingrad had created three crises for the Germans. The Red Army's advance from Stalingrad had thrown Manstein's Army Group Don (renamed South on 12 February) back on to Rostov, the 'gateway' of the southern front. The enforced withdrawal of Kleist's Army Group A from the Caucasus had carried it to the shore of the Sea of Azov, leaving a gap a hundred miles wide between his front and Manstein's. Moreover, the continuation of Vatutin's attack on the Hungarians defending Voronezh, north-west of Stalingrad, threatened after 14 January to detach Manstein's northern flank from contact with Kluge's Army Group B (Centre after 12 February). The opening stages of the Stavka's offensive augured well for its success. Between 2 and 5 February Russian pressure on the lower Don was so intense that Hitler, at Manstein's insistence, was forced to agree to the abandonment of Rostov, while a simultaneous advance from Voronezh on the upper Don brought Vatutin's South-West Front to Kharkov on 14 February. A bitter battle for the city erupted, in which the population took part, and, despite the efforts of the elite I SS Panzer Corps (*Leibstandarte Adolf Hitler* and *Das Reich* divisions), the Germans were defeated and forced to abandon it on 16 February. As a result a gap nearly 200 miles wide yawned between Army Groups South and Centre.

The Stavka had, however, made two fatal miscalculations. One was to overestimate the Red Army's capabilities. The other was to underestimate Manstein. 'Both the Voronezh and South-West Fronts', comments Professor John Erickson, the leading Western historian of the Great Patriotic War, 'had done some prodigious fighting and covered great stretches of ground, following nothing less than a trail of destruction as retreating German units blew up bridges, buildings and airfields, tangled railway lines and damaged the few roads as much as possible.' However, by mid-February the Popov spearhead, which had begun the offensive with only 137 tanks (no more than a single German Panzer division normally fielded), could put only fifty-three into action, while the so-called Third Tank Army of the Voronezh Front could find only six.

Vatutin's decision on 12 February to 'broaden' the offensive, in accordance with the Stavka's general directive, would therefore have been incautious even against a normally competent opposing commander who retained a modicum of tank reserves. Against Manstein – a supreme master of what both the German and the Russian armies called the 'operational' level of command – the broadening of the offensive was foolhardy. Even before the height of the crisis had been reached, Hitler had ordered seven divisions from France to his front, where he himself arrived to confer with Manstein on 17 February. The pretext was to oversee the unleashing of a counterstroke by Army Group South and to rally the *Ostheer* to the concept of 'total war', which Goebbels proclaimed to the German people in an inflammatory speech at the Berlin Sports Palace the next day. 'The outcome of a crucial battle depends on you,' Hitler wrote in an order of the day. 'A thousand miles from the Reich's frontiers the fate of Germany's present and future is in the balance. . . . The entire German homeland has been mobilised. . . . Our youth are manning the anti-aircraft defences around Germany's cities and workplaces. More and more divisions are on their way. Weapons unique and hitherto unknown are on the way to your front. . . . That is why I have flown to you, to exhaust every means of alleviating your defensive battle and to convert it into ultimate victory.' In reality the counterstroke was not Hitler's conception but Manstein's. Not only had he extracted permission to launch it from Hitler during an urgent visit to Rastenburg on 6 February. He had also found the necessary armoured striking force – of a strength to make Popov's look insignificant – by concentrating all available Panzer reserves under his reconstituted Fourth Panzer Army and positioning it alongside the First

Panzer Army, in the neck of ground between the Donetz and the Dnieper across which Vatutin's South-West Front was seeking to break its way into the German rear.

So dangerous was Vatutin's manoeuvre, threatening as it did to cut off Army Group A in its bridgehead on the Asiatic shore of the Sea of Azov, that Hitler had actually granted permission for troops to be airlifted from it to join Manstein. Over 100,000 were to be sent in that way; but before they or any of the divisions alerted in the west arrived Manstein had struck. On 20 February his two Panzer armies mounted convergent attacks on the flanks of Popov's Front Mobile Group, still advancing to the crossings over the Dnieper less than fifty miles away. The Russian higher command failed altogether to grasp the gravity of the changed situation. It urged Popov onward and on 21 February the General Staff even ordered Malinovsky's Southern Front on Vatutin's flank to join more actively in the offensive: 'Vatutin's troops are speeding on at an extraordinary pace . . . the hold-up on his left is due to the absence of active operations on the part of your front.' In fact Popov was already threatened with encirclement, had begun to run out of fuel and was stopped in his tracks. By 24 February, when despite reinforcements he had only fifty tanks left, over 400 German tanks were operating against his left flank alone. By 28 February, when German tanks reached the banks of the Donetz, his group and much of the rest of Vatutin's South-West Front were surrounded, and such units as escaped did so only because the river was still frozen.

Manstein renews the offensive

The collapse of Popov's offensive now allowed Manstein to unleash the second phase of his plan, for the recapture of Kharkov. The Fourth Panzer Army had now begun to receive the reinforcements sent from the west, including the SS *Totenkopf* Division (originally formed from concentration camp guards) which went to join *Leibstandarte* and *Das Reich* in I SS Panzer Corps. Their loss of Kharkov the previous month rankled savagely with these ideological warriors who, in formidable strength, opened their attack to retake the city on 7 March. By 10 March the northern suburbs were the scene of savage fighting, and two days later the city was effectively surrounded, together with numbers of Soviet units struggling to sustain the defence. Now the Germans threatened the Red Army's centre with envelopment at exactly the spot from which they had hoped to begin the encirclement of Hitler's. So dangerous did the situation suddenly appear to the Stavka that rather than send reinforcements to help their beleaguered formations at Kharkov they rushed them instead to the neighbouring Voronezh Front, south of Kursk, where they succeeded in holding a sector which was to become the southern face of what would soon be called the Kursk salient. With the commitment of these troops to the defensive rather than the offensive, the Soviet spring offensive of 1943 could be seen to have failed, like that which followed victory in the Battle of Moscow the year before. Some Russians had already foreseen that outcome. As Golikov, the commander of the Voronezh Front, had signalled to a subordinate at the height of the Red Army's effort: 'There are 200-230 miles to the Dnieper and to the spring *rasputitsa* there are 30-35 days. Draw your own conclusions.'

The *rasputitsa*, the twice-yearly wet season caused by the autumn rains and the spring thaw, which turns the dirt roads to quagmires and the surrounding steppe to swamp, had worked to Germany's disadvantage in 1941 and 1942, delaying the advance on Moscow, into the Ukraine and on Stalingrad. Now it brought a welcome breathing space. With all the *Ostheer*'s reserves concentrated in the south, the Red Army was able to reopen a land route to Leningrad and to move against the force isolated since the Battle of Moscow in the northern Demyansk pocket – though not to prevent its escape. It was also able to sustain sufficient pressure on the Vyazma salient west of Moscow to persuade Hitler to sanction an uncharacteristic withdrawal to a short front, prepared in advance and called the 'Buffalo Line'. However, while the wet season lasted, and despite the immense losses it had inflicted on the enemy – 185,000 among the Italians, 140,000 among the Hungarians, 250,000 among the Romanians and, by the Wehrmacht's own reckoning, nearly half a million among the Germans – it could not find the strength to resume the offensive on any major sector.

Operation Citadel

Despite the hair's breadth by which the *Ostheer* had escaped disaster on the southern front during the Stalingrad winter, Hitler and his generals were nevertheless turning to a resumption of the offensive at precisely the moment that the Red Army was admitting defeat. 'The real struggle is only beginning', Stalin warned in his message to his soldiers on Red Army Day, 23 February 1943. He and the Stavka knew that it had exhausted its current strength, and that until the awaited donations of Lend-Lease aid and output from the relocated Urals factories had been received, until the next inflow of young conscripts and older 'comb-outs' had been trained, Russia could not form the reserve of force which would safely allow her generals to go over to the attack. The German calculation was precisely contrary. Because the *rasputitsa* and the Red Army's exhaustion had granted the *Ostheer* a breathing space, it must attack as soon as possible, or suffer the consequences of inactivity.

The question was: where? It was an issue which, for the last time during the war, the generals were largely to settle between themselves. Hitler's confidence, his sense of personal credibility in the eyes of his commanders, had been so shaken by the outcome of his insistence on holding Stalingrad as a 'fortress' that he had temporarily lost the will to dictate strategic terms to his subordinates. During his visit to Manstein's headquarters on 17-19 February, before the Kharkov counterstroke, he had listened to a review of the opportunities which might flow from its success. The discussion between Kleist (Army Group A), Jodl, Zeitzler, the new army chief of staff, and Manstein was far more free-ranging than any he permitted on home territory at Rastenburg. Towards the end of the three-day meeting, conducted at times to the sound of Russian gunfire, he had intervened decisively to quash a typically bold proposition by Manstein for a 'one step backward, two steps forward' manoeuvre north of the Crimea, since it entailed the temporary surrender of ground, something to which he was temperamentally opposed. The alternative proposition, for a concentric attack on the developing salient of Kursk, he did not reject but left to Zeitzler and the generals of the *Ostheer* to put into executive form.

During the lull imposed by the *rasputitsa* in March and April, the longest the soldiers of both sides on the Eastern Front were to enjoy throughout the war, the staffs of both the German and the Red armies busied themselves with detailed planning for the great battle which summer must bring, while their overlords, in an uncanny convergence of mutual doubt, sought to modify their proposals, even to temporise with the inevitability of action. Stalin seemed unable to follow the logic of his generals' strategic analysis, believing that the whole Soviet front was threatened but particularly the sector opposite Moscow, and argued for using available strength in a 'spoiling' attack which would at least ensure that the Germans did not win a third summer victory in 1943. In a meeting with his senior commanders on 12 April he agreed that the construction of deep defences in the Kursk salient should be given priority, but also insisted that defences be constructed on all the main axes of advance open to the Germans. Stalin's outlook diverged from the opinion of such now highly experienced generals as Vatutin and Zhukov, who had concluded that Kursk was certainly the sector on which the *Ostheer* would attack, that the correct Soviet response was to fortify that front as strongly as possible to absorb the Panzer offensive, but that reserves accumulated by the Stavka should not be committed exclusively to Kursk but be apportioned to provide a *masse de manoeuvre* with which the Red Army might subsequently unleash a counterstroke on its own account. As Zhukov put it to Stalin on 8 April, 'An offensive on the part of our troops in the near future aimed at forestalling the enemy I consider to be pointless. It would be better if we grind down the enemy in our defences, break up his tank forces and then, introducing fresh reserves, go over to a general offensive to pulverise once and for all his main concentrations.'

Hitler, though committed in principle to the concept of an attack on the Kursk salient, was changeable about date and still oddly indecisive about the form of the attack. Although on 15 April he signed the order committing Army Groups South and Centre to an attack on the Kursk bulge on 3 May, he almost immediately had second thoughts and proposed to Zeitzler that the attack be launched against the nose of the salient. The suggestion was in defiance of all military orthodoxy – which holds that troops in a salient must always be cut off rather than attacked frontally – and Zeitzler was able to talk him out of it on 21 April. Then Model, who was to command one of the two armies consigned to the convergent attacks, persuaded him that the observed strength of the Russian defences would require more time for penetration than the plan allowed unless he got extra tanks. Hitler accordingly approved a postponement of some days while Guderian, his new Inspector of Panzer Troops, found the extra tanks. With the involvement of Guderian (by title a mere administrator) in operational planning, delays began to lengthen. Guderian was well informed about both the quantity and the quality of Soviet tank production, and it was his purpose to raise Germany's production to match it. On 2 May he outlined to Hitler a schedule of tank deliveries which made postponement look advisable. He promised not only more tanks – over 1000 a month on a rising scale, ten times Germany's annual output in 1939 – but better tanks, including the new Panther Mark V and the 'family' of 88-mm gun-carriers, Hornets, Tigers and Ferdinands, which were believed to be invincible on the battlefield; but he – not Speer – warned that the Panther, on which Hitler counted heavily for the success of Kursk, had not yet shed development 'bugs'. On 4 May, after yet another conference with his leading generals at Munich, Hitler accordingly postponed the Kursk attack, now codenamed Citadel, until mid-June.

Soviet industry, however, was not only continuing to turn out tanks at twice the German rate but in addition to the outstanding T-34 was now producing heavier models, including the KV-85, with an 85-mm gun, the first mark of the super-heavy Joseph Stalin, which would eventually mount a 122-mm gun, and various equivalents of the turretless assault guns which the Germans favoured. Russian production of anti-tank weapons was even more impressive. Over 200 reserve anti-tank regiments, equipped with powerful 76-mm guns, had been formed, while 21,000 lighter anti-tank guns had been issued to infantry units. 'By the summer of 1943', Professor John Erickson judges, 'the Soviet infantryman [was] equipped as no other for anti-tank fighting.' As well as armoured and anti-armoured resources the Red Army now had enormous quantities of artillery. 'Artillery is the god of war,' Stalin had said. It had always been the leading arm of Russian armies, and by the summer of 1943 the Red Army's artillery was the strongest in the world, in both quality and quantity of equipment. During 1942 whole divisions of artillery had been formed – an entirely novel military concept – and equipped with the new 152-mm and 203-mm guns. They included four divisions of Katyusha rocket-launchers; with this revolutionary weapon each division could fire 3840 projectiles weighing 230 tons in a single salvo. The Katyusha, which the Germans were hastily to copy, became one of the most feared weapons of the eastern battlefront, dazing and disorientating infantrymen who were not directly disabled by its tremendous blast effect.

This re-equipment of the Red Army, made possible by the regeneration of production in the factories transported eastward behind the Urals during the terrible Barbarossa months, spelt great danger to the *Ostheer*. More ominously, in accordance with the appreciation made by the Stavka on 12 April, huge quantities of material were poured into the Kursk salient during April and May, including 10,000 guns, anti-tank guns and rocket launchers. The civilian population of the salient – about 60 by 120 miles in area – was mobilised to dig entrenchments and anti-tank ditches, while army engineers laid mines in a density of over 3000 to each kilometre of front. The troops defending it, the Centre Front (Rokossovsky) and the Voronezh Front (Vatutin), laid out their own defensive positions, each consisting of a forward line three miles deep and two rearward positions. Eventually, with 300,000 civilians labouring in the rear, the Kursk salient was to contain eight defensive lines, echeloned to a depth of 100 miles. Nothing like it had ever been seen on a battlefield, not even on the Western Front at the height of trench warfare.

Hitler's prevarication over choosing a date for Operation Citadel reflected his doubts about the feasibility of the operation – and those of commanders committed to carrying it out, like Model of the Ninth Army. Model had originally asked for the plan to allow two days for his armour to penetrate the northern face of the salient. On 27 April, however, he arrived at Berchtesgaden, where Hitler was holidaying from the forest dankness of Rastenburg, with air photographs of the Russian defences at Kursk and a request for more tanks and more time. 'When Model told me,' Hitler recalled a year later, 'that he would need three days – that is when I got cold feet.' Cold feet or not, Hitler took no decisive action to cancel Citadel. His self-confidence remained weakened, while Zeitzler's was strong. The fighting infantry subaltern of the First World War was bent on doing 'something' during 1943, and for him that meant fighting a battle on the Eastern Front, his sole area of responsibility. Hitler also agreed that something had to be done, if the Red Army was not to grow unchecked in strength for a major offensive in 1944. However, besides the persisting depression caused by Stalingrad, he had other things on his mind: not only the worsening situation in Tunisia, ended by the surrender of the German-Italian army in May, but the increasingly precarious position of Mussolini, the uncertainty over where the Allies would strike next in the Mediterranean, and the growing civil defence crisis in Germany, where the British Bomber Command and the US Army Air Force were making heavier and deeper strikes each week. Three times during June he postponed Citadel again: on 6 June, when Guderian demanded more time to accumulate tank reserves, on 18 June, and again on 25 June, when Model raised more objections. Finally on 29 June he announced that he would return to Rastenburg and that Citadel would begin on 5 July. As he explained to his staff when he arrived on 1 July, 'The Russians are biding their time. They are using their time replenishing for the winter. We must not allow that or there will be fresh crises. . . . So we have got to *disrupt* them.'

The demand for 'disruption' was a far cry from the trumpet call to *Blitzkrieg* uttered in the summers of 1941 and 1942. It revealed how much Hitler had narrowed his horizons during the two years of the Russian war, how strong the

Red Army remained despite the devastation he had inflicted on it and how weakened the *Ostheer* was by the relentless programme of offensives and 'standfasts' to which he had committed it in the previous two years. The Red Army numbered 6.5 million at the beginning of July 1943, an actual increase since the outbreak of the war, despite the loss of over 3 million men as prisoners alone; the *Ostheer*, by contrast, fielded 3,100,000, a net decrease of 200,000 since 22 June 1941. The number of its divisions was static at about 180, but all establishments, of both men and equipment (except in the favoured SS divisions), were below strength. In the Red Army divisional establishments were also low, about 5000 men each, but the number of its divisions equalled the German, was rising, and was complemented by large numbers of 'non-divisional' units, including the specialised artillery formations. Moreover, while the German army was dependent exclusively on the output of home industry to supply its needs, the Russians were now the beneficiaries of a growing tide of Lend-Lease aid, including vast numbers of vital motor supply vehicles; no less than 183,000 modern American trucks had arrived by mid-1943 alone. Meanwhile war was destroying the *Ostheer*'s means of transport, the horse; by the spring of 1942 it had lost a quarter of a million horses, half those with which it had entered Russia, and losses had continued at an equivalent rate ever since.

The issue of manoeuvre, however, was not central to Citadel, where battering power alone was to count. German battering power was considerable. It was distributed between Model's Ninth Army, which was to attack the northern face of the Kursk salient, and Hoth's Fourth Panzer Army, which was to attack the southern. Together they disposed of some 2700 tanks supported by 1800 aircraft, the largest concentration of force against such a confined area yet seen on the Eastern Front. Model controlled eight Panzer and Panzergrenadier divisions, with seven infantry divisions in support, Hoth eleven Panzer, one Panzergrenadier and seven infantry divisions. The plan was straightforward. Model and Hoth were to cut into the 'neck' of the Kursk salient, between Orel and Kharkov, join hands, and then envelop and destroy Vatutin's and Rokossovsky's sixty divisions.

Into the furnace

The attack began at 4.30 am on 5 July, a date of which Stalin's 'Lucy ring' in Switzerland had apparently given him warning. Erickson has described the battle:

> Within twelve hours both sides were furiously stoking the great glowing furnace of the battle for Kursk. The armour continued to mass and move on a scale unlike anything seen anywhere else in the war. Both commands watched this fiery escalation with grim, numbed satisfaction: German officers had never seen so many Soviet aircraft, while the Soviet commanders . . . had never before seen such formidable massing of German tanks, all

blotched in their green and yellow camouflage. These were tank armadas on the move, coming on in great squadrons of 100 and 200 machines or more, a score of Tigers and Ferdinand assault guns in the first echelon, groups of 50-60 medium tanks in the second and then the infantry screened by the armour. Now that the Soviet tank armies were moving up into the main defensive fields, almost 4000 Soviet tanks and nearly 3000 German tanks and assault guns were being steadily drawn into this gigantic battle, which roared on hour after hour, leaving ever greater heaps of the dead and the dying, clumps of blazing or disabled armour, shattered personnel carriers and lorries, and thickening columns of smoke coiling over the steppe.

Rokossovsky counter-attacked Model on 6 July, trying to recapture the ground lost on the first day, but his troops were rolled back by the advancing German divisions. On 7 July the 18th, 19th, 2nd and 20th Panzer Divisions approached the high ground at Olkhovatka, thirty miles from the start-line, from which they would be able to look down on Kursk from the north and dominate Soviet lines of communication within the salient. The Soviet defenders were wiped out, but reserves arrived just in time to deny the Germans their prize. Meanwhile on the southern sector Hoth, who had three SS Panzer divisions under command, *Leibstandarte, Das Reich* and *Totenkopf,* as well as the 3rd and 11th Panzer Divisions and the powerful *Grossdeutschland* Panzergrenadier Division, was also making dogged progress. Vatutin contemplated launching a counter-attack on 6 July but, in view of the strength the Germans deployed, decided to remain on the defensive. By the evening of 7 July Hoth's Panzer 'fist' had smashed through the Soviet defensive crust to within twelve miles of Oboyan, which defended Kursk from the south. The junction of the northern and southern Panzer thrusts, on which the logic of Citadel depended, now seemed near to realisation.

The Russian defences, however, were proving extremely costly to penetrate. The whole front was crisscrossed by earthworks, while the Soviet anti-tank batteries were organised as single units, which discharged concentrated salvoes of shot at single tanks in the German spearheads. During 10 July Hoth was obliged to bring up his armoured reserve, the 10th and SS Viking Divisions, to sustain progress on the southern sector, but the pace of advance began to slow none the less. Moreover, Zhukov and Vasilevsky, who assumed direct control of the battle from Stalin and the Stavka on 11 July, were now about to unleash the Soviet reserves in a general counter-attack. During 11 July they committed the Bryansk Front (Popov), on Rokossovsky's right, to a drive into Model's flank. More importantly, on 12 July they brought forward the tank reserves held under Konev's Steppe Front to engage Hoth's Fourth Panzer Army south of Kursk. This decision was to precipitate perhaps the greatest tank battle of the Second World War. Erickson writes: 'In the area of Prokhorovka two great bodies of armour, Soviet and German, rushed into a huge swirling tank battle

with well over a thousand tanks in action. The two groups of German armour
. . . mustered some 600 and 300 tanks respectively; Rotmistrov's Fifth Guards
Army [from Konev's reserve] just under 900 tanks – approximate parity, except
that the Germans were fielding about 100 Tigers.' The battle blazed all day

> at point-blank range as the Soviet T-34s and a few KVs raced into the
> German formation, whose Tigers stood immobile to deliver their fire; once
> at close range with scores of machines churning about in individual
> engagements, front and side armour was more easily penetrated, when the
> tank ammunition would explode, hurling turrets yards away from the
> shattered hulls or sending up great spurts of fire. . . . With the coming of the
> deep night, when thunderclouds piled up over the battlefield, the gunfire
> slackened and the tanks slewed to a halt. Silence fell on the tanks, the guns
> and the dead, over which the lightning flickered and the rain began to rustle.
> The *Prokhorovskoe poboische*, the 'slaughter at Prokhorovka', was
> momentarily done, with more than 300 German tanks (among them 70
> Tigers) . . . lying wrecked on the steppe . . . more than half the Soviet Fifth
> Guards Tank Army lay shattered in the same area. Both sides had taken and
> delivered fearful punishment. The German attack from the south and west,
> however, had been held. At Oboyan the attack had been halted.

It was not only on Hoth's southern sector that Operation Citadel failed to
reach its objectives. 'On the broad slopes of the Sredne-Russki heights on
Rokossovsky's Central Front the attack on Kursk from the north [by Model's
Ninth Army] had been halted also, and Rokossovsky had considerable reserves
in hand.'

No one was readier to admit defeat than Hitler. 'That's the last time I will
heed the advice of my General Staff,' he told his adjutants after a meeting with
Manstein and Kluge on 13 July to decide the future of the operation, and he
ordered Citadel to be closed down. Manstein was sure that he could still cut off
the salient, if only he were given the last armoured reserve on the Eastern Front.
Hitler would have none of it. His generals had persuaded him that the defences
of the Kursk salient, despite their unparalleled depth and strength, could be
penetrated by armoured assault, and they had been proved wrong. Citadel
flickered on until 15 July but the decision had gone to the Russians, if at terrible
cost; over half their tank fleet used in the battle was gone. The cost to the
Germans, however, had also been very high: the 3rd, 17th and 19th Panzer
Divisions, for example, now had only a hundred tanks between them, instead of
the 450 with which they began. Moreover, these were strategic losses. German
tank output, for all Guderian's – and Speer's – efforts, did not approach the
thousand per month scheduled for 1943; it averaged only 330. More tanks than
that had been lost on several days during Citadel, 160 out of Fourth Army's
Panzers having simply broken down on the battlefield. As a result, the central

armoured reserve on which the *Ostheer* had always thitherto been able to call in a crisis was now dissipated and could not be rebuilt out of current production, which was committed to the replacement of normal losses. The Red Army, thanks to the burgeoning output of heavy industry beyond the Urals, was producing tanks at a rate which would approach 2500 a month in 1944, far greater than the rate at which tanks were lost, and so sufficient to increase its net complement of armoured formations. The main significance of Kursk, therefore, was that it deprived Germany of the means to seize the initiative in the future and so, by default, transferred it to the Soviet Union.

The Russian exploitation of the Kursk victory was at first clumsy and tentative. A Russian attack towards Orel, north of the Kursk salient, involved Soviet armour in a heavy battle with four Panzer divisions which tried to block its advance, although the Germans suffered losses. A simultaneous drive on Belgorod, south of the salient, organised by a recently arrived commander, Tolbukhin, was counter-attacked and the troops committed forced to retire on 1 August. However, these Russian attacks, by drawing off Kluge's and Manstein's remaining reserves, had exposed Kharkov to a renewed offensive which Stalin had approved on 22 July. It was launched on 3 August, with devastating effect. A single German infantry division, the 167th, was first subjected to bombardment by the massed artillery of the Soviet Sixth Guards Army belonging to Vatutin's Voronezh Front. After several hours, when its sector had been pulverised, a Russian tank column broke through. On 5 August it took Belgorod and by 8 August had opened up a gap on the flank of the Fourth Panzer Army which led directly to the crossings of the Dnieper a hundred miles away.

Manstein now informed Hitler that he must either receive a reinforcement of twenty divisions from the west or yield the Donetz basin, with all the mineral and industrial resources which were so valuable to the German and the Russian war effort. Hitler responded to this ultimatum not by conceding one or the other condition but by proposing a third option. Far from being able to offer reinforcements, he was actually withdrawing divisions, including the elite *Leibstandarte*, from Russia to Italy to protect his position there which was increasingly under threat. However in view of the deepening crisis in the east, he now conceded the desirability (which hitherto he had consistently rejected) of constructing an 'East Wall' to match the West Wall along the Rhine, behind which the *Ostheer* could defend the territory captured in 1941-2. It was to run from the shore of the Sea of Azov northward to Zaporozhe, Army Group South's headquarters, and then along the lines of the rivers Dnieper and Desna via Kiev and Chernigov, north to Pskov and Lake Peipus until it reached the Baltic at Narva.

He ordered work on this line, which was also to be a 'stop' position behind which the *Ostheer* was not to retreat, to begin at once. In fact both the manpower and the resources to construct it were lacking, while the Red Army would not concede the time necessary to undertake the work. Mounting simultaneous drives all along the southern sector of the front, the Red Army took Kharkov, the most fought-over city in the Soviet Union, on 23 August (it was to remain in Russian hands thereafter) and crossed the Donetz and its short tributary the Mius at the same time. These drives threatened to envelop Kleist's Army Group A, still holding its bridgehead beyond the Crimea, and compromised the position of the Sixth Army, the southernmost formation of Manstein's Army Group South, above the Crimea itself. On 31 August Hitler sanctioned further withdrawals in the south. But Army Group Centre's defences had now also been penetrated in three places, and the whole lower sector of the *Ostheer*'s front was crumbling under the weight of the Red Army's might. By 8 September the Russian vanguard was within thirty miles of the Dnieper and by 14 September was threatening Kiev. Kluge's Army Group Centre was unable to sustain its defence of the Desna, designated only a month earlier as part of the 'East Wall', and on the same day Sokolovsky's West Front began a drive against Smolensk in Army Group Centre's sector, focus of the great encirclement battle of 1941 in the heyday of the *Ostheer*'s Russian triumphs. Next day Hitler gave permission for a retreat to the line of the Dnieper, Sozh and Pronya rivers, roughly that reached in the great *Blitzkrieg* of July 1941; but the instruction came too late to permit an ordered withdrawal. It developed into a race for the river positions which many German formations lost, so that by 30 September the Red Army had five bridgeheads over the Dnieper – some seized by parachute assault – including a large lodgement to the immediate south of the Pripet Marshes.

For the *Ostheer* this was a disastrous outcome of the summer's fighting, since the Dnieper, with its high western scarp slope, was the strongest defensive position in southern Russia. During five weeks of continuous combat it had been forced back 150 miles along a 650-mile front and, although Hitler had decreed that it should conduct a 'scorched earth' retreat, in which factories, mines, power stations, collective farms and railways were destroyed, his demolition teams had not been able to obliterate the road network along which the Red Army made its way forward. Moreover, the fortification he had decreed had made no progress at all. The 'East Wall' remained a line on the map, nowhere transformed into earthworks, minefields or obstacle belts.

The growing strength of the Red Army

For the Red Army, by contrast, the summer fighting had been a triumph. It had regained all the objectives laid down for it by Stalin and the Stavka in the aftermath of the Kursk victory and though its human losses and material expenditures continued to run at a high rate – it had expended the astonishing total of 42 million rounds of artillery ammunition in July and August – its strength and therefore its offensive capabilities continued to grow. By October its strength stood at 126 rifle corps (of two to three divisions each), 72 independent rifle divisions, five tank armies (of three to five divisions), 24 tank corps (of two to three divisions), 13 mechanised corps (of two to three divisions), 80 tank brigades, 106 independent tank regiments, and a vast array of artillery formations – 6 artillery corps, 26 artillery divisions, 43 regiments of self-propelled guns, 20 artillery brigades and 7 divisions of Katyusha rocket-launchers. To mark the advances the Red Army had made, moreover, its fronts were now renamed. The Voronezh, Steppe, South-West and South Fronts became the First, Second, Third and Fourth Ukrainian Fronts in the first week of October, as they paused to regroup for the next stage of their offensive. Those to their north would shortly be retitled the First and Second White Russian and the First and Second Baltic Fronts. The Red Army was on the march.

Winter was its favoured time for attack. It was a season to which the Russian soldier was more accustomed and for which he was better equipped than his German counterpart. The German infantryman's feet froze in his 'diceboxes', as the army boots were called. The Red Army man, shod in felt boots, of which 13 million pairs were manufactured to Soviet specifications in the United States during the war and shipped back under Lend-Lease arrangements, resisted frostbite; he also knew the tricks, learned painfully by the Wehrmacht, of keeping motor vehicles running at sub-zero temperatures – mixing petrol with lubricating oil was one of them – and of caring for draught animals when frost formed icicles round their nostrils. Not until this third winter, when Hitler was at last prepared to admit that victory was not imminent, did the *Ostheer* receive adequate supplies of cold-weather clothing (in the first winter men had stuffed their uniforms with torn-up newspaper); Soviet soldiers received sheepskins and furs as normal issue.

As the first frosts of winter descended, the newly named Ukrainian fronts began their attacks across the lower Dnieper. The Red Army's recent accretion of numbers and material did not suffice for an offensive along the whole front, so for the next eighteen months it was to proceed by a sequence of advances, first on the right or southern sector, then on the left or northern sector of the front; this autumn manoeuvre was to be the first of its left-hand strokes. The target presented itself. By far the most vulnerable formation in the *Ostheer* was the Seventeenth Army, which occupied the Crimea and its approaches. Hitler attached disproportionate importance to the possession of the Crimea, both because he had fought so hard to acquire it in the summer of 1942 and because he was obsessed by the belief that it provided the best point for aerial attack on the Ploesti oilfields. When the Fourth Ukrainian Front (Tolbukhin) opened a major attack towards it on 27 October, his first thought was to request reinforcements from the Romanians, who he believed would share his perception of the developing danger; when the Romanian leader General Ion Antonescu refused to raise his stake in the Eastern Front, Hitler simply decreed that the Seventeenth Army must hold on and fight it out. Under heavy Soviet pressure its nearest neighbour, the Sixth Army, was quickly driven back beyond the land neck which links the Crimea to the mainland (the Perekop isthmus), while landings were made from the Asiatic shore on the Kerch peninsula. By 30 November not only were 210,000 German soldiers isolated in the Crimea; they were also threatened with battle on the territory they were defending.

Meanwhile the other three Ukrainian fronts had gone over to the offensive along the whole length of the lower Dnieper, with results which threatened the flanks of Manstein's Army Group South. The Third and Second Ukrainian Fronts first seized a large bridgehead near Krivoy Rog on Manstein's southern flank; then, on 3 November, the First Ukrainian Front broke across the Dnieper below the Pripet to recapture Kiev, in the most spectacular reversal of fortunes on the Eastern Front since the encirclement of Stalingrad.

During November the White Russian and Baltic Fronts also moved into action north of the Pripet, advancing from Bryansk to recapture Smolensk – a place of agony for the Red Army in 1941 – and threaten Vitebsk. They were now mobile on Napoleon's route to Moscow of 1812, but in the opposite direction, and giving Hitler cause to fear for the safety of the Baltic states and the approaches to the 1939 frontier of eastern Poland.

Unseasonably mild weather in December, which left unfrozen the network of waterways and small lakes above the Pripet, temporarily spared the *Ostheer* from the difficulty of defending the Smolensk-Minsk route westward across the upper Dnieper. However, Hitler had announced in Führer Directive No. 51 (3 November) that he imminently expected an Anglo-American invasion in the west and that he could 'no longer take responsibility for further weakening in the west, in favour of other theatres of war'. Indeed, he had 'therefore decided to reinforce its defences'; the decision meant that the *Ostheer* could no longer count on reinforcements from the quieter, so-called OKW sectors – France, Italy and Scandinavia – but must fight its battles with the strength it had available and such replacements as the Home Army could find.

'The vast extent of territory in the east', Hitler conceded in Führer Directive No. 51, 'makes it possible for us to lose ground, even on a large scale, without a fatal blow being struck to the nervous system of Germany.' This admission implied that he might be ready to accept the submissions of his eastern marshals, Manstein foremost among them, that the most profitable way of fighting the Red Army was to employ a strategy of withdrawing from territory before mounting a further attack. The implication was not to be borne out in practice. During the winter of 1943-4 the Red Army came on in even greater strength than before; but Hitler's reluctance to concede territory proved as fixed as ever – nowhere more so than on the southern front. Hitler not only clung to the hope of retaining the mines at Nikopol and Krivoy Rog but constantly emphasised the danger of allowing the Crimea to become a Soviet air base for attack on the Romanian oilfields and – a particular obsession – argued that its loss would encourage Turkey to enter the war on the Allied side.

Hitler orders retreat

Manstein, who had withdrawn his command post to Hitler's old summer headquarters at Vinnitsa in the Ukraine, travelled to Rastenburg twice during January to argue the case for withdrawal, but on both occasions it was refused. His Army Group South, moreover, fought stoutly to hold its front together against relentless attacks by the First and Second Ukrainian Fronts, now under Zhukov's direct command, and at first gave ground less slowly than Kleist's Army Group A. Assaulted by the Third and Fourth Ukrainian Fronts on 10 January, Army Group A was nearly encircled in its efforts to retain Nikopol and Krivoy Rog, and after Hitler issued formal permission for a retreat which could be no longer avoided it eventually escaped, at the expense of abandoning most of its artillery and transport. By mid-February, however, Army Group South was also in severe straits; two of its corps were encircled by Vatutin between the Dnieper and Vinnitsa west of Cherkassy and were rescued on 17 February only by concentrating all the available armour to help them break out. This operation by the First and Fourth Panzer Armies, which had previously halted a menacing thrust by Konev's Second Ukrainian Front towards Uman, caused Manstein's tanks to be wrongly placed to check a subsequent onset by the First Ukrainian Front south of Pripet. By 1 March the First Ukrainian Front had crossed the 1939 frontier of Poland, was menacing Lvov and was less than 100 miles from the Carpathians, southern Europe's only mountain barrier against an invasion from the east.

There was a crisis too on the northern front, where Army Group North, now commanded by General Georg von Küchler, was attacked on 15 January. In a whirlwind advance three Soviet fronts, the Leningrad, Volkhov and Second Baltic, had moved to the assault and by 19 January had breached Army Group North's defences in three places, widened the narrow corridor connecting Leningrad to the rest of Russia and liberated the city after a thousand days of siege; the blockade, which had starved a million Leningraders to death, was formally declared at an end on 26 January, when the city's entire artillery fired a twenty-four-gun salute. Behind Leningrad, however, ran the only length of the projected 'East Wall' which had been brought to a state of completion. During the early stages of the Leningrad offensive Hitler withheld permission for a withdrawal into it, demanding that Küchler, whom he reproached with having the strongest army in the east, should fortify an intermediate position on the Luga river. As it became clear that time and resources lacked, however, on 13 February he was obliged to sanction a retreat to the 'Panther' line, as the East Wall along the line from Narva to Lake Peipus and Lake Pskov was denominated. The retreat, like the prospect of abandoning the Crimea, caused him acute political misgivings, since he believed – with reason – that it would encourage Finland to open secret negotiations with the Russians for a separate peace.

Hitler's current difficulties remained military, however, not political.

Zeitzler, his chief of staff, fed him with assurances in late February that 18 million Russians of military age had now been eliminated and that Stalin disposed of only 2 million in his manpower reserve; in mid-October Colonel Reinhard Gehlen, head of the Foreign Armies East section, had warned, by contrast, that the Red Army would in future 'surpass Germany in terms of manpower, equipment and the field of propaganda'. Gehlen was right, Zeitzler wrong. The Red Army had now assembled exactly the sort of central armoured reserve which Hitler had allowed OKH to cast away in the cauldron of Kursk and was able to move it about the front as opportunity for breakthrough offered. By mid-February the Stavka had concentrated five of its tank armies opposite Army Group South; the sixth arrived at the end of the month. On 18 February Stalin issued orders for them to attack at the beginning of March: the First Ukrainian Front was to open the offensive on 4 March; the Second and Third were to join it on 5 and 6 March. Between them they outnumbered the Germans opposite by two to one in infantry and more than two to one in armour.

There was a last-minute impediment. Vatutin, commanding the First Ukrainian Front, fell into an ambush mounted by Ukrainian separatist partisans – combatants in a shadow war between Russians, Germans and Poles for local dominance over the Pripet borderlands which the onset of the Red Army was now making irrelevant – and was mortally wounded on 29 February. He was a grave loss to the Soviet high command; but his place was immediately taken by Zhukov, who, as at Stalingrad, exercised direct control of the coming offensive. It opened with one of the devastating bombardments which had become the signature of the Red Army's operational methods ever since the enormous wartime expansion of the artillery. The First Ukrainian Front quickly opened a gap between the flanks of the First and Fourth Panzer Armies and rolled forward; the Fourth Panzer Army was encircled at Kamenets near Lake Ilmen and forced to break out. The Second and Third Ukrainian Fronts made even faster progress against the weaker forces of Army Group A. By 15 April they had broken and crossed all three of the river lines on which the Germans might have hoped to stand, the Bug, the Dniester and the Prut, had retaken Odessa and had left the Seventeenth Army isolated in the Crimea to their rear. On 8 April moreover, Tolbukhin's Fourth Ukrainian Front suddenly enlarged its bridgehead on the Crimean Kerch peninsula, rolled forward, and surrounded the survivors of the Seventeenth Army in a small pocket around Sevastopol, exactly as the French and British had done to the Russians in 1854 – and Rundstedt to the Red Army in 1941. The Russian defenders of Sevastopol had held out heroically in the eight-month siege from November 1941 to July 1942. In early May 1944 Hitler conceded that he could not sustain the city's defence and it was evacuated in four nights between 4 and 8 May; over 30,000 German soldiers were nevertheless abandoned within the perimeter and made captive when the Russians liberated the city on 9 May.

The spring offensive on the southern front had been a triumph for the Red Army. Between March and mid-April it had advanced 165 miles, overrun three potential defensive positions, recovered, if in gravely devastated condition, some of the most productive territory in the Soviet Union, deprived Hitler of his cherished strategic outpost and inflicted irreparable damage on Army Groups A, South and Centre. The Seventeenth Army, the garrison of the Crimea, had disappeared altogether, with the loss of over 100,000 German and allied Romanian troops.

The débâcle in the south had already prompted Hitler to impose a cosmetic change on the *Ostheer* there. On 30 March he had summoned Manstein and Kleist to Rastenburg, told them that what the southern front needed was 'a new name, a new slogan and a commander expert in defensive strategy' and announced that they were relieved. Manstein was to be replaced by Model, the general who had stabilised the Leningrad front on the 'Panther Line', and Kleist by General Ferdinand Schörner, an even more fanatical devotee of the Nazi regime and a consummate promoter of his own reputation. Thus at a stroke the two great Panzer breakthrough experts were removed, to be replaced by men whose capacity was for the ruthless subjection of their soldiers to orders and slavish obedience to the Führer's authority. Zeitzler, also a disciplinarian, retained enough integrity to offer his own resignation at the news. Hitler refused it, with the warning, 'A general cannot resign'.

In a wholly empty gesture, Hitler also ordered a few days later that Army Groups South and A were to be renamed North and South Ukraine respectively, in token of his stated determination to recapture that territory, of which none now remained in his possession. However, not only did he lack the means to mount any offensive or even to find the reserves to shore up a defensive battle; in the spring of 1944 he was faced with the threat of seaborne invasion in France and the reality of an Allied breakthrough in Italy. In the east it was Stalin who retained the strategic initiative and who, on the heels of his Ukrainian triumph, was planning a further offensive which would clear the *Ostheer* from the soil of Russia once and for all.

During May Stalin commissioned two of his senior staff officers, S. M. Shtemenko, General Staff chief of operations, and Timoshenko, representing the Stavka, to examine each sector of the Soviet front – 2000 miles long between the Baltic and Black Seas – and report on possibilities. Their analysis was as follows: to persist in the advance towards the Carpathians, despite the political advantages of heightening the threat to Romania, Bulgaria, Hungary and ultimately Yugoslavia, was dangerous because it would lengthen the flank presented to Army Group Centre. To advance from Leningrad down the Baltic coast would menace East Prussia but not the German heartland, and would also risk a counterstroke by Army Group Centre. By elimination, therefore, the desirable strategy was to wipe out Army Group Centre itself, which still occupied the most important sector of historically Russian territory and also guarded the route to Warsaw, on the high road to Berlin. To do so would require organisational changes, notably the reinforcement and division of the White Russian and Baltic Fronts which opposed it. However, given the Red Army's new-found ability to concentrate strength rapidly on an altered axis, such a redeployment was feasible.

Operation Bagration

During April the 'western' theatre of operations was reorganised. The two White Russian and Baltic Fronts each became three. New generals were appointed, and senior commanders – Vasilevsky and Zhukov – were nominated to supervisory roles. Tank reinforcements and artillery reserves were concentrated on the White Russian fronts. Diversionary moves were co-ordinated at the extreme southern and northern ends of the whole theatre of operations – the latter not merely diversionary, since a subsidiary component of the summer offensive was intended to be a surprise attack designed to drive Finland out of the war. Finally, the First Ukrainian Front, south of the Pripet, commanded by the experienced Konev, was filled out with tank armies drawn from the other Ukrainian fronts to mount a long-range encircling manoeuvre round the Pripet Marshes into the flank of Model's Army Group North and eventually against that of Army Group Centre itself. The operation was to be the most ambitious the Red Army had ever staged. All it lacked was a name; on 20 May, when Stalin received the detailed plan from the General Staff, he announced that it would be called Bagration, after the general mortally wounded at Borodino on the route between White Russia and Moscow during Napoleon's invasion of 1812.

The attack on Finland by the Leningrad Front began on 9 June and, though mounted only with marginal force, soon consumed the tiny Finnish army's reserves. On 28 July the Finnish President asked leave to transfer his office to the national leader, Marshal Mannerheim, who at once began negotiations for a separate peace. His approaches were to be answered at the end of August.

Meanwhile Stalin had set a date for the opening of Bagration. At Tehran the previous November he had assured Churchill and Roosevelt that the operation would be timed to coincide with D-Day. The date he chose was 22 June, the third anniversary of Hitler's unprovoked and surprise attack on the Soviet Union. In the three preceding nights the Russian partisan groups based in Army Group Centre's rear area busied themselves laying demolition charges on the rail lines which supplied its logistic needs; over 40,000 charges were exploded on 19, 20 and 21 June. Both OKH and OKW nevertheless discounted the possibility that these attacks supplied evidence of an offensive in preparation. Since early May the Eastern Front had been quiet, and Gehlen's Foreign Armies East section insisted that such signs as there were indicated the preparation of a new offensive against Army Group North Ukraine; it was what Gehlen called 'the Balkans solution', precisely what the Stavka had rejected.

The Luftwaffe, through its own intelligence service and reconnaissance flights, took a contrary view; it had established that 4500 Soviet aircraft were concentrated against Army Group Centre. Warning of this, which came to Hitler on 17 June, alarmed him and prompted him to order IV Air Corps, his last intact air striking force in the east, to undertake spoiling attacks. However, the Russian concentration was too large for any attack to be effective, and it was now too late to move ground forces to stand behind Army Group Centre's front.

At 4 am on 22 June Bagration opened with a short artillery bombardment, behind which infantry reconnaissance units moved to the attack. Zhukov was anxious that the assault should not waste itself on empty positions. The real offensive developed next day, when heavier infantry waves supported by dense formations of aircraft pressed up to the main German defences, opening the way for the tanks in their rear. They were the vanguard of 166 divisions, supported by 2700 tanks and 1300 assault guns, against which Army Group Centre, on an 800-mile front, could oppose only thirty-seven divisions, weakly supported by armour.

The first German formation to suffer disaster was the Ninth Army, holding the southern sector of the army group's front. It was threatened with encirclement by the First and Second White Russian Fronts on the second day of the offensive, and when, on 26 June, Hitler gave it permission to fall back east of Minsk to the river Berezina (on which Napoleon's Grand Army had been savaged in 1812) it was too late. Its neighbour, the Fourth Army, was the next to suffer. Although also granted permission to withdraw on 26 June, it was trapped in a wider encirclement by the First and Third White Russian Fronts and devastated east of Minsk by 29 June. At the end of the first week of Bagration the three German armies which had stood in its immediate path had lost between them nearly 200,000 men and 900 tanks; the Ninth Army and the Third Panzer Army were shadows, each with only three or four operational divisions, and the Fourth Army was in full retreat. Hitler was now confronted with the prospect of a vast gap opening on the Eastern Front, and he was troubled simultaneously by other crises, great and small. The Allied landing in France had secured a foothold, Finland was crumbling, and his favourite general, Dietl, who had saved the situation in northern Norway in 1940, had been killed in an air crash while flying back to secure the German position in the Arctic on 22 June. On 28 June he replaced Field Marshal Ernst von Busch with the general who was emerging as his 'fireman', Model, at Army Group Centre, also leaving him Army Group North Ukraine as a source of reserves.

However, not even Model's firefighting abilities could quell the conflagration which was destroying his army group. By 2 July he concluded that there was no hope of getting the Fourth Army back intact to Minsk, since it was now pinned against the Berezina, which the Second White Russian Front had already crossed at Lepel. He therefore concentrated his efforts on trying to hold open escape routes on each side of the city, but the rapid advance of Soviet armour – Rotmistrov's Fifth Tank Army made thirty miles in a day down the Minsk-Moscow highway on 2 July – quickly quashed that plan. Minsk fell on 3 July while the Fourth Army stood encircled to the east; 40,000 of its 105,000 troops died trying to break out. After a last attempt to airdrop supplies on 5 July, the survivors began to surrender; the commander of XII Corps offered a formal surrender on 8 July; and by 11 July there was no further resistance in the pocket.

The encirclement of the Fourth Army effectively concluded the Battle of White Russia; the victory was formally celebrated on 17 July, when 57,000 captives were marched through silent crowds in the streets of Moscow to the prison camps. By then the spearheads of the Soviet attacking fronts were already far to the west of the ground where the captives had been made prisoner. On 4 July the Stavka had designated new objectives for each of them, on an arc which ran from Riga in Latvia to Lublin in southern Poland and touched the frontier of East Prussia. For the first time in the war the territory of the Reich itself lay under threat. During the second week of July they drove on; by 10 July Vilna, the capital of Lithuania, was in Soviet hands and the Third White Russian Front had set foot on German soil. On 13 July Konev's First Ukrainian Front, with 1000 tanks and 3000 artillery pieces, opened its offensive by attacking Lvov, the old Austro-Hungarian bastion of eastern Galicia. A difficult objective, and strongly defended, Lvov fell on 27 July; by then Rokossovsky's First White Russian Front had swung southward round the edge of the Pripet Marshes to reach out towards it. At the end of the first week in August the two fronts had reached the line of the river Vistula and its tributary, the San, south of Warsaw, while the First Baltic and the other White Russian Fronts had crossed the Niemen and the Bug, the Vistula's northern tributary, to menace Warsaw from the other flank.

A sense of treachery oppressed Hitler at every hand. The common German soldiers remained loyal to him, but his officers, whom he had never trusted as a class, had begun to throw up traitors. The 'Seydlitz' group, formed from those who had surrendered at Stalingrad and gone over to the 'anti-fascist cause', had been active with radio and leaflet propaganda against Army Group Centre on the eve of Bagration; sixteen of the generals captured in its course were prevailed upon by the Russians to issue an 'appeal' on 22 July which alleged that its 350,000 lost soldiers had been 'sacrificed in a game of chance'. Two days earlier Colonel Claus von Stauffenberg had tried to kill him in his own headquarters, as the opening move in a conspiracy designed to replace the National Socialist regime with a government acceptable to the West. In the course of that day Zeitzler disappeared from his post as chief of staff; whether he was dismissed or simply fled has never been clarified. Then on 1 August the Polish Home Army seized the centre of Warsaw.

The motivation for the Polish Home Army's rising in Warsaw was complex; so too was the explanation for the failure of the Red Army – by 1 August close at hand on the eastern bank of the Vistula – to come to the patriots' rescue. On 29 July Radio Kosciuszko in the Soviet Union, run by Polish communists under Soviet control, had broadcast an appeal for an uprising and promised that Russian help was close at hand. The Home Army itself was caught in a dilemma: according to Erickson, 'not to act meant being stigmatised a virtual collaborator with the Nazis or else being written off as the nonenity which Stalin insisted the Polish underground was.' Stalin had his own satellite Polish army, the People's Army, currently fighting in Operation Bagration. He also had an alternative Polish government, known in the West as the 'Lublin Committee', from the recently captured Polish border city where it had proclaimed itself. On 3 August Stalin met Stanislaw Mikolajczyk, the Polish Prime Minister of the government in exile, who had come to Moscow for talks about future relationships with the Soviet regime. Stalin at first professed to be uninformed about the rising, then warned that he could not tolerate disunity between the 'London' and the 'Lublin' Poles, consistently refused facilities for British-based planes to supply arms to the insurgents, regretted that German counter-attacks near Warsaw made it impossible for the First White Russian Front (nearest to the city) to continue its advance, but finally assured Mikolajczyk, on 9 August, that 'we shall do everything possible to help'.

By then it was too late. Model had indeed scraped together enough armour to mount a holding attack on 29 July against Rokossovsky's First White Russian Front near Praga, the Warsaw suburb across the Vistula. Then Himmler, whom Hitler had entrusted with authority to put down the Warsaw rising, brought up troops, including the Dirlewanger brigade of German criminals and the Kaminski brigade of Russian turncoats, both recently formed by the SS for internal security operations judged to demand particular ruthlessness. Within twenty-four hours of the outbreak they opened a reign of terror against the city's inhabitants – combatant or not – which, through fighting, massacre or area bombardment, would bring death to 200,000 of them before the rising was over.

TWENTY-SIX

Resistance and Espionage

'Now set Europe ablaze' was Winston Churchill's instruction to Hugh Dalton on appointing him to direct the Special Operations Executive set up at Churchill's express wish on 22 July 1940, to foment and sponsor resistance to Hitler's rule of occupied Europe. No event in the history of occupied Europe's defiance of Hitler more exactly matched Churchill's expectation of what a 'setting ablaze' might mean than the Warsaw uprising of August 1944. It confronted Hitler with an acute internal military crisis; it threw down a challenge to oppressed peoples elsewhere within his empire to do likewise; it endorsed the message Churchill had proclaimed to the English-speaking nations throughout the war – that the defeated were ready to rise against tyranny when the moment offered; and it validated the 'parallel' war of subversion and sabotage sponsored by the British and American 'special' agencies inside occupied Europe during four years of conventional war against its periphery.

So it appeared on the surface. Historically, however, it must be recognised that for all the bravery and suffering of the Polish Home Army in seven weeks of combat against Hitler's security troops – 10,000 fighters killed, perhaps as many as 200,000 civilians killed also – the Warsaw rising was not a spontaneous reaction to the brutality of occupation. Nor, by any objective valuation, did it seriously undermine Hitler's ability to maintain order within Poland at large while continuing to sustain an effective defence against the Red Army, which had halted on the far side of the Vistula at the moment the rising broke out. On the contrary: the rising was precipitated by the Home Army's calculation that the Germans' defeat in the Battle of White Russia presented it with a not to be repeated chance of seizing Poland's capital city for the government in exile before the arrival of the Red Army led to the installation of Stalin's puppet Polish regime; but its calculation was invalidated by the failure of the Russians to maintain military pressure on the Germans, who, in turn, found the means to fight – and eventually defeat – the insurgents without drawing on their front-line strength.

Far from demonstrating what an earlier uprising in a series of similar insurrections might have contributed to the defeat of Hitler, therefore, Warsaw stood as an awful warning of how dangerous it was, even at that late stage of the war, for any of his subject peoples to take up arms against him on territory which remained under the Wehrmacht's control. If Warsaw were not sufficient proof, the point was reinforced by the experience of the Maquis in southern France in June and of the Slovaks in July. In France, on D-Day itself, the Maquis of the Grenoble region set up the standard of revolt on the plateau of Vercors, from which it began to raid German troops using the Rhône valley route; by July there were several thousand Maquisards at Vercors, most of them fugitives from the forced labour programme. Army Group G, which was being troubled by pinprick attacks all over its area of responsibility, then decided to make an example of this isolated and vulnerable resistance base, cordoned the plateau, landed SS troops by glider on the summit and between 18 and 23 July brutally killed everyone they found there. On a small scale it anticipated the action German security troops were simultaneously taking against the Slovak rebels in eastern Czechoslovakia. Elements of the satellite Slovak army rose against the occupation forces in the expectation of imminent Russian intervention, but were not rescued and were put to the sword. Hitler had to deal with no further uprising – except the carefully timed outbreak of insurrection in Paris during the week of its liberation – while his armies still stood on conquered territory.

What made the Vercors massacre all the more dispiriting for those committed to Churchill's policy of 'setting Europe ablaze' was that the resistance had been supported and supplied by the Special Operations Executive; one of SOE's liaison teams (codenamed 'Jedburghs') had been parachuted to the support of the Vercors Maquisards, who on 14 July had received a drop of 1000 loads of weapons and ammunition flown to the plateau by the USAAF. This supply had availed them not at all. Although led by a French regular officer, the resistance fighters wholly lacked the experience and training to engage professional troops – who were in any case indoctrinated to put down resistance with pitiless brutality.

Warsaw, Slovakia and Vercors, coming so late in the course of the Second World War, were key events in the history of Hitler's Europe against which all other resistance and subversion to his rule between 1939 and 1945 must be set in the balance. (The partisan wars behind German lines in Russia and Yugoslavia are exceptions and require separate consideration.) If the three uprisings typify in their outcome the unintended effect of the programme of subversion, sabotage and resistance which Churchill, later abetted by Roosevelt, and the European governments in exile so ardently supported after June 1940, the programme must be adjudged a costly and misguided failure. All failed at the price of very great suffering to the brave patriots involved but at trifling cost to the German forces that put them down; as a result, all the lesser and preliminary activities of the resistance forces which they crowned must be seen, by any objective reckoning, as irrelevant and pointless acts of bravado. If that is a fair verdict on European resistance and the Allied efforts to plan and support it, what is the explanation for its failure?

US Marines in the hellish landscape of Iwo Jima, a speck in the Pacific
defended to the death by a Japanese garrison of 23,000 men.

A Japanese aircraft falls blazing from the sky in the fight for the Marianas chain. The capture of the Marianas, in June–August 1944, provided bases from which Allied naval and air forces could cut the lines of communication to Japan's southern empire and launch a sustained bombing campaign against the Japanese home islands by B-29 Superfortresses of XXI Bomber Command. The bombing campaign began on 24 November 1944 when 111 B-29s took off from Saipan to attack the Musashi aero-engine plant on the outskirts of Tokyo.

The escort carrier *USS St Lô* takes a direct hit from a Japanese kamikaze aircraft at Leyte Gulf, 25 October 1944. She sank less than an hour later.

Amphibious landing craft churn towards Okinawa, 1 April 1945, while the 16-in guns of a battleship plaster the shoreline. In taking the island the Americans suffered nearly 50,000 casualties, over 12,000 of them dead,. In contrast the Japanese sustained 117,000 casualties, of whom 110,000 were killed.

Left: The flight deck of the carrier *USS Bunker Hill* after direct hits by two kamikaze aircraft within the space of one minute on 11 May 1945 during the fighting for Okinawa. In the two months that the US fleet stood off Okinawa the Japanese flew 1900 kamikaze missions, sinking 38 warships, mostly smaller types, and damaging dozens more. They also sacrificed the battleship *Yamato*, which was despatched on a suicide mission, with fuel for a one-way journey, only to be sent to the bottom off Okinawa by 300 US aircraft on 7 April.

Below: Transferring the wounded from the *USS Bunker Hill*.

Above: US Marines cautiously await the results of a satchel charge tossed into a Japanese strongpoint in the murderous terrain of Okinawa, which saw some of the most savage fighting of the Pacific War.

Below: B-29s over Yokohama, 29 May 1945, a daylight raid on which their P-51 escorts shot down 26 Japanese fighters.

Facing page: Ground zero – the blasted landscape of Hiroshima after the atomic bomb raid of 6 August 1945. A new military era had dawned.

MacArthur watches the Japanese surrender, Tokyo Bay, 2 September 1945.

At the root of Churchill's misapprehension of what resistance could achieve against ideological tyranny, a misapprehension shared by hundreds of intelligent and energetic men and women among his fellow countrymen, was a total misunderstanding of the role of public opinion in the politics of conquest. Britain's history is suffused both by conquest and by resistance to it. In Churchill's own lifetime the boundaries of the British Empire had been greatly extended by military force, in South, West and East Africa, in the Middle East, in Arabia and in south-east Asia. However, the tide of British imperialism had always been tempered by extraneous factors: the continuing influence of anti-imperialism, domestic and foreign, and the British empire-builders' own ethos of equity and trusteeship. Confronted by rebellion and atrocity in India during the Great Mutiny of 1857-8, the mid-Victorians reacted with a ruthlessness from which Hitler's security forces could have learned little. Their successors were raised in a more equitable philosophy of empire. 'Eventual self-rule' became the principle on which colonial governments were founded in Africa in the late-Victorian and post-Edwardian era; 'trusteeship' was the concept on which Britain administered the African and Arabian mandates granted by the League of Nations; 'self-rule as soon as possible' informed the regime which Britain imposed on the Afrikaner republics in the wake of the Boer War; and the same spirit transfused British rule in India in the years after the First World War.

At the heart of Britain's self-imposed moderation of its right to rule over its enormous twentieth-century empire lay deference to its own democratic beliefs and concern for the good opinion of other peoples, particularly Americans, who shared those beliefs. Churchill, though he had isolated himself from his own party in the 1930s by his opposition to the devolution of government in India, was emotionally, if not intellectually, as committed to the principle of self-determination as the most doctrinaire liberal. Moreover, through his experience in fighting the Afrikaners in the Boer War, he had learned how deep the urge to freedom could drive, and how difficult it was for an occupying power to persist in imposing alien rule on any people inspired by faith in their right to independence. Churchill's personal experience was reinforced by his wide reading in modern history, which abounded in examples of the success of popular resistance to foreign conquest and domination – for instance, resistance by the Spaniards and Prussians against Napoleon, and by the American colonists against George III.

Hitler's philosophy of empire

A wider mismatch between the philosophies of empire held by Churchill and Hitler could scarcely be imagined. Imperialistic though he was, Churchill believed in the dignity of man; Hitler held 'the dignity of man' to be a bourgeois vacuity. As recognised by those in the Anglo-Saxon world who had read *Mein Kampf* – they were still only a tiny handful in 1940 – he rejected with contempt the idea of self-rule for those who did not belong to the Germanic race. For purposes of expediency he was prepared to make common cause with the Japanese; out of loyalty he included Mussolini ('a descendant of the Caesars') and the Italians in the Germanic confraternity; he had an ideologically soft spot for the modern Greeks, whom he identified with the defenders of Thermopylae against the Asiatic hordes and esteemed as dogged warriors; the Scandinavians he recognised as racial cousins, a title he yearned for the British to accept, and which he also extended to those Dutchmen and Belgian Flemings who identified with his cause; he was prepared at a pinch to include the Finns and Balts among his approved minorities; and, as long as they fought on his side, he excepted the Hungarians, Romanians, Slovaks or Bulgarians from racial stigma. For the rest of the inhabitants of Europe who by the end of 1941 had fallen under his sway he reserved nothing but contempt. They belonged either to those groups, like the French, which were tainted by their subjugation to Roman rule (Hitler's political memory was long) or to the Slavonic 'riff-raff', Poles, Serbs, Czechs and above all Russians, whose history was one of subjection to superior empires.

In consequence Hitler was not at all affected by the moral reservations which so easily touched Anglo–Saxon attitudes to empire. He positively exulted in the ease with which he had extinguished autonomous governments in Poland, Czechoslovakia and Yugoslavia; and he measured the rectitude of the authority with which he had replaced these legitimate regimes purely in terms of expediency: if the successor administrations worked, with the minimum of vexation to his occupying forces, he was content to leave them in office undisturbed. Thus he devolved authority in Norway to the Quisling regime from February 1942 (Vidkun Quisling was the local Nordic authoritarian), conceded continuing rights of parliamentary government to the Danes, who conducted a democratic election as late as 1943, in which 97 per cent of the candidates returned were patriots, and left Pétain to embody the appearance if not the reality of sovereign French head of state even after he had extended German military occupation to the whole of France in November 1942.

The complexity of Hitler's occupation policies was reflected in the complex pattern of resistance to his occupation regimes in both western and eastern Europe. However, the pattern of resistance was determined not only by the nature of the regime Hitler chose to impose in any particular occupied territory. Three other factors operated: the first was the attitude of the left; the second was the degree of assistance which the British (and, after December 1941, the Americans) were able to bring to local resistance organisations; the third was geography.

Geography, being a constant, is best dealt with first. The degree of success of any movement of resistance to enemy occupation is directly determined by the difficulty of the terrain in which it operates – with this proviso: difficult terrain, mountain, forest, desert or swamp, is bereft of the resources necessary to support an irregular military force, and external supplies are therefore required. Most of German-occupied Europe, however, was either topographically unsuited to irregular operations or too distant from Allied bases of support for irregular forces operating there to be easily and regularly supplied. For example, Denmark, in which the spirit of resistance was strong (despite the existence of military and political groups sympathetic to Hitler's anti-Bolshevik propaganda), lends itself badly to partisan activity, being flat, treeless and densely inhabited; the same conditions characterise most of Holland, Belgium and northern France. In all those areas clandestine activity was readily monitored by the police – and throughout occupied Europe the domestic police forces accepted the authority and direction of the conqueror from the outset – and reprisal punishments were as readily inflicted. The ease and ruthlessness with which reprisals were carried out, either by the Germans or by their satellite security forces, such as the Vichy *Milice*, proved a sufficient deterrent for much of the war. Moreover, fear of reprisal – on a scale which ran from curfew through arrest, hostage-taking and transportation to exemplary execution – encouraged informing, which in turn directly heightened the efficiency of German control. Most resistance organisations, when they began to form, were obliged to devote a high proportion of their energy to combating informers, nowhere with complete success.

The only part of occupied western Europe in which the terrain favoured resistance activity was Norway, north of Oslo; but there the population was so sparse and the density of German occupation troops was so high that all guerrilla activity had to be organised outside the country. The infiltration of Norwegian resistance fighters from Scotland (who in February 1943 destroyed the heavy-water plant at Vermork, thus crippling the German atomic weapons programme), reinforced by the programme of British commando raids against German military outstations, had the highly desirable effect of persuading Hitler grossly to over-garrison Norway throughout the war; but the internal resistance itself was of negligible strategic significance.

Certain regions of eastern and south-eastern Europe were topographically favourable to partisan activity, notably Carpathian Poland, the Bohemian Forest in Czechoslovakia, much of Yugoslavia, the mountainous parts of the Greek mainland and its larger islands, and the Italian Alps and Apennines. The growth of resistance in Italy, however, had to await the fall of Mussolini in July 1943, while Czechoslovakia was too distant from bases of external support for resistance to take root. The Czech government in exile ran the most efficient of Allied-oriented intelligence organisations to operate inside Europe during the war, but SOE's only serious sponsorship of resistance activity inside the country, the assassination of Reinhard Heydrich, SS 'Deputy Protector of Bohemia-Moravia', in May 1942, provoked so terrible a reprisal (the extinction of the population of the village of Lidice) that the effort was not repeated; it reveals much about the efficiency of Hitler's occupation policies that the assassins were betrayed by one of their own number, who made himself known to the Gestapo as soon as he was parachuted into his homeland. In Greece, where SOE set up an extensive network of agents as early as the autumn of 1942, many of them Oxford- and Cambridge-educated classical scholars inspired by the philhellenes (Byron foremost among them) who had aided the Greeks in their struggle against the Turks in the 1820s, the Germans responded to partisan activity with such pitiless cruelty that the British officers soon found themselves obliged actually to dissuade activists from initiating attacks against the occupiers.

In Poland – again partitioned after 1939 so that its western province became German, its eastern Russian and only its centre, the 'General-Government', remained a separately administered entity – the 'Home Army', under the direction of the government in exile in London, abstained from provocative military action against the occupier until it unleashed the Warsaw rising of August 1944. Though the Poles ran an intelligence network second in efficiency only to that of the Czechs (one of its triumphs was to supply the British government with key parts of crashed German pilotless weapons which had made rogue flights), they decided from the outset that the national interest lay in preserving the strength of the Home Army against the moment when the collapse of Germany would allow it to strike for the recovery of independence. The military efforts of the Home Army were also restricted, however, by its difficulty in acquiring arms. Its lack of weapons was a factor in its non-intervention against the Germans during their destruction of the Warsaw ghetto in April 1943, when its heroic Jewish resistance groups were systematically overwhelmed in a street-by-street battle conducted by SS troops and militia under the command of SS General Jürgen Stroop. Until 1944 SOE lacked aircraft with sufficient range to reach central Poland; even after the acquisition of bases in Italy in 1943, flights were still lengthy and dangerous. The Soviet Union, which occasionally granted the Western air forces refuelling facilities for bombing raids against Germany, refused to do so for arms-dropping missions to Poland. It also refused to supply arms to the Home Army itself.

Russia's attitude was determined by its political differences with the government in exile and the Home Army, which persisted even after the signing of the agreement in August 1941 which released Polish prisoners held in Russia to join the British armies in the Middle East. Stalin had politically identified the Home Army as a potential opponent of the Polish Communist Party, through which he began to sponsor his own army in exile in the Soviet Union after June 1941. This was the only negative effect of Barbarossa on the development of resistance to German occupation inside Hitler's Europe. Almost everywhere else the efforts were positive. The European communist parties, through the persisting control of the Comintern, had been restrained from joining in resistance to occupation as long as the Molotov-Ribbentrop Pact remained in force. As soon as it was broken, all European communist parties were ordered to initiate subversive activity, with a marked increase in the efficiency of resistance groups of whatever political colour. This effect was due either to collaboration by the communists, whose habits of secrecy were far superior to those of recently formed clandestine groups, with the non-communist resistance, as notably in Holland, or to creative competition between left and right, as in France: there de Gaulle, alarmed by the prospect that 'Free France' might fall under communist leadership on home territory, succeeded in creating a pan-resistance 'Secret Army', commanded by a National Resistance Council under his authority. The marriage it imposed between communist and non-communist groups was one of convenience. The French Communist Party privately reserved the intention to operate to its own political advantage as soon as opportunity offered, and it did indeed institute local reigns of terror against its committed opponents during the interregnum which followed liberation in August 1944; but from June 1941 to July 1944 the marriage worked to unify and strengthen the resistance as a whole.

Objectively, however, it must be recognised that the principal achievement of resistance in western Europe during the years of Hitler's strength was psychological rather than material. The most visible symbol of resistance was the underground newspaper (120 separate imprints were circulating in Holland in 1941) and the most seditious activity the transmission of intelligence, of varied value, via clandestine networks to London. Some of these networks fell into enemy hands and were 'turned'; the North Pole network, for example, was 'run' by the Germans between March 1942 and April 1944. Such setbacks did little harm to the Allied war effort but resulted in numbers of brave men and women (SOE judged that women made better agents than men) being parachuted straight into the hands of the Gestapo. The publication of underground newspapers and the running of intelligence networks, whose subsidiary activities included the smuggling of crashed aircrew out of occupied territory, occasional acts of sabotage and sporadic assassinations, did a great deal to sustain national pride during the occupation years, but none of the activities shook the German system of control, which was both efficient and remarkably economic. Historians of the resistance are naturally reluctant to put figures to the size of the German security forces (civilian *Sicherheitsdienst*, military *Feldgendarmerie*) which were the resistance groups' enemies, but it is probable that their total strength in France did not exceed 6500 at any stage during the war; the German police garrison of Lyon, the second largest city of France, comprised about 100 secret policemen and 400 security troops in 1943. The divisions of the German army stationed in France (sixty in June 1944) took no part whatsoever in security duties, and, since they were almost exclusively stationed in coastal districts, they were not in a position to do so. Against the German security forces the resistance deployed at most 116,000 armed men, a figure established in July 1944 when the arrival of the Allied liberation armies raised their strength to its maximum. During the occupation proper, the number and size of armed groups were small and their activities consonantly limited; in the first nine months of 1942 the total number of assassinations of German security officers was 150, while major acts of sabotage throughout the war did not exceed five (interference with the railway network was extensive, but was largely confined to the months before and during the D-Day landings).

The popular idea of western Europe 'ablaze' under German occupation, first promulgated in John Steinbeck's inspirational novel *The Moon is Down* (1942) and fed by an army of authors since, must therefore be recognised as a romantic, if understandable, myth. Western Europe's urban and pastoral regions, where the population was so vulnerable to reprisals, were quite unsuited to the sort of sustained partisan activity which, when supplied and supported by external regular forces, is the only form of guerrilla warfare which constrains a conqueror to divert appreciable military effort from the battlefront. During the whole course of Hitler's war, he was confronted with such effective guerrilla resistance in only two areas of operations: in the rear of the Eastern Front, where Stalin, after an initial hesitation, supported, supplied and eventually reinforced partisan formations centred on the impenetrable Pripet Marshes; and in Yugoslavia.

The Soviet partisan formations were initially based on fragments of regular divisions isolated by the German advance through White Russia and the Ukraine in the summer of 1941, survivors who retained the will and some of the means to fight on after they had been cut off from their higher headquarters and sources of supply. For recruitment, however, they depended upon volunteers from the White Russian and Ukrainian populations, both suspect in Stalin's eyes as undependable minorities and as potential collaborators with the occupation authorities. From the outset he put the partisan formations under NKVD (secret police) control; the command structures, infiltrated through German lines to the partisan bands, were known as *orgtroika* (tripartite organisations) consisting of state, party and NKVD officers. As late as the summer of 1943, their members in the Ukraine did not exceed 17,000. In January 1944, when the partisans were returned to Red Army control, thirteen partisan brigades in the Ukraine numbered 35,000; on the eve of Operation Bagration in June 1944, when partisans carried out 40,000 railway demolitions, their numbers were 140,000. They had grown as a result of Soviet support, despite ferocious German repression. From the spring of 1944 onwards, specialist SS anti-partisan units, to which German formations 'resting' from operations at the front were regularly attached, carried out sweeps through 'band-infested' areas, burning and killing without pity; 'kills' of up to 2000, including women and children as well as men, were regularly reported for each operation. Post-war investigations by historians with access to German records suggest that such sweeps were extremely effective, that Soviet estimates of the achievements of partisans were wildly exaggerated, and that the losses inflicted by partisans, whether on the personnel or the material of the Wehrmacht, were a fraction of those claimed by the Soviet authorities. The Soviet estimate that 147,835 German soldiers were killed by partisans in the Orel region, west of the River Don, has been challenged by a Western scholar, J. A. Armstrong, who suggests a figure of 35,000 killed and wounded.

The Yugoslavian Partisans

It is to Yugoslavia that historians ultimately turn in arguing for the effectiveness of partisan warfare and in estimating the contribution of resistance forces to the defeat of the Wehrmacht in Europe in the Second World War. Yugoslavia is unquestionably a special case. Its mountainous terrain, intersected by deep valleys and bounded by a coastline that gave easy access to SOE's air and sea supply units, is ideally suited to irregular warfare. Its Serbian population was accustomed by resistance to the Turks and by the Austrian invasion of 1914-15 to fighting on its home territory. Hitler's aggression of April 1941 had outraged the national pride and by its suddenness left hundreds of military units in possession of weapons and ground which provided the basis for irregular operations. The first to raise the standard of revolt were Serbian monarchists commanded by a Serbian regular officer, Draza Mihailović. His Chetniks, so called from the Serbian word for opponents of the Turkish occupation, were at odds from the outset with the Croatian Ustashi who made common cause with the Italian occupation forces in Slovenia and Croatia; they also understandably opposed the Hungarian, Bulgarian and Albanian appropriations of Yugoslav border areas on the northern and eastern frontiers of the kingdom. Properly, however, their quarrel was with the Germans, who had imposed a puppet government on the territory of historic Serbia and against whom they initiated partisan warfare as early as May 1941.

The Special Operations Executive made contact with the Chetniks in September 1941 and began to supply them with weapons and money in the summer of 1942. However, SOE's original emissary to Mihailović, Captain D. T. Hudson, had also come across groups of anti-monarchist guerrillas who called themselves 'Partisans' and were led by an experienced Comintern agent, Josip Broz who used the *nom de guerre* Tito. Hudson early formed the impression that Tito was a more serious opponent of the Axis occupiers than Mihailovicü, whom he suspected of wishing to build the Chetniks into a Serbian 'Home Army' on the Polish model and preserve its strength against the day when external circumstances would allow him to liberate the country from within. His suspicion did Mihailović less than justice, for the Chetniks were conducting a guerrilla war against the Germans in 1942 and (as Ultra revealed) were regarded by them as troublesome enemies as late as 1943. It was undoubtedly the case, however, that Mihailović was an extreme Serb nationalist, that he refused to co-operate with Tito in creating a national resistance movement, that his Chetniks had begun to fight the Partisans for control of western Serbia in November 1941, and that he early entered into local truces with the Italians to acquire arms for the prosecution of this burgeoning civil war.

A principal motive of Mihailović's policy was to spare the Serb population from reprisal and atrocity at the hands of the occupiers – an estimable aim in view of the appalling consequences of the internal war which none the less ensued, costing as it did the lives of nearly 10 per cent (1,400,000) of the pre-war population. Tito made no such reservations. In the classic tradition of revolution, he committed the Partisans to waging war against the occupier to the bitter end. By late 1943 he had established himself in the eyes of SOE (whose Yugoslav section was dominated by officers with left-wing views) as the most effective of the Yugoslav guerrilla leaders. From the spring of 1944 onwards all British aid was sent to Tito's Partisans and withdrawn from Mihailovicü. Although some officers of the American Office of Strategic Services remained in contact with the Chetniks, their abandonment by the British had the effect of driving them into closer co-operation with the Germans, with whom Mihailović agreed to a local armistice in November 1943 as a means of continuing the civil war against Tito, thus confirming the Allied prejudice against them which Hudson had voiced at the outset.

Tito meanwhile had been building up his army and instituting increasingly ambitious attacks against the Germans in central and southern Yugoslavia. When these attacks began to threaten the Germans' exploitation of the country's mineral resources and their line of communication with Greece, Hitler was forced to commit sizeable forces and mount large-scale pacification operations against them. Until the collapse of Italy in September 1943 twenty Italian divisions were permanently stationed in Yugoslavia and Albania (where SOE also sponsored a minor guerrilla movement), together with six German divisions. After the dissolution of the Italian occupying force, the German was reinforced with an additional seven divisions, together with four from the Bulgarian army. A Partisan offensive at the Neretva river in Bosnia in February 1943, defeated at some cost to the Italians and Germans, prompted them to launch Operation *Schwarz* in the following May. It involved over 100,000 German and satellite troops and drove Tito out of Montenegro, where he had retreated. Similar offensives cleared western Bosnia in December, while in May 1944 Operation Knight's Move in southern Bosnia was so successful that Tito was obliged to seek rescue at British hands and fly to Bari in Italy – even though at the time of the September armistice he had acquired large quantities of Italian weapons which allowed him to raise the number of armed men he kept in the field to about 120,000.

The Royal Navy quickly returned Tito to Yugoslavia, though only as far as the island of Vis, where it had established a base to support Partisan operations. Meanwhile the British Balkan Air Force, set up at Bari in June, was flying vast quantities of (largely American) weapons to the Partisans in the interior of the country. In August Tito left Vis to visit Stalin, who until February 1944 had been tepid in his support for Tito's campaign; in Moscow Tito granted 'permission' for Soviet troops to enter the country and they began to cross the border from Romania on 6 September. Their arrival, and Hitler's decision to evacuate Greece in October, transformed the Partisans' position. Army Group F, outflanked in the Balkans by the Red Army and along the Adriatic coast by Allied Armies Italy, immediately beat a hasty retreat into central Yugoslavia. Belgrade, the capital, fell to a joint force of the Red Army and the Partisans on 20 October. Stalin, at his August meeting with Tito in Moscow, had given a guarantee that the Red Army would evacuate Yugoslavia as soon as its presence was no longer militarily necessary, and his promise was indeed kept after the German surrender of May 1945.

Mihailović ended the war a tragic figure. Tito's ascendancy had driven him deeper into complicity with the Germans; his belated efforts to reingratiate himself with the Allies totally failed, and after having hidden from Tito's troops in the mountains of central Serbia for over a year he was caught in March 1946, tried in Belgrade in June and executed by firing squad on 17 July. His plea of exculpation, 'I wanted much, I began much, but the gale of the world swept away me and my work', has entered into the memorabilia of the Second World War. 'Destiny', he said, had been 'merciless' to him, and hindsight, by which many of his judgements have been forgiven, accords weight to that view. His tragedy was to have been a nationalist leader in a state composed of minorities, whose differences Hitler cynically exploited in order to divide and rule.

Hindsight has also greatly diminished Tito's achievement. At the end of the war he was widely hailed as the only European resistance leader to have liberated his country by guerrilla effort. Many strategic commentators further credited him with having diverted such numbers of German and satellite troops from the eastern and Mediterranean battlefields as to have materially influenced the outcome of the war in those theatres. Realistically, it is now accepted that the liberation of Yugoslavia was the direct result of the arrival of Russian troops in the country in September 1944. What now seems most surprising about the Tito era is that Stalin should have so unwisely agreed to remove the Red Army from Yugoslav territory at the moment of victory – a misjudgement which robbed Soviet post-war control of eastern Europe of consistency from the outset. Strategically, estimates of Tito's diversion of force from Hitler's main centres of operation are now seen to be exaggerated. The principal army of occupation in Yugoslavia was always Italian. After the Italian collapse Hitler was indeed obliged to double the number of German divisions deployed in Yugoslavia from six to thirteen; but few were suitable for use against the Red Army or Allied Armies Italy. Only one, the 1st Mountain Division, brought from Russia in the spring of 1943, was first class; the rest, including the SS Prinz Eugen and Handschar Divisions and the 104th, 117th and 118th Divisions, were composed either of ethnic Germans from central Europe or of locally enlisted non-German minorities, including a high proportion of Balkan Muslims from Bosnia and Albania. They were quite unsuitable for war against Russian, British or American mechanised formations; their presence in Yugoslavia, even their existence, was in itself evidence that fighting there partook more closely of the character of civil rather than international war. In a sense, Hitler's cunning in setting Serb against Croat and monarchist against communist rebounded on itself; for, though his only real interest in the country lay in the exploitation of its resources and the free use of its lines of communication to southern Europe, he eventually became a party to its internal quarrels. In objective military terms, his involvement cost him little, but it would have simplified his politico-military arrangements if he had taken the trouble, after his whirlwind victory of April 1941, to establish a pan-Yugoslav satellite administration, charged with maintaining order within the country, rather than cynically bribing Yugoslavia's neighbours with portions of its territory to impose occupation policies which rapidly proved ineffectual.

The Special Operations Executive, though puffed by a powerful lobby of historians, some of whom were its former officers, largely fails in its claim to have contributed significantly to Hitler's defeat, since its achievement in Yugoslavia, its principal theatre of operations, was ambiguous. The same verdict holds true for the activities of the American Office of Strategic Services (OSS), set up in June 1942. Through an agreement allocating responsibilities between OSS and SOE signed on 26 June 1942, OSS took the major role in supporting the Italian partisans and the Johnny-come-lately resistance movements in Hungary, Romania and Bulgaria. Italian resistance activity discommoded the Germans very little, the Hungarian, Romanian and Bulgarian subversives scarcely at all. SOE's and OSS's parallel effort in psychological warfare (sponsored in Britain by SOE's parent organisation, the Ministry of Economic Warfare, through its Political Warfare Executive) afforded high excitement to the journalists and intellectuals who staffed it; its effect on opinion in the occupied countries was marginal and on German civilian morale negligible. 'Black propaganda', transmitted by radio stations purporting to operate within the boundaries of the Reich, understandably convinced no German who could daily witness the absolute control the Gestapo exercised over German society. The only non-military manifestation of internal resistance to Nazi rule, the Catholic Bavarian White Rose group, was pitilessly liquidated almost as soon as it appeared in February 1943. Allied efforts at economic warfare were equally unavailing; the principal success, the purchase of future production of Swedish ball-bearings, was negotiated so late in the war (mid-1944) that victory had been won by conventional military means before it could take effect.

The 'indirect' offensive encouraged and sustained by the Allies against Hitler – military assistance to partisans, sabotage and subversion – must therefore be judged to have contributed materially little to his defeat. Among his army of 300 divisions deployed across Europe on 6 June 1944, the last moment in the war when Hitler exercised unchallenged control over the greater part of the territory he had conquered in 1939-41, fewer than twenty can be identified as committed to internal security duty. Outside central Yugoslavia, parts of western Russia far behind the Wehrmacht's lines and tiny pockets of defiance in mountain Greece, Albania and southern France, all peripheral to his conduct of the larger war, occupied Europe lay inert under the jackboot. The 'dawn of liberation', so seductively promised by Churchill, Roosevelt and the governments in exile to the conquered populations, was signalled only by the flicker of gunfire at the military boundaries of the Wehrmacht's zone of operations.

The bacillus of espionage

If the structure of Hitler's empire was penetrated and fissured by clandestine activity, that took a form quite different from the 'setting ablaze' Churchill had so optimistically demanded in July 1940. Resistance may have been a gnat on the hide of the Wehrmacht; espionage was a bacillus debilitating its vital system. The real triumph of the Allies' indirect campaign against Hitler between 1939 and 1945 was won not by the brave and often foolhardy saboteur or guerrilla warrior but by the anonymous spy and the chairborne cryptographer.

Of the two, spies were by far the less important. Popular imagination invests 'Humint' (human intelligence, in the jargon of the trade) with a significance even greater than it gives to the resistance fighter; certainly far greater than is given to 'Sigint' (signals intelligence). Moreover, governments to a considerable degree conform to popular estimations of the worth of a spy. The notion that the inner councils of the enemy are penetrated by an 'agent in place' who transmits their deliberations and decisions swiftly and directly to his master in the friendly camp is beguilingly attractive to any war leader; Churchill, Roosevelt, Stalin and Hitler were all beguiled in this way during the Second World War. Hitler, for example, was led to believe by the Wehrmacht intelligence service, the Abwehr, that it maintained an extensive network of agents in Britain who were able to report after June 1944 on the accuracy of pilotless-weapons strikes on London. Churchill and later Roosevelt were supplied through the Czech intelligence service with significant information of German capabilities and intentions by Agent A-54, probably the Abwehr officer Paul Thümmel. Stalin, who benefited from the dedication of the international communist movement's members to the Soviet cause, drew on the Swiss-based 'Lucy ring' for information of German activities in occupied Europe, on Richard Sorge's network in Japan for warning of German military intentions, on the 'Red Orchestra' for day-by-day intelligence of the German order of battle and, during 1941-2, on the Schulze-Boysen Luftwaffe network (a Red Orchestra component) for technical data; through the 'Cambridge Comintern', composed of British intelligence officers, Anthony Blunt, Guy Burgess and

Kim Philby, Stalin also kept his finger intermittently on the pulse of Anglo-American higher strategy.

All these sources were, however, to some degree unsatisfactory or compromised. Agent A-54's transmissions, for example, were too sporadic to provide a coherent picture of German strategy. The Red Orchestra was wrongly placed to monitor key German military activity, the Schulze-Boysen network was insecure and quickly penetrated (117 of its members were hanged), Sorge was too detached from the Comintern network to be always believed (though he undoubtedly influenced Stalin's decision to transfer troops from Siberia to Moscow in the winter of 1941), while the Lucy ring was probably transmitting information tailored and supplemented by the Swiss intelligence service for its own purposes; another interpretation is that the Lucy ring derived its information from Bletchley, either directly or through Allied agents in Swiss intelligence. The Cambridge Comintern was perhaps the most influential of all Stalin's networks, though in a reverse direction; Philby, by deliberately maligning his circle's reliability, may have been decisive in dissuading the British government from lending support to Stauffenberg's anti-Hitler conspiracy, which Stalin undoubtedly judged inimical to his long-term plans for the post-war control of Germany, since the conspirators were avowedly anti-communist. Least satisfactory of all the networks was the Abwehr's in Britain; it was turned as early as 1939 by the capture of one of its spies, and after that all the information it transmitted to Germany was compiled by the British officers of the 'Double Cross' organisation which controlled subsequently captured agents. One of the controllers' achievements was to persuade the German staff of the pilotless-weapons force progressively to shorten the range of their launched missiles, so that the majority fell south of London.

The role of Ultra

The contribution of 'Humint' to the direction of strategy in the Second World War looks all the more marginal and patchy when compared with that of 'Sigint'. Signals intelligence is concerned with the interception, decryption and interpretation of the enemy's secure messages, however transmitted, and the protection of one's own from his interception services. In practice the majority of material so intercepted in the Second World War was radio traffic protected by elaborate mathematical ciphers, though some traffic was sent by the older system of codes, constructed from secret codebooks; much tactical intelligence, known to the British as 'Y', was gleaned from messages sent in 'low-grade' ciphers or even *en clair* (not in cipher) between units in the heat of the action.

All five major combatants, Germany, Britain, the United States, the Soviet Union and Japan, devoted costly and extensive efforts to protecting their radio signals traffic and seeking to penetrate that of the enemy (telex, telegraph and telephone traffic sent along landlines – which carried, for example, 71 per cent of all German naval traffic in 1943 – proved generally impenetrable, but had limited use). Minor combatants, notably the Poles, French and Italians (whose codes and ciphers were exceptionally secure), also played a major role in the radio warfare. In particular it was the Poles' early success in attacking Germany's military machine cipher (Enigma), subsequently revealed to the French, which eventually allowed the British to break Enigma on a regular and rapid basis, thus laying the foundations for the Ultra organisation, which disclosed information of truly war-winning value to the Allies from late 1940 onwards.

The extent of the Ultra triumph is now so well known (the equally important American 'Magic' penetration of the Japanese naval and diplomatic ciphers less so) that consideration of both ought to wait upon an evaluation of the cryptographic successes of other combatants, which were more significant than received opinion admits. It is probable, for example, that the Russians had their own success against Enigma; so Professor Harvey Hinsley, the historian of Ultra, guardedly indicates. Russia's own high-grade ciphers were certainly of the best quality, and this suggests a capacity to read others; they had not yielded to foreign intelligence services' attack for some years before 1941 (Churchill forbade the British Government Code and Cipher School from making the attempt after 22 June 1941). Russian medium- and low-grade ciphers, however, were much read by the Germans in 1941-2 and perhaps later; the decrypts yielded valuable tactical intelligence. The Germans also succeeded in breaking the American military-attaché code in late 1941, and their decrypts of messages originating with the US liaison officer in Cairo supplied Rommel with important information about the Eighth Army in the desert.

The most important German success, however, was in breaking the British naval book ciphers, to which the Admiralty clung long after the army and air force had gone over to more secure ciphers. These ciphers (nos. 1, 2, 3 and 4) were more properly book codes: the letters of a message were translated into figures by reference to standard tables and then 'super-enciphered' by mathematical techniques, which were altered at regular intervals. The weakness of the system was that the book itself might be reconstructed if sufficient radio traffic was collected and analysed by the enemy, who could then apply normal calculations of mathematical probability to break the super-encipherment. This was exactly the achievement of the German navy's Observation Service (*Beobachtungs-* or *B-Dienst*); by April 1940 it was reading 30-50 per cent of traffic encoded in naval cipher no. 1; when a change was made to cipher no. 2 it again broke this traffic on a large scale between September 1941 and January 1942; it had less success with the replacement, no. 4; but between February 1942 and June 1943, with short interruptions, it was reading as much as 80 per cent of no. 3, often in 'real time'. 'Real time', a cryptographer's term, means that messages sent are intercepted and broken by the enemy at the same speed as they are received and decrypted (or decoded) at the denominated station. What made the B–Dienst's success against naval cipher no. 3 so disastrous for the Allies was that it was the code used to carry information between London and Washington about transatlantic convoys; as a result the B–Dienst 'was sometimes obtaining decrypts about convoy movements between 10 and 20 hours in advance' of their departure, according to Harold Hinsley. Such information was the key to the success of Dönitz's U-boat wolf packs, which, alerted by such intelligence, could be deployed across the track of an east- or west-bound convoy in numbers that frequently overwhelmed their escorts. It was not until the Admiralty accepted a combined cipher machine (CCM), employed also by the American and Canadian navies and brought into general use in 1943, that the German navy's penetration of convoy ciphers in the Battle of the Atlantic was ended. By then, however, the balance of advantage in the battle had been shifted to the Allies by conventional military means.

In the land and air war, the Allied armies and air forces conceded no such advantage to the enemy as the British Admiralty did by its arrogant persistence in the use of book codes; they used machine cipher systems from the outset and in consequence resisted German attack on their secure transmissions. The German services were also committed to a machine cipher system, but, having adopted theirs ten years before the British and Americans, found themselves – unwittingly – at the outbreak of the war equipped with a semi-obsolete system. Hence the success of the British Government Code and Cipher School (GCCS), located at Bletchley, between Oxford and Cambridge (from which it drew so much of its talent), at breaking into the transmissions encrypted on the Enigma machine the Germans used.

The Enigma machine outwardly resembled a portable typewriter but the depression of a key worked an internal system of gears which allotted any letter input an alternative letter not logically to be repeated before 200 trillion subsequent depressions. The Germans therefore understandably regarded Enigma transmissions as unbreakable in 'real time', indeed in any sort of human time whatsoever. Unfortunately for them they were deceived; because of the need to indicate to a receiving station the way in which the sending Enigma machine had been geared to transmit, the operator was obliged to preface each message with a repeated sequence of the same letters. This established a pattern which a trained mathematician could use as a 'break' into the message and so into the whole of its meaning. As the mathematicians recruited by Bletchley included Alan Turing, the author of universal computing theory, Gordon Welchman, a principal pioneer of operational analysis, and Max Norman and Thomas Flowers, designer and builder respectively of the first electronic computer ('Colossus', so called at Bletchley because of its enormous size), such 'breaks' were rapidly exploited to yield complete readings of German messages quite quickly after interception – and eventually in 'real time'.

There were important exceptions to Bletchley's success. Luftwaffe 'keys' – the different methods of enciphering Enigma traffic used by separate German service branches – proved easier to break than army and naval keys, and some naval keys were never broken; significantly, nor was the Gestapo key, though it was not changed from 1939 until the end of the war. Enigma security depended heavily upon the experience and skill of the German senders; mistakes in procedure made by inexperienced, tired or lazy operators provided Bletchley with the majority of their 'breaks' into traffic. Gestapo operators were meticulous; so too were the naval officers who used the 'Shark' key controlling U-boats in the North Atlantic. For most of 1942, while the B-Dienst was regularly reading naval cipher no. 3 in real time, Shark resisted all Bletchley's efforts; during those months (February-December) the Germans were masters of the radio war in the Battle of the Atlantic. In consequence hundreds of thousands of tons of Allied shipping were sunk.

The Shark episode was, however, an exception to the general rule that the British dominated radio warfare in the West, as the Americans did in the Pacific, where they matched Bletchley's achievement by breaking both the Japanese naval (JN 25b) and diplomatic (Purple) machine ciphers before the outbreak of the war. The joint Allied triumph, successfully concealed from both enemies throughout the war, naturally prompts the query why, if the Allies enjoyed such direct access to the most secret internal messages sent by their opponents, they were nevertheless on occasion suprised by large-scale enemy initiatives – Pearl Harbor, Crete and the Ardennes offensive being the obvious examples. The answer is that there are limits to the usefulness of even the best intelligence system, and this is precisely demonstrated in different forms by each of these episodes. The Japanese, for example, disguised their intentions before Pearl Harbor by hiding their fleet in the remoteness of the Pacific and imposing absolute radio silence on its units, which moved to their attack positions by preordained plan; alert though the Americans were, they were thereby denied the intelligence which would have allowed them to anticipate attack. Before the Ardennes the Germans also imposed radio silence on their attack units. Nevertheless, primarily through troop movements, they unwittingly betrayed sufficient warning to the Allies of the attack they planned for a truly sensitive intelligence organisation to have detected the danger and alerted higher authority. Both the intelligence service and higher headquarters had, however, persuaded themselves that the Germans were too weak to launch an offensive on the Ardennes front during December 1944; accordingly they discounted the evidence to the contrary and so were discountenanced.

The case of Crete reveals a third and highly frustrating limitation in the use of intelligence: the inability to act on clear warning because of disabling weakness. Before the German parachute landings in May 1941, Ultra – the organisation which evaluated and distributed the raw decrypts produced by Bletchley – had identified from Enigma intercepts both the German order of battle and the German plan. However, Freyberg, the British commander on the island, lacked not only the troops but also (more ironically) the transport that would have enabled him to concentrate counter-attack forces swiftly against points of danger. As a result the Germans, despite heavy initial losses, seized a vital airfield which allowed them to fly in reinforcements and swamp the defences.

There is a fourth and universal limitation on the usefulness of intelligence: the need to protect a source. It has been widely alleged, for example, that Churchill 'allowed' Coventry to be bombed in November 1940 because to have taken extraordinary defensive measures against the attack would have revealed to the Germans the 'Ultra secret'. It is now known that this interpretation is false; although Churchill did indeed have advance warning via Ultra of the Coventry raid, it was too short to enable defensive measures to be taken – which he would certainly have done, at whatever the risk of compromising Ultra, had time been available. A more telling accusation is that in the weeks before Barbarossa the British did not validate their warnings to the Russians of the imminence of the German attack by revealing the authenticity of the source. However, in view of Stalin's wishful thinking to the contrary and his desire to placate Hitler at any cost, the betrayal to him of the Ultra secret would have plumbed the depths of insecurity. In this case, as in every other where such a calculation had to be made, Churchill was unquestionably right to put the long-term security of the source above current advantage.

Despite the intrinsic and artificial limitations to the usefulness of Allied access to the enemy's secret traffic, both Ultra and the American 'Magic' organisation were undoubtedly responsible for major, even crucial, strategic success in the Second World War. The first and most important was the victory of Midway, where knowledge of Japanese intentions allowed the Americans to position their inferior fleet of carriers in such a way as to destroy the much larger enemy force. Midway, the most important naval battle of the Second World War, reversed the tide of advantage in the Pacific and laid the basis for America's eventual triumph. In the European theatre, Ultra supplied Montgomery with vital intelligence of Rommel's strength and intentions both before and during the battle of Alamein and later provided Alexander in Italy with timely warnings of the German intention to counter-attack at the Anzio bridgehead – 'one of the most valuable decrypts of the whole war', according to Ralph Bennett in his account of the role of Ultra in the Mediterranean. Ultra intelligence also allowed Alexander correctly to time his subsequent break-out from Anzio and enabled General Jacob Devers to undertake his headlong pursuit of Army Group G up the Rhône valley after the landing in Provence in August 1944, safe in the knowledge that this would not be opposed.

Ultra's greatest contribution to the war in the West, however, occurred during the Battle of Normandy, when Bletchley provided Montgomery with information of day-to-day German strengths at the battlefront, of the effect of Allied air-strikes, such as that which destroyed the headquarters of Panzer Group West on 10 June, and eventually of Hitler's order to counter-attack at

Mortain against the flank of Patton's break-out into Brittany – a disclosure which led to the destruction of Army Group B's armoured reserve and to the climactic encirclement of the *Westheer* in the Falaise pocket. The Mortain decrypts were certainly the most important which came to any general on any front throughout the course of the Second World War.

Whether Ultra 'shortened the war', as is sometimes suggested, or even materially altered its course, is more difficult to argue. There was no single Ultra triumph as great as the American codebreakers' success in identifying Midway as the target for the Japanese fleet in June 1942, a genuinely tide-turning intelligence operation. Although the breaking of the Shark key in December 1942 very greatly contributed to the winning of the Battle of the Atlantic in the following spring, against it must be set the cost of the B-Dienst's concurrent success in reading the British naval convoy codes. Ultra did not much influence the course of the war in the air, despite the insecurity of the Luftwaffe keys, and in the ground fighting between the Germans and the Western Allies it can never be said to have given the advantage consistently to the eavesdropping side. That was because, as Clausewitz's famous and accurate observation on combat reminds us, on the battlefield 'friction' always intervenes between the intentions and achievements of even the best-informed general: accident, misunderstanding, delay, disobedience inevitably distort an enemy's plans so that, whatever advance knowledge his opponent may have of them, he can never so predisposition his troops and responses as to be sure of frustrating the enemy's actions; nor, because of 'frictions' working against him, can he count on smoothly carrying out his own counter-measures. Ultra reduced friction for the Allied generals; but it did not abolish it.

If we shift the focus and ask whether in the spectrum of clandestine warfare cryptanalysis was more or less valuable to the Allies than the activity of the resistance, the answer is simple. Cryptanalysis was consistently and immensely more valuable indeed. The Second World War in the West could have been won without either the resistance or Ultra; but the cost of the former was heavy, and its material, as opposed to psychological significance was slight. The cost of Ultra, by contrast, was trivial – the whole apparatus employed only 10,000 people, including clerks and cryptographers – while its material value was considerable and its psychological significance inestimable. The proof of that comes from the German as well as the Allied side. Ultra sustained the confidence of the very few Western decision-makers who were privy to its secret in a way nothing else could have done. Twenty years after the war was over, when their German opponents discovered that their most secret correspondence had been read daily by the British and Americans, they were struck speechless.

TWENTY-SEVEN

The Vistula and the Danube

The destruction of Army Group Centre had cast Germany's strategic position on the Eastern Front into ruins. The military implications were grave enough. The remnants of Army Group Centre now stood on the line of the Vistula less than 400 miles from Berlin, with the great Polish plain at its back and no obstacle but the river Oder between it and the capital. On the Baltic coast Army Group North, now commanded by Ferdinand Schörner, one of Hitler's chosen 'standfast' generals, was threatened by the Baltic fronts' thrust to Riga with encirclement in northern Latvia and Estonia – the 'Courland pocket' as it was called by OKH. From there the army group could be supplied only by sea, but Hitler would not allow the position to be abandoned because he insisted on preserving free use of the Baltic to train the crews of his new U-boats. The physical damage the army had suffered in the summer battles was staggering. Between June and September the number of dead on the Eastern Front rose to 215,000 and of missing to 627,000; when losses in the west were added in and the number of wounded included, the total rose to nearly 2 million, or as many casualties as the army had suffered from the beginning of the war until February 1943, including those of Stalingrad. By the end of 1944 106 divisions – a third of those in the order of battle – had been disbanded or rebuilt, more than the army had fielded on the eve of its victory era in September 1939.

Hitler resisted striking divisions from the order of battle. His solution to the massacre that had taken place, therefore, was to decree the formation of new divisions with the same number as the old, but now to be designated 'people's grenadier' *(Volksgrenadier)* divisions. Despite the rising output of the German arms industry, which Speer raised to unprecedented heights in September 1944, the Volksgrenadier divisions were only 10,000 strong (divisions had contained 17,000 men in 1939), lacked anti-tank guns and mounted their reconnaissance battalions on bicycles. Even so only sixty-six Volksgrenadier divisions altogether could be formed to replace the seventy-five infantry divisions lost in the west and east during 1944. They were raised within the Home Army, command of which Hitler had given to the SS chief, Heinrich Himmler, in the aftermath of the July Plot. After 23 July the military salute was also abolished; instead all servicemen were required to give the 'Heil Hitler' with outstretched arm. Guderian, who replaced Zeitzler as chief of staff after 20 July, accepted this and the institution of military 'courts of honour' to strike suspects from the officer corps before they were tried by the people's courts.

The political implications of the outcome of Operation Bagration were even more menacing than the military ones. The Russian triumph threatened the integrity of the whole complex Balkan alliance Hitler had so painstakingly constructed through the Tripartite Pact between August 1940 and March 1941. On 20 August the Second and Third Ukrainian Fronts opened an offensive against Army Group South Ukraine and burst across the river Prut to the delta of the Danube in five days. The weight of the attack fell on the Romanian Third and Fourth Armies and the Romanians were panicked into changing sides. On 23 August King Michael staged a palace revolution in Bucharest, arrested Ion Antonescu, Hitler's Romanian collaborator, and replaced his government with one of 'national unity', which included communists. When Hitler responded by bombing Bucharest on 24 August, the King declared war on Germany. In demonstration of the country's change of heart, but also to avenge a national grievance, the surviving elements of the Romanian army at once invaded Hungary, still in Hitler's camp, to recover the province of Transylvania which had been transferred to Hungary under the terms of the Tripartite Pact of August 1940. The Russians would not at first accept the move as an act of co-belligerency. Having already overrun Ploesti and its oilfields, the jewel in the crown of Hitler's economic empire, they entered Bucharest as conquerors on 28 August. Not until 12 September did they concede an armistice, allowing Romania to retain Transylvania but taking back the provinces of Bessarabia and northern Bukovina which had been their local share of the spoils of the Ribbentrop-Molotov Pact.

Romania's defection provoked Bulgaria's. The Bulgarians, traditionally the most Russophile of Slavs, had been careful not to allow their accession to the Tripartite Pact in March 1941 to commit them to war against Russia. They had granted the German army basing and transit facilities; they had taken their share of Yugoslavia and also sent occupation troops to Greece; but no Bulgarian soldier had fought against the Red Army. Indeed the existence since 1943 of a small anti-German partisan movement inside the country made a change of sides easier to arrange. With the death of King Boris in August 1943 Hitler had been robbed of his most dependable Bulgarian supporter; the successor government, though rebuffed by the Western Allies when it explored the possibility of changing sides, knew it must disentangle itself from the German alliance. On the approach of the Red Army, however, a 'Fatherland Front' proclaimed a national uprising on 9 September, supplanted the government – which had already asked Russia for a truce on 5 September – and took power. On 18 October the Red Army marched into Sofia and the 150,000-strong Bulgarian army went over to its side.

The collapse of the German front in the north had already forced Finland to reconsider its position. The Finns had never been ideological allies of Hitler. They were fiercely democratic – they had succeeded in retaining their national parliament even under tsarist rule – and their quarrel with Russia was a territorial one. Once they had regained during the Barbarossa campaign the territory they had been forced to concede at the end of the Winter War (together with land east of Lake Ladoga to which they had an historic claim) they had halted. As early as January 1944 they had made approaches to the Allies through Washington but had been warned that the Russian price for a separate peace would be high: a return to the 1940 frontier, the cession of Petsamo, centre of Finland's mineral industry and its outlet to the Arctic in the Far North, and a large financial indemnity. The terms had then seemed too harsh; as Bagration developed they came to look more attractive. During June the Finnish president, Risto Ryti, was personally confronted with conflicting demands from Hitler and Stalin: he must either formally reject a separate peace with Russia (the German ultimatum) or capitulate (the Russian ultimatum). Under pressure from Marshal Mannerheim, effectively Finland's leader, Ryti gave Ribbentrop his assurance that Finland would not make a separate peace. However, Mannerheim had privately resolved to use this assurance to fight for time. During July he managed to blunt the Russian attack on Finland's fortified frontier until the retreat of Army Group North had drawn the Russian northern fronts westward into the Baltic states. Then on 4 August he assumed the office of President, revoked Ryti's commitment and opened direct negotiations with Moscow. On 2 September he broke relations with Germany and on 19 September signed a treaty with Russia whose terms were much as they had been in January. The most important differences were a halving of the size of the indemnity, offset by the grant of a naval base on the Porkala peninsula, near Helsinki, and a Finnish undertaking to disarm the German Twentieth Mountain Army in Lapland. As Mannerheim privately recognised, the Finnish army had neither the strength nor the the will to drive the Twentieth Army out of the country into Norway, from which it was supplied, and the operation was not completed until April 1945, and then only with Russian help.

Despite the Finns' excellence as soldiers (they were alone among the Wehrmacht's allies in regarding themselves and being regarded, man for man, as the Germans' military equals, even superiors), by late 1944 Finland was peripheral to Hitler's strategic crisis. Hungary, on the far southern flank and next in the firing line after Romania and Bulgaria, was by contrast central to the defence of the Reich's outworks. The Hungarians too had a reputation as excellent soldiers which had been won in the service of the Habsburg emperors – and in rebellion against them. However, Admiral Horthy, the Hungarian dictator, had made the mistake of committing them to Operation Barbarossa against the Red Army, which they were not equipped to fight once the shield of Wehrmacht protection had been withdrawn from them. The collapse of Army Group South Ukraine (renamed Army Group South in early September) and the defection of the Romanians now exposed them to a Soviet thrust which they lacked the power to repel, even from their strong positions in the Carpathians.

Horthy hoped that he would be saved from choosing between the Germans and the Russians by an Anglo-American advance from Italy into Yugoslavia. Not only had the Allies been deflected by internal disagreement from such a manoeuvre; the Americans, with whom he was in contact through their ambassador in Switzerland, informed him in August that he must make his own arrangements with the Russians. As soon as the Romanians attacked his army in Transylvania, he had no option but to do so. A Hungarian delegation arrived in Moscow at the end of September to negotiate terms for a change of sides. Horthy, however, had unilaterally undermined the chances of doing so successfully by allowing Hitler in March to station German troops on Hungarian soil. When the Russians heightened their pressure on Horthy's delegates in Moscow by launching an attack into eastern Hungary towards Debrecen on 6 October, the occupation army – reinforced by three Panzer divisions – counter-attacked and blunted the advance. Hitler, moreover, had by now got wind of Horthy's impending treachery. He was aware that Horthy had issued orders to his First and Second Armies, which were still fighting with Army Group South, to make a unilateral retreat; he also suspected that Horthy was on the point of announcing a change of sides. On 15 October, therefore, he authorised Skorzeny, his expert in such operations, to kidnap Horthy's son and then confronted the dictator with a demand that he transfer power to a pro-German replacement. Early on 16 October Horthy abdicated as Regent, and German troops took control of the whole of Budapest. The Second Ukrainian Front was by then only fifty miles from the Hungarian capital, but it was to remain safe in German hands for several months.

Revolt in the Balkans

Hitler had meanwhile also quashed another attempt at defection among his eastern satellites. Slovakia, ruled since October 1939, in the aftermath of Czechoslovakia's dismemberment, by Joseph Tiso, a signatory of the Tripartite Pact and a co-belligerent in the war against Russia, had been seething with internal discord since the spring. While the 'London' Czechs, legitimately the government in exile, looked to a post-war settlement to restore them to power, the dissident Slovaks, through the underground Czechoslovak Communist Party, were in contact with Moscow, which sponsored a small army in exile stationed on Russian territory. Part of the Slovak army of Monsignor Tiso's puppet state remained under German control on the Eastern Front; the rest, stationed at home, fell increasingly under patriot influence. A pro-Soviet partisan movement was also active in eastern Slovakia, towards which Operation Bagration had drawn the Fourth Ukrainian Front at the beginning of August. At the end of August the pro-Soviet partisans precipitated action. Liaising directly with the Red Army and bypassing both the London Czechs and the dissidents' 'Slovak National Council', on 25 August they initiated a national uprising, in which they were joined by the home-based Slovak army, and looked for support to the Russians beyond the Carpathians. Their response was far more positive than it had been to the Polish Home Army in Warsaw. They at once sent liaison officers and initiated an offensive by the First and Fourth Ukrainian Fronts to come to the insurgents' rescue. They also airlifted parts of the Czech army in exile from Russia into Slovakia and embodied the rest in the Ukrainian fronts fighting to cross the Slovak passes through the Carpathians. However, pressure from without and within was not strong enough to overcome the response Hitler organised to preserve his position in Slovakia. Two German corps, XXIV Panzer and XI, were sent to man the Carpathian position, including the key Dukla Pass. At the end of September the Soviet Thirty-Eighth Army, assisted by the I Czechoslovak (exile) Corps, was still battering against the pass, and it did not fall until 6 October. Meanwhile the security troops which were so experienced in anti-partisan operations in the eastern theatre were being earmarked for commitment. Two SS divisions formed from ethnic minorities, the 18th *Horst Wessel* (racial German) and the 14th Galizian (Ukrainian), were concentrated for a counter-offensive, together with five German army divisions; by 18 October the Dirlewanger and Kaminski brigades had also been brought down from Warsaw to turn their murderous talents against the Slovaks. Between 18 and 20 October 'free Slovakia' was assaulted at eleven points and by the end of the month the insurrection was extinct. The Soviet Thirty-Eighth Army and the I Czechoslovak Corps (commanded by General Ludwik Svoboda, whom the Russians would install as Dubcúek's successor after the 'Prague Spring' of 1968) suffered 80,000 casualties in the effort to come to the insurgents' rescue; almost all the insurgents who did not escape into the hills died in combat or in concentration camps.

In the extreme south of his Balkan theatre, Hitler was to prove less successful at shoring up a defence than in Hungary and Slovakia. The occupation of Greece had been crumbling since the capitulation of the Italians in September 1943, when at least 12,000 of their weapons had fallen into the hands of the resistance. The Greek partisans had been fighting bravely and doggedly against the occupiers, as their forefathers had done in the war to liberate their homeland from the Turks 120 years earlier. Many of the British liaison officers whom SOE infiltrated into the Greek islands and mainland were touched by a Byronic afterglow, seeing themselves as successors to the philhellenes who had fought at the patriots' side in the War of Liberation in the 1820s. However, German reprisals against the villages near which the resistance attacks took place were ferocious; at the Nuremberg trial a prosecution lawyer was to testify that 'in Greece there are a thousand Lidices [Lidice was the Czech village obliterated after the assassination of Heydrich], their names unknown and their inhabitants forgotten.' Therefore much SOE effort was devoted to restraining rather than encouraging the partisans; but the British liaison officers had not been able to check violence between the right and left wings of the resistance movement, which, as in Yugoslavia, obeyed different authorities – ELAS the Greek

Communist Party, EDES the Greek government in exile in Cairo. The Germans restored and maintained order after the surrender of the Italians – whom they treated with almost as much brutality as any partisans they caught – but, as their Balkan position began to collapse, they evacuated the Greek islands (except for Crete and Rhodes) from 12 September and then on 12 October the whole of Greece. As they left and the British began to arrive, the first round of a civil war between ELAS and EDES broke out; it was to be quelled, at a tragic cost in British lives, by the intervention of the 2nd Parachute Brigade and other formations against ELAS at Christmas.

Army Group E, the German command in Greece and Albania, had a single hope of salvation, which was to find its way through the Ibar and Morava valleys to link up with Army Group F in Yugoslavia. The sudden onset of the Third Ukrainian Front, now supported by the Bulgarian Army, forced it to fight a desperate rearguard action. Meanwhile Army Group F was confronted by a Soviet assault on its eastern flank aimed at the Yugoslav capital, Belgrade. The Third Ukrainian Front had crossed the Yugoslav border on 6 September, prompting Tito to fly to Moscow from a British airfield on the Adriatic island of Vis to discuss the terms on which the Red Army would operate on Yugoslav territory. In a remarkable exercise in negotiation from weakness, Tito persuaded Stalin by 28 September to agree to lend troops from the Third Ukrainian Front for a joint assault on Belgrade but to withdraw them, leaving civil administration in Tito's hands, once the operational task was complete. The battle for Belgrade opened on 14 October and ended on 20 October; 15,000 German soldiers were killed and 9000 taken prisoner in the defence of the city. Tito paraded his Partisans through the streets as victors on 22 October; of his 'Belgrade battalion', which had fought the three-year partisan war, only two of its original members were still in the ranks.

The rest of Yugoslavia now lay open to an extension of the Soviet offensive; Army Group E, which had incorporated F, was holding an indefensible north-south flank which ran from the outskirts of Belgrade to the Albanian frontier. However, in mid-October Stalin had agreed with Churchill in Moscow a strange division of 'spheres of influence' in the Balkans which gave Britain a 50 per cent share in Yugoslavia. An odd streak of legalism in Soviet diplomacy lent this agreement force; but Stalin also had other fish to fry. Hitler's successful coup against Horthy in Budapest had destroyed the chance of making a quick advance, by a negotiated armistice, into the Hungarian plain. The approach to Vienna, up the Danube valley, would now have to be fought for; the force it would require meant that the Red Army could not afford to dissipate its strength in the mountains of central Yugoslavia, where conditions would put even the battered formations of Army Groups E and F on equal terms. On 18 October, therefore, the Stavka had ordered Tolbukhin to halt the Third Ukrainian Front west of Belgrade and turn its formations back to the Danube to take part in the coming battle of Hungary.

Hungary, however, had now been reinforced and parts of the Hungarian army (First and Second Armies) hijacked to fight on the German side. On 19 October Army Group South counter-attacked, and when Malinovsky's Second Ukrainian Front began its assault on 29 October, at Stalin's express orders 'to take Budapest as quickly as possible', it found twelve German divisions in its path. The Russian advance reached the eastern suburbs on 4 November but was then halted; when the assault was resumed on 11 November a sixteen-day battle ensued which left much of the city in ruins but still in German hands. By then the German front line, though withdrawn 100 miles since mid-October, rested from west to east on the strong defences of the river Drava, Lake Balaton and the flanks of the Carpathians. Vienna, the prize which Stalin sought, remained secure 150 miles away along the Danube.

The campaign in Hungary thereafter took on a logic of its own and proceeded quite separately from the Red Army's preparations on the far side of the Carpathians for the final advance into Germany. On 5 October Malinovsky's Second Ukrainian Front began an offensive designed to encircle Budapest from the north-west, while Tolbukhin's Third Ukrainian Front made a feint to the south of the city between Lake Balaton and the Danube. By 31 January the Third Ukrainian Front was within seven miles of the city centre and emissaries were sent forward to offer terms for a capitulation. The city was completely surrounded, the suffering of its inhabitants was intense, and the situation of the German and Hungarian defenders appeared hopeless. Hitler, however, had decided upon a Stalingrad-style Panzer rescue. He had dismissed the commander of the Sixth Army, Maximilian Fretter-Pico, on the Budapest front, to replace him with Hermann Balck, another 'standfast' general of the Model type, and in late December had brought IV Panzer Corps from Army Group Centre to stage a counter-attack in concert with III Panzer Corps, which was already on the scene. The attack by IV SS Panzer began on 18 January 1945 and during the next three weeks IV and III Panzer Corps fought savagely, switching from one axis to another by road and rail, in an awful warning to Malinovsky and Tolbukhin of what damage experienced German tank soldiers could still inflict on Soviet formations operating on stereotyped and predictable fixed lines of advance. By 24 January IV Panzer Corps had driven to within fifteen miles of the German perimeter in Budapest, and the defenders could have broken out to safety had that been Hitler's wish. As during Manstein's winter thrust to Stalingrad in December 1942, however, he wanted the city to be recaptured, not evacuated. This vain hope collapsed when IV Panzer Corps, after three weeks of frantic operations, ran out of steam.

Within the perimeter, meanwhile, the Russians had brought up dense concentrations of 152-mm guns and 203-mm howitzers to reduce the German positions block by block in Pest, the northern half of the city. Its garrison began to surrender *en masse* on 15 January, when they were trapped with their backs to the Danube. In Buda, Pest's twin city on the south bank, resistance held up fiercely until 5 February, when Malinovsky ordered a final assault. For a week the Germans stuck it out, taking to the sewers to frustrate the Russian advance, but by 13 February they had no more room for manoeuvre and were overwhelmed. The Stavka claimed to have killed 50,000 German and Hungarian soldiers and taken 138,000 prisoners since 27 October; it is known that only 785 Germans and some 1000 Hungarians escaped from Budapest. The Red Army's own undisclosed losses in killed and wounded may have equalled those of the enemy.

There was to be one more battle fought in Hungary, at Lake Balaton, from which Hitler drew his last supplies of non-synthetic oil. By the time it opened on 15 February, however, a far greater battle was in preparation for the ultimate objective of the war itself: Berlin. Since early February Zhukov's First White Russian and Konev's First Ukrainian Fronts had been poised astride the river Oder, ready to launch themselves into the climactic offensive as soon as the Stavka defined the attack plan and made available the requisite forces. On 15 January Hitler had left the western headquarters in the Eifel mountains (*Amt* 500) from which he had overseen the Ardennes offensive to return to the Reich Chancellery. For all his talk of secret weapons that were still to bring victory, he sensed the approach of the final struggle and was resolved to be present on the field in person.

The road to Berlin

Hitler's Rastenburg headquarters in East Prussia, from which he had directed most of the war, was now in Russian hands. The Red Army's offensive north of the Carpathians had begun on 15 September 1944 when the three Baltic and the Leningrad Fronts had opened an attack on Schörner's Army Group North, designed to cut it off in the Baltic states from contact with Army Group Centre and its lines of communication into Germany through East Prussia. Schörner commanded some thirty divisions, disposed in well-fortified terrain, but lacked mobile forces with which to counter-attack. Though his army group was able to slow the Russian advance, therefore, it was not able to disrupt it and on 13 October, after an eight-day battle, Riga fell to Bagramyan's First Baltic Front. This breakthrough to the coast completed the encirclement of Army Group North (shortly to be renamed Courland) in the 'Courland pocket', where it would linger in pointless isolation until the end of the war; six separate battles were fought by the Red Army against it. Finland's defection in September allowed Schörner (before he was removed to command Army Group Centre in January) to improve its position by abandoning Estonia and concentrating his forces in Latvia. Its four dependent divisions in the port of Memel, between East Prussia and Lithuania, were also surrounded in October and held out until January 1945.

The Baltic front's clearance of the approaches to East Prussia (which a unit of the Third White Russian Front had actually entered on 17 August) now laid it open to major assault. Plans for the great offensive had been laid by the Stavka in early November and allotted the greater effort to the two fronts which lay most directly astride the route to Berlin, Konev's First Ukrainian and the First White Russian, command of which Stalin conferred directly on Zhukov, in testimony of his proven strategic achievements. Each front now greatly exceeded any German army group in strength. Between them they controlled 163 rifle divisions, 32,000 guns, 6500 tanks and 4700 aircraft, or one-third of all current Soviet infantry strength and half the Red Army's tanks. Together they outnumbered the German formations opposite, Army Groups Centre and A, over twofold in infantry, nearly fourfold in armour, sevenfold in artillery and sixfold in airpower. For the first time in the war the Red Army had achieved both the human and material superiority that thitherto the Wehrmacht had only faced in the west. Army Groups Centre and A, now commanded by new generals, Hans Reinhardt and Josef Harpe, disposed between them of seventy-one divisions, 1800 tanks and 800 aircraft; all their formations were under strength, and their defensive capabilities depended greatly on the 'fortresses' which the Prussian and Silesian border towns – Königsberg, Insterburg, Folburg, Stettin, Küstrin, Breslau – had now been so designated by Hitler.

The Stavka plan was for Zhukov to lead off down the Warsaw-Berlin axis, while Konev aimed for Breslau. Both offensives were to be direct power-drives against the German defences, eschewing manoeuvre, in what had now become the Red Army's distinctive, brutal and terrifying means of making war. Over a million tons of supplies were brought up to Zhukov's front alone in the days before the attack; they were carried in 1200 trains and 22,000 of the American-supplied six-by-six trucks which were the backbone of the Soviet logistic system. Almost equal quantities were stockpiled behind Konev's front. The daily requirement of each front was 25,000 tons, less fuel and ammunition.

Konev's offensive opened first on 12 January 1945 behind a barrage fired by guns disposed at a density of 300 to each kilometre of front – an earthquake concentration of artillery power. By the evening of the first day his tanks had broken the front of the Fourth Panzer Army to a depth of twenty miles, in exactly the same sector as the Germans and Austrians had made their great breakthrough in the Gorlice-Tarnow battle against the tsar's army in 1915, but in the opposite direction. Cracow, the great Polish fortress-monastery city, was threatened; beyond it the way lay open to Breslau and the industrial regions of Silesia, where Speer had concentrated clusters of German armaments factories out of range of the Anglo-American bomber force.

Zhukov's offensive on the Warsaw-Berlin axis began two days later, behind another pulverising bombardment, from the Vistula bridgehead south of Warsaw. The city was quickly encircled, and inevitably decreed a 'fortress' by Hitler, but it fell on 17 January before the reinforcements he had allotted it could reach the defenders. On 20 January he announced, to the despair of his commanders both in the west and the east, that he was transferring the Sixth SS

Panzer Army, just extricated from the débâcle of the Ardennes offensive, to the east: 'I'm going to attack the Russians where they least expect it. The Sixth SS Panzer Army is off to Budapest! If we start an offensive in Hungary, the Russians will have to go too.' This wild diversion of precious defensive resources demonstrated how little he grasped both the Wehrmacht's growing debility and the imperviousness of the Russians to subsidiary manoeuvres; the Ukrainian fronts, as events would prove, could deal adequately with the Sixth SS Panzer Army's intervention, and Zhukov and Konev were not at all deflected from their drive on Berlin.

On 21 January, again clutching at straws, Hitler decreed the creation of a new army group, Vistula, command of which he gave to Himmler (also head of the Home Army), though he was quite unfitted to exercise military command, in the belief that loyalty to the Führer might prove a substitute for generalship. Army Group Vistula, positioned behind the threatened front, had almost no troops except *Volkssturm* units – the militia of Germans too young or too old to serve in the army which Hitler had set up on 25 September under Martin Bormann, the Nazi Party secretary.

The advance to the Oder

The *Volkssturm* would shortly be fighting for German territory. On 22 January Konev's First Ukrainian Front crossed the Oder at Steinau; Rokossovsky's Second White Russian Front, which had attacked across the Narew on 14 January, was by then deep into East Prussia. The arrival of the Red Army *en masse* on German soil provoked a stampede of refugees towards any tenuous outlet to safety. It was as if the submerged knowledge of what the Wehrmacht had done in the east had suddenly come to the surface, seized whole populations with terror and flung them on to the snowbound roads in an agony of urgency to put themselves beyond the reach of the Red Army's columns. In a few days 800 years of German settlement in the east were ended as 2 million East Prussians left homes, farms, villages and towns in a frantic trek towards the German interior or the coast; 450,000 were evacuated from the port of Pillau in the next few weeks, while another 900,000 sought rescue at Danzig, many of them trudging across the frozen waters of the Frisches Haff lagoon to reach it. Many escaped, many did not. As Professor John Erickson, no enemy to the Red Army, has described this terrible episode:

> Speed, frenzy and savagery characterised the advance. Villages and small towns burned, while Soviet soldiers raped at will and wreaked an atavistic vengeance in those houses and homes decked out with any of the insignia or symbols of Nazism . . . some fussily bedecked Nazi Party portrait photograph would be the signal to mow down the entire family amidst their tables, chairs and kitchenware. Columns of refugees, combined with groups of Allied prisoners uprooted from their camps, and slave labour no longer enslaved in

farm or factory, trudged on foot or rode in farm carts, some to be charged down or crushed in a bloody smear of humans and horses by the juggernaut Soviet tank columns racing ahead with assault infantry astride the T-34s. Raped women were nailed by their hands to the farm carts carrying their families. Under these lowering January skies and the gloom of late winter, families huddled in ditches or by the roadside, fathers intent on shooting their own children or waiting whimpering for what seemed the wrath of God to pass. The Soviet Front command finally intervened, with an order insisting on the restoration of military discipline and the implementation of 'norms of conduct' towards the enemy population. But this elemental tide surged on, impelled by the searing language of roadside posters and crudely daubed slogans proclaiming this and the land ahead 'the lair of the Fascist beast', a continuous incitement to brutalised ex-prisoners of war now in the Soviet ranks or to the reluctant peasant conscripts dragged into the Red Army in its march through the Baltic states, men with pity for no one.

None of the German army groups north of the Carpathians could stem this onrush; the only impediment to Zhukov's and Konev's uninterrupted advance on Berlin was provided by the attenuation of their own supplies, which the enormous artillery preparations consumed at the rate of 50,000 tons for each million shells fired, losses in the ranks – divisional strengths in the two fronts averaged only 4000 at the end of January – and the resistance of the 'Führer fortresses'. On Rokossovsky's front Memel held out until 27 January, Thorn until 9 February, Königsberg until mid-April; on Zhukov's and Konev's, Posen (Poznan) held until 22 February, Küstrin until 29 March, Breslau until the day before the end of the war. The loss of other places brought the Red Army great propaganda sensation: on 21 January Rokossovsky's Second White Russian Front took Tannenberg, where a 'miracle' battle had saved East Prussia from the tsar in 1914, and from which the retreating Germans just managed to save the remains of the victor of that battle, Field Marshal Hindenburg, and the colours of the regiments he had commanded (they hang now in the hall of the Bundeswehr Officer Cadet School at Hamburg), before blowing up his memorial tomb. On 27 January Konev's First Ukrainian Front stumbled on the extermination camp of Auschwitz, chief place of the Holocaust, from which its operatives had not succeeded in removing the pathetic relics of the victims – clothes, dentures, spectacles and playthings. Meanwhile the strong places of Germany's eastern frontier, so many of them fortresses of the Teutonic knights who had once pushed the tentacles of *Germantum* eastward into the Slav lands, held out to block or menace the lines of advance which the Soviet fronts were punching westward towards Berlin.

By the beginning of February, however, as the Allied leaders gathered for the last great conference of the European war at Yalta in the Crimea, Zhukov's and Konev's fronts were firmly established on the line of the Oder, ready to begin their final advance on Berlin. The German army groups opposite them – now reorganised as Vistula and Centre, the latter commanded by the Führer-dedicated Schörner – were shadows of their former selves. In East Prussia the Third Panzer Army was still active, and was to launch a brief counter-attack against the flank of the Russian concentration on 15 February; on 17 February the Sixth SS Panzer Army opened Hitler's promised diversionary offensive against Tolbukhin's Third Ukrainian Front to the east of Lake Balaton in Hungary. But the sands were now running out fast for the Wehrmacht. On 13 February Dresden, the last undevastated city of the Reich and packed with refugees, but also stripped of anti-aircraft guns to bolster the anti-tank screen on the Oder front, was overwhelmed by a British bomber assault and burnt to the ground, with appalling loss of life. Although the figure sometimes quoted of 300,000 dead is grossly exaggerated, at least 30,000 were killed in the raid. The consequences of this attack, for which the champions of the strategic bombing have never been able to advance a convincing military justification, quickly became known throughout Germany and gravely depressed civilian morale in the last months of the war. The Lake Balaton offensive, though mounted with the last 600 tanks at Hitler's disposal as an uncommitted force, soon ran into immovable Russian defensive lines. Army Group E in Yugoslavia was meanwhile bending its front back towards the bastion of pro-German Croatia. The remnants of Army Group South gathered what strength they had left to bar the approaches to Vienna. But the crisis of the war hovered between Küstrin and Breslau where, along the Oder and the Neisse, Zhukov's and Konev's fronts stood ready to race the last forty-five miles to Berlin.

TWENTY-EIGHT

City Battle:
The Siege of Berlin

The siege of cities seems an operation that belongs to an earlier age than that of the Second World War, whose campaigns appear to have been exclusively decided by the thrust of armoured columns, the descent of amphibious landing forces or the flight of bomber armadas. Cities, however, are as integral to the geography of war as great rivers or mountain ranges. An army – however well mechanised, indeed precisely because it is mechanised – can no more ignore a city than it can the Pripet Marshes or the defile of the Meuse. On the Eastern Front the three 'cities of Bolshevism' – Leningrad, Moscow and Stalingrad – which Hitler had marked out as the targets of the *Ostheer*'s advance had each brought one of his decisive campaigns to grief. His own designation of cities as fortresses – Calais, Dunkirk and the Ruhr complex in the west, Königsberg, Posen, Memel and Breslau in the east – had severely hindered the progress of his enemies' armies towards the heartland of the Reich. Capital cities, with their maze of streets, dense complexes of stoutly constructed public buildings, labyrinths of sewers, tunnels and underground communications, storehouses of fuel and food, are military positions as strong as any an army can construct for the defence of frontiers, perhaps stronger indeed than the Maginot Line or the West Wall, which merely tried to replicate in artificial form the features that capital cities intrinsically embody. Hitler's return to Berlin on 16 January 1945, and his decision by default not to leave it thereafter, ensured that the last great siege of the war, shorter than Leningrad's but even more intense than Stalingrad's, would be Berlin's. The final moment at which he might have left Berlin, and over which he deliberately prevaricated, was his birthday, 20 April. 'I must force the decision here', he told his two remaining secretaries on his birthday evening, 'or go down fighting.'

Berlin was a stout place for a last stand. It was unique among German cities in being large, modern and planned. Hamburg, densely packed around its port on the Elbe, had burned as if by spontaneous combustion in July 1943; the fragile and historic streets of Dresden had gone up like tinder in February 1945. Berlin, though heavily and consistently bombed throughout the war, was a tougher target. A complex of nineteenth- and twentieth-century apartment blocks standing on strong and deep cellars, and disposed at regular intervals along wide boulevards and avenues which served as effective fire-breaks, the city had lost about 25 per cent of its built-up area to Bomber Command during the Battle of Berlin between August 1943 and February 1944. Yet it had never suffered a firestorm, as Hamburg and Dresden had done, nor had its essential services been overwhelmed, and new roads had since been constructed. While the destruction of their dwellings had driven many Berliners into temporary accommodation or out of the city, the ruins left behind were as formidable military obstacles as the buildings left standing.

At the heart of the city, moreover, beat the pulse of Nazi resistance. Hitler's bunker had been constructed under the Reich Chancellery at the end of 1944. The bunker was a larger and deeper extension of an air-raid shelter dug in 1936. It contained eighteen tiny rooms, lay 55 feet under the Chancellery garden, had independent water, electricity and air-conditioning supplies and communicated with the outside world through a telephone switchboard and its own radio link. It also had its own kitchen, living quarters and copiously stocked storerooms. For anyone who liked living underground, it was competely self-sufficient. Although Hitler had spent extended periods of the war in spartan and semi-subterranean surroundings, at Rastenburg and Vinnitsa, he felt the need for fresh air; his after-dinner walks had been favourite occasions for his monologues. On 16 January, however, he descended from the Chancellery into the bunker and, apart from two excursions, on 25 February and 15 March, and occasional prowls about his old accommodation upstairs, he did not leave it for the next 105 days. The last battles of the Reich were conducted from the bunker conference room; so too was the Battle of Berlin.

Berlin did not have its own garrison. Throughout the war, except for the brief period of uneasy peace between the French armistice and Barbarossa, the German army had been at the front; the units of the Home Army which remained within the Reich performed recruitment or training functions. Inside the capital, the only unit of operational value was the Berlin Guard Battalion, out of which had grown the *Grossdeutschland* Division. It had figured largely in the suppression of the July Plot and was to fight in the siege of Berlin. However, the bulk of Berlin's defenders was to be supplied by Army Group Vistula as it fell back from the Oder on the capital. Its strength at the beginning of the siege was about 320,000, to oppose nearly 3 million men in Zhukov's, Konev's and Rokossovsky's fronts, and it comprised the Third Panzer and Ninth Armies. The most substantial force within Army Group Vistula was LVI Panzer Corps, containing the 18th Panzergrenadier and SS *Nordland* divisions, as well as fragments of the 20th Panzergrenadier and 9th Parachute Divisions and the recently raised Müncheberg Division; Müncheberg belonged to a collection of 'shadow' formations, based on military schools and reinforcement units, without military experience. To them could be added a motley of *Volkssturm*, Hitler Youth, police, anti-aircraft and SS units; among the latter was the Charlemagne Assault Battalion of French SS men and a detachment of the SS Walloon Division, formed from pro-Nazi French Belgians commanded by the fanatically fascist Léon Degrelle, the man Hitler is alleged to have said he would have liked for a son, and who would lead it in a fight to the end over the ruins of the Reich Chancellery.

During the last weeks of March and the first of April Zhukov's and Konev's fronts assembled the force and supplies they would need for the assault on the city. Zhukov accumulated 7 million shells to supply his artillery, which was to be massed at a density of 295 guns to each attack kilometre; Konev, who needed to capture assault positions across the river Neisse from which to launch his offensive, had concentrated 120 engineer and thirteen bridging battalions to seize footholds, and 2150 aircraft to cover the operation.

While Zhukov and Konev were preparing for the great assault, Tolbukhin and Malinovsky opened the drive out of central Hungary on Vienna. On 1 April their tank columns began their race northward across the wide Danubian plain, brushing aside German armoured brigades which could put no more than seven to ten tanks in the field. By 6 April Tolbukhin's spearheads had entered the western and southern suburbs of Vienna and on 8 April there was intense fighting for the city centre. Local SS units fought fanatically, with total disregard for the safety of the monuments they made their strongpoints. Point-blank artillery duels broke out around the buildings of the Ring, there was fierce fighting in the Graben and the Körtnerstrasse in the heart of the old city which had resisted the Turkish siege of 1683, and the Burgtheater and the Opera House were totally burnt out. Miraculously the Hofburg, the Albertina and the Kunsthistorischemuseum survived; but when the survivors of the German garrison eventually dragged themselves northward over the Danube across the Reichsbrücke on 13 April one of the great treasure-houses of European civilisation lay burning and devastated in acres behind them.

Crossing the Rhine

In the west too the great cities of the Reich were now falling to Allied attack. Eight armies were aligned along the west bank of the Rhine at the beginning of March, from north to south the Canadian First, Allied First Airborne, British Second, American Ninth, First, Third and Seventh and French First, the last facing the Black Forest on the far bank of the river. Patton's Third and Patch's Seventh Armies were still separated from the Rhine by the difficult terrain of the Eifel, but both succeeded in driving deep corridors to the river by 10 March. Eisenhower's plan for the Rhine crossing consisted of a deliberate assault on a wide front, with the heaviest effort to be made in the north by the Canadian, British and American Ninth and First Armies, aimed at encircling the great industrial region of the Ruhr. The British Second and American Ninth Armies' operations, codenamed respectively Plunder and Grenade, were vast and spectacular offensives involving large numbers of amphibious craft, massive air and artillery preparations and the dropping of two divisions of the Allied Airborne Army behind the German defences on the east bank of the river. They began on 23 March and were lightly opposed; the Allied Liberation Army now contained eighty-five divisions and numbered 4 million men, while the real strength of the *Westheer* was only twenty-six divisions.

The evolution of Eisenhower's plan, however, had already been altered by a chance event. On 7 March spearheads of the US 9th Armoured Division, belonging to the First Army, had found an unguarded railway bridge across the Rhine at Remagen below Cologne and had rushed it to establish a bridgehead on the far side. It could not at first be exploited, but on 22 March Patton's Third Army established another bridgehead by surprise assault near Oppenheim. The German defences of the Rhine were therefore broken at two widely separated places, in the Ruhr and at its confluence with the river Main at Mainz, thus threatening the whole Wehrmacht position in the west with envelopment on a large scale. On 10 March Hitler had relieved Rundstedt of supreme command in the theatre (it was the old warrior's third and last dismissal) and replaced him with Kesselring, brought from Italy where he had so successfully contained the Anglo-American drive up the peninsula; but a change of commanders could not now deflect the inevitable penetration of Germany's western provinces by the seven Allied armies. While the British and Canadian armies pressed on into northern Germany, aiming towards Hamburg, the US Ninth and First Armies proceeded with the encirclement of the Ruhr and completed it on 1 April, forcing the surrender of 325,000 German soldiers and driving their commander, Model, to suicide. At the same time Patton's Third Army was embarking on a headlong thrust into southern Germany which at the beginning of May would have carried it to within thirty miles of both Prague and Vienna.

On the evening of 11 April the US Ninth Army reached the river Elbe, designated the previous year as the demarcation line between the Soviet and Western occupation zones in Germany. At Magdeburg the 2nd Armoured Division seized a bridgehead across the Elbe and next day the 83rd Division established another at Barby; their soldiers believed they were going to Berlin, since the 83rd Division was only fifty miles away after enlarging its bridgehead on 14 April. Word swiftly came down the line, however, that they were misled. Eisenhower was bound by the inter-Allied agreement, according to which his American forces in the central sector would stay where they were, while the British and Canadians continued to clear northern Germany and the southernmost American and French armies overran Bavaria and occupied the territory in which Allied intelligence suggested the Germans might be organising a 'national redoubt'. The capture of Berlin was to be left exclusively to the Red Army.

It was not, however, to be a simple operation of war, but a race between military rivals. In November 1944 Stalin had promised Zhukov – who as his personal military adviser, senior army staff officer and operational commander was the principal architect of the Red Army's victories – that he should have the privilege of taking Berlin. Then on 1 April, at a Stavka meeting in Moscow devoted to ensuring that the Soviets and not the Western powers would be the first into the Reich capital, General A.I. Antonov of the General Staff posed the question how the demarcation line between Zhukov's and Konev's fronts should be drawn. To exclude Konev from the drive on Berlin would be to make the final operation more difficult than it need be. Stalin listened to the argument and then, drawing a pencil line on the situation map, designated their approach routes to within forty miles of the city. Thereafter, he said, 'Whoever breaks in first, let him take Berlin.'

The fall of Berlin

The two fronts jumped off across the Oder on 16 April. On Zhukov's front the honour of leading the assault went to Chuikov's Eighth Guards Army (formerly the Sixty-Second Army, which had defended Stalingrad), whose soldiers had sworn an oath to fight without thought of retreat in the coming battle. German resistance was particularly strong in their sector, however, and at the end of the day it was Konev's front which had made greater progress. On 17 April Konev continued to make the faster advance, closing on the Spree, Berlin's river, and persuaded Stalin by telephone that he was now better placed to open the assault on the city from the south, rather from the direct eastern route on which Zhukov's armoured columns were labouring against fierce opposition by German anti-tank teams. Zhukov now lost patience with his subordinate commanders and demanded that they lead their formations against the German defences in person; officers who showed themselves 'incapable of carrying out assignments' or 'lack of resolution' were threatened with instant dismissal. This warning produced a sudden and notable increase in the pace of advance through the Seelow heights. By the evening of 19 April Zhukov's men had cracked all three lines of defences between the Oder and Berlin and stood ready to assault the city.

Rokossovsky's Second White Russian Front was now aiding Zhukov's advance by pressing the German defenders of the lower Oder, where their defences still held, from the north. Zhukov was more concerned by the urgent advance of Konev's front through Cottbus, on the Spree, to Zossen, the headquarters of OKH, since it threatened to take the capital's fashionable suburbs from the south. On the evening of 20 April, when Konev ordered his leading army 'categorically to break into Berlin tonight', Zhukov brought up the guns of the 6th Breakthrough Artillery Division and began the bombardment of the streets of the capital of the Third Reich.

On 20 April Hitler celebrated his fifty-sixth birthday with bizarre solemnity in the bunker, leaving it briefly to inspect an SS unit of the *Frundsberg* Division and to decorate a squad of Hitler Youth boys, orphans of the Allied bombing raid on Dresden, who were defending the capital. This was to be his last public appearance. His power over the Germans nevertheless remained intact. On 28 March he had dismissed Guderian as chief of staff of the German army and replaced him with General Hans Krebs, once military attaché in Moscow and now installed in the bunker at his side; soon the Führer would dismiss others who had managed to make their way to the bunker to offer their congratulations on his birthday, including Goering, as head of the Luftwaffe, and Himmler, as head of the SS. There would be no lack of Germans willing to carry out these orders; more impressively, there was no lack of Germans, whether or not intimidated by the 'flying courts martial' which had begun to hang deserters from lampposts, ready to continue the fight for the Nazi regime. Keitel and Jodl, intimates of every one of his command conferences throughout the war, left the bunker on 22 April to take refuge at Fürstenberg, thirty miles north of Berlin and conveniently close to Ravensbruck concentration camp, where a group of so-called *Prominenten*, well-connected foreign prisoners, were held as hostages. Dönitz, the Grand Admiral, went to Plön, near Kiel on the Baltic, immediately after his last interview with the Führer on 21 April; he had transferred naval headquarters there during March. Speer, chief of war industry, came and went on 23 April; other visitors included Ribbentrop, still his Foreign Minister, his adjutant Julius Schaub, his naval representative Admiral Karl-Jesko von Puttkamer, and his personal physician Dr Theodor Morell, whom many in the inner circle believed had secured his privileged place by dosing Hitler with addictive drugs.

A few others actually overcame great danger to make their way to the bunker, including Goering's successor as commander of the Luftwaffe, General Robert Ritter von Greim, and the celebrated test pilot, Hanna Reitsch, who succeeded in landing on the East-West Axis in a training aircraft, while outside the bunker the garrison of Berlin kept up a ferocious struggle against the encroaching Russian formations throughout the week beween 22 April, the day on which Hitler definitively announced his refusal to leave – 'Any man who wants may go! I stay here' – and his suicide on 30 April.

On the morning of 21 April, Zhukov's tanks entered the northern suburbs, and the units following them were regrouped for siege warfare: Chuikov, who had fought the Battle of Stalingrad, knew what was necessary. Assault groups were formed from a company of infantry, supported by half a dozen anti-tank guns, a troop of tank or assault guns, a couple of engineer platoons and a flamethrower platoon. According to the theory of siege warfare, assault weapons were used to blast or burn down resistance in the city blocks, into which the infantry then attacked. Overhead the heavy artillery and rocket-launchers threw crushing salvoes to prepare the way for the next stage, house-to-house fighting. Medical teams stood close in the rear; street fighting produces exceptionally heavy casualties, not only from gunshot at short range but also from falls between storeys and the collapse of debris.

On 21 April, Zossen fell into the hands of Konev's front, its elaborate telephone and teleprinter centre still receiving messages from army units all over what remained of unconquered Germany. The next day Stalin finally delineated the thrust lines for the advance into central Berlin. Konev's sector was aligned on the Anhalter railway station, a position which ensured that his vanguard would be 150 yards away from the Reichstag and Hitler's bunker. Zhukov, whose troops were already dug deep into the city's streets, was to be the 'conqueror of Berlin' after all, as Stalin had promised the previous November.

However, German resistance was still stiffening. From his bunker Hitler constantly demanded the whereabouts of the two surviving military formations nearest the city, General Walther Wenck's Twelfth and General Theodor Busse's Ninth Armies. Although he railed at their failure to come to his rescue, both were fighting hard from the west and south-east to check or throw back the Soviet advance. Nevertheless by 25 April Konev and Zhukov had succeeded in encircling the city from south and north and were assembling unprecedented force to reduce resistance within it. For the final stage of the assault on the centre, Konev massed artillery at a density of 650 guns to the kilometre, literally almost wheel to wheel, and the Soviet 16th and 18th Air Armies had also been brought up to drive away the remnants of the Luftwaffe still trying to fly munitions into the perimeter, either via Tempelhof, the inner Berlin airport, or on to the great avenue of the East-West Axis (by which Greim and Reitsch made their spectacular arrival and eventual departure) in the city centre.

On 26 April 464,000 Soviet troops, supported by 12,700 guns, 21,000 rocket-launchers and 1500 tanks, ringed the inner city ready to launch the final assault of the siege. The circumstances of the inhabitants were now frightful. Tens of thousands had crowded into the huge concrete 'flak towers', impervious to high explosive, which dominated the centre; the rest, almost without exception, had taken to the cellars, where living conditions rapidly became as squalid. Food was running short, so too was water, while the relentless bombardment had interrupted electrical and gas supplies and sewerage; behind the fighting troops, moreover, ranged those of the second echelon, many released prisoners of war with a bitter personal grievance against Germans of any age or sex, who vented their hatred by rape, loot and murder.

By 27 April, when a pall of smoke from burning buildings and the heat of combat rose a thousand feet above Berlin, the area of the city still in German hands had been reduced to a strip some ten miles long and three miles wide, running in an east-west direction. Hitler was demanding the whereabouts of Wenck; but Wenck had failed to break through, as had Busse's Ninth Army, while the remnants of Manteuffel's Third Panzer Army were withdrawing to the west. Berlin was now defended by remnants, including shreds of foreign SS units – Balts, and Frenchmen from the Charlemagne Division, as well as Degrelle's Walloons, whom the chaos of fighting had tossed into the environs of the bunker. On 28 April these last fanatics of the National Socialist revolution found themselves fighting for its government buildings in the Wilhelmstrasse, the Bendlerstrasse and near the Reich Chancellery itself. Professor John Erickson has described the scene:

The *Tiergarten*, Berlin's famous zoo, was a nightmare of flapping, screeching birds and broken, battered animals. The 'cellar tribes' who dominated the life of the city crept and crawled about, but adding to the horror of these tribalised communities clinging to life, sharing a little warmth and desperately improvised feeding, when the shelling stopped and the assault troops rolled through the houses and across the squares, there followed a brute, drunken, capricious mob of rapists and ignorant plunderers. . . . Where the Russians did not as yet rampage, the SS hunted down deserters and lynching commands hanged simple soldiers on the orders of young, hawk-faced officers who brooked no resistance or excuse.

On the same day the German defenders of the central area around the Reich Chancellery and the Reichstag tried to hold off the northern Russian thrust into this 'citadel', as it had been designated, by blowing the Moltke bridge over the river Spree. The demolition damaged but did not destroy it, and it was rushed early next morning under cover of darkness. There then followed a fierce battle for the Ministry of the Interior building 'Himmler's house', as the Russians dubbed it – and shortly afterwards for the Reichstag. Early on 29 April the fighting was less than a quarter of a mile from the Reich Chancellery, which was being demolished by heavy Russian shells, while 55 feet beneath the surface of the cratered garden Hitler was enacting the last decisions of his life. He spent the first part of the day dictating his 'political testament', enjoining the continuation of the struggle against Bolshevism and Jewry, and he then entrusted copies of this to reliable subordinates who were ordered to smuggle them through the fighting lines to OKW headquarters, to Field Marshal Schörner and to Grand Admiral Dönitz. By separate acts he appointed Schörner to succeed him as commander-in-chief of the German army and Dönitz as head of state. Doónitz's headquarters at Ploón thus became the Reich's temporary seat of government, and he would remain there until 2 May, when he transferred to the naval academy at Mürwik, near Flensburg, in Schleswig-Holstein. Hitler also dismissed Speer, for recently revealed acts of insubordination in refusing to carry out a 'scorched earth' policy, and expelled Goering and Himmler from the Nazi Party, the former for daring to anticipate his promised succession to Hitler's place, the latter for having made unauthorised peace approaches to the Western Allies. He had already appointed Ritter von Greim commander of the Luftwaffe and specified eighteen other military and political appointments to Dönitz in the political testament. He also married Eva Braun, who had arrived in the bunker on 15 April, in a civil ceremony performed by a Berlin municipal official hastily recalled from his *Volkssturm* unit defending the 'citadel'.

Hitler had not slept during the night of 28/29 April and retired to his private quarters until the afternoon of 29 April. He attended the evening conference, which began at ten o'clock, but the meeting was a formality, since the balloon which supported the bunker's radio transmitting aerial had been shot down that morning and the telephone switchboard no longer communicated with the outside world. General Karl Weidling, the 'fortress' commander of Berlin, warned that the Russians would certainly break through to the Chancellery by 1 May, and urged that the troops remaining in action be ordered to break out of Berlin. Hitler dismissed the possibility. It was clear that he was committed to his own end.

During the night of 29/30 April he took his farewells first from the women – secretaries, nurses, cooks – who had continued to attend him in the last weeks, then from the men – adjutants, party functionaries and officials. He slept briefly in the early morning of 30 April, attended his last situation conference, at which the SS commandant of the Chancellery, Wilhelm Mohnke, reported the progress of the fighting around the building, and then adjourned for lunch with his two favourite secretaries, Gerda Christian and Traudl Junge, who had spent the long months with him at Rastenburg and Vinnitsa. They ate noodles and salad and talked sporadically about dogs; Hitler had just had his cherished Alsatian bitch, Blondi, and four pups destroyed with the poison he intended to use himself, and inspected the corpses to assure himself that it worked. Eva Braun, now Frau Hitler, remained in her quarters; then about three o'clock she emerged to join Hitler in shaking hands with Bormann, Goebbels and the other senior members of the entourage who remained in the bunker. Hitler then retired with her into the private quarters – where Frau Goebbels made a brief and hysterical irruption to plead that he escape to Berchtesgaden – and after a few minutes, measured by the funeral party which waited outside, together they took cyanide. Hitler simultaneously shot himself with a service pistol.

An hour earlier soldiers of Zhukov's front, belonging to the 1st Battalion, 756th Rifle Regiment, 150th Division of the Third Shock Army, had planted one of the nine Red Victory Banners (previously distributed to the army by its military soviet) on the second floor of the Reichstag, chosen as the point whose capture would symbolise the end of the siege of Berlin. The building had just been brought under direct fire by eighty-nine heavy Russian guns of 152 mm and 203 mm; but its German garrison was still intact and fighting. Combat within the building raged all afternoon and evening until at a little after ten o'clock a final assault allowed two Red Army men of the 1st Battalion of the 756th Regiment, Mikhail Yegorov and Meliton Kantaria, to hoist their Red Victory Banner on the Reichstag's dome.

· The bodies of Hitler and his wife had by then been incinerated by the funeral party in a shell crater in the Chancellery garden. Once the flames, kindled with petrol brought from the Chancellery garage, had died down the remains of the bodies were buried in another shell crater nearby (from which they were to be disinterred by the Russians on 5 May). Shells were falling in the garden and in the Chancellery area, and fighting was raging in all the government buildings in the 'citadel'. Goebbels, appointed Reich Chancellor at the same time as Hitler nominated Dönitz to succeed him as head of state, nevertheless felt it important to make contact with the Russians to arrange a truce so that preparations could be made for peace talks, which, in the deluded atmosphere prevailing in the bunker, he believed were possible. Late in the evening of 30 April a colonel was sent as emissary to the nearest Russian headquarters, and early on the morning of 1 May General Krebs, since 28 March the army chief of staff, but formerly military attaché in Moscow (at the time of Barbarossa) and a Russian-speaker, went forward through the burning ruins to treat with the senior Soviet officer present. It was Chuikov, now the commander of the Eighth Guards Army, but who two years earlier had commanded the Russian defenders in the siege of Stalingrad.

A strange four-sided conversation developed. Chuikov heard Krebs out and was then connected by telephone to Zhukov, who in turn spoke to Stalin in Moscow. 'Chuikov reporting,' the general said. 'General of Infantry Krebs is here. He has been authorised by the German authorities to hold talks with us. He states that Hitler ended his life by suicide. I ask you to inform Comrade Stalin that power is now in the hands of Goebbels, Bormann and Admiral Dönitz. . . . Krebs suggests a cessation of military operations at once.' Krebs, however, like Bormann and Goebbels, remained deluded by the belief that the Allies would be ready to treat with Hitler's successors as if they were legitimate inheritors of the authority of a sovereign government. Stalin tired quickly of the conversation, declared abruptly that the only terms were unconditional surrender and went to bed. Zhukov persisted a little longer but then announced that he was sending his deputy, General Sokolovsky, and broke off communication. Sokolovsky and Chuikov between them engaged in interminable parleys with Krebs, who had difficulty in establishing his credentials, so murky were recent developments in the bunker (with which he communicated twice by runner). Eventually Chuikov's patience ran out. In the early afternoon of 1 May he told Krebs that the new government's powers were limited to 'the possibility of announcing that Hitler is dead, that Himmler is a traitor and to treat with three governments – USSR, USA and England – on *complete capitulation*'. To his own forces Chuikov sent the order: 'Pour on the shells . . . no more talks. Storm the place.' At 6.30 pm on 1 May every Soviet gun and rocket-launcher in Berlin opened fire on the unsubdued area. The eruption was signal enough to those remaining in the bunker that hopes of arranging a succession were illusory. About two hours later Goebbels and his wife – who had just killed her own six children by the administration of poison – committed suicide in the Chancellery garden close to Hitler's grave. Their bodies were more perfunctorily cremated and buried nearby. The rest of the bunker party, underlings as well as grandees like Bormann, now organised themselves into escape parties and made their way through the burning ruins towards what they hoped was safety in the outer suburbs. Meanwhile the Soviet troops – understandably reluctant to risk casualties in what were clearly the last minutes of the siege of Berlin – pressed inward behind continuous salvoes of artillery fire. Early on the morning of 2 May LVI Panzer Corps transmitted a request for a cease-fire. At 6 am Weidling, the commandant of the Berlin 'fortress', surrendered to the Russians and was brought to Chuikov's headquarters, where he dictated the capitulation signal: 'On 30 April 1945 the Führer took his own life and thus it is that we who remain – having sworn him an oath of loyalty – are left alone. According to the Führer's orders, you, German soldiers, were to fight on for Berlin, in spite of the fact that ammunition had run out and in spite of the general situation, which makes further resistance on our part senseless. My orders are: to cease resistance forthwith.'

In John Erickson's words: 'At 3 pm on the afternoon of 2 May Soviet guns ceased to fire on Berlin. A great enveloping silence fell. Soviet troops cheered and shouted, breaking out the food and drink. Along what had once been Hitler's parade route, columns of Soviet tanks were drawn up as for inspection, the crews jumping from their machines to embrace all and sundry at this new-found cease-fire.' The peace which surrounded them was one of the tomb. About 125,000 Berliners had died in the siege, a significant number by suicide; the suicides included Krebs and numbers of others in the bunker party. Yet probably tens of thousand of others died in the great migration of Germans from east to west in April, when 8 million left their homes in Prussia, Pomerania and Silesia to seek refuge from the Red Army in the Anglo-American occupation zones. By one of the most bizarre lapses of security in the entire war, the demarcation line agreed between Moscow, London and Washington had become known to the Germans during 1944, and the last fight of the Wehrmacht in the west was motivated by the urge to hold open the line of retreat across the Elbe to the last possible moment. Civilians too seem to have learned where safety lay and to have pressed on ahead of the Red Army to reach it – but at terrible cost.

The cost to the Red Army of its victory in the siege of Berlin had also been terrible. Between 16 April and 8 May, Zhukov, Konev and Rokossovsky's fronts had lost 304,887 men killed, wounded and missing, 10 per cent of their strength and the heaviest casualty list suffered by the Red Army in any battle of the war (with the exception of the captive toll of the great encirclement battles of 1941). Moreover, the last sieges of the cities of the Reich were not yet over. Breslau held out until 6 May, its siege having cost the Russians 60,000 killed and wounded; in Prague, capital of the 'Reich Protectorate', the Czech National Army resistance group staged an uprising in which the puppet German 'Vlasov army' changed sides and skirmished against the SS garrison in the hope of delivering the city to the Americans – a vain hope, for which Vlasov's men paid a terrible price in blood when the Red Army entered it on 9 May.

By then, however, the war in what remained of Hitler's empire was almost everywhere over. A local armistice had been arranged in Italy, through the SS General Karl Wolff, on 29 April, scheduled for announcement on 2 May. On

3 May Admiral Hans von Friedeburg surrendered the German forces in Denmark, Holland and North Germany to Montgomery. On 7 May Jodl, dispatched by Dönitz from his makeshift seat of government at Flensburg in Schleswig-Holstein, signed a general surrender of German forces at Eisenhower's headquarters at Reims in France. It was confirmed at an inter-Allied meeting in Berlin on 10 May. Norway, which the Russians had fractionally penetrated only at the very north of the country from Finland in October 1944, was surrendered by its intact German garrison on 8 May. The Courland pocket capitulated on 9 May. Dunkirk, La Pallice, La Rochelle and Rochefort, the last of the 'Führer fortresses' in western Europe, surrendered on 9 May, as did the Channel Islands, Lorient and Saint-Nazaire on 10 May. The final surrender of the war in the West was at Heligoland on 11 May.

Peace brought no rest to the human flotsam of the war, which swirled in hordes between and behind the victorious armies. Ten million Wehrmacht prisoners, 8 million German refugees, 3 million Balkan fugitives, 2 million Russian prisoners of war, slave and forced labourers by the millions – and also the raw material of the 'displaced person' tragedy which was to haunt Europe for a decade after the war – washed about the battlefield. In Britain and America crowds thronged the streets on 8 May to celebrate 'VE Day'; in the Europe to which their soldiers had brought victory, the vanquished and their victims scratched for food and shelter in the ruins the war had wrought.

PART VI
THE WAR IN
THE PACIFIC
1943–1945

THE ALLIED ADVANCES IN THE PACIFIC AND ASIA, MARCH 1945

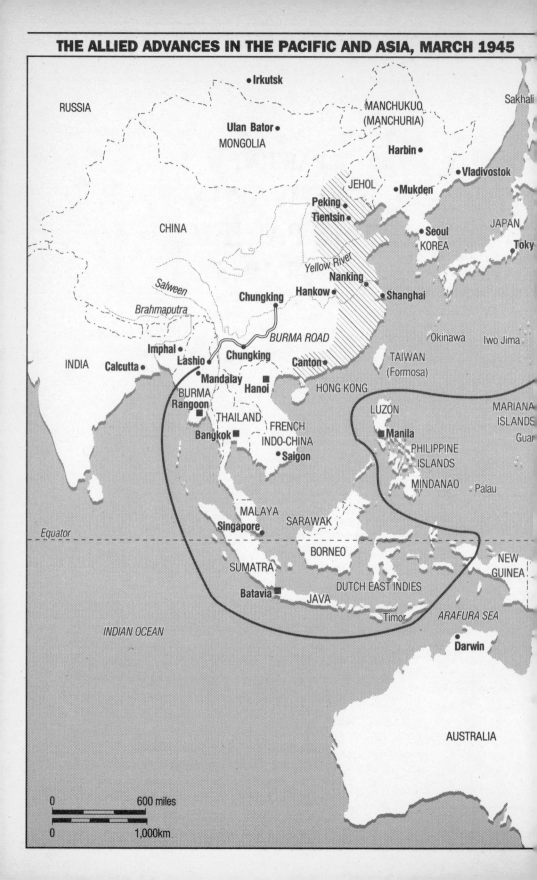

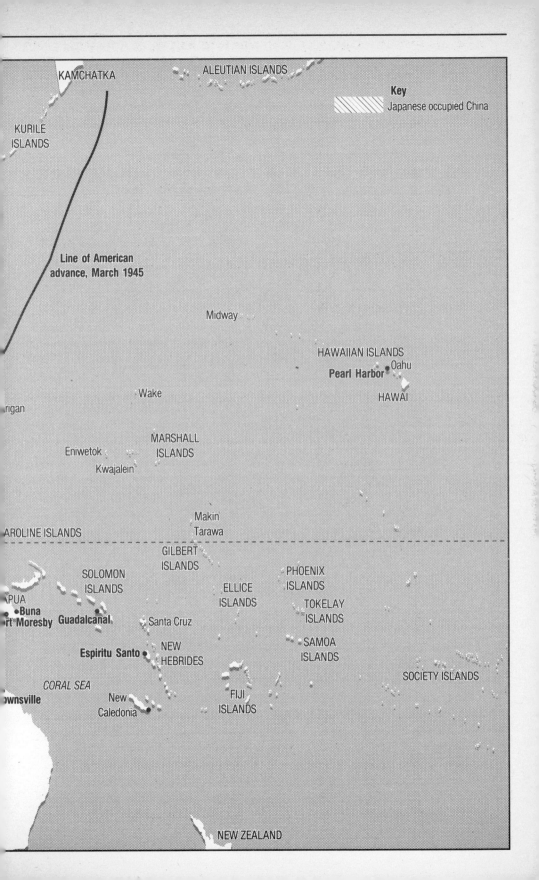

KAMCHATKA

ALEUTIAN ISLANDS

Key
Japanese occupied China

KURILE
ISLANDS

**Line of American
advance, March 1945**

Midway

HAWAIIAN ISLANDS
Oahu
Pearl Harbor

HAWAI

Wake

rigan

MARSHALL
ISLANDS

Eniwetok

Kwajalein

Makin

AROLINE ISLANDS

Tarawa

GILBERT
ISLANDS

SOLOMON
ISLANDS

PHOENIX
ISLANDS

PUA

ELLICE
ISLANDS

TOKELAY
ISLANDS

•Buna
rt Moresby **Guadalcanal**

Santa Cruz

SAMOA
ISLANDS

Espiritu Santo•

NEW
HEBRIDES

SOCIETY ISLANDS

CORAL SEA

New
Caledonia

FIJI
ISLANDS

wnsville

NEW ZEALAND

TWENTY-NINE

Roosevelt's Strategic Dilemma

The news of Roosevelt's death on 12 April 1945 had stirred a flicker of optimism in the Berlin bunker. Hitler had sustained his spirits during the last year of the war by two beliefs: that his secret weapons would break the will of the British; and that the contradictions of an alliance between a decadent capitalist republic, a moribund empire and a Marxist dictatorship must inevitably lead to the disintegration of that alliance. By March 1945, when his V-2 had been driven beyond the last sites from which Britain could be hit, he knew that his secret weapons had failed. Thereafter he clung all the more desperately to the hope of dissension among the Allies. Goebbels, the political philosopher of his court, had explained to some intimates in early April how such a falling-out might occur. According to the historian Hugh Trevor-Roper, he had 'developed his thesis that, for reasons of Historical Necessity and Justice, a change of fortune was inevitable, like the Miracle of the House of Brandenburg in the Seven Years War.' When Frederick the Great of Prussia faced defeat by the combined armies of Russia, Austria and France in the Seven Years War, the tsarina Elisabeth had unexpectedly died, to be succeeded by a tsar who was Frederick's admirer; the alliance then collapsed and Frederick's Prussia survived. In April 1945, on hearing the news of the President's passing, Goebbels exclaimed, 'the tsarina is dead', and telephoned Hitler 'in an ecstasy' to 'congratulate' him. 'It is the turning-point,' he said, 'it is written in the stars.'

Hitler himself was briefly moved to share Goebbels's euphoria. Throughout the latter years of the war he had come to identify closely with Frederick the Great and was even ready to believe that the evolution of his fortunes might mirror those of the Prussian king. He was particularly ready to believe that Roosevelt's death would produce the disabling crack in the alliance that he predicted, since one of his fundamental misappreciations was that the American people were unwarlike and had been drawn into the conflict by the machinations of their President. 'The arch-culprit for this war', he had told a Spanish diplomat in August 1941, 'is Roosevelt, with his freemasons, Jews and general Jewish-Bolshevism.' He said, whether he believed it or not, that he had proof of Roosevelt's 'Jewish ancestry'. He was certainly obsessed by the number of Jews in American government, including Henry Morgenthau, the Secretary of the Treasury, whose plan to reduce defeated Germany to a nation of cultivators and pastoralists had been leaked and republished in the German press in September 1944, to the great benefit of Goebbels's propaganda for a 'total war' effort.

Hitler's understanding of America's commitment to the war was almost exactly contrary to the truth. Isolationism was certainly a powerful force in American politics before December 1941, while America's parents remained naturally reluctant to see their sons depart to a foreign war up to the moment of Pearl Harbor. Few Americans, however, objected to the measures of rearmament enacted in 1940, which doubled the size of the fleet, allocated funds for an air force of 7800 combat aircraft – three times the size of the Luftwaffe – and increased the size of the army from 200,000 to one million men, to be raised by conscription. When war came, moreover, the nation reacted enthusiastically. The sense of being 'out of things' had waxed powerfully in the United States during the eighteen months of the Blitz and the Battle of the Atlantic; so too had hostility to Hitler, as a paradigm of everything against which American civilisation stood. As in Europe in 1914, the coming of war was ultimately almost a relief, since Americans had been oppressed by indecision and inactivity and were untainted by any fear of defeat.

Roosevelt too saw Hitler as a tyrant and a malefactor. However, Hitler's belief that Roosevelt dragged his people to war reluctantly behind him is at variance with the facts; more accurately, the facts of Roosevelt's war policy, in the months before Pearl Harbor, defy objective arrangement or analysis. Roosevelt's attitude to United States entry into the Second World War remains profoundly ambiguous, as do the aims and objectives of his war-making in the three and a half years in which he acted as commander-in-chief of the United States forces.

Roosevelt is by far the most enigmatic of the major figures of 1939-45. Stalin, though devious, double-dealing and treacherous in his methods, steadfastly pursued a quite limited set of aims: while determined to sustain his position as head of government, party and army, whomever he had to dismiss or even kill to maintain his power, he wanted, first, to save the Soviet Union from defeat, second, to expel the Wehrmacht from Soviet territory and, third, to extract the largest possible benefit – territorial, diplomatic, military and economic – from the Red Army's eventual victory. Hitler, however mysterious the workings of his psyche, also held to a clear-cut if wildly over-ambitious strategy: he wanted revenge for Versailles, then German mastery of the continent, followed by the subjugation of Russia and the eventual exclusion of the Anglo-Saxon powers from any influence in European affairs. Churchill was transparently a patriot, a romantic and an imperialist: victory was his first and last desire; only secondarily did he consider how victory might be gained in a way that secured British interests in Europe and the survival of the British Empire overseas. His 'naturally open and unsuspicious nature', as his wife described it, automatically revealed his motives to all who treated with him during the war. Captious and contrary though he often was, he had no capacity for sustained dissimulation, grasped eagerly at the semblance of generosity in the statements of others and was as powerfully swayed as his listeners by the force and nobility of his own oratory.

Roosevelt too was a magnificent speaker; his range, indeed, was far greater than Churchill's, for he was the master of not only the high-flown set-piece – his proclamation of the 'Four Freedoms' to Congress in January 1941, for example, or his 'Day of Infamy' speech after Pearl Harbor – but also the intimate radio appeal to families and individuals in his 'fireside chats', a medium of political communication which he himself invented, the *ad hominem* stump speech of the political campaign, subtly varied from place to place and audience to audience, the disingenuously frank news conference, the personal telephone call, above all the face-to-face conversation, flattering, funny, discursive, beguiling and ultimately almost wholly baffling to the interlocutor who sat mesmerised by the flow of words. Roosevelt was a magician with words. According to his biographer James McGregor Burns, he sent visitors away from the Oval Office entranced by his 'expansiveness, openness, geniality'; but they rarely took back with them any answer to the problems or questions they had brought. For Roosevelt talked perhaps above all 'to find bearings and moorings in his own experiences and recollections'. Roosevelt had dozens of attitudes and a few deeply held values, which were precisely those of Americans of his class and time: he believed in human dignity and freedom, in economic opportunity, in political compromise, he felt deeply for the hardships of the poor, and he detested recourse to violence; but he had few policies, either for peace or for war, while war itself he found utterly distasteful.

Hence his profoundly ambiguous attitude towards American involvement. Churchill had sustained his own spirits during the darkest hours of 1940 and 1941 by the belief that the New World would eventually come forth to redress the balance of the Old. Roosevelt had given him every reason to believe that such would be the outcome. He had erected an American armed neutrality against the Axis almost from the moment of Hitler's opening of the war, selling arms to Britain and France which would certainly have been refused to Germany, then authorising unrestricted 'cash and carry' arms shipments and progressively extending American protection to Britain-bound convoys in the Atlantic. He first defined a neutrality zone which effectively denied the U-boats access to American waters, then in April 1941 extended the zone to the mid-ocean line and allowed American warships to act as convoy escorts, while in July he dispatched American Marines to replace British troops in the garrison of Iceland, which Britain had summarily occupied after the fall of Denmark in 1940. On 11 March 1941 Congress, at his persuasion, passed the Lend-Lease Act, which effectively allowed Britain to borrow war supplies from the United States against the promise of later repayment; in February he had sponsored Anglo-American staff talks in Washington (the ABC-1 conference) which agreed on most of the strategic fundamentals, including 'Germany First', which would in practice be implemented after December.

By every outward sign, therefore, Churchill had reason to believe that the President was leading his nation to intervention on Britain's side; certainly Hitler was acutely aware of that danger and laid his U-boat commanders under strict orders not to provoke the Americans in any way at all – even after Roosevelt authorised the freezing of all German assets in the United States in June 1941. Churchill, in his private conversations with the President on the transatlantic scrambler telephone (erratically intercepted by the Germans), was given even more strongly to understand the warmth of the President's commitment, while from the Placentia Bay meeting of August 1941 he brought back the agreement that the United States Navy would protect all ships in a convoy which included one American ship, in effect a means of defying Dönitz to sink an American warship. On his return from Placentia Bay, Churchill told the war cabinet that Roosevelt was 'obviously determined that they should come in'; his concluding message had been that 'he would wage war but not declare it, and that he would become more and more provocative'. If the Germans 'did not like it', Roosevelt said, 'they could attack American forces.'

Churchill's Chiefs of Staff were more suspicious and formed a different impression. A staff officer, Ian Jacob, noted in his diary that the United States Navy 'seem to think that the war can be won by our simply not losing it at sea', and that the army 'sees no prospect of being able to do anything for a year or two'. He observed that not 'a single American officer had shown the slightest keenness to be in the war on our side. They are a charming lot of individuals but they appear to be living in a different world from ourselves.' Moreover, when on 31 October the Germans committed the ultimate provocation by sinking the destroyer USS *Reuben James* in the Atlantic with the loss of 115 American lives, Roosevelt chose not to regard it as a *casus belli* – though it was a far more flagrant act of aggression than, for example, the 'Gulf of Tonkin' incident used by President Johnson to authorise American military intervention in Vietnam in 1964.

Roosevelt's inaction over the sinking of the *Reuben James* may be taken as the key to the 'strategic enigma' he remained during 1941, as his biographer James MacGregor Burns has characterised him.

Roosevelt was following a simple policy: all aid to Britain short of war. This policy was part of a long heritage of Anglo-American friendship; it was a practical way of blocking Hitler's aspirations in the west; it could easily be implemented by two nations used to working with one another; it suited Roosevelt's temperament, met the needs and pressures of the British, and was achieving a momentum of its own. But it was not a grand strategy . . . it did not emerge from clear-cut confrontation of political and military alternatives. . . . Above all this strategy was a negative one in that it could achieve its full effect – that is, joint military and political action with Britain – only if the Axis took action that would force the United States into war. It was a strategy neither of war nor of peace, but a strategy to take effect (aside from war supply to Britain and a few defensive actions in the Atlantic) only in the event of war. . . . [Roosevelt] was still waiting for a major provocation from Hitler even while recognising that it might not come at all. Above all, he was trusting to luck, to his long-tested flair for timing. . . . He had no plans. 'I am waiting to be pushed into the situation,' he told Morgenthau in May – and clearly it had to be a strong shove.

Trusting to luck and waiting to be shoved were to characterise Roosevelt's conduct as commander-in-chief from Pearl Harbor almost to the very end of his life. Revisionist historians have argued that he was playing a deep game both before the United States entry into the war and during the years thereafter: that he saw in Britain's isolation and desperate need for arms on any terms a means of liquidating her overseas investments (as they were indeed liquidated by 'cash and carry' sales), and thus of reducing the mistress of the world's greatest empire – an institution he disliked as strongly as he did industrial trusts and financial cartels within his own country – to a state where she could not resist American pressure to divest herself of her colonies. This is surely to endow Roosevelt with a Machiavellianism he did not possess. War, Machiavelli said, is the only proper study for a prince; and Roosevelt was indeed princely in a distinctively Renaissance style, transacting much of his business through a court favourite, Harry Hopkins, permitting no official – not even the implacable Marshall – to establish himself as indispensable to him, dispensing charm and empty flattery with lordly largesse, operating a political oubliette for those who incurred his displeasure, maintaining a private country palace as a refuge from the heats and longueurs of Washington (no Camp David for FDR), even formally maintaining a mistress in the White House and treating his cousin-wife of thirty years as the honoured spouse of a dynastic marriage of convenience. None the less Roosevelt was not Machiavellian in strategy, for the simple reason that the wealth, power and ethos of the New World had liberated its rulers from the Old World's narrow needs to dissemble and traduce. The United States had been founded on the principle of 'no entangling alliances'; it had grown up to riches which absolved it from the temptation to pursue cheap and temporary advantages over weaker states.

As a result, Roosevelt was able to hold aloof from the business of directing war, an activity alien to his temperament. Such an aloofness was not granted to any of the other leaders. Churchill, of course, revelled in high command, dedicated his days (and nights) to war-making, had rooms, suites, even whole houses adapted to his needs as a wartime Prime Minister, preferred his 'siren suit' to any other garb (though he also kept handy his uniforms of an honorary air commodore and an honorary colonel of the Cinque Ports Battalion), demanded a constant diet of Ultra intelligence intercepts and lived in hour-to-hour intimacy with his military advisers. Hitler turned himself into a military hermit after the opening of Barbarossa, seeing few but his generals, even though he found their company grating. Stalin's wartime routine conformed strangely in pattern to Hitler's – secretive, nocturnal, troglodyte. Roosevelt scarcely altered his pattern of life at all after Pearl Harbor. Unthreatened by air attack, he continued to live at the White House, occasionally vacationing at Hyde Park, and there pursued a timetable that drove the methodical and purposeful almost to distraction. Marshall's day was measured to the minute: his only relaxation was to visit his wife in his official quarters for lunch, which was served as he stepped on to the veranda from his staff car. Roosevelt lunched off a tray brought into the Oval Office, did not begin work until ten in the morning and took few telephone calls at night. According to Burns, there were a few fixed points in his week:

> He saw the congressional Big Four – the Vice-President, the Speaker and the majority leader of each chamber – on Monday or Tuesday; met with the press on Tuesday afternoons and Friday mornings; and presided over a Cabinet meeting on Friday afternoons. [Otherwise] there seemed to be no pattern at all in the way that Roosevelt did his work. Sometimes he hurried through appointments on crucial matters and dawdled during lesser ones. He ignored most letters altogether. . . . He took many phone calls, refused others, saw inconsequential and dull people, and ignored others of apparently greater political or intellectual weight – all according to some mystifying structure of priorities known to no one, perhaps not even to himself.

This pattern, or lack of it, persisted from 7 December 1941 to 12 April 1945. Unlike Churchill, who was constantly on the move – to Paris (before the fall of France), to Cairo, to Moscow, to Athens (where he spent Christmas Day 1944 while the sound of gunfire between British troops and ELAS rebels rocked the city), to Rome, Naples, Normandy, the Rhine – Roosevelt travelled little. His mobility was, of course, limited by his physical disability, which was the result of poliomyelitis and which a discreet press disguised from its readership almost completely. Nevertheless he travelled when he chose, but during the war his travels took him only to Casablanca in January 1943, Quebec twice (August 1943 and September 1944), Hawaii and Alaska in the summer and Cairo and Tehran at the end of 1944 and Yalta, in the Russian Crimea, in February 1945. He saw nothing of the war at first hand, no bombed cities, no troops at the front, no prisoners, no after-effects of battle, and probably did not choose to; he directed American strategy as he had directed the New Deal – by lofty rhetoric and by rare but decisive strikes at the conjunctions of power.

There were effectively four decisive actions in all. The first was his endorsement of the 'Germany First' decision, advanced by Admiral Stark, Chief of Naval Operations, in November 1940, adopted by the Anglo-American ABC-1 conference of February–March 1941, agreed with Churchill at Placentia

Bay in August, but enshrined as national policy only after Pearl Harbor, when Roosevelt, who might with his political heart so easily have yielded to the popular demand for vengeance on Japan, let his strategic head dictate that the greater should be beaten before the lesser enemy. The second was his settlement of the dispute between Marshall and the British in London in July 1942 on terms which authorised the Torch landing in North Africa, with all the dubious consequences that flowed from that expedition. The third was his insistence on the proclamation of 'unconditional surrender' at Casablanca in January 1943, a high-minded re-echoing of the terms on which the United States had conducted its war against the Confederacy. The last was his decision to distance himself from Churchill at the Yalta conference in February 1945 and deal directly with Stalin on the future of Europe.

There had been anticipations of Roosevelt's Yalta initiative both at Placentia Bay, when Churchill had reluctantly accepted the more liberalising provisions of the Atlantic Charter – which in effect committed the British Empire to granting independence to its colonies – and at the Cairo conference, where Roosevelt had shown a typically 'China lobby' over-tenderness to Chiang Kai-shek. At Cairo the British had been persuaded to surrender their historic rights of extraterritoriality in China as a token of commitment to their belief in the nominal equality of Chiang's leadership with that of the Western democracies.

Chiang Kai-shek was to let Roosevelt down. Contrary to the President's expectations, he neither went through the motions to reform China's political and economic structures – how could he have done, a realist might have asked, with the more productive half of his country in the hands of the enemy? – nor utilised American aid and American advice, supplied so liberally first by Stilwell and then, after Chiang had tired of Stilwell's lecturing, by Wedemeyer, to maximise China's fighting power.

By the time of Yalta, therefore, Roosevelt had privately written off Chiang; for form's sake, China was elevated to permanent membership of the Security Council of the United Nations Organisation, whose institution and structure was decided at Yalta, but Chiang was accorded no fruits of the victory he had done so little to advance, certainly not the annexation of Indo-China he had been offered at Cairo. Poland too was written off at Yalta, though it had fought every day of the war since 1 September 1939, maintaining an army in exile which stood fourth in size among those opposed to the Wehrmacht, after the Russian, American and British; its eastern provinces, over-generously delimited in 1920, were permanently transferred to Russia at Yalta, though this Roosevelt-Stalin deal was an act less of political treachery than of political reality, since the Red Army already occupied the whole of Poland's territory.

However, the most important of all decisions taken at Yalta, agreed directly between Roosevelt and Stalin, concerned the future conduct of the war in the Pacific. Roosevelt's willingness to barter away the future of Poland and to finalise a division of Germany which accorded the Soviet Union an over-generous allocation of occupation territory was ultimately determined by his anxiety to engage the Red Army in the battle to defeat Japan. At the time of Yalta, the United States had neither yet assured itself that its nuclear-research programme would result in the successful test explosion of an atomic bomb nor advanced its forces to a point from which the land invasion of Japan might be undertaken. The amphibious assault on Iwo Jima was in preparation but had not been launched; the devastating fire-bombing of Japan had not begun. The Red Army's commitment in Europe, on the other hand, was clearly almost at an end, and from western Russian the Trans-Siberian railway led directly to the border of Manchuria, where in 1904–5 Tsar Nicholas II's army had suffered a humiliating defeat. The opportunity to avenge it stood high on the list of Stalin's wartime priorities. When he might take the opportunity, however, was what preoccupied the American President. To ensure that he did so sooner rather than later motivated almost all Roosevelt's initiatives at Yalta. The price he paid in the end was to discredit Churchill in the eyes of their joint Polish allies, to concede Russia rights over territory in sovereign China which were not America's to grant, but ultimately to assure that the repossession of Japan's conquests in the Pacific would not be bought at the cost of American lives alone. To a nation which had watched the heroic advance of the United States Navy, Marine Corps and MacArthur's army divisions from New Guinea to the Philippines, the diplomatic price paid at Yalta – when the cost to a distant European state's territory and to Britain's good name was balanced against further American casualties – seemed a small one to pay.

THIRTY

Japan's Defeat in the South

In the six months of 'running wild' between Pearl Harbor and the expulsion of the British from Burma between December 1941 and May 1942, the Japanese had succeeded in what five other imperial powers – the Spanish, Dutch, British, French and Russians – had previously attempted but failed to achieve: to make themselves masters of all the lands surrounding the seas of China and to link their conquests to a strong central position. Indeed, if China is included among the powers with imperial ambitions in the western Pacific, Japan had exceeded even her achievement. The Chinese had never established more than cultural dominance over Vietnam, and their power had failed altogether to penetrate the rest of Indo-China, the East Indies, Malaya or Burma. In mid-1942 the Japanese had conquered all those lands, were preparing to establish puppet regimes in most of them, were also the overlords of thousands of islands which were *terrae incognitae* in Peking, and had joined their maritime and peripheral annexations to the broad swathes of mainland territory in Manchuria and China which they had seized since 1931.

In crude territorial terms the extent of Japanese power even in mid-1944 was one and a half times greater than the area Hitler had controlled at the high tide of his conquests in 1942 – 6 million against 4 million square miles. However, Hitler held down his empire by brute force of manpower, deploying over 300 German and satellite divisions at the battlefront and in the occupied lands. Japan, by contrast, deployed an army only one-sixth the size, with only eleven divisions available for mobile operations. The rest were committed to the interminable, enervating and (apparently) ultimately irresoluble war against Chiang Kai-shek in the Chinese hinterland. This state of affairs left Japan in a fundamentally unbalanced strategic position. Though the map represented her situation as strong, since she occupied that 'central position' in the theatre of war which all military theorists have argued is the most desirable to hold, logistics pointed to a different conclusion. Intercommunication between many of the Japanese strongholds, particularly southern China, Indo-China and Burma, had always been difficult if not impossible by land because of the mountain chains which define their frontiers. Intercommunication by sea was wearisome and increasingly perilous because of the bold and effective depredations of the American submarine captains. Intercommunication between the Pacific and East Indian islands was menaced both by submarines and by American airpower, land- or carrier-based. Finally the Japanese army in China itself was effectively immobilised by the size of the country, its units committed to pacification or occupation – in which they were assisted by thousands of so-called 'puppet' Chinese troops belonging to the bogus government of Wang Ching-wei, set up in 1939 – and only rarely freed to undertake offensive operations against the Chinese armies proper.

Those armies belonged to two hostile camps, the army of the legitimate Kuomintang government commanded by Chiang Kai-shek, and the communist army of Mao Zedong. By a pre-war truce they had agreed to fight the Japanese instead of each other, but the truce was often broken, while the communist army was certainly more interested in letting Chiang's troops exhaust themselves in battle with the foreign enemy than in helping them to victory. Their actions were quite unco-ordinated, in any case, for Mao's base was in the distant north-west, around Yenan in the great bend of the Yellow River beyond the Wall where rivals to the central government had traditionally established themselves, while Chiang had been driven into the deep south, around his emergency capital of Chungking, 500 miles away. Between the two seethed the remnants of the warlord armies which had carved out their territories after the collapse of the empire in 1911; the Japanese made accommodations with them and also recruited from them puppet troops.

To both the warlord and puppet armies Chiang's was militarily superior – but only barely so. In 1943 it was theoretically 324 divisions strong and therefore the largest army in the world, but in reality it consisted of only twenty-three properly equipped divisions, and those were small ones of only 10,000 men. For their equipment and supplies, moreover, they depended entirely on the Americans, who in turn depended upon the British to provide them with facilities to fly transport aircraft from India into southern China over 'the Hump', the mountain chain 14,000 feet high between Bengal in India and the province of Szechuan. These supplies had previously been delivered via the 'Burma Road' from Mandalay; but since the fall of Burma to the Japanese in May 1942 that route was closed. Chiang was dependent on the Americans not only for armament and subsistence but also for training and air support – provided by the few dozen aircraft of General Claire Chennault's Flying Tigers, originally the 'American Volunteer Group' of pilots and machines supplied to China by the United States in 1941. He was, moreover, dependent on the Americans for his armies' cutting edge, since the most effective element in his command was the American brigade-size 5307th Provisional Regiment, to become famous as Merrill's Marauders. The man he had accepted as his nominal chief of staff, 'Vinegar Joe' Stilwell, displayed an impatience with the Chinese that was exceeded in degree only by his rudeness towards the British with whom he was co-operating.

The Japanese army in China, twenty-five divisions strong, was so successful in keeping Mao pinned in his 'liberated area' of the north-west and Chiang backed against the mountains of Burma in the south that for the first two and a half years of the Second World War in the East it was not under an obligation to mount mobile operations. It already controlled the most productive parts of the country, Manchuria and the valleys of the Yellow and Yangtse rivers, as well as enclaves around the ports of the south, Foochow, Amoy, Hong Kong and Canton, together with the key island of Hainan in the South China Sea. It was taking what it wanted from China, particularly rice, coal, metals and Manchurian industrial goods, was scarcely discommoded either by 'resistance' – from which any sensible Chinese held aloof – or by the operations of Chiang's and Mao's armies and, above all, continued to exercise by its presence in the country all the advantages of occupying the strategic 'central position'.

The Ichi-Go and U-Go offensives

The Pacific Fleet's sudden advance into the central Pacific dissipated Japanese complacency. Nimitz's thrust was aimed like an arrow at the heart of Japan's central position. Ultimately it threatened their control of the South China Sea – the Pacific 'Mediterranean' which washes the shores of China, Thailand, Malaya, the East Indies, Formosa and the Philippines – and that control was essential to Japan's maintenance of its empire in the 'Southern Area'. On 25 January 1944, therefore, imperial headquarters in Tokyo issued orders to General Iwane Matsui, the chief of staff in China, to undertake a large-scale offensive. The last offensive in China had occurred in the spring of 1943, when the North China Army had cleared the area west of Peking in Shansi and Hopei provinces. Now the plan was to occupy more territory in the south, with the object of both opening a direct north-south rail route between Peking and Nanking and clearing the south of American airfields in Chiang's area, from which Chennault's air force, which had reached a strength of 340 aircraft including strategic bombers, was harassing the Japanese Expeditionary Army throughout China.

This Ichi-Go offensive was to open on 17 April 1944. Earlier in the year an associated offensive, U-Go, had opened in Burma. Curiously the two Japanese plans were not co-ordinated in time, objectives or aims – except in the general and favoured Japanese aim of confronting the enemy with a complexity of thrusts – whereas the Allied campaigns in southern China and Burma did in fact interconnect. For one thing, Chiang's armies based on Chungking were dependent on supply via the 'Hump route'; secondly, Chinese troops, effectively commanded by Stilwell, were operating in southern China with the object of reopening the Burma Road; and, thirdly, Chinese troops were being trained in India as a means to improving the quality of Chiang's army. Nevertheless imperial headquarters did not order General Renya Mutaguchi, commanding the Fifteenth Army in Burma, to make an attack up the Burma Road to lend assistance to the Ichi-Go offensive. Instead it directed him to undertake nothing less than a full-scale invasion of India, in an entirely different direction.

U-Go was an operation to which Mutaguchi was wholeheartedly committed. Between November 1942 and February 1943 his predecessor, Iida, had successfully turned back a British offensive into Burma down the Arakan coast on the Bay of Bengal. A subsequent irregular operation, mounted by the long-range penetration Chindit forces led by their creator, the messianic Orde Wingate, had also been defeated between February and April 1943. However, Mutaguchi had been rightly impressed by the success of Wingate's troops in penetrating the Japanese front on the mountainous and roadless terrain of the Indo-Burmese frontier. He feared that where Wingate's tiny penetration force had marched larger Allied armies might follow. He also saw that Wingate's route was one his own hardy soldiers could take in the opposite direction, as the best means of defending Burma, interrupting the Allied efforts to reopen the Burma Road (on which American engineers were working from a roadhead in India at Ledo), quashing Stilwell's increasingly intrusive thrusts from southern China, and so indirectly assisting Ichi-Go in China proper.

Mutaguchi's offensive spirit was justified by the principle that the best form of defence is attack. South-East Asia Command, which had come into existence on 15 November 1943 with the dynamic Admiral Lord Louis Mountbatten at its head, was indeed planning offensives of its own designed to re-establish Allied power in Burma. Among the operations planned was another offensive in the Arakan, a major offensive across the Indo-Burmese border from Assam to the river Chindwin, gateway to the Burmese central plains, two Chinese offensives into north-eastern Burma from the province of Yunnan, one of which was to be mounted by Stilwell's Chinese troops with the support of Merrill's Marauders, the other to be a Chindit operation into the Japanese rear at Myitkyina, at which Merrill's Marauders were going to strike.

Mutaguchi's operation was therefore not merely an offensive; it was also a pre-emptive attack. For this operation the whole Burma Area Army, commanded by General Count Terauchi, had been reinforced, in part with troops from Thailand, in part with the 1st Division of the Indian National Army, raised by Subhas Chandra Bose from 40,000 of the 45,000 Indians captured in Malaya and Singapore who had shown themselves sympathetic to his cause. However, Mataguchi's spoiling attack was itself preceded by another one, for in November 1943 the British had resumed their attempt to penetrate the steamy Arakan. On 4 February, therefore, the Japanese 55th Division was launched into the British lines in the Arakan, with a mission to disrupt the advance. Only with the greatest difficulty was the 55th Division dispersed and driven back to its departure point at the end of the month. Meanwhile the Japanese 18th Division was dealing harshly with Stilwell's advance towards Myitkyina, behind which Wingate's second Chindit expedition was due to descend by glider in March.

It was in a highly disturbed northern Burma, therefore, that Mutaguchi opened his U-Go offensive on 6 March, when his three divisions crossed the

Chindwin river to invade India, the 31st heading for Kohima, the 15th and 33rd for Imphal.

These tiny places in the high hills of Assam had been centres of the tea-growing industry before the war. They provided no facilities for the basing of the large British-Indian army which now occupied the front and were poorly connected by road to the rest of India. Moreover, General William Slim, commanding the British Fourteenth Army, was preparing to go over to the offensive and was not in a position to receive attack. The Fourteenth Army, under his inspired leadership, had been transformed from the low state it had reached after the agonising retreat from Burma in the spring of 1942 and the humiliating withdrawal from the Arakan eight months later. It had not yet, however, fought a full-scale battle against Japanese troops at the peak of their aggression.

Slim had nevertheless scented a Japanese offensive in the offing and was not wholly surprised by it. He therefore persuaded Mountbatten to coax sufficient air transport out of the Americans to fly the 5th Indian Division, one of the most experienced in the British-Indian forces, up from the Arakan front between 19 and 29 March, and he himself sent forward supplies and reinforcements from the resources he had been gathering for his own offensive to the defenders on the border. He also gave his subordinate commanders strict instructions not to withdraw without permission from higher authority. Since the British defenders stood fast at the key points on the mountainous Indo-Burmese frontier, without attempting to defend its whole length, the Japanese succeeded in their object of encircling Imphal and Kohima, but could not take possession of the frontier roads that lead down into the Indian plain. Kohima was surrounded on 4 April, Imphal the following day. The fighting that ensued was among the most bitter of the war, as the two sides battled it out often at ranges no wider than the tennis court of the district commissioner's abandoned residence which formed part of no man's land on Kohima ridge. The British were supplied by airlift erratically at Kohima, rather more regularly at Imphal. The Japanese were not supplied at all; diseased and emaciated, they persisted in their attacks even after the coming of the monsoon. On 22 June, however, after over eighty days of siege, Imphal was relieved, and four days later Mutaguchi was forced to suggest to Terauchi that the Fifteenth Army ought now to retreat. In early July imperial headquarters gave its approval, and the survivors struggled off down roads liquefied by the tropical rains to cross the river Chindwin and return to the Burmese plains. Only 20,000 of the 85,000 who had begun the invasion of India remained standing; over half the casualties had succumbed to disease. The 1st Division of the Indian National Army, mistrusted as turncoats and therefore mistreated by the Japanese commanders, had ceased to exist.

The focus of the fighting in Burma now shifted to the north-eastern front, where the Japanese were holding their own with tenacity against both Stilwell and the Chindits; Slim meanwhile began to prepare the Fourteenth Army for its delayed offensive across the Chindwin to recover Mandalay and Rangoon. However, with the defeat of Mutaguchi's U-Go offensive, Burma itself ceased to be a major preoccupation of imperial headquarters. Though the Ichi-Go offensive was proceeding satisfactorily in southern China – so much so that the American government had begun to entertain fears of Chiang Kai-shek's imminent collapse – the situation in the southern and (more critically) central Pacific continued to worsen. In New Guinea, the fall of the Vogelkop peninsula in July was followed by the capture of the island of Morotai, midway between New Guinea and the southern Philippines island of Mindanao, on 15 September; the fall of Guam and Saipan was followed, also on 15 September, by the invasion of Peleliu, in the Palau islands, the closest point to the Philippines the Americans had yet reached on the central Pacific front. The invasion of the Philippines, which gave access to China, Indo-China and Japan itself, was now at hand.

A timetable for the Leyte landings

The extent and rapidity of MacArthur's and Nimitz's success had, however, so surprised the Joint Chiefs of Staff and their planners in Washington that the exact nature of the invasion was now once again a matter for debate. As in the European theatre, where in 1943 the chiefs of staff to the Supreme Allied Commander Designate had laid down a timetable for the advance to the German border which the actual pace of events then overtook with unanticipated speed, all sorts of operations which had once seemed important now faded into insignificance. In Europe events had made irrelevant the capture of the Atlantic ports, as points of supply from the United States for an American army fighting in central France, as well as the invasion of southern France. In the Pacific it was the capture of ports on the south China coast to supply Chennault's air bases, the invasion of Formosa and the occupation of the southern Philippines island of Mindanao that lost their significance. Two of these projects cancelled themselves. The success of Ichi-Go in southern China had led to the loss of most of Chennault's airfields near the coast, thereby making the capture of nearby ports irrelevant; and the invasion of Formosa, an island twice the size of Hawaii and defended by the highest sea cliffs in the world, was calculated to require so many troops that it could not be undertaken until the war in Europe was over. The attack on Mindanao was abandoned on 13 September after Halsey's carriers encountered only light resistance in the area. He urged instead that a landing be made first in Leyte, in the centre of the archipelago, and that the troops proceed thereafter to the northernmost island, Luzon, in December, two months ahead of schedule. As this timetable suited MacArthur, who thought the approach of the Chief of Naval Operations, Admiral King, over-ponderous, it carried his support. The debate had run between president, Joint Chiefs of Staff and operational commanders since 26

July; it was ended on 15 September, when the joint chiefs authorised MacArthur to begin landings on Leyte on 20 October.

The Japanese in the Philippines were ill prepared to withstand invasion. Indeed, the Japanese forces as a whole were now suffering the consequences of their own earlier success. Having passed what Clausewitz calls 'the culminating point of the offensive', they found themselves in possession of more territory than they could closely defend and were confronted by an enemy who was on the rampage and whose resources were growing by the month. Though the manpower available to the US Army in the Pacific and to the Marine Corps was limited by the demands of the war in Europe, the USAAF had been acquiring more and better aircraft throughout 1944, particularly the B-29 Superfortress, which had the range to bomb the Japanese home islands from the old bases in southern China and the new bases on Saipan. The United States Navy, whose particular theatre was the Pacific, enjoyed almost an embarrassment of riches; it had new battleships, cruisers and destroyers, fast attack transports, landing craft large and small, but above all new carriers: twenty-one Essex-class carriers had come into service since 1941 or were about to do so, and the total carrier fleet provided flight-deck space for over 3000 aircraft, an embarked naval air force three times the size of the Japanese at its largest.

Japan, by contrast, had already passed the high point of its war production. Its army had been fully mobilised since 1937 and was stuck at a size of about fifty divisions. Its navy had been continuously in action since 1941, had suffered heavy losses and could not make them good from the output of its shipyards. Only five fleet aircraft carriers were launched between 1941 and 1944. Losses in Japan's merchant fleet were far higher and threatened the collapse of the Japanese system. Because Japan could not feed itself or supply its own raw material needs, free use of the western Pacific seas was essential to the running of its economy; it was also necessary to the sustenance, reinforcement and movement of garrisons within the Southern Area. During 1942 American submarines had sunk 180 Japanese merchant ships, totalling 725,000 tons deadweight, of which 635,000 tons was replaced by new building; the tanker tonnage actually increased. In 1944, however, because the skill of American submarine captains had increased and they were operating from bases much further forward in New Guinea, the Admiralties and the Marianas, the total of sinkings increased to 600, or 2.7 million tons, more than had been sunk in the years 1942 and 1943 combined. By the end of 1944 half Japan's merchant fleet and two-thirds of her tankers had been destroyed, the flow of oil from the East Indies had almost stopped, and the level of imports to the home islands had fallen by 40 per cent.

The destruction of the merchant fleet obliged the navy to use destroyers instead of merchantmen to ship and provision units, and this seriously impeded the movement of troops between threatened spots, thus affecting Japan's defence of the Philippines. Imperial headquarters had correctly divined that the

Americans planned to invade first the southernmost island of Mindanao, from New Guinea, and then the northernmost island of Luzon, as a stepping-stone to Japan; but they had not anticipated that the Americans would change their plan in the light of events. In consequence, Leyte was left even more weakly garrisoned than Mindanao. Although on 20 October 1944 there were 270,000 Japanese troops in the Philippines, Tomoyoku Yamashita, the conqueror of Singapore and commander of the Northern Area, had only the weak 16th Division on Leyte itself. With only 16,000 men, it was no match for the four divisions of General Walter Krueger's Sixth Army, which began to go ashore in Leyte Gulf that morning.

Although the Japanese army was unprepared for the Leyte landing, the Japanese navy was not. It was now divided into two halves, the remaining carriers and their escorts being kept in home waters, the battleships – of which there were still nine, including the *Yamato* and *Musashi*, the largest in the world, of 70,000 tons and mounting 18-inch guns – lying at Lingga Roads, near Singapore, to be near their supply of East Indies oil which could not be shipped to the home islands. Both sections of the fleet had sensibly held back from involvement in the latest of the American island landing operations, the descent on Peleliu in the Palaus on 15 September – an uncancelled operation of the original central Pacific strategy which had lost its point (though it inflicted agony on the veteran 1st Marine Division). The home fleet did not evade involvement in the pre-Leyte air offensives on Formosa, Okinawa and Luzon, during which the American Third Fleet destroyed over 500 Japanese carrier- and land-based aircraft between 10 and 17 October; but Admiral Jisaburo Ozawa's carrier force risked none of its ships, while the bulk of the Combined Fleet at Lingga remained intact.

It was in these circumstances that imperial headquarters decided to launch a decisive naval offensive, codenamed Sho-1, against the American Third and Seventh Fleets covering the Leyte landings. Of great complexity, as large Japanese offensives always were, in essence it was diversionary: Ozawa's carriers, brought down from Japan's Inland Sea, were to lure Halsey's Third Fleet away from the Leyte beaches; then the battleships and heavy cruisers, divided into the 1st and 2nd Attack Forces and Force C, were to attack the transports and landing craft in Leyte Gulf and destroy them. The 1st Attack Force was to approach through the San Bernardino Strait to the north of Leyte, the 2nd Attack Force and Force C through the Surigao Strait to the south.

What followed was the largest naval battle in history, larger even than Jutland, but, like Jutland, confused by misreportings and misunderstandings. First into the fray was Vice-Admiral Takeo Kurita's 1st Attack Force, which had sailed from Lingga. It was intercepted and damaged by American submarines en route but reached the western approaches to the San Bernardino Strait on 24 October. The land-based aircraft which supported it inflicted heavy damage on one of the American carriers from Halsey's Third Fleet, USS

Princeton, which eventually sank, but they lost more heavily themselves against the American Hellcat fighters; and as the day developed American torpedo-bombers took Kurita's own battleships under attack. During the afternoon the *Musashi* suffered nineteen torpedo hits, more even than its enormous bulk could absorb, and at 7.35 in the evening rolled over and sank. Kurita decided he could not risk *Yamato,* his two other battleships and his ten heavy cruisers in the confined waters of the San Bernardino Strait, without the assurance of support from Ozawa's carriers (of which he had heard nothing), and so turned back to retreat to Lingga.

At the moment he did so, however, the Sho-1 plan was on the point of success, for Halsey in the Third Fleet's flagship *New Jersey,* stationed off the southern tip of Luzon, had just received news that Ozawa's carriers had been sighted 150 miles to the north. Halsey had been offended by whispered allegations that he had let the Japanese escape too easily in the Battle of the Philippines Sea the previous June, and he was determined to make Ozawa fight. He therefore extemporised plans to leave behind part of his force, designated Task Force 34, to guard the San Bernardino Strait while he raced his heavy units northward to seek and destroy the Japanese carriers.

Two changes of mind now supervened to alter the course of the battle. The first was Kurita's. Shamed by urgings from the Combined Fleet that he was shrinking from the chance of victory, he reversed course to pass through the San Bernardino Strait after all on the night of 24/25 October and sailed onward towards Leyte Gulf. The second was Halsey's. Excited by reports of how vulnerable Ozawa's carriers were, he decided not to leave any part of his force to guard the San Bernardino Strait but to take those ships which would have formed Task Force 34 with him northward to attack them.

Sho-1 was suddenly after all on the point of success. Kurita's 1st Attack Force was about to appear off Leyte Gulf, where the landing force was protected only by a fragile fleet of destroyers and escort carriers. Vice-Admiral Kiyohide Shima's 2nd Attack Force and Vice-Admiral Shoji Nishimura's Force C were meanwhile heading for the Surigao Strait to take the Leyte landing force in the rear from the south. While Halsey proceeded northwards to a putative encounter with the Japanese carriers, unknown to him the American invasion of Leyte was threatened with disaster.

All that stood between the two advancing Japanese forces and disaster were three tiny escort carriers in the San Bernardino Strait and Admiral Oldendorf's six battleships in the Surigao Strait. Oldendorf's battleships were a spectre from the past, since all predated the Second World War and five had been raised from the bottom of Pearl Harbor. In the intervening years, however, they had been refurbished and re-equipped, particularly with modern radar. In the darkness of the night of 24/25 October, the images of Nishimura's ships appeared distinct on Oldendorf's radar screens. His destroyers crippled the battleship *Fuso* as it approached; his own battleship salvoes then finished her off and sank the other Japanese battleship *Yamashiro* as well. The survivors of Force C beat a retreat, not alerting the 2nd Attack Force as it passed them to the danger that lurked in the Surigao Strait. It too suffered damage, hastily reversed course and followed in Nishimura's wake.

The Battle of the Surigao Strait was a lucky escape for the Americans. The second round in the San Bernardino Strait promised not to be. Kurita's 1st Attack Force greatly outgunned any American force which stood between it and the landing fleet, while the United States Navy's heavy metal was far away. As Halsey cruised in search of Ozawa, he was pursued by messages which included the notorious 'Where is Task Force 34 the whole world wonders'; the last four words were a misunderstood piece of security padding, but to Halsey they were eternally galling. In the meantime Kurita had fallen among the landing fleet's protecting warships. Those he found first were a puny group of five escort carriers, converted merchant ships with little speed and few aircraft, which were equipped for anti-submarine rather than torpedo strikes. The five nevertheless rose to the occasion with aplomb and superb bravery. While Admiral Clifton Sprague manoeuvred Task Force 3 at all available speed to escape 16- and 18-inch salvoes, his pilots flew off their aircraft to launch anti-submarine bombs at the battleships. One of the carriers, *Gambier Bay*, was hit and left on fire. The rest, to which another group of 'baby flattops' from Task Force 2 lent assistance, managed to cover their retreat by air strikes and torpedo attacks launched by their own escorting destroyers. In the face of this Tom Thumb defiance, and dispirited by the non-appearance of Ozawa's carriers, Kurita decided to break off action and retreat through the San Bernardino Strait. It was 10.30 on the morning of 25 October.

To the south, Oldendorf's battleships were steaming to the rescue from the Surigao Strait but were still three hours away; to the north, Halsey had reversed course from his pursuit of Ozawa but would take even longer to reach the scene. Halsey's aircraft had nevertheless taken their toll. In an early-morning strike they left the light carriers *Chitose* and *Zuiho* sinking. A second strike destroyed the carriers *Chiyoda* and Ozawa's flagship *Zuikaku*, a veteran of Pearl Harbor; though they had come to the battle with only 180 aircraft embarked, their loss virtually completed the extinction of the great Japanese naval air force. To their loss, moreover, had to be added that of three battleships, six heavy cruisers, three light cruisers and ten destroyers, in total a quarter of the losses the Imperial Japanese Navy had suffered since Pearl Harbor.

Leyte Gulf was therefore not only the largest but also one of the most decisive battles of naval history, even though for the Americans it had been a close-run thing. The battle for Leyte itself was a more long-drawn-out affair. The Japanese, recognising that their hold on the Philippines stood or fell by the defence of Leyte, rushed in reinforcements by destroyer from elsewhere in the islands – the 8th, 26th, 30th and 102nd Divisions, as well as the elite 1st Division from the dwindling general reserve in China. The Americans too reinforced the four divisions with which they had made their initial landing, so that by November they deployed six of their own – the 1st Cavalry, 7th, 11th Airborne, 24th, 32nd, 77th and 96th Divisons. Fighting during the next month was bitter, and on 6 December the Japanese launched a counter-attack to take the main American airfield complex on Leyte. When the attack failed, the campaign for the island was effectively at a close. It had cost the Japanese 70,000 and the Americans 15,500 losses.

On 9 January 1945 Krueger's Sixth Army moved from Leyte – and the nearby islands of Mindoro and Samar, which had also been cleared – to invade Luzon, where the Philippines capital, Manila, was located. In the far south the Australian First Army was mopping up Japanese resistance in New Guinea, New Britain and Bougainville. In Burma, while Slim's Fourteenth Army opened its offensive into the plains of Burma by its capture of Kalewa on the Chindwin on 2 December, Chiang's troops, with American assistance, were also making progress on the north-eastern front. They were no longer commanded by the vitriolic Stilwell, who had definitively fallen out in turn with the British, the Chinese and ultimately President Roosevelt. After Stilwell's removal on 18 October, his role was divided between Generals Albert C. Wedemeyer, the architect of the American 'Victory Plan' of 1941, and Daniel Sultan. The former had taken over as the American commander in China; the latter now commanded Merrill's Marauders (renamed Mars Force) and the Indian-trained Chinese forces in Burma.

In China, Chiang's armies, strengthened by two Indian-trained divisions brought from Burma, at last managed to halt the Ichi-Go offensive at Kweiyang, after it had threatened to drive a corridor from the Japanese-held coastal areas to Chiang's capital at Chungking itself. Ichi-Go had achieved a subsidiary object in opening a continuous corridor from northern Indo-China to Peking, but it had not brought about the destruction of Chiang's army. Indeed, in January 1945 the best of his troops (under Sultan's command) finally succeeded in breaking across the mountainous north of Burma through Myitkyina, which Stilwell had taken in August, to join up with the so-called Y-Force of Chiang's China forces advancing from Yunnan. On 27 January the two reopened the Burma Road, thus assuring a direct source of land supply from the Anglo-American base in India to the Kuomintang heartland around Chungking. The Japanese nevertheless remained the dominant force in southern China. British strength was aligned towards the plains of Burma, into which the Fourteenth Army was making its advance, and neither Wedemeyer's, Sultan's nor even Chiang's troops were powerful enough to stem any determined Japanese operation south of the Yangtse. In the spring of 1945, as in every year since 1941, the future of the war in China was closely bound to the outcome of the main battle between the Imperial Japanese and United States Navies' fleets and their amphibious forces in western Pacific waters.

THIRTY-ONE

Amphibious Battle: Okinawa

With the fall of the Philippines and the capture of the Marianas, the war in the Pacific approached its climactic amphibious phase. Ground fighting was to continue throughout 1945 at a score of places inside or close to the Japanese 'defensive perimeter' of 1942; in Burma and the northern Philippines, where Manila would become a ghost city as devastated as Warsaw, fighting was to be very heavy indeed. The character of the Pacific war, however, now underwent a radical change. No longer would there be two separate and competitive American strategies, with the navy bringing overwhelming force to the landing of individual Marine divisions on tiny, remote atolls, while the army moved by shorter hooks in greater strength to seize large land masses in the Indies. Navy and army would now combine to mount large-scale amphibious operations against the outlying islands of Japan itself, involving several divisions at a time, enormous fleets and naval air forces as well as dense concentrations of embarked troops. The success of these operations would depend entirely on the combined amphibious skills of sailors, soldiers, airmen and Marines.

The Joint Chiefs of Staff had confidence in the outcome of the projected operations, of which the most important was to be the landing on Okinawa in the Ryukyu islands, only 380 miles from Kyushu, the southernmost of the large Japanese home islands. American amphibious skills were now very high but had taken time to develop – indeed, they had been developing throughout the Pacific campaign. Credit for their conception, however, belonged above all to the United States Marine Corps, which had seen the need to learn how troops could best be transferred from ship to shore twenty years before the Second World War began. The United States Marine Corps put forward the idea that transit between ship and shore must be essentially a tactical movement. The idea, so arrestingly simple, had been grasped by none of the oceanic powers before. Neither the British nor the French, though they had built great empires by projecting military through naval force, had perceived that there was more to landing troops than putting them in ships' boats and debarking them at the water's edge. When in 1915 they jointly mounted the great amphibious landing at Gallipoli the result was catastrophic. Hastily adapted lighters, towed to shore by steam pinnaces, were grounded under Turkish machine-guns and the soldiers on board were massacred in the water. After the First World War the US Marine Corps determined that such would not be its men's fate. It had, admittedly, an institutional reason for wishing to make amphibious landing tactics its own particular specialism, for it entertained the fear – common to small organisations which operate between the margin of two larger ones – of being absorbed by either the army or the navy; but there was more to it than that. The Marines foresaw the danger of a Pacific war with Japan. They also saw that it could be won only with specialised methods and specialised equipment, and they set about developing both.

The architect of the Marines' amphibious warfare doctrine was Major Earl Ellis, who in 1921 first proposed the concept of landing as a 'ship-to-shore tactical movement'. He emphasised the need for landing troops to be covered with the heaviest available firepower as they left the ship, to debark on the run and to take up their first positions not on the beach itself but on dry ground inland. Sea and beach, in short, were to be regarded as a no man's land. The fighting would commence in or beyond the enemy's first defensive line well above the high-water mark. The realisation of such a concept required not only special training but also purpose-built equipment. One item was a dive-bomber, operating from a carrier but flown if possible by pilots of the Marines' own air arm; dive-bombing was an essential means of delivering pinpoint firepower on to enemy beach strongpoints. The other was a 'dedicated' landing craft, with the power to cross the danger zone between ship and shore at high speed, and with the build to enable it to beach, debark and back off without waiting for tides. With time, the US Marine Corps perceived the need for two and eventually three types of landing craft. The first was a tracked amphibian or amphtrac, armoured if possible, which could actually drive out of the water and across the beach before its occupants debarked; a prototype was produced in 1924 by Walter Christie, the astonishingly creative American tank pioneer (who also fathered the T-34). The second was a larger beaching craft to carry the second wave; the successful model, the Higgins boat, was based on a civilian design built by the Higgins Company of New Orleans for use in the Mississippi delta. The third was a ship capable of beaching tanks; the sketch for the first of more than a thousand Landing Ships Tank (LST) built during the war was roughed out in a few days in November 1941 by John Niedermair of the US Navy's Bureau of Ships. All three types could, of course, also be used to tranship the supplies which the landing troops needed once ashore.

By early 1945 the Pacific Fleet possessed all three types and many variations, in enormous numbers; the United States Coast Guard had specialised in the role of manning the Marines' landing craft. In addition it possessed large numbers of fast 'attack transports', on which landing troops and craft were embarked, and which could keep pace with the destroyers and carriers of an amphibious task force. It also had numbers of dedicated command ships from which admirals and generals could jointly direct operations.

Plans for the advance to the Ryukyu islands had been laid as early as July 1944, before the Leyte landing, when Admiral Raymond Spruance, commanding the Fifth Fleet, had suggested that intermediate positions, particularly Formosa, should be bypassed and a giant stride taken to Japan's doorstep. Admiral King, Chief of Naval Operations, at first thought the scheme over-ambitious. By September, however, when it became clear that the persistence of the war in Europe (through Hitler's evident determination to stand on the West Wall) and MacArthur's deep involvement in the Philippines precluded the release of more army formations, King relented. With six Marine divisions and five army divisions under his command, Nimitz now had an independent force of sufficient size for him to mount large-scale operations of his own. On 29 September 1944, therefore, King, Nimitz and Spruance, meeting at San Francisco, agreed to make Okinawa the principal target for amphibious operations in the following year. Because a main aim of the advance to the Ryukyus was to secure better air bases for the preparatory bombardment of Japan and to drive an 'air corridor' between the home islands and the Japanese airfields on Formosa and Luzon, it was also agreed that a subsidiary base should be seized on a smaller island nearby, which could be taken more quickly, to provide a staging post and emergency landing field for B-29s. Iwo Jima in the Bonin islands seemed the best choice. On 3 October the Joint Chiefs of Staff issued a directive for Iwo Jima to be attacked in February and Okinawa in April.

The Ten-Go plan

Meanwhile the Japanese were revising their own plans for the future conduct of the war. In September 1943 they had accepted that the 1942 defensive perimeter was untenable and had defined a new Absolute National Defence Zone, enclosing the Kuriles to the north of the home islands, the Bonins, Marianas and Carolines in the central Pacific, and western New Guinea, the East Indies and Burma in the south-west. Subsequently the American advance in 1944 had so deeply penetrated this zone that plans based on its defence were abandoned, and its architects withdrew from government; in July Tojo resigned as Prime Minister, to be replaced by the more moderate Kuniaki Koiso, though the inclusion of representatives of the war and navy ministries in the cabinet ensured that it remained under military control. By the spring of 1945 the situation had so gravely deteriorated everywhere except in China (where the Ichi-Go offensive proceeded) that imperial headquarters had to think again. It formulated a plan codenamed Ten-Go for the defence of the most vulnerable points on what remained of Japan's defensive cordon, which included the island of Hainan between China and Indo-China, the China coast itself, Formosa and, lastly, the Ryukyus. The sub-plan for the defence of the Ryukyus, of which Okinawa was recognised to be the island most at risk, was codenamed Ten-Ichigo, and 4800 aircraft based on Formosa and the home islands were allotted to its execution. Because of the shortage of fuel, which limited the number of sorties that could be flown and severely restricted the pilots' training hours,

Ten-Ichigo was to be a new sort of offensive. The aircraft would be loaded with high explosive and would fly one-way missions to crash themselves on American ships in what the Americans would learn to call 'kamikaze' ('divine wind') suicide strikes.

The Americans had already experienced a foretaste of kamikaze tactics on the last day of the Battle of Leyte Gulf, but fortunately those suicide missions had been hastily improvised. Ten-Ichigo was more methodically prepared and was not ready for launching when the 3rd, 4th and 5th Marine Divisions assaulted Iwo Jima on 19 February. That was the only mercy granted the Americans at Iwo Jima; heavily gunned and garrisoned, honeycombed with tunnels, its bedrock of basalt covered with a deep layer of volcanic dust, the island subjected the Marines to their worst landing experience of the Pacific war. Amphtracs lost traction and ditched on the beaches, to be destroyed by salvoes from close-range artillery which three days of battleship bombardment had not destroyed; riflemen dug trenches which collapsed as soon they were deep enough to give cover; the wounded were wounded again as they lay out on the beaches awaiting evacuation. Robert Sherrod, the correspondent who had been at Tarawa and most island landings in between, thought it the worst battle he had ever seen: men died, he said, 'with the greatest possible violence'. When Iwo Jima was finally secured on 16 March, 6821 Americans had been killed and 20,000 wounded, over a third of those who had landed; the 21,000 Japanese defenders died almost to a man.

Okinawa, the last battle

Iwo Jima provided an awful warning of what lay in store for the American divisions assigned to Okinawa – the 1st, 6th and 7th Marine, and the army's 7th, 27th, 77th, 81st and 96th Divisions. Because of the casualties taken at Iwo Jima on the first day, it was decided to make the preparatory bombardment the heaviest yet delivered on to a Pacific island. It lasted from 24 to 31 March, and when it was over nearly 30,000 heavy-calibre shells had impacted on the landing area. On 1 April, from an armada of 1300 ships including eighteen battleships, forty carriers and 200 destroyers, the 1st and 6th Marine and 7th and 96th Divisions raced to shore in their amphtracs and Higgins boats to seize the central waist of the island, where its airfields lay, and then reduce resistance in the two halves.

Okinawa is a large island nearly eighty miles long. The American scheme for its capture was based on the supposition that, as at all but one landing so far, the Japanese would resist tenaciously at the water's edge and then be beaten back inland, to increasingly untenable positions, by the weight of American air and naval firepower. The Japanese, anticipating American expectations, had adopted a contrary scheme for Okinawa's defence. They were to let the Marine and army divisions land unopposed, then draw them into battle against what they regarded as impregnable defence lines within the island, meanwhile turning the weight of the kamikaze against the ships offshore. The ultimate aim was to drive the fleet away, leaving its landbound half to be destroyed at leisure.

The Japanese forces on the island numbered some 120,000 against 50,000 Americans landed on the first day – a figure that eventually rose to nearly a quarter of a million in the US Tenth Army. These Japanese troops were organised into the 24th and 62nd Divisions and together with a large number of non-divisional units formed the Thirty-Second Army, commanded by General Mitsuru Ushijima. He was more realistic than the staff officers in imperial general headquarters, since he recognised that victory on Okinawa was unattainable; nevertheless he intended to inflict the largest possible toll of casualties on the invaders, and had made preparations accordingly. The island was honeycombed with tunnels and firing positions, many of which concealed large-calibre weapons; and the fighting positions formed a series of lines which extended from the beaches where he had correctly judged the Americans would land into the high ground to the south and north.

The Americans landed virtually without loss on 1 April. The 1st and 6th Marine Divisions (whose volunteer soldiers were for the first time in the war diluted with conscripts) then turned north to clear the top of the island before joining the army's 7th and 96th Divisions in the battle for the more mountainous south. By 6 April, as casualties mounted, both were in contact with the Machinato Line covering the southern cities of Shuri and Naha. It was on that day that the Japanese air and sea offensive against the offshore fleet began.

The Americans had already had a taste of how fiercely the Japanese intended to defend Okinawa when Task Force 58, still commanded by the resolute Admiral Mitscher, had raided the Inland Sea on 18-19 March as a preliminary to the landing. Although American carrier aircraft had destroyed some 200 Japanese aircraft, the task force had also suffered heavily itself. The carrier *Wasp* was badly damaged by a kamikaze and only saved by rapid firefighting, a technique at which the American now excelled all other navies. Another carrier, *Franklin*, was hit by two bombs which almost incinerated the ship; 724 of her crew died, the highest fatal casualty toll suffered by any surviving American ship in the Pacific war.

On 6 April kamikazes attacked in dense waves; at the same time, far to the north, Japan's last operational surface force, the giant battleship *Yamato* escorted by a cruiser and eight destroyers set sail from Japan. *Yamato* had taken on board the last 2500 tons of fuel available at her Japanese home port to make the one-way trip. Her mission was to penetrate the screen around the Okinawa beaches and inflict unacceptable damage on the amphibious force. She was detected long before she got within range, however, and at noon on 7 April was taken under attack by 280 aircraft of Task Force 58. Between noon and two o'clock she suffered six torpedo hits, lost speed and steering, became a sitting duck to successive waves of American aircraft and at 2.23 pm rolled over and sank with almost all the 2300 sailors on board. Her cruiser and four of her seven destroyer escorts were also sunk. This 'Special Surface Attack Force' had launched the Imperial Japanese Navy's last sortie of the war.

The kamikazes proved far more difficult to repel. About 900 aircraft, of which a third were on suicide missions, attacked the amphibious fleet on 6 April and by the end of the day, although 108 were shot down, three destroyers, two ammunition ships and an LST had been sunk. The attacks were repeated on 7 April when a battleship, a carrier and two destroyers were all hit by kamikaze strikes. The American response was to thicken the screen of radar-picket destroyers, lying off Okinawa up to ranges of 95 miles, which gave early warning of attacks. There were soon sixteen on station, eleven of which lay in the semicircle between the north-eastern and south-western azimuths, nearest Japan and Formosa. As the British task force at the Falklands was to rediscover forty years later, however, a screen of radar pickets may give the large units of a fleet early warning of attack; but its mission is a sacrificial one, for the incoming enemy strikes readily choose its ships as targets. That was to be the American destroyers' fate. Between 6 April and 29 July fourteen American destroyers were sunk by suicide pilots, together with another seventeen LSTs, ammunition ships and assorted large landing craft lying within the screen. Over 5000 American sailors died as a result of the Okinawa kamikaze campaign – the heaviest toll the US Navy had suffered in any episode of the war, including Pearl Harbor.

Between 6 April and 10 June, besides many smaller missions the kamikaze corps mounted ten mass attacks by 50-300 aircraft, which damaged battleships and aircraft carriers as well as destroyers; the venerable *Enterprise* and the newer carriers *Hancock* and *Bunker Hill* were all kamikaze victims, and *Bunker Hill*, Spruance's flagship, lost 396 of her crew killed. American carriers, which were horizontally armoured above the engine room but below the flight deck, burned all too easily when a kamikaze landed aboard. A principal advantage of the four British carriers of Task Force 57, which joined the American force off Okinawa in March, was that they were armoured on their flight decks as a precaution against the shellfire likely to be encountered in narrower European waters, and therefore survived kamikaze strikes without serious damage.

Ultimately the kamikaze attacks could not go on, for the Japanese began to run out of both pilots and aircraft; the number of raids was heavier in April than in May and far heavier in May than in June, when only four ships were sunk. However, the pickets were bound to remain in place – and so expose themselves to damage or sinking, at almost unbearable cost to their crews' nerves – as long as the army and Marines battled ashore. As the campaign protracted, Nimitz grew increasingly impatient with the Tenth Army commander, General Simon Bolivar Buckner, complaining that he lost 'a ship and a half a day' at the pace at which the front was moving. Buckner, son of the general who had fought Ulysses S. Grant in the American Civil War in 1862, resolutely defended his methodical tactics. Successive ridge lines imposed delay on every offensive mounted. The lines were washed by constant rain, which bogged tanks trying to give support, and they were fanatically defended by Japanese who, whether trained infantrymen or wholly inexperienced naval shore personnel, fought literally to the death. Not until the end of June did resistance cease, and some 4000 Japanese surrendered in the last days. All the Japanese senior officers, including Ushijima, committed ritual suicide, as did many of their subordinates and some civilian Japanese. The Okinawan population, 450,000 strong at the outset, had suffered terribly; at least 70,000 and perhaps as many as 160,000 died in the course of the fighting. Thousands took refuge in the island's numerous caves, which the garrison subsequently occupied as strongpoints, and were killed when the American infantry attacked them with flamethrowers and high explosive.

For the fighting troops Okinawa had been the grimmest of all Pacific battles. The American army divisions lost 4000 killed, the Marine Corps 2938; 763 aircraft were destroyed and 38 ships sunk. The Japanese lost 16 ships and an almost incredible total of 7800 aircraft, over a thousand in kamikaze missions. The Japanese servicemen on the island – shore-based sailors as well as front-line riflemen, clerks, cooks, Okinawan labour conscripts – found ways of dying almost to the last man. The American total of prisoners, including men too badly wounded to commit suicide, was 7400; all the others, 110,000 in number, died refusing to surrender.

THIRTY-TWO

Super-weapons and the Defeat
of Japan

Okinawa left an awful warning of what awaited the American forces as the Pacific war drew in towards the perimeter of the Japanese home islands. It was the first battle for a large island on the approaches to the empire's heartland, and its cost and duration hinted at far worse ordeals to come once the United States Navy advanced to land soldiers and Marines on the shores of the Inland Sea. From a source never satisfactorily identified, the figure of 'a million casualties', even 'a million dead', had begun to circulate among American strategic planners as the number of losses to be expected in an invasion of Japan. It cast a terrible shadow over their discussions of how the victorious campaign in the Pacific was to be brought to an end without a national tragedy.

So far – and this implies no slur on the courage, dedication and self-sacrifice of the American sailors, Marines and soldiers who fought and died in the front line – the Pacific war had been a small war. The number of major ships engaged exceeded that deployed in any other theatre: with a dozen battleships, fifty aircraft carriers, fifty cruisers, 300 destroyers and 200 submarines, the Pacific Fleet in 1945 was not only the largest navy in the world but the largest navy that had ever existed; it had extinguished the Imperial Japanese Navy, whose few units still afloat lacked the fuel for them to put to sea. The American naval air force, 3000 strong, was also the largest in being; in addition the navy and the US Army Air Force had tens of thousands of shore-based aircraft, including the B-29 Superfortress, 250 of which had begun to operate regularly against Japanese cities since March, with devastating effect.

The Pacific war had been an enormous war in its geographical scope, encompassing over 6 million square miles of land and ocean. In terms of human numbers, however, the war had been quite small compared to that fought in Europe. There the Soviet Union had mobilised 12 million men against Germany's 10 million, and the theatre had also engaged most of Britain's 5 million and about a quarter of the United States' 12 million. In the Pacific, by contrast, although the Japanese had mobilised 6 million men, five-sixths of those deployed outside the home islands had been stationed in China; the number committed to the fighting in the islands had perhaps not exceeded that which America had sent. Between 1941 and 1945 a million and a quarter United States servicemen were posted to the Pacific and China-Burma-India theatres; of these, however, only 450,000 belonged to army or Marine divisions, and of those twenty-nine divisions only some six army and four Marine divisions were involved in regular periods of prolonged combat. Compared to the European theatre, where in mid-1944 300 German and satellite divisions confronted 300 Russian and seventy British and American divisions, the 'ground combat' dimension of the Pacific war was small indeed – if one sets aside the appalling casualties suffered by the Japanese island garrisons.

In the aftermath of Okinawa, its scale suddenly threatened to swell exponentially. The surrender of Germany meant that all of the ninety divisions the United States had mobilised and most of the British Empire's sixty could be made available for the invasion of Japan, together with whatever proportion of the Red Army Stalin decided to allot as soon as he declared war (as he had undertaken to do, once Germany was defeated, at Tehran in November 1943). According to the Okinawan experience, however, even numbers such as these could not guarantee that the defeat of the Japanese on their home territory would be quick or cheap. Okinawa and Japan were similar in terrain, but Japan offered a defender a vast succession of ridge, mountain and forest positions from which to hold an invader at bay. The prospect appalled the United States' decision-makers. Admiral William Leahy, chairman of the Joint Chiefs of Staff, pointed out to President Truman at a meeting on 18 June that the army and Marine divisions had suffered 35 per cent casualties on Okinawa, that a similar percentage could be expected in an attack on Kyushu, the first of the Japanese home islands selected for invasion, and that, with 767,000 men committed to the operation, the toll of dead and wounded would therefore amount to 268,000, or about as many battle deaths as the United States had suffered throughout the world on all fronts so far.

Truman's comment was that he 'hoped there was a possibility of preventing an Okinawa from one end of Japan to the other'. The Joint Chiefs' plan, worked out in Washington at the end of May 1945, called for an invasion of Kyushu (codenamed Olympic) in the autumn of 1945 and an assault on the main island of Honshu (codenamed Coronet) in March 1946. It had been agreed with difficulty. The army, whose view had been largely fixed by MacArthur, insisted that only an invasion would definitively finish the war. The navy, to which the US Army Air Force commanders lent unspoken support, argued that the seizure of bases on the coast of China from which close-range strategic air bombardment could be mounted would reduce Japanese resistance without the need to risk American lives in an amphibious landing. Strategic bombing, however, had thus far inflicted little damage on the home islands and had had insignificant effect upon its government's will to war. MacArthur's view therefore prevailed.

The destruction of Japan's cities
Before the Joint Chiefs of Staff issued their directive for Olympic and Coronet, however, the strategic bombing campaign had taken a different turn. Like the

British bomber chiefs in 1942, the Americans had been constrained to abandon the belief – which they had held much more dogmatically than the British – that the bomber was a precision tool and to accept that it had to be used as a blunt instrument. They had been driven to that change of doctrine by the success of the Japanese (in imitation of Speer's programme in Germany in 1943-4) in dispersing production of weapon components away from the main industrial centres to new factories which could not be easily located or hit by the Twentieth Air Force. In February 1945 General Curtis LeMay arrived in the Marianas, which had become the main base for the Superfortresses of XXI Bomber Command, to implement new bombing tactics. Targets were to be subjected not to precision high-level daylight strikes by high explosive but to low-level drenching by incendiary bombs at night, exactly the method by which 'Bomber' Harris had made his 'thousand-bomber raids' an instrument of terror in 1942 and created firestorms in one German city after another. The incendiary bomb LeMay's aircrew used, however, being filled with jellified petrol, was a far more efficient agent of conflagration than the RAF's; more important, Japan's flimsy wood-and-paper cities burned far more easily than European stone and brick.

On 9 March Bomber Command attacked Tokyo with 325 aircraft armed exclusively with incendiaries, flying at low altitude under cover of darkness. In a few minutes of bombing the city centre took fire and by morning 16 square miles had been consumed; 267,000 buildings burned to the ground, and the temperature in the heart of the firestorm caused the water to boil in the city's canals. The casualty list recorded 89,000 dead, half as large again as the number of injured survivors treated in the city's hospitals. Losses to the bombers were below 2 per cent and were to decline as the campaign gathered force. LeMay's command soon rose in strength to 600 aircraft and brought one city after another under attack; by mid-June Japan's five other largest industrial centres had been devastated – Nagoya, Kobe, Osaka, Yokohama and Kawasaki – 260,000 people had been killed, 2 million buildings destroyed and between 9 and 13 million people made homeless.

The destruction continued relentlessly, at virtually no loss to the American bomber crews but at appalling cost to Japan; by July 60 per cent of the ground area of the country's sixty larger cities and towns had been burnt out. As MacArthur and other military hardheads had argued, however, the devastation did not seem to deflect the Japanese government from its commitment to continuing the war. In early April, after failing to draw China into a separate peace, Koiso had been replaced as Prime Minister by a moderate figurehead, the seventy-eight-year-old Admiral Kantaro Suzuki; Tojo, though a deposed Prime Minister, nevertheless retained a veto over cabinet decisions through his standing in the army, and he and other militarists were determined to fight it out to the end. This determination exacted sacrifices which even Hitler had not demanded of the Germans in the closing months of the war. The food ration was reduced below the 1500 calories necessary to support life, and more than a million people were set to grubbing up pine roots from which a form of aviation fuel could be distilled. On the economic front, reported a cabinet committee instructed by Suzuki to examine the situation, the steel and chemical industries were on the point of collapse, only a million tons of shipping remained afloat, insufficient to sustain movement between the home islands, and the railway system would shortly cease to function. Still no one dared speak of peace. Tentative openings made in May through the Japanese legation in Switzerland by the American representative, Alan Dulles, were met with silence; over 400 people were arrested in Japan during 1945 on the mere suspicion of favouring negotiation.

The search for revolutionary weapons

In midsummer the American government began both to lose patience at Japan's intransigence and to yield to the temptation to end the war in a unique, spectacular and incontestably decisive way. They were aware through Magic intercepts that the Suzuki cabinet, like Koiso's before it, was pursuing backdoor negotiations with the Russians, whom it hoped would act as mediators; they were also aware that a principal sticking-point in Japan's attitude to ending the war was the 'unconditional surrender' pronouncement of 1943, which all loyal Japanese recognised as a threat to the imperial system. However, since the Russians mediated in no way at all, and since the Potsdam conference following the surrender of Germany indicated that unconditional surrender need not extend to the emperor's deposition, America's willingness to wait attenuated during the summer. On 26 July the Potsdam Proclamation was broadcast to Japan, threatening 'the utter destruction of the Japanese homeland' unless the imperial government offered its unconditional surrender. Since 16 July President Truman had known that 'utter destruction' lay within the United States' power, for on that day the first atomic weapon had been successfully detonated at Alamagordo in the New Mexico desert. On 21 July, while the Potsdam meeting was in progress, he and Churchill agreed in principle that it should be used. On 25 July he informed Stalin that America had 'a new weapon of unusually destructive force'. Next day the order was issued to General Carl Spaatz, the commander of the Strategic Air Forces, to 'deliver its first special bomb as soon as weather will permit visual bombing after about 3 August 1945 on one of the targets: Hiroshima, Kokura, Niigata and Nagasaki'. The attempt to bring the Second World War to an end by the use of a revolutionary super-weapon had been decided.

The search for a revolutionary weapon was one of the most immediate and persistent outcomes of the industrialisation of war in the mid-nineteenth century, and both a logical and an inevitable extension of the revolution in war which preceded it. Until the fifteenth century, warfare was a muscular activity, and decision on the battlefield went to the side which could sustain muscular effort longer than the other. The invention of gunpowder changed that; by allowing energy to be stored in chemical form, it made the weak man the equal of the strong and transferred advantage in war to the side which possessed superior intellectual quality and morale. The first attempts to draw on the products of industrialisation for military purpose therefore took the form of multiplying the power of chemical energy by accelerating the rate at which projectiles could be discharged; breech-loading and magazine rifles and then the machine-gun were the result. Their purpose was to nullify morale and intellectual quality by weight of metal.

When human resilience and adaptability demonstrated that the fighting man of the industrial age could survive even quantum leaps in firepower, military inventors changed their tack. They began to apply their inventiveness not to the problem of killing or disabling warriors *en masse* but to attacking and destroying the protective systems in which they took shelter – on land, fortifications; at sea, armoured ships. Human ingenuity had sought a means of destroying ships by stealth even before the industrial age, and the idea of the submarine and the torpedo had found primitive forms in sailing days. Between 1877 and 1897 both the torpedo and the submarine emerged as practicable weapons and did indeed transform the nature of naval warfare. The tank, which appeared in 1916, promised a comparable transformation of land warfare.

The promise, however, proved illusory. Tank and submarine, though they appeared to be strategic weapons in essence, were more or less quickly revealed to be tactical; that is to say, they were susceptible to counter-measures at the point of encounter and they struck at the products, not the structure, of an enemy's war-making system. However great the human losses and material damage they inflicted at the battlefront, the enemy, as long as he could replace those losses and repair that damage from his internal resources, might continue to wage war. The production of tanks and submarines, as those committed to battle were destroyed and had to be replaced from current output, itself became a charge on industrial capacity and therefore merely raised instead of reducing the price of victory.

This perception was one of the most important military legacies of the First World War. It was to lead to the formulation of the theory of strategic bombing. In the years after the war, both the British and the American air forces were converted to the belief that the heavy bomber, by carrying the high explosive which had proved so ineffectual against trench systems to the industrial heartland of the enemy, could quickly and finally destroy his means of making war and so win victory without the need for armies or navies to fight 'decisive' battles at all. The British further persuaded themselves that if such a 'strategic bombing' campaign were carried out at night it would spare the bombing force appreciable loss, while the Americans independently arrived at the conclusion advanced by the Italian dogmatist, Giulio Douhet, that a large, heavily armed day-bomber could be made self-defending: the Flying Fortress was the result.

As we have seen, the experience of war proved the theory of strategic bombing to be ill founded. A major cause of its failure lay in the realisation of one of the war's greatest scientific endeavours, the development of radar. Invented by the British before the outbreak, in 1940 it provided an effective though static chain of early-warning stations which allowed Fighter Command to be directed quickly and accurately against incoming Luftwaffe raids during the Battle of Britain. The British invention of the cavitron valve, which became operational in 1942, enabled radar to function at 'centimetric' wavelengths on a directional arc. These developments, which greatly reduced the bulk of radar sets, increased the definition of the image received and allowed the operator to search a chosen sector of air space, meant that effective search radars could be mounted in night-fighters; a further application of the cavitron valve was the miniaturised radar proximity fuse, introduced in August 1944, which exploded an anti-aircraft shell at a range lethal to an aircraft. It was used with considerable success against the V-1s. When 'centimetric' radar was developed by the Germans, however, they began to inflict a heavy toll on Bomber Command during its night raids on the Reich; had they also discovered the secret of miniaturising radar fusing for anti-aircraft shells, the American formations bombing by daylight would have suffered proportionately.

By 1944 it had become clear to all but the dogmatists in Bomber Command and the US Eighth Air Force that strategic bombing would not win the war in Europe (just as in mid-1945 it seemed clear that the fire-bombing of the home islands would not beat Japan). The strategic bomber, like the submarine and the tank, had been revealed to be a weapon susceptible to counter-measures, a system that required expensive 'dedicated' defences to protect it and a victim of attritional losses which imposed a heavy and continuing charge on war production. If there was such a thing as a revolutionary war-winning weapon, the search for it lay in another direction.

Hitler's 'revenge weapons'

In one field of the search, the Germans had made greater progress than any of the other combatants. They were on the point of deploying a ballistic missile. German pilotless weapon research had an extended history, much of it intertwined with the life stories of two individuals, Werner von Braun and Walter Dornberger. Von Braun was a professional technologist whose youthful enthusiasm for the idea of space travel had translated itself by the late 1920s into practical experimentation with rockets. Dornberger was a regular gunner officer who had served with the heavy artillery in the First World War and in 1930 was charged with rocket development at the Army Weapons Office. Circumstances brought the two into contact and in 1932 they began to experiment together with rocket firings. Braun supplied the technical expertise, Dornberger defined the practical criteria a successful rocket would have to meet. 'I had been a heavy gunner,' he wrote. 'Gunnery's highest achievement to date had been the huge Paris Gun', which fired 'a 21-cm shell with about 25 lb of high explosive about 80 miles. My idea of a first big rocket was something that would send a ton of high explosive over 160 miles.' He also 'stipulated a number of military requirements, among others that for every 1000 feet of range a deviation of only 2 or 3 feet [from the chosen impact point] was acceptable.' He finally 'limited the size of the rocket by insisting that we must be able to transport it intact by road and that it must not exceed the maximum width laid down for road vehicles.'

Dornberger's prescription revealed both the institutional roots of his thinking and, at the same time, an astonishing prescience of the rocket's potential. His insistence on road transportability went back to the characteristics of the 305-mm and 420-mm guns with which the German artillery had devastated the Belgian forts in 1914; it ensured that future German ballistic missiles would be weapons of the artillery arm. His requirements for range, accuracy and size of warhead, on the other hand, cast German rocket research far into the future. What he demanded, in effect, was a prototype for the transportable ballistic missile which has become the principal strategic weapon of the superpowers in the late twentieth century. When, later, he was to insist that the successful production model (known to the Germans as the A-4, to the Allies as the V-2) should move on a vehicle (the *Meillerwagen*) which was also its launcher, he ensured the appearance of the 'transporter-erector' which in our own time has made the Soviet SS-20 and the American Pershing 2 instruments of strategic power so 'survivable' that their existence has produced the world's first ever categorical agreement of disarmament between leading military powers.

The German army's decision to invest in rocket development was motivated by the provisions of the Versailles treaty, which forbade it to possess heavy artillery but did not proscribe rockets. By 1937, however, when work on the V-2's predecessors was sufficiently advanced for Braun and Dornberger to have secured funds to establish a testing station on the Baltic island of Peenemünde, Hitler had already breached the Versailles treaty at every point. The rocket team's current preoccupation was to sustain funding to continue research. The army favoured the programme, since the V-2 was to be an army weapon, and provided finance. In October 1942 a successful test firing was staged, in December Speer, the Armaments Minister, authorised mass production, and on 7 July 1943 Hitler, after viewing a film of a missile launch, designated it 'the decisive weapon of the war' and announced that 'whatever labour and materials [Braun and Dornberger] need must be supplied instantly.'

By 1943, however, the British were already aware that the German ballistic missile programme was well advanced. Warning from a still-unidentified German wellwisher, received in Norway in 1940 and known as the 'Oslo Report', had alerted London to the existence of a missile research programme. The trail had then gone cold, but had revived when new evidence suggested in

December 1942 that a ballistic missile was under development in Germany and, in April 1943, that the Luftwaffe was also experimenting with a pilotless aircraft. Both clues came from 'Humint' (human intelligence, or the word of agents' contacts), one of its few successes of the war. By June both German programmes had been identified as centring on Peenemünde, where in fact the Luftwaffe was developing the FZG-76 (V-1 flying-bomb) at one end of the island while Dornberger and Braun worked on the V-2 rocket at the other. On 29 June Churchill personally ordered Bomber Command to take Peenemünde under heavy attack and on the night of 16/17 August it was attacked by 330 aircraft and devastated.

The Peenemünde raid so gravely set back the German pilotless weapons programme that it was not until 12 June 1944 that the first flying-bomb landed in Britain; 8 September was the date of the first successful V-2 rocket attack. By then the Luftwaffe's 155 Regiment had been driven back from the positions whence its V-1s could reach England; as a result, out of the 35,000 produced, only 9000 were fired against England and of these over 4000 were destroyed by anti-aircraft fire or fighter attack. The V-2s were never fired from their chosen launch sites in northern France; from Holland they could just reach London, on which 1300 impacted, and after October an equal number were directed at Antwerp, which by then was the Allied Liberation Armies' main logistic base.

The V-2s killed 2500 Londoners between 8 September 1944 and 29 March 1945, when their launch positions were finally overrun by the 21st Army Group. Britain had had a lucky escape – and perhaps also America, for Braun and Dornberger had already written the specifications for a missile, designated the A10 and utilising the V-2 (A-4) as its second stage, which would have had a range of 2800 miles and been launched across the Atlantic. Under other circumstances, moreover, these missiles, to which the Allies had not even the beginnings of a counterpart and no counter-measure whatsoever, would have carried a warhead as revolutionary in nature as the missiles were themselves. For Germany too had its atomic weapons programme.

It was the crowning mercy of the Second World War that it came to nothing. For a complex of reasons, which included Nazi Germany's self-deprivation of significant scientific talent by its persecution of the Jews, but also the inefficient multiplication of research programmes by as many as a dozen agencies which all hoped to win the Führer's favour by bringing him news of the successful development of the super-weapon, the American atomic intelligence team which ransacked Germany in May 1945 found that 'they were about as far as we were in 1940, before we had begun any large-scale work on the bomb at all'. In the last months of his life, Hitler, whose enthusiasm for nuclear weapons, as for ballistic missiles, developed too late in the war to ensure their decisive operational deployment, attempted to revitalise those about him with promises of unanswerable vengeance on his enemies. However, the evidence showed that, 'although [he] had been advised of the possibility of an atomic weapon in

1942, the Germans had failed to separate U 235 [the essential fissile element] and that, while they had apparently started separation on a small scale by means of a centrifuge and were constructing a uranium pile, they had only recently succeeded in manufacturing uranium metal . . . and had not by August 1944 taken their experiments to the point at which they were aware of the difficulties they would have to overcome before the pile would function.'

In short, the Germans were years from manufacturing an atomic bomb at the time when the Allied atomic weapons programme was already close to fulfilment. In October 1939 Albert Einstein, then the most famous man of science in the world and an émigré to the United States, had nevertheless been prompted by two younger physicists to write to President Roosevelt warning that Germany might be bent on an atomic weapons programme and suggesting that the United States should study the possibility itself; Roosevelt set up a 'Uranium Committee', which reported in July 1941 that the project was feasible and, if so, would be 'determining'. In 1942 the British, who had been pursuing their own researches with excellent manpower but insufficient funds, amalgamated their efforts with those of the Americans in the United States. By 1945 120,000 people were employed by the Manhattan Project, which had succeeded in separating uranium 235 and the synthetic element plutonium and in developing mechanisms to explode both as warheads of bomber-borne weapons.

It was the uranium 235 version of this atomic bomb that the B-29 *Enola Gay* dropped over Hiroshima on the morning of 6 August 1945; a few hours later, while 78,000 people lay dead or dying in the ruins, a White House statement called on the Japanese to surrender or 'they may expect a rain of ruin from the air'. No word being received, on 9 August another B-29 flew from Tinian to bomb the city of Nagasaki, killing 25,000. The United States thus temporarily exhausted its supply of nuclear weapons and awaited the outcome of the damage done.

On 8 August, following a warning it had issued in April that it would repudiate its 1942 non-aggression treaty, the Soviet Union declared war on Japan and opened a vast offensive into Manchuria the following day. This offensive had been promised to the Western Allies, but the Americans had grown decreasingly enthusiastic for it as the moment to launch their atomic strike approached. Stalin had shown little surprise when told by Truman at Potsdam of America's 'secret weapon'; as we now know, the treachery of certain Western scientists, in particular the German communist émigré Klaus Fuchs, had revealed its existence to the Soviets already. Marshall, the American chief of staff, was particularly insistent that Russian intervention was no longer necessary to the success of the Allied cause and would win them advantages in the Far East which the United States would find cause to regret. He equally admitted that there was no means of deterring the Russians from their offensive, which had been in preparation ever since the German surrender.

Three Far-Eastern Army Groups had been formed from the best-equipped and most experienced veterans of the European campaign, the third under the famous Marshal R. Y. Malinovsky. They were highly mechanised, the Japanese Kwantung Army was not. Though 750,000 strong, and regarded as the best formation in the imperial army, it had little recent experience of fighting. It bitterly defended the approaches to the central Manchurian plain, but when the Soviet Sixth Guards Tank Army broke out into open country on 13 August large sections of it were rapidly enveloped. The remainder was driven back across the river Yalu into northern Korea, where fighting continued until a final Japanese collapse on 20 August.

By then the Japanese forces everywhere else within the Pacific war zone had made their surrender to whichever Allied troops were at hand. On 15 August Emperor Hirohito, in the first public speech a Japanese sovereign had ever made, broadcast to his soldiers, sailors and people to announce that his government had decided to treat with the enemy. Explaining that the war had 'turned out not necessarily to Japan's advantage' and that the enemy had begun 'to employ a new and most cruel bomb', he called upon them, in a series of strange and obscure phrases which never mentioned surrender, to accept the coming of peace. A few intransigents disobeyed and attempted briefly to continue the fight; a few irreconcilables committed ritual suicide. The rest of the emperor's seventy million subjects relapsed instantly into the posture of defeat. On 28 August MacArthur arrived at Yokohama to institute the American occupation and reconstitution of Japan. On 2 September, aboard the battleship *Missouri* lying in Tokyo Bay, in the presence of representatives of Britain, the Soviet Union, China, France, Australia, New Zealand and Canada, MacArthur and the Japanese Foreign Minister, chief of staff and chief of naval operations signed the instrument of surrender. The Second World War was over.

EPILOGUE

THIRTY-THREE

The Legacy of the Second World War

The war was over, but the return of peace to the peoples who had fought it would prove patchy and erratic. In some places the war had touched – Greece, Palestine, Indonesia, Indo-China, China itself – peace was scarcely to return at all. In Greece, where the ELAS guerrillas, despite their defeat by the British in Athens at Christmas 1944, retained bases in the northern mountains, their communist leaders resolved in February 1946 to resume the civil war. The war dragged on until August 1949, at cruel cost to the rural population, 700,000 of whom fled to the cities and towns under government control; many families were bereft of their children, who had been kidnapped in thousands to be raised as future guerrilla fighters across the border in states under communist control.

In Palestine the British sponsors of the Jewish National Home, who were also the territory's rulers under a League of Nations (then United Nations) Mandate, soon found themselves in conflict with the Zionist settlers. Fearful of damaging relations with the native Arabs, the British refused to raise the limit they had set on further Jewish immigration, fixed at 75,000 in 1939, even when Washington petitioned London to allow 100,000 survivors of the concentration camps to be given refuge. Haganah, the semi-official Zionist militia, was shortly driven to side with the radical Jewish terrorist organisations against the Mandate government. In October 1945 Haganah initiated a sabotage campaign, setting off 500 explosions, and by the spring of 1946, when 80,000 British troops were deployed in Palestine, the territory trembled on the brink of open insurrection, which threatened to become a communal war should the Palestinian Arabs judge that the British intended to permit large-scale Jewish immigration or abandon the Mandate.

In Indonesia and Indo-China the British also found themselves caught between the fires of a local nationalism and an alien presence. In Indonesia, as the Dutch East Indies were shortly to be called, the Javanese set upon their former masters when the internees were released from prison camp, and it took the deployment of the whole of the 5th Indian Division in nineteen days of fighting in November 1945 to restore order. The Indian sepoys and their British officers were assisted by Japanese troops, whom Major-General E. C. Mansbridge released from captivity, rearmed and kept under control as long as the struggles against the Japanese-trained Indonesian army lasted.

Released Japanese prisoners were also used by the commander of the 17th

Indian Division when it was sent to reoccupy southern Indo-China in September 1945. The embryo Viet Minh party and the army of Ho Chi Minh had taken power in the vacuum left by Japan's surrender. In the north, which the great powers had agreed at Potsdam should temporarily be garrisoned by Chinese nationalist forces, the arriving Chinese general established a co-existence with Ho Chi Minh. In the south the British conceived it their duty under the Potsdam directive to wrest control of the civil administration from the Viet Minh, and they found they needed the help of rearmed Japanese soldiers to do so. In October a division of French troops arrived, led by Leclerc, the Gaullist hero who had liberated Paris in August 1944. His title to re-establish French authority was disputed, but he did so none the less, at the cost of beginning the 'war of the ricefields' which, in one form and another, was to drag on for the next thirty years.

In China the war between communists and nationalists, first begun in the 1920s, had only been interrupted by the Second World War. Both sides deployed large armies: Mao Zedong had nearly half a million men under arms, Chiang Kai-shek over 2 million. In 1937 they had agreed a truce, to hold as long as both were engaged in war against the Japanese invader. The defeat of Japan brought to China 50,000 American Marines and General George C. Marshall, the wartime chief of staff, with a mission to prolong the truce. In January 1946 an extension of the truce was indeed agreed; but its basis was unstable. Chiang Kai-shek's principal concern was to re-establish his position in Manchuria, overrun the previous August by the Russians, who were busy stripping the province (the richest in China) of its industrial plant, which they claimed was due to them as war reparations from Japan. Chiang lacked the power to check the depredations; but he was determined to see that the Russians, who had agreed to evacuate Manchuria by 1 February 1946, should not allow Mao Zedong's troops to succeed them as occupiers. While the truce was being negotiated, therefore, he was busily transferring units from his area of control in the south of China into Manchuria, even though these troop movements inevitably provoked local clashes with Mao's soldiers. Despite the best efforts of the American mediators, sporadic clashes were destined to swell into outright conflict and by July 1946 into full-scale civil war. An American attempt to bring hostilities to a close by denying the nationalists military aid merely enhanced the chances of the communists, who returned to the offensive when General Marshall was recalled by President Truman in January 1947. They were shortly to carry the war to the valley of the Yellow River as well as Manchuria, reviving the agony which had left 50 million Chinese homeless and 2 million orphaned as result of Japanese occupation.

The Allies brought to trial over 5000 of the Japanese who had waged the Pacific War and the 'China Incident' and executed 900 of them, in most cases for their mistreatment of Allied prisoners of war. At the Tokyo trial of major war criminals, however, twenty-five of Japan's leaders were arraigned for general war crimes and seven were condemned to death; they included Tojo and Koiso (his successor as Prime Minister) and might have included Konoye, had he not evaded arrest by taking poison. The Tokyo trial was inspired by the much larger and more widely publicised Nuremberg Tribunal, before which the Nazi leaders were tried between November 1945 and October 1946. There were twenty-one defendants at Nuremberg, one defendant (Bormann) tried *in absentia* and five corporate accused – the Reich Cabinet, the Leadership Corps of the Nazi Party, the SS/SD, the Gestapo and the General Staff. Of the individual defendants, who were charged with one, other or all of (a) crimes against peace, (b) war crimes, (c) crimes against humanity, two were acquitted, eight sentenced to terms of imprisonment varying from life to ten years and eleven condemned to death. The last included Goering, who managed to acquire poison and commit suicide on the eve of his execution; Kaltenbrunner of the SS (Himmler having committed suicide on capture); three governors of the occupied territories and the administrator of the forced labour regimes, Frank, Rosenberg, Seyss-Inquart and Sauckel; the two generals from Hitler's operations staff, Keitel and Jodl, whose endorsement of the 'Commando Order' of 1942 directing raiders in uniform to be murdered ensured their condemnation; Ribbentrop; Frick, the author of the Nuremberg decrees against the Jews; and Streicher, Nazism's principal mouthpiece of anti-Semitism. At a series of subsequent trials of lesser war criminals, another twenty-four were executed, largely for the perpetration of atrocities, thirty-five were acquitted and 114 imprisoned. Numbers of other war criminals were also later arrested, tried and sentenced by national courts in the countries where they had committed their offences.

The legal philosophy of the Nuremberg system continues to be debated by academic lawyers; but both at the time of the trials and thereafter the natural justice of the proceedings and of the verdicts has been universally accepted by the citizens of the states against which Germany and Japan waged war. Some 50 million people are estimated to have died as a result of the Second World War; it is in the nature of war-making that an exact figure can never be established. By far the most grievous suffering among the combatant states was borne by the Soviet Union, which lost at least 7 million men in battle and a further 7 million civilians; most of the latter, Ukrainians and White Russians in the majority, died as a result of deprivation, reprisal and forced labour. In relative terms, Poland suffered worst among the combatant countries; about 20 per cent of her pre-war population, some 6 million, did not survive. About half of the war's Polish victims were Jewish, and Jews also figured large in the death tolls of other eastern European countries, including the Baltic states, Hungary and Romania. Civil and guerrilla war accounted for the deaths of a quarter of a million Greeks and a million Yugoslavs. The number of casualties, military and civilian, were far higher in eastern than in western Europe – an index of the intensity and ferocity of war-making where Germans fought and oppressed Slavs. In three

European countries, however, France, Italy and the Netherlands, casualties were heavy. Before June 1940 and after November 1942 the French army lost 200,000 dead; 400,000 civilians were killed in air raids or concentration camps. Italy lost over 330,000 of whom half were civilians, and 200,000 Dutch citizens, all but 10,000 of them civilians, died as a result of bombing or deportation.

The Western victors suffered proportionately and absolutely much less than any of the major allies. The British armed forces lost 244,000 men. Their Commonwealth and imperial comrades-in-arms suffered another 100,000 fatal casualties (Australia 23,000, Canada 37,000, India 24,000, New Zealand 10,000, South Africa 6000). About 60,000 British civilians were killed by bombing, half of them in London. The Americans suffered no direct civilian casualties, although a Japanese balloon bomb killed a woman and five children of a Sunday School class picnicking in Oregon on 5 May 1945; their military casualties, which contrast with 1.2 million Japanese battle deaths, were 292,000, including 36,000 from the navy and 19,000 from the Marine Corps.

Germany, which had begun the war and fought it almost to Hitler's 'five minutes past midnight', paid a terrible price for war guilt. Materially her cities and towns stood up to bombing more stoutly than the flimsy Japanese population centres. Nevertheless, Berlin, Hamburg, Cologne and Dresden had effectively been reduced to rubble by 1945, and many smaller places had been brutally damaged.

When the cultural losses of the Second World War are reviewed, most can be seen to have occurred on German territory. Forethought had assured the preservation of the Great European libraries and art collections; the treasures of the Kaiser Wilhelm collection had been stored in the Berlin Zoo flak tower, and the pictures from the British National Gallery had spent the war in caves in Wales. Architectural treasures, by their nature, could not be protected. Fortunately, the course of the fighting, except in Italy, spared most of Europe's most beautiful creations. Berlin was devastated, but it was largely a nineteenth-century city; much of London's pre-eighteenth-century fabric was burnt in the Blitz; classical Leningrad suffered under bombardment and delights like Tsarkoe Selo (now, thankfully, completely restored) were burnt to the ground; baroque Dresden was burnt out; the Old City of Warsaw destroyed block by block (again miraculously re-created since 1945 by reference to the paintings of Bernado Belotto); the Old City of Vienna badly damaged in the fighting of 1945; Budapest on both banks of the Danube ravaged; the centre of Renaissance Rotterdam incinerated; William the Conqueror's medieval Caen laid flat. Yet historic Paris, Rome, Athens, Florence, Venice, Bruges, Amsterdam, Oxford, Cambridge, Edinburgh and almost all the other great European temples of architecture remained untouched.

In Germany, by contrast, not only the large but also the small historic cities suffered fearful destruction, including Potsdam, the Versailles of the Prussian kings, Jülich, Freiburg-im-Breisgau, Heilbronn, Ulm, Freudenstadt,

Würzburg and Bayreuth, the centre of the Wagner festival. In the west the twenty-eight towns which make up the industrial centre of the Ruhr and its environs all came under heavy attack: Stuttgart, the capital of south Germany, was bombed out; and Breslau, the largest German city in the east, was effectively destroyed during its defence against the Russian advance in the spring of 1945.

The German people paid a greater human than material price for initiating and sustaining war against their neighbours between 1939 and 1945. Over 4 million German servicemen died at the hands of the enemy, and 593,000 civilians under air attack. Although more women than men were killed by Allied bombing – a ratio of 60:40 – the numbers of women in the Federal Republic in 1960 still exceeded those of men by a ratio of 126:100. The male-female disproportion among the 'lost generation' was not as severe as in the Soviet Union, where women outnumbered men by a third after the war; but not even in Russia did the population undergo the horrors of forced migration which defeat visited on the Germans in 1945.

The uprooting of the Germans from the east comprised two phases, both tragic in their effect: the first was a panic flight from the Red Army; the second a deliberate expulsion of populations from regions of settlement where Germans had lived for generations, in some places for a thousand years. The flight of January 1945 was an episode of human suffering almost without parallel in the Second World War – outside the concentration camps. Terrified at the thought of what the Red Army would do to the first Germans it encountered on home territory, the population of East Prussia, already swollen by refugees from the areas of German settlement in Poland and the Baltic states displaced by the Bagration offensive, left home *en masse* and, in bitter winter weather, trekked to the Baltic coast. Some 450,000 were evacuated from the port of Pillau during January; 900,000 others walked along the forty-mile causeway to Danzig or crossed the frozen lagoon of the Frisches Haff to reach the waiting ships – one of which, torpedoed by a Russian submarine with 8000 aboard, became a tomb for the largest number of victims ever drowned in a maritime disaster. The Wehrmacht put up a fight of almost demented bravery to cover the rescue of refugees; Richard von Weizsäcker, son of the state secretary of Hitler's Foreign Ministry and Ex-President of the Federal German Republic, won the Iron Cross First Class in the battle of the Frisches Haff.

It seems possible that a million Germans died in the flight from the east in the early months of 1945, either from exposure or mistreatment. In the winter of 1945 most of the remaining Germans of eastern Europe – who lived in Silesia, the Czech Sudetenland, Pomerania and elsewhere, numbering some 14 million altogether – were systematically collected and transported westward, largely into the British zone of occupation in Germany. The transportees who arrived were destitute and often in the last stages of deprivation. Of those who failed to complete this terrible journey, it is calculated that 250,000 died in the course of the expulsion from Czechoslovakia, 1.25 million from Poland and 600,000 from elsewhere in eastern Europe. By 1946 the historic German population of Europe east of the Elbe had been reduced from 17 million to 2,600,000.

The expulsions, often conducted with criminal brutality, were not illegal under the settlement the victors had agreed between themselves at the Potsdam Conference of July 1945. Article 13 of its protocol stated that the 'transfer to Germany of Germans remaining in Poland, Czechoslovakia and Hungary will have to be undertaken'; at Potsdam, moreover, the Western Allies agreed to a realignment of the German frontier, giving half of East Prussia to Poland (the other half went to the Soviet Union), together with Silesia and Pomerania. These readjustments, balanced by the enforced cession by Poland of its eastern province to Russia, had the cartographic effect of moving Poland a hundred miles westward; demographically, they ensured that post-war Poland would be wholly Polish, at the expense of displacing the German populations of its new western borderlands.

The Potsdam agreement, to a far greater extent than that of Yalta, determined the future of European government in the post-war years. The concessions made to the Soviet Union by Britain and the United States at Yalta have been widely condemned by Western politicians and polemicists in the aftermath as a 'betrayal', particularly of the anti-communist Poles. As Roosevelt and Churchill recognised at the time, the Red Army's victorious advance into Poland made Stalin's plans for the most important country in eastern Europe a *fait accompli*. It ensured that the 'London Poles' would have no effective role in the post-war Warsaw administration, which would be dominated by the communist puppet 'Lublin committee'. Potsdam took post-war arrangements far further than that. By endorsing the resettlement westward of eastern Europe's Germans – both those of the borderlands of *Deutschstum* in Poland and Czechoslovakia and the more scattered settlements of German commercial, agricultural and intellectual enterprise in the Slav and Baltic states – it returned ethnic frontiers in Europe largely to those that had prevailed at the creation of Charlemagne's empire at the beginning of the ninth century, solved at a stroke the largest of the 'minority problems', and ensured Soviet domination of central and eastern Europe for two generations to come.

The Soviet Union's subsequent refusal to co-operate in the staging of free elections throughout the zones of occupation in post-1945 Germany had the additional effect of consolidating the 'Iron Curtain' between communist and non-communist Europe identified by Winston Churchill in his Fulton speech in 1946. The post-war settlement of 1918, by creating self-governing 'successor states' out of the tsarist, Hohenzollern and Habsburg empires which had dominated the eastern half of the continent before 1914, greatly diversified its political complexion. Potsdam ruthlessly simplified it. Post-1945 Europe west of the Elbe was to remain a polity of democratic states; east of the Elbe it was to relapse into autocracy, conforming to a single political system dictated and dominated by Stalinist Russia.

The imposition of Stalinism east of the Elbe after 1945 solved 'the German problem', which had transfixed Europe since 1870. It did not solve the problem of how to establish a lasting peace, either in Europe or in the wider world. The United Nations, which the United States, Britain and the Soviet Union had agreed to establish as a more effective successor to the League of Nations at Tehran in 1943, and which came into being at San Francisco in April 1945, was intended to be an instrument of international peace-keeping, with its own general staff commanding forces contributed by the member states under the authority of its Security Council (comprising representatives of Britain, the United States, the Soviet Union, France and China as permanent members). The Soviet Union's opposition to the establishment of the general staff, and its subsequent use of its veto to block peace-keeping resolutions, quickly emasculated the Security Council's authority. Stalin's foreign policy, which may be interpreted either as a resumption of Bolshevik commitment to the fomentation of revolution in the capitalist world or, more realistically, as an effort to entrench the Soviet victory of 1945 by keeping the anti-communist states of western Europe under threat of military attack, did not directly challenge the United Nations' role. His sponsorship of an anti-democratic coup in Czechoslovakia and his institution of the Berlin blockade in 1948 apart, in the post-war years he took no step which directly threatened the stability of Europe as constituted at Yalta and Potsdam. His challenge to the Western position in the world was to be laid elsewhere – in the Philippines, in Malaya and, above all, in Korea, where he was to endorse an aggression by the communist north against the non-communist south in June 1950.

The Soviet Union, indeed, demobilised its military forces in Europe as quickly, if not as completely, as did the United States and Britain theirs after August 1945. By 1947 the size of the Red Army had been reduced by two-thirds; the remaining force sufficed to outnumber the occupation forces of the Americans and the British many times – the British Army of the Rhine numbered only five divisions in 1948, the American army in Bavaria only one – but, though its continuing preponderance was to drive the North Americans and Western Europeans into a North Atlantic alliance in 1949, the disparity did not tempt the Soviet leadership to risk extending its power west of the Elbe.

There are many explanations for this. One is that Soviet foreign policy, for all its coarseness and brutality, was directed by a distinct legalism, which constrained Russia to the spheres of influence defined at Yalta and Potsdam. Another is that the American monopoly of nuclear weapons, persisting in its strict form until 1949 but effectively for a decade thereafter, deterred the Soviet Union from foreign policy adventures. A third, and contestably the most convincing, is that the trauma of the war had extinguished the will of the Soviet people and their leadership to repeat the experience.

The legacy of the First World War was to persuade the victors, though not the vanquished, that the costs of war exceeded its rewards. The legacy of the Second World War, it may be argued, was to convince victors and vanquished alike of the same thing. 'Every man a soldier', the principle by which the advanced states had organised their armies, and in large measure their societies, since the French Revolution, achieved its culmination in 1939-45 and, in so doing, inflicted on the countries which had lived by it a tide of suffering so severe as to banish the concept of war-making from their political philosophies. The United States, least damaged and most amply rewarded by the war – which left it in 1945 industrially more productive than the rest of the world put together – would be able to muster sufficient national consent to fight two costly, if small, wars in Asia, in Korea and Vietnam. Britain, which had also come through the war relatively unscathed in terms of human if not material loss, would preserve the will to fight a succession of small colonial wars, as France, another country comparatively untouched by severe loss of life, would do as well. By contrast, the Soviet Union, for all the fierce face it showed its putative enemies in the post-war era, eschewed confrontations which put its soldiers at direct risk; its recent venture into Afghanistan, costing a quarter of the number of lives lost by the United States in Vietnam, appears to reinforce, not vitiate, that judgement. Not a single German soldier, despite the Federal Republic's resumption of conscription in 1956, has been killed by enemy action since May 1945, and the likelihood of such a death grows more, not less, remote. Japan, the most reckless of the war-makers of 1939-45, is today bound by a constitution which outlaws recourse to force as an instrument of national policy in any circumstances whatsoever. No statesman of the Second World War was foolish enough to claim, as those of the First had done, that it was being fought as 'a war to end all wars'. That, nevertheless, may have been its abiding effect.

Bibliography

Fifty Books on the Second World War

Bibliographies of the Second World War abound. None is comprehensive, nor is that surprising, since 15,000 titles in Russian alone had appeared by 1980. Excellent working bibliographies may be found, nonetheless, in most good general histories of the war, such as the revised edition of *Total War* by P. Calvocoressi, G. Wint and J. Pritchard (Lodon, 1989).

Rather than supply an equivalent of such bibliographies, I have decided to offer a list of fifty books available in English which together provide a comprehensive picture of the most important events and themes of the war, which are readable and from which the general reader can derive his own picture of the war as a guide to deeper reading. The list inevitably reflects my own interests and prejudices and is certainly not complete; it does not, for example, contain a title on the Polish campaign of 1939 or on the Scandinavian or Italian campaigns; it is thin on the war at sea in western waters and on the war in the air; and it is biased towards the fighting in Europe rather than in the Pacific. These distortions are, however, in most cases caused by gaps in the literature. There are still no books which meet the criteria I set myself on the Polish or Italian campaigns. If this judgement seems a depreciation of the remarkable work of the American, British and Commonwealth Official Historians, may it please be noted that I have nevertheless included several volumes which appear in those series, and have omitted others purely for reasons of space. I have included no books in foreign languages, though I would have dearly liked to include the war diary of the *Oberkommando der Wehrmacht*, the daily record of Hitler's operations staff. Its full title is: P. Schramm, *Kriegstagebuch des OKW der Wehrmacht*, vols 1–8, Munich, 1963. The place of publication of the titles cited is London, unless otherwise stated, and the edition, including those in English translation, is the most recent.

An indispensable guide to the campaigns is Colonel Vincent J. Esposito's *The West Point Atlas of American Wars*, vol 2, New York, 1959; the atlas contains meticulous maps of the main theatres of fighting, whether American troops were engaged or not, complemented by clear narratives on the facing page.

The best biography of Hitler, whose personality stands at the centre of the Second World War, is still that of Alan Bullock: *Hitler, a Study in Tyranny*, 1965. Complementing it as a picture of how he directed Germany's war effort is David Irving's *Hitler's War*, 1977, which has been described as 'the autobiography Hitler did not write' and is certainly among the half-dozen most important books on 1939–45. Robert O'Neill's *The German Army and the Nazi Party*, 1966, is an essential portrait of both institutions and their relationship in the pre-war years. Two books on the relationship between Hitler and German government and army in the war years which will always be read are: W. Warlimont, *Inside Hitler's Headquarters*, 1962, by one of his operations officers, and A. Speer, *Inside the Third Reich*, 1970; Speer was Hitler's armaments minister from 1942 and a technocrat of brilliant intelligence who nevertheless allowed himself to become a court favourite. H. Trevor-Roper is the author of two indispensable works: *Hitler's War Directives*, 1964, and his eternally fascinating classic, *The Last Days of Hitler*, 1971.

Contentious though it is, A. J. P. Taylor's *The Origins of the Second World War*, 1963, cannot be bettered as an introduction to that subject. On the beginning of the war in the west an outstanding work of historical drama is Alistair Horne's *To Lose a Battle*, 1969; Guy Chapman's *Why France Fell*, 1968, meticulously analyses that persisting conundrum. Some of the consequences are described in Robert Paxton's too little known *Parades and Politics at Vichy*, Princeton, 1966, a study of 'the French officer corps under Marshal Pétain', which is also a brilliant dissection of the dilemmas of resistance and collaboration. The best account of the aftermath of Hitler's victory in the west is Telford Taylor's *The Breaking Wave*, 1967, which is also an account of his defeat in the Battle of Britain.

Whether or not Hitler had ever seriously contemplated invading Britain, by the autumn of 1940 his thoughts were turning eastwards. Martin van Creveld, in *Hitler's Strategy, the Balkan Clue*, Cambridge, 1973, describes the stages through which his thinking proceeded and provides one of the most original of all analyses of strategy and foreign policy in the historiography of the war. A brilliant monograph on a critical aspect of the Balkan campaigns is *The Struggle for Crete*, 1955, by I. M. G. Stewart, the medical officer of one of the British battalions overwhelmed by the German airborne descent. The fighting in the Western Desert, for the Germans an appendix to their advance to the Mediterranean, has been much written of, but nowhere better than in Correlli Barnett's *The Desert Generals*, 1983.

The Balkans was the prelude to Hitler's attack on Russia. Overtowering all other writers in English on the war in the east (probably in Russian also) is John Erickson, who has published three magisterial works: *The Soviet High Command*, 1962, *The Road to Stalingrad*, 1975 and *The Road to Berlin*, 1983; the last two are over-complex at the operational level but magnificent in their portrayal of the Red Army and the Soviet peoples at war. The reality of the war waged by the Germans, and of its self-defeating nature, is conveyed in A. Dallin's scholarly *German Rule in Russia*, New York, 1957. A slight but vital monograph on how devastated Russia's resistance was sustained is Joan Beaumont's *Comrades in Arms*, 1980, which, though devoted to British aid to Russia, also tells much of the far greater American aid effort.

Hitler's embroilment in Russia, together with America's entry into the war which shortly followed it, cast the strategic initiative for the first time to the Allied side. Two key monographs which outline Britain's efforts to make strategy on its own acount are Michael Howard's *The Continental Commitment*, 1972, and *The Mediterranean Strategy in the Second World War*, 1968; the latter frankly acknowledges British reluctance to meet American enthusiasm for a direct assault on North-West Europe. Splendid documentary surveys of joint Anglo-American strategic decision-making from the moment of American entry are provided in two volumes of the great American Official History, E. Snell's *Strategic Planning for Coalition Warfare, 1941-2*, Washington, 1953, and M. Matloff, same title but for 1943-4, Washington, 1959. An associated volume, investigating how particular strategic choices (not all Allied) were made, is *Command Decisions*, Washington, 1960, edited by K. R. Greenfield.

Because we now know that the making of Allied strategy – sometimes of tactics – was guided by Britain's ability to read German secure communication (Ultra) and the Americans' ability to read the Japanese (Magic), it is inevitable that this list should contain several titles on both activities. By far the most important is the first volume of the Official History, by F. H. Hinsley (and others), *British Intelligence in the Second World War*, 1979; it contains the essential information on the breaking of Enigma, the German cipher system, and on the establishment and early use of Ultra, the intelligence derived from it. Additional but vital technical details are supplied by Gordon Welchman, a pioneer at the cipher-breaking centre at Bletchley, in *The Hut Six Story*,

1982. Ronald Lewin provided broad but highly reliable accounts of the influence both of Ultra and Magic in *Ultra Goes to War*, 1978, and *The American Magic*, New York, 1982; the latter also explains how the Americans complemented Bletchley's achievement by breaking the Japanese ciphers. Two detailed studies of Ultra in action are P. Beesly's *Very Special Intelligence*, 1977, about the Battle of the Atlantic, and R. Bennett's *Ultra in the West*, 1979, about the North-West Europe campaign.

American's war in the Pacific has produced an enormous literature. The most illuminating introduction, for a westerner, is Richard Storry's *A History of Modern Japan*, 1960, by a scholar who served as an intelligence officer with the British army in South-East Asia and had taught in Japan before its disastrous decision to make the surprise attack on Pearl Harbor. H. P. Willmott's *Empires in the Balance*, 1982, surveys the strengths and strategies of the Pacific antagonists before and during the first year of the war and is particularly well-informed on the Japanese side. The best general history of the war in the Pacific, which also finds room for accounts of events in China and Burma, is Ronald Spector's *Eagle against the Sun*, 1988, enthrallingly written and brilliantly compressed. It would be unfair not to include a volume from Samuel Eliot Morison's Official History of United States Naval Operations in World War II; in fact his fourth, *Coral Sea, Midway and Submarine Operations*, Boston, 1949, provides superb and moving accounts of those two crucial battles and is a justification in itself of the official historiography programme. The most important survey of the politics of the Pacific War, which is also a monument of diplomatic history, is Christopher Thorne's *Allies of a Kind*, 1978, subtitled 'The United States, Britain and the War against Japan, 1941-5' which exactly describes its content.

Japan's defeat ultimately derived from the disparity between its economic resources and those that the United States could deploy, as Admiral Yamamoto had warned the Imperial government would be the case. An essential survey of the economic factors underlying the course of the war is Alan Milward's *War, Economy and Society, 1939-45*, 1977, which encapsulates his many monographs on national wartime economies. A separate large monograph to which I frequently turn for illumination of how economies adapt to the particular needs of war-making is a volume in the British Official Histories, *The Design and Development of Weapons*, 1965, by M. M. Postan and others; it does not, however, deal with the British contribution to the atomic weapons programme, nor is there, indeed, any single book which satisfactorily covers the development and use of the atomic bomb in the Second World War. The effort to destroy economies by conventional bombing has produced an enormous literature; I particularly value Max Hastings' *Bomber Command*, 1987, for its study of the effects of the campaign both on the Germans and the crews who took part. Germany's reciprocal effort to attack the Allied war economy through its U-boat campaign has also been amply recounted; Peter Padfield's biography of the admiral who created and directed the U-boat fleet, *Dönitz, The Last Führer*, 1984, is an outstanding study, as well as a riveting 'Portrait of a Nazi War Leader'.

I have chosen only one book among the thousands written on the North-West Europe campaign, Chester Wilmot's *The Struggle for Europe;* I use the original 1952 edition, though there has been a re-issue. Wilmot, a war correspondent, effectively invented the modern method of writing contemporary military history, which combines political, economic and strategic analysis with eye-witness accounts of combat. Though many of his judgements have been challenged, and some demolished, his book remains for me the supreme achievement of Second World War historiography, combining a passionate interest in events with a cool dissection of the material realities which underlay them. It was the book which first awoke my interest in the war as history and which I come to admire more rather than less as time passes.

Wilmot correctly perceived that the war was one of 'the big battalions', an important corrective to the already burgeoning Anglo-Saxon interest in clandestine operations. That interest has swollen since, to a point where irregular and resistance campaigns assume a greater significance than Stalingrad or Normandy. Resistance forms, nevertheless, an essential ingredient of the story of the war. The best general survey is H. Michel's *The Shadow War*, 1972, and the best particular study of the most important resistance campaign, that in Yugoslavia, is F. W. Deakin's *The Embattled Mountain*, 1971. W. Rings has provided a highly original account of the other side of the story, the German effort to run a European empire, in *Life with the Enemy*, 1982. The horrors of the blackest side of that empire were first objectively obsessed by G. Reitlinger in *The Final Solution*, 1953; though the historiography of the Holocaust has since been greatly elaborated, and while his book is largely concerned with the Jews, rather than the many other groups systematically massacred by the Nazi extermination apparatus, it retains, for me at least, a power to shock, to instruct and to warn that later publications lack.

Finally there are the personal memoirs of the war. Among the thousands of soldiers' stories, I am haunted by one from the Pacific War, *With the Old Breed*, Novato, California, 1981. E. B. Sledge, now a professor of biology, fought the campaign with the 1st Marine Division. His account of the struggle of a gently raised teenager to remain a civilised human being in circumstances which reduced comrades – whom he nevertheless loved – to 'twentieth-century savages' is one of the most arresting documents in war literature, all the more moving because of the painful difficulty someone who is not a natural writer found in re-creating his experience on paper. A brilliant literary achievement, by contrast, is *Wartime*, 1977, by M. Djilas, the Yugoslav intellectual who belonged to Tito's entourage, negotiated with Stalin, fought as a Partisan but eventually fell out with his master and rejected the 'heroic' ethos which had driven so many men of passion and ability to create the tragedy of the Second World War. The last two books I have chosen relate the experience of women, that half of the wartime generation whose fate was to bear so much of the tragedy it brought. The *Berlin Diaries* of Marie Vassiltchikov, 1985, the memoirs of an Anglophile white Russian whom circumstances cast into the heart of Nazi Germany at the outbreak of the Second World War, present an extraordinary picture of human resilience under bombing attack, of the strange normalities that persisted even as the shadows drew in and of the high-spirited disdain for the clods of Nazi bureaucracy that a beautiful girl of noble birth could openly display throughout the wartime years. Christabel Bielenberg, an Englishwoman married to one of the July conspirators against Hitler, felt the same disdain; her account, in *The Past is Myself*, first published in 1968, of her brave and eventually successful effort to rescue her husband from the Gestapo, shows how narrowly an enemy of the regime, even if a woman, had to measure disdain against deference in preserving her loved ones from destruction.

This list might have been decupled in length; but at fifty books I cut it short. With this extension: in *Armed Truce, The Beginnings of the Cold War, 1945-6*, 1986, Hugh Thomas has written what is not only an essential guide to the war's aftermath but also a great work of modern history, meticulous in its use of sources and enthralling in the sweep of its narrative. No history of the war itself, and certainly not mine, can match it in quality or authority.

Index